Paleoecology: Past, Present and Future

Paleoecology: Past, Present and Future

Edited by Anne Offit

SYRAWOOD
PUBLISHING HOUSE
New York

Published by Syrawood Publishing House,
750 Third Avenue, 9th Floor,
New York, NY 10017, USA
www.syrawoodpublishinghouse.com

Paleoecology: Past, Present and Future
Edited by Anne Offit

International Standard Book Number: 978-1-68286-546-0 (Hardback)

Cataloging-in-Publication Data

Paleoecology : past, present and future / edited by Anne Offit.
p. cm.
Includes bibliographical references and index.
ISBN 978-1-68286-546-0
1. Paleoecology. 2. Ecology. 3. Paleobiology. I. Offit, Anne.
QE720 .P35 2018
560.45--dc23

Printed in China.

TABLE OF CONTENTS

PREFACE

The world is advancing at a fast pace like never before. Therefore, the need is to keep up with the latest developments. This book was an idea that came to fruition when the specialists in the area realized the need to coordinate together and document essential themes in the subject. That's when I was requested to be the editor. Editing this book has been an honour as it brings together diverse authors researching on different streams of the field. The book collates essential materials contributed by veterans in the area which can be utilized by students and researchers alike.

Paleoecology studies ancient organisms and their relationship with the environment they inhabited. As a field, it provides key clues into the processes of habitat adaptation among organisms as well as to biotic and abiotic stress factors. This book on paleoecology highlights the most up-to-date concepts in this field. Paleoecology recreates the past environment for precise interpretation of data. Significant changes have occurred in our environment over the past centuries which are studied under this field. For someone with an interest and eye for detail, this book covers the most significant topics in the field of paleoecology. It aims to serve as a resource guide for students and experts alike and contribute to the growth of the discipline.

Each chapter is a sole-standing publication that reflects each author´s interpretation. Thus, the book displays a multi-facetted picture of our current understanding of application, resources and aspects of the field. I would like to thank the contributors of this book and my family for their endless support.

Editor

1

Rebuilding Biodiversity of Patagonian Marine Molluscs after the End-Cretaceous Mass Extinction

Martin Aberhan[1]*, Wolfgang Kiessling[1,2]

1 Museum für Naturkunde, Leibniz Institute for Evolution and Biodiversity Science, Berlin, Germany, 2 GeoZentrum Nordbayern, Paläoumwelt, Universität Erlangen–Nürnberg, Erlangen, Germany

Abstract

We analysed field-collected quantitative data of benthic marine molluscs across the Cretaceous–Palaeogene boundary in Patagonia to identify patterns and processes of biodiversity reconstruction after the end-Cretaceous mass extinction. We contrast diversity dynamics from nearshore environments with those from offshore environments. In both settings, Early Palaeogene (Danian) assemblages are strongly dominated by surviving lineages, many of which changed their relative abundance from being rare before the extinction event to becoming the new dominant forms. Only a few of the species in the Danian assemblages were newly evolved. In offshore environments, however, two newly evolved Danian bivalve species attained ecological dominance by replacing two ecologically equivalent species that disappeared at the end of the Cretaceous. In both settings, the total number of Danian genera at a locality remained below the total number of late Cretaceous (Maastrichtian) genera at that locality. We suggest that biotic interactions, in particular incumbency effects, suppressed post-extinction diversity and prevented the compensation of diversity loss by originating and invading taxa. Contrary to the total number of genera at localities, diversity at the level of individual fossiliferous horizons before and after the boundary is indistinguishable in offshore environments. This indicates an evolutionary rapid rebound to pre-extinction values within less than *ca* 0.5 million years. In nearshore environments, by contrast, diversity of fossiliferous horizons was reduced in the Danian, and this lowered diversity lasted for the entire studied post-extinction interval. In this heterogeneous environment, low connectivity among populations may have retarded the recolonisation of nearshore habitats by survivors.

Editor: Donald J. Colgan, Australian Museum, Australia

Funding: The study was supported by the Deutsche Forschungsgemeinschaft (grant number: Ab 109/8-1; www.dfg.de). The funders had no role in study design, data collection and analysis, decision to publish, or preparation of the manuscript.

Competing Interests: The authors have declared that no competing interests exist.

* Email: martin.aberhan@mfn-berlin.de

Introduction

The study of mass extinction events can reveal patterns of selective survival and the dynamics of ecosystem recovery over evolutionary time scales [1–3]. Apart from developing models of the processes involved in recovery dynamics [2], [4], such information can be gleaned from empirical studies of the geological past.

The mass extinction at the Cretaceous-Palaeogene (K-Pg) boundary is among one of the best studied Phanerozoic bioevents [5]. Geographically, the scope of studies encompasses local, regional, and global scales [6–8]. Yet, our understanding is still limited about how global signals of diversity dynamics arose as a result of the accumulation of events at local and regional scales. A general challenge is the reciprocal relationship between the geographical extent of a study and its temporal resolution. The 11-myr binning resolution of global occurrence compilations of marine macrofauna [8] is too coarse to uncover the dynamics of post-extinction diversity rebounds, whereas local-scale analyses may lack general validity.

Here we construct high-resolution diversity trajectories from benthic marine molluscs at two fossiliferous K-Pg boundary sites in Patagonia and incorporate information on the evolutionary and biogeographical characteristics of the fauna and on species abundances. The two studied sites represent marginal to shallow marine (nearshore) and middle shelf environments (offshore). We focus on evolutionary patterns and longer-term diversity trends across the K-Pg interval. In particular, we test two null hypotheses: Within the first *ca* 4 million years (myr) after the extinction event (1) the diversity of mollusc assemblages is indistinguishable from that of pre-extinction faunas; (2) the transient loss of diversity through global extinctions and local extirpations was compensated for by gains via immigrating and newly originating taxa. We demonstrate that hypothesis (1) is only partly supported, and hypothesis (2) can be rejected by empirical evidence.

Materials and Methods

No permits were required for the described study, which complied with all relevant regulations.

Sampling localities

We analysed two stratigraphic sections in the Argentine part of Patagonia. The first one is a composite section from the adjacent localities Bajada del Jagüel (BJG, 38°6′S, 68°23′W) and Opaso (38°8′S, 68°24′W), located in the Neuquén Basin north of the city of Neuquén. The second section, close to Estancia San Ramón in the middle Chubut River valley (42°41′S, 69°51′W), is located in the Cañadón Asfalto Basin. During latest Cretaceous–earliest

Palaeocene times the sea in both areas formed narrow embayments, which opened to the Atlantic Ocean to the east [9].

The sections at BJG and Opaso consist of monotonous, middle shelf siliciclastic mudstones of the Jagüel Formation, representing open marine sedimentation below, or close to, the storm wave base [10]. In the following they are referred to as 'offshore'. Because the section at Opaso virtually lacks a Maastrichtian record to which the Danian samples could be compared, the faunal data from BJG and Opaso were arranged chronologically according to their distance from the K-Pg boundary and integrated into a single succession (i.e. our offshore section). We thereby assumed equal sedimentation rates at these close-by localities. Faunal samples were derived from 11.4 metres below the K-Pg boundary to 27.5 metres above the boundary.

In San Ramón, faunal sampling extends over 188 m of varied sedimentary rocks of the Lefipán Formation. The Maastrichtian to early Palaeocene succession comprises siliciclastic mudstones, siltstones, sandstones, and conglomerates. Litho- and biofacies indicate deposition in a marginal to shallow marine, tide-dominated deltaic setting with tidal bars, prodelta, and delta plain environments, which are represented in both the Maastrichtian and the Danian [9]. In this paper, the aforementioned deltaic setting is referred to as 'nearshore'. Salinity was an important factor affecting the composition and ecological structure of the fauna in these environments. The uppermost Maastrichtian is characterized by virtually monospecific assemblages of the bivalve genus *Corbicula* owing to oligohaline to mesohaline salinity, whereas more fully marine conditions existed below and above this interval [9]. These very low-diversity brackish-water assemblages were excluded from our analyses because they represent a different environment.

Biostratigraphy and interval lengths

In BJG, the K-Pg boundary is at the erosive base of a coarse-grained, tuffaceous sandstone bed [10], [11]. Calcareous nannofossils prove the presence of the first Cenozoic nannofossil zone (NP1 or NNTp1) in the upper part of the sandstone layer, and younger subzones in the mudstones overlying the sandstone [10]. The mudstones below the sandstone are late Maastrichtian in age [10], [11]. According to Keller et al. [11], the 11.4 metres of section below the K-Pg boundary, which yielded the late Maastrichtian quantitative samples in our study, are characteristic of the nannofossil zone CC26 and the *A. mayaroensis* and *P. hariaensis* foraminiferal zones (zonal names according to [12]). In Opaso, the K-Pg boundary sandstone is exposed about half a metre above the base of the sampled section. Micro- and nannofossils indicate the presence of early (NP1-NP3?/P1a-b) to late Danian (NP4/P1c) biozones [13]. Applying the ages given in Gradstein et al. [12], the samples of BJG and Opaso extend from ~2 myr before the K-Pg boundary to ~4 myr after the boundary. Because the earliest Danian planktonic foraminiferal zones P0 and Pα, the latest Maastrichtian *P. hantkeninoides* foraminiferal zone, and impact tracers are missing, the sections at BJG and Opaso are stratigraphically incomplete [10], with possibly up to 500 kyr of the earliest Danian and 300 kyr of the latest Maastrichtian missing [10], [11]. In summary, the amount of time represented by the Danian strata is about twice that of the Maastrichtian strata.

In the nearshore San Ramón section, a distinct boundary layer is not preserved. However, the presence of latest Maastrichtian and earliest Danian dinoflagellate markers and age-diagnostic palynomorphs indicate a very minor hiatus [14]. A more detailed biostratigraphical subdivision is not available in this very shallow-water setting. Because depositional environments are the same below and above the boundary, and because the sampled Maastrichtian succession (100 metres) is similar in thickness to that of the Danian (88 metres), we tentatively assume roughly equal time spans for the studied pre- and post-extinction intervals.

Analytical methods

We quantitatively sampled 157 single beds of uniform lithology for macroinvertebrates, comprising a total of 5,547 individuals. Bivalves and gastropods dominate the macrofauna. Taxa were identified as precisely as possible, mostly at species level. Our analyses focus on genus diversity of bivalves and gastropods. We elevated subgenera to genus rank and omitted all taxa not resolvable at least to the genus level. Because most genera are represented by a single species at the studied localities, the number of taxa is almost identical at genus and species level. Following standard procedures [15] the number of individuals of bivalves was determined as the sum of articulated specimens plus the larger number of right or left valves of disarticulated specimens plus one-half the number of fragments which could not be assigned to either right or left valves but make up more than 50% of a valve. In one special case – *Entolium membranaceum,* an equivalved and equilateral pectinoid with symmetrical auricles – distinction of the right and left valves was impossible, thus numbers of specimens were halved to yield numbers of individuals. Gastropods were counted when more than 50% of the shell was preserved.

To avoid large variations in sample size, we pooled samples of successive fossiliferous beds until at least 100 individuals were reached or until the lithology changed markedly. In the offshore section, this resulted in 24 pooled samples with a mean size of 129.1±7.1 (standard error) individuals. The 28 pooled nearshore samples had a mean of 63.1±5.5 individuals.

We focus on two measures of diversity. The first one determines the standardised number of genera for each of the pooled samples. We refer to this measure as sample-level diversity [16]. We standardised sample-level diversity using the 'shareholder quorum' subsampling method developed by Alroy [8], [17]. This method involves random sampling of individuals to equal frequency coverage, the shareholder quorum. In our analyses the quorum was set to 0.8 and our values are averages of 100 iterations. The shareholder quorum is superior to traditional subsampling methods (e.g., rarefaction) in that variations in evenness are explicitly taken into account. Therefore, it is less biased by different abundance distributions and represents a nearly unbiased estimate of richness.

Our second measure of diversity aims at comparing the total number of Danian genera in a section with that of Maastrichtian genera in the same section. This measure is termed total (Maastrichtian or Danian) diversity. To this end, the genera of the pooled and standardised samples were summed up over consecutively younger samples for the Maastrichtian and for the Danian part of the section, respectively. Genera are added to the count of total diversity at the sampling level at which they first appeared in the succession and are considered to persist up to the K-Pg boundary (for Maastrichtian total diversity) or to the youngest Danian sample (for Danian total diversity). This procedure constructs sample-based accumulation curves [18], whereby the sequence of samples is ordered in time via stratigraphy. This approach assumes that diversity accumulates during both the Maastrichtian and the Danian. While confidence intervals on stratigraphic ranges of genera support this approach, we emphasize that this procedure is thought to provide theoretical 'genus accumulation curves', which allow for a comparison of Maastrichtian and Danian diversity at various sampling levels.

With reference to the K-Pg boundary, we assigned each genus to one of five categories: (1) going extinct globally; (2) disappearing in Patagonia but continuing elsewhere; (3) surviving; (4) immigrating; and (5) originating. The reference area for categories 2–5 is Patagonia because geographical restriction to localities may produce strong sampling biases. We categorised genera as immigrants when they occurred in the Danian sections of our study area, had no Cretaceous record from Patagonia, and had been reported outside of Patagonia either before the Danian or within a part of the Danian which is stratigraphically older than our own field records. We categorised genera as originating in Patagonia when they are endemic to this region and first appeared in the Danian, or when a Danian occurrence in our sections is stratigraphically older than records from anywhere else in the world. For each pooled and standardised sample we calculated proportional diversity of the various categories. For the purpose of assigning genera to the five evolutionary/biogeographical categories, geographical and stratigraphical range data were retrieved from the Paleobiology Database (http://paleobiodb.org), Sepkoski's compendium [19], and numerous monographs on Cretaceous and Palaeogene molluscs. The main sources of information and specific references that were used for the categorisation of taxa are provided in Tables 1 and 2. A full documentation of primary data is presented in Tables S1 and S2.

Extinction intensities are expressed as percent extinctions of genera using the data presented in Tables 1 and 2. For comparison with our Patagonian sites, percent global extinctions of molluscan genera were also calculated for three other well-studied K-Pg boundary localities: Brazos River, Texas [20]; Braggs, Alabama [21]; and Stevns Klint, Denmark [22]. These calculations used the primary data provided in the respective publications.

Finally, we compared the relative abundances of the most common Maastrichtian species with their abundances in the Danian and vice versa at both sections. We summed up the total abundances of each species in the Danian and Maastrichtian parts of our respective sections and calculated their proportional abundance in each pooled sample. The proportional abundance of the most common species (those in the upper 90th percentile of all proportional abundances) was then ordered by decreasing proportional abundance and compared with the proportions in the stratigraphically older (for Danian abundances) or younger (for Maastrichtian abundances) parts of the sections. The bivalve *Corbicula*, which is very abundant in the low-diversity brackish-water assemblages, was excluded from this analysis.

All analyses were performed in the R programming environment (www.r-project.org). The field-collected Patagonian specimens are housed at the Departamento de Ciencias Geológicas, Universidad de Buenos Aires, Argentina.

Results

Genus-level extinction rates

Intensities of end-Cretaceous extinctions may vary markedly across local, regional, and global scales. In our nearshore section the percentage of global extinctions of benthic molluscan genera is 21.7%. In the offshore section 32.1% of genera became extinct. These values are comparable to those of well-studied Northern Hemisphere boundary sections such as Braggs, Alabama (30.5% extinction of bivalves and gastropods [21]) or Stevns Klint, Denmark (22.5% extinction of bivalves [22]), but markedly higher than at Brazos River, Texas (10.7% extinction of bivalves and gastropods [20]). Local extirpations (60.9% in our nearshore section, 53.6% in the offshore section) are much greater than global extinctions. These higher values are the cumulative result of local extirpations, global extinctions, and sampling effects.

Maastrichtian and Danian diversity compared

Sample-level diversity in offshore environments fluctuated without a consistent temporal trend (Figure 1). The diversity of even the first quantitative Danian samples is mostly within the range of Maastrichtian assemblages. In contrast, Danian sample-level diversity in nearshore habitats is, on average, distinctly lower than that of pre-extinction assemblages (Figure 1). This pattern is evident in both fine-grained and coarse-grained lithologies and persists until the end of the studied time interval, suggesting contrasting diversity patterns between nearshore and offshore environments. Despite comparable extinction intensities, recovery of sample-level diversity was only achieved offshore.

Total diversity in the Danian is significantly and consistently lower than total Maastrichtian diversity at the same sampling level (Figure 2). This pattern holds for both nearshore and offshore environments, and points to a distinct difference in diversity structure between Maastrichtian and Danian faunas.

The contributions of surviving, immigrating, and originating genera

In both environments, the most diverse components of Maastrichtian assemblages are genera that survived the K-Pg boundary event (Figure 3). Immigrating and newly evolving genera play only subordinate roles in the composition of Danian samples. Their occurrence is often sporadic and without obvious trends (Figure 3). The reduced total diversity in the Danian (Figure 2) indicates that immigrants and newly evolved taxa could not compensate for the loss of Maastrichtian diversity at the K-Pg boundary.

Abundance shifts of dominant species across the K-Pg boundary

In both environmental settings, the dominant Maastrichtian species became extinct at the K-Pg boundary (Figure 4). In the nearshore environment, this affected the five most abundant Maastrichtian species. Offshore, three of the four most abundant species did not survive the boundary, only the oyster *Pycnodonte vesicularis*, the second most common Maastrichtian species in the offshore environment, survived. However, with only a single specimen found in the Danian, it can be considered as ecologically extinct. It is best classified as a 'Dead Clade Walking', a term coined to characterise lineages that survived mass extinctions only to remain marginal or decline in the aftermath [23].

A second intriguing abundance pattern is seen in species that dominate the Danian assemblages. These were usually rare in the Maastrichtian part of the respective stratigraphic section (Figure 5). The three most abundant nearshore species, collectively making up almost 60% of Danian individuals, are survivors with low abundances during the Maastrichtian. In the offshore environment, three of the five most abundant Danian species are survivors that also changed their relative abundances from rare to common. The two dominant Danian offshore species – the oyster *Pycnodonte burckhardti* and the pectinid *Delectopecten neuquenensis* – are Patagonian endemics that originated in the early Danian.

Discussion

Incumbency effects and recovery with minor origination

The persistent dominance of survivors (Figure 3) points to the effectiveness of survival mechanisms [24] and contrasts with recovery models in which a short-term survival phase is followed by diversification owing to origination of new taxa [25]. At any

Table 1. Categorisation of molluscs at San Ramòn.

Class	Genus	Range in section	Category	Additional notes and references on records in the	
				Cretaceous	Palaeogene
Bivalvia	*Acesta*	Maa	2		
Gastropoda	*Acteon*	Maa	3		[50]
Bivalvia	*Anomia*	Maa	3		See Table 2
Bivalvia	*Aphrodina*	Danian	3	Valanginian of Argentina [51], [52]	
Bivalvia	*Arcomytilus*	Danian	4		
Bivalvia	*Astarte*	Danian	3	Hauterivian of Argentina [53]	Palaeocene of Argentina [54]
Bivalvia	*Australoneilo*	Maa – Danian	3		
Gastropoda	*Austroaporrhais*	Maa	3		Danian of Patagonia [54], [55]
Bivalvia	*Austrotrigonia*	Maa	1		
Bivalvia	*Bicorbula*	Maa – Danian	3		
Bivalvia	*Brachidontes*	Maa	3		Eocene of Patagonia [56], [57]
Bivalvia	*Camptochlamys*	Maa	1	Casadío et al. [58]	
Bivalvia	*Corbicula*	Maa – Danian	3		
Gastropoda	*Cryptorhytis*	Maa	3		See Table 2
Bivalvia	*Cucullaea*	Maa	3		Danian of Patagonia (unpublished data of authors)
Bivalvia	*Cucullastis*	Danian	4	Late Cretaceous of New Zealand [59]	
Bivalvia	*Cuspidaria*	Danian	4		
Bivalvia	*Eriphyla*	Maa – Danian	3		
Bivalvia	*Etea*	Maa	2		
Gastropoda	*Euspira*	Maa – Danian	3		
Bivalvia	*Glycymeris*	Danian	3	Maastrichtian of Argentina (unpublished data of authors from Sierra Huantraico)	
Gastropoda	*Heteroterma*	Maa – Danian	3		
Bivalvia	*Inoperna*	Maa	1		
Bivalvia	*Lahillia*	Maa	3		Eocene of Patagonia [60]
Bivalvia	*Ledina*	Maa – Danian	3		
Bivalvia	*Leionucula*	Maa – Danian	3		
Bivalvia	*Linearia*	Maa	2		Eocene of USA [61]
Bivalvia	*Lyriochlamys*	Maa	1		
Bivalvia	*Meretrix*	Maa – Danian	3		
Bivalvia	*Mesocallista*	Maa	1		
Bivalvia	*Modiolus*	Maa – Danian	3		
Gastropoda	*Nacella*	Danian	3	See Table 2	
Bivalvia	*Neilo*	Maa	3		Danian [62], Eocene [60] and Oligocene [50] of Patagonia
Bivalvia	*Nucula*	Maa	3		Danian of Patagonia [26]
Bivalvia	*Ostrea*	Maa – Danian	3		
Bivalvia	*Pachymya*	Maa	1		
Bivalvia	*Pacificor*	Maa – Danian	3		
Bivalvia	*Pacitrigonia*	Maa	1		
Bivalvia	*Panopea*	Maa – Danian	3		
Gastropoda	*Parasyrinx*	Maa	2		
Bivalvia	*Pholadomya*	Maa	2		
Bivalvia	*Phygraea*	Maa	3		See Table 2
Bivalvia	*Pinna*	Maa – Danian	3		
Gastropoda	*Priscaphander*	Maa	3		See Table 2

Table 1. Cont.

Class	Genus	Range in section	Category	Additional notes and references on records in the Cretaceous	Additional notes and references on records in the Palaeogene
Bivalvia	*Protagelus*	Maa	1		
Gastropoda	*Pseudamaura*	Danian	4		[54]
Bivalvia	*Pseudolimea*	Maa	3		See Table 2
Gastropoda	*Pseudotylostoma*	Danian	5		Genus endemic to Patagonia
Bivalvia	*Pteromyrtea*	Maa - Danian	3		
Gastropoda	*Pyropsis*	Maa	2		
Bivalvia	*Rinetrigonia*	Maa	1		
Bivalvia	*Saccella*	Danian	3	Maastrichtian of Argentina (unpublished data of authors from Sierra Huantraico)	
Bivalvia	*Solyma*	Maa	1		
Bivalvia	*Spineilo*	Danian	5		Danian records from Patagonia are from the lower part of the Danian (this study; foraminiferal zone P1b of [63]), whereas oldest records from New Zealand [64] are of late Danian age [26]. Palaeocene records from Russia appear to be late Danian or younger [65]; see also [66]
Gastropoda	*Struthioptera*	Maa - Danian	3		
Bivalvia	*Tikia*	Maa - Danian	3		
Gastropoda	*Turritella*	Maa - Danian	3		

Alphabetical list of genera (including subgenera) of bivalves and gastropods from the late Maastrichtian and Danian at San Ramón, western Argentina, and their assignment to evolutionary/biogeographical categories.
Evolutionary/biogeographical categories: 1 = going extinct globally in the late Maastrichtian; 2 = disappearing regionally from Patagonia in the late Maastrichtian; 3 = surviving the K-Pg boundary in Patagonia; 4 = immigrating to Patagonia in the Danian; 5 = originating in Patagonia during the Danian. Maa = Maastrichtian.

given sampling level, originating and invading genera could not compensate for the loss of genera caused by extinctions (Figure 2). A similar observation was reported for Antarctic molluscs at the K-Pg boundary, where few new genera were observed, and only later in the Danian did the number increase [26].

Delayed recovery is frequently explained by unfavourable environments, either as protracted impact-related effects, or as specific regional effects [27–29]. However, local environmental stress fails to explain the Danian *versus* Maastrichtian difference in total diversity in the offshore section where there is no evidence for a shift to more unfavourable environments. Similarly, in the nearshore section the range of sedimentary environments and overall environmental heterogeneity in the Danian and in the Maastrichtian match well [9]. Furthermore, previous work from our working group that used the same raw data from the same nearshore section showed that even Danian assemblages from stable, subtidal environments, characterised by diverse trophic and life habit groups, did not reach Maastrichtian diversity levels [9]. This is exemplified by comparing the Danian *Struthioptera-Venericardia* faunal association with the Maastrichtian *Struthioptera-Panopea* association of Scasso et al. [9]. Both associations are dominated by infaunal guilds but also have relatively high proportions of epifaunal molluscs. In both associations, trophic resources have been subdivided among suspension-feeders, deposit-feeders, and carnivores, suggesting biologically controlled communities [9]. Yet, the diversity of samples forming the Danian association remains distinctly below the diversity of the Maastrichtian samples. Thus, unfavourable habitats cannot be held responsible for the differences among pre- and post-extinction diversity at either of the two studied sections.

An alternative explanation for the delayed recovery would be low immigration rates caused either by geographic isolation of the basins or by a general shortage in the supply of taxa from outside Patagonia. However, there is no evidence of isolation because the Patagonian marine basins opened to the South Atlantic Ocean to the east, which was a large open ocean with pelagic sedimentation during K-Pg boundary times [9], [10]. Based on general physical oceanographic principles, wind driven surface currents, tidally induced amphidromic circulation, and thermohaline circulation are assumed to have been well developed in the studied areas, all of which would have contributed to the exchange of waters with surrounding regions. The presence in the Neuquén Basin of diverse planktonic foraminifera [11], a plankton group only found in high diversity in marine waters of normal salinity, provides direct evidence for such exchange with open marine waters. Palaeogeographic reconstructions of the marine basins in Patagonia for the Maastrichtian–Palaeocene interval show direct shallow water connections between the basins [9]. Also, with the closely located South American shelf to the north, the Antarctic shelf to the south, and refugia along the margin of the southern-Pacific [26], several potential areas were available as a source for new immigrants. Therefore, geographic isolation and a shortage in the supply of larvae are unlikely scenarios for the delayed recovery in diversity.

We suggest that the numbers of new taxa and of taxa invading from outside Patagonia were low due to the presence of survivors

Table 2. Categorisation of molluscs at Bajada del Jagüel and Opaso.

Class	Genus	Range in sections	Category	Additional notes and references on records in the	
				Cretaceous	Palaeogene
Gastropoda	*Acteon*	Maa	3		[50]
Bivalvia	*Ambigostrea*	Maa	1		
Bivalvia	*Amphidonte*	Maa	1		
Bivalvia	*Anomia*	Maa – Danian	3		
Bivalvia	*Aphrodina*	Danian	3	Valanginian of Argentina [51]	
Bivalvia	*Arcomytilus*	Danian	4		
Bivalvia	*Australoneilo*	Maa – Danian	3		
Bivalvia	*Bicorbula*	Danian	3	See Table 1	
Bivalvia	*Brachidontes*	Maa	3		Eocene of Patagonia [56], [57]
Bivalvia	*Camptochlamys*	Maa	1	[58]	
Bivalvia	*Cosmetodon*	Maa – Danian	3		
Gastropoda	*Cryptorhytis*	Maa – Danian	3		
Bivalvia	*Cucullastis*	Danian	4	Late Cretaceous of New Zealand [59]	
Bivalvia	*Delectopecten*	Danian	4	Early Cretaceous of Greenland [67]	Danian of Patagonia [62], [68]
Bivalvia	*Disparilia*	Maa – Danian	3		
Bivalvia	*Entolium*	Maa	1		
Bivalvia	*Glycymeris*	Danian	3	Maastrichtian of Argentina (unpublished data of authors from Sierra Huantraico)	
Bivalvia	*Gryphaeostrea*	Danian	3	Aguirre-Urreta [69]	del Río et al. [62], [66]
Gastropoda	*Heteroterma*	Danian	3	See Table 1	
Bivalvia	*Jagolucina*	Danian	5		Thanetian of Australia [70]; Eocene of France [71]
Bivalvia	*Leionucula*	Danian	3	See Table 1	[66]
Bivalvia	*Lima*	Maa	2		
Bivalvia	*Linearia*	Maa	2		Eocene of USA [61]
Bivalvia	*Lyriochlamys*	Maa	1		
Bivalvia	*Mesocallista*	Maa	1		
Bivalvia	*Modiolus*	Maa – Danian	3		
Gastropoda	*Nacella*	Maa – Danian	3		
Bivalvia	*Neilo*	Maa	3		Eocene [60] and Oligocene [50] of Patagonia
Bivalvia	*Nicaniella*	Danian	4		
Bivalvia	*Nuculana*	Maa – Danian	3		
Bivalvia	*Ostrea*	Maa – Danian	3		
Bivalvia	*Pacificor*	Danian	3	See Table 1	
Bivalvia	*Panopea*	Danian	3	See Table 1	
Gastropoda	*Perissodonta*	Danian	5		Danian records from Patagonia are from the lower part of the Danian (this study; foraminiferal zone P1b of [63]), whereas oldest Palaeocene records from New Zealand [64] are late Danian [26]
Bivalvia	*Phelopteria*	Maa	1	[72]	
Bivalvia	*Phygraea*	Maa – Danian	3		
Bivalvia	*Pinna*	Danian	3	See Table 1	
Gastropoda	*Priscaphander*	Danian	3	See Table 1	
Bivalvia	*Protagelus*	Maa	1		
Bivalvia	*Pseudolimea*	Maa – Danian	3		

Table 2. Cont.

Class	Genus	Range in sections	Category	Additional notes and references on records in the Cretaceous	Palaeogene
Bivalvia	*Pteromyrtea*	Maa – Danian	3		
Gastropoda	*'Pyropsis'*	Danian	4		Species belonging to a probably new genus referred to as *'Pyropsis'* have been discussed in [26], [73], [74]; oldest records are from Antarctica
Bivalvia	*Roudairia*	Maa	1		
Bivalvia	*Saccella*	Danian	3	Maastrichtian of Argentina (unpublished data of authors from Sierra Huantraico)	
Bivalvia	*Spineilo*	Danian	5		See Table 1
Bivalvia	*Spondylus*	Maa	2		
Gastropoda	*Struthioptera*	Danian	3	See Table 1	
Gastropoda	*Taioma*	Danian	4	Late Cretaceous of Antarctica [74]	[74], [75]
Bivalvia	*Tikia*	Maa – Danian	3		
Gastropoda	*Tornatellaea*	Danian	4		Palaeocene of Argentina [54], [75]

Alphabetical list of genera (including subgenera) of bivalves and gastropods from the late Maastrichtian and Danian at Bajada del Jagüel and Opaso, western Argentina, and their assignment to evolutionary/biogeographical categories.
Evolutionary/biogeographical categories: 1 = going extinct globally in the late Maastrichtian; 2 = disappearing regionally from Patagonia in the late Maastrichtian; 3 = surviving the K-Pg boundary in Patagonia; 4 = immigrating to Patagonia in the Danian; 5 = originating in Patagonia during the Danian. Maa = Maastrichtian.

within Patagonia impeding their progress. This phenomenon is known as incumbency or priority effects, in which existing organisms prevent the successful establishment of newly evolving or immigrating species [30], [31]. Resistance to invasions or to the establishment of newly evolved taxa occurs because these 'would-be usurpers' are initially less well adapted than established species, and the latter keep incipient populations of new species from taking hold [32], [33]. A closely related concept is that of pre-emptive competition [34], emphasizing the importance of being 'first in the field'. Incumbency can be expected to play a role when resources are limited and communities are saturated with species [15]. Mass extinctions are generally thought to reduce or remove the incumbency effect [35]. Our analyses, however, suggest that despite significant extinctions, the removal of incumbents did not entail an increase in immigration or origination rates to counterbalance diversity reduction. Rather, the number of

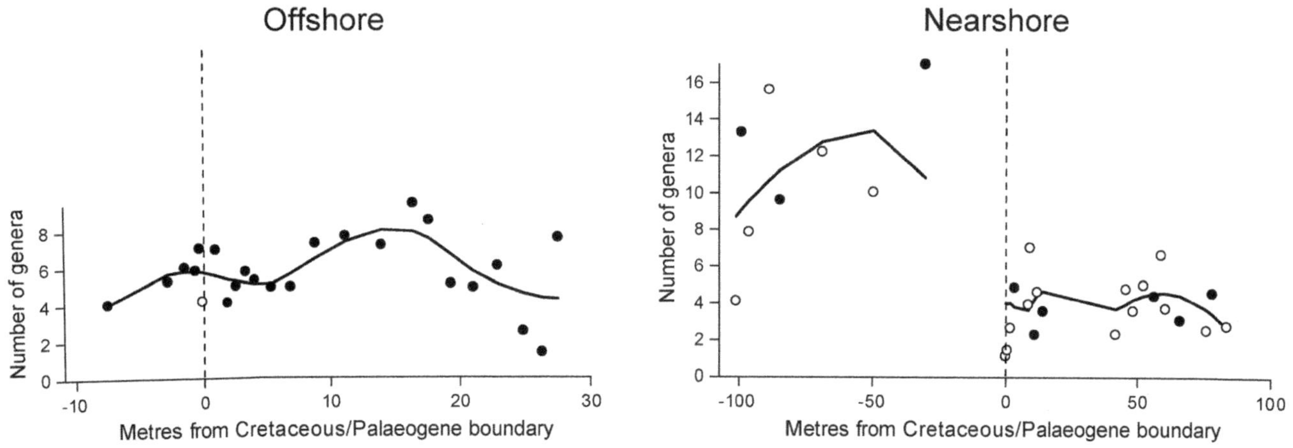

Figure 1. Sample-level diversity of benthic marine molluscs in two Cretaceous-Palaeogene boundary sections in Patagonia. Standardised genus diversity is based on the shareholder quorum subsampling method [17]. Dashed vertical lines indicate position of the Cretaceous-Palaeogene boundary; values on x-axis refer to metres before (Cretaceous, negative values) or after the boundary. Curved solid line is LOESS smoothing with a smoothing parameter of 0.5. Black dots indicate samples from fine-grained sedimentary rocks (siliciclastic mudstones and siltstones), open circles indicate samples from coarse-grained sedimentary rocks (sandstones and conglomerates). Nearshore, the uppermost *ca* 26 metres of the Cretaceous are characterised by low-diversity brackish-water assemblages, which are not included in the figure.

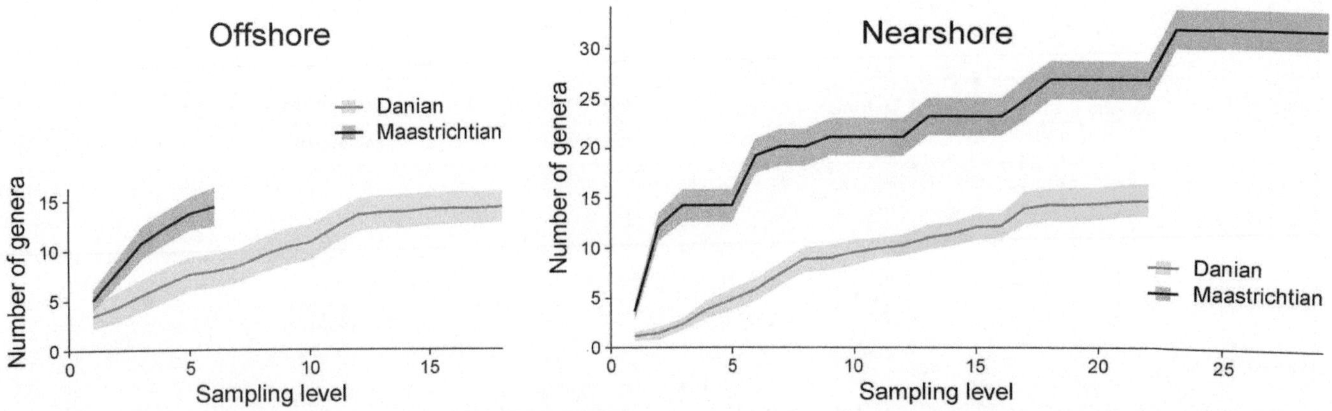

Figure 2. Comparing Cretaceous (Maastrichtian) and Palaeogene (Danian) accumulation curves for the estimated number of molluscan genera. Sampling level refers to consecutively numbered fossiliferous samples, arranged from stratigraphically oldest to youngest, and shows the cumulative number of genera at the respective level (see text for details on pooling and standardising of samples). Grey areas indicate the standard deviations around the means. In both offshore and nearshore habitats total diversity in the Danian is consistently lower than total Maastrichtian diversity at any given sampling level.

survivors seems to have remained sufficiently high to impose constraints on the establishment of immigrants and originating taxa, consistent with incumbency. We consider incumbency as an

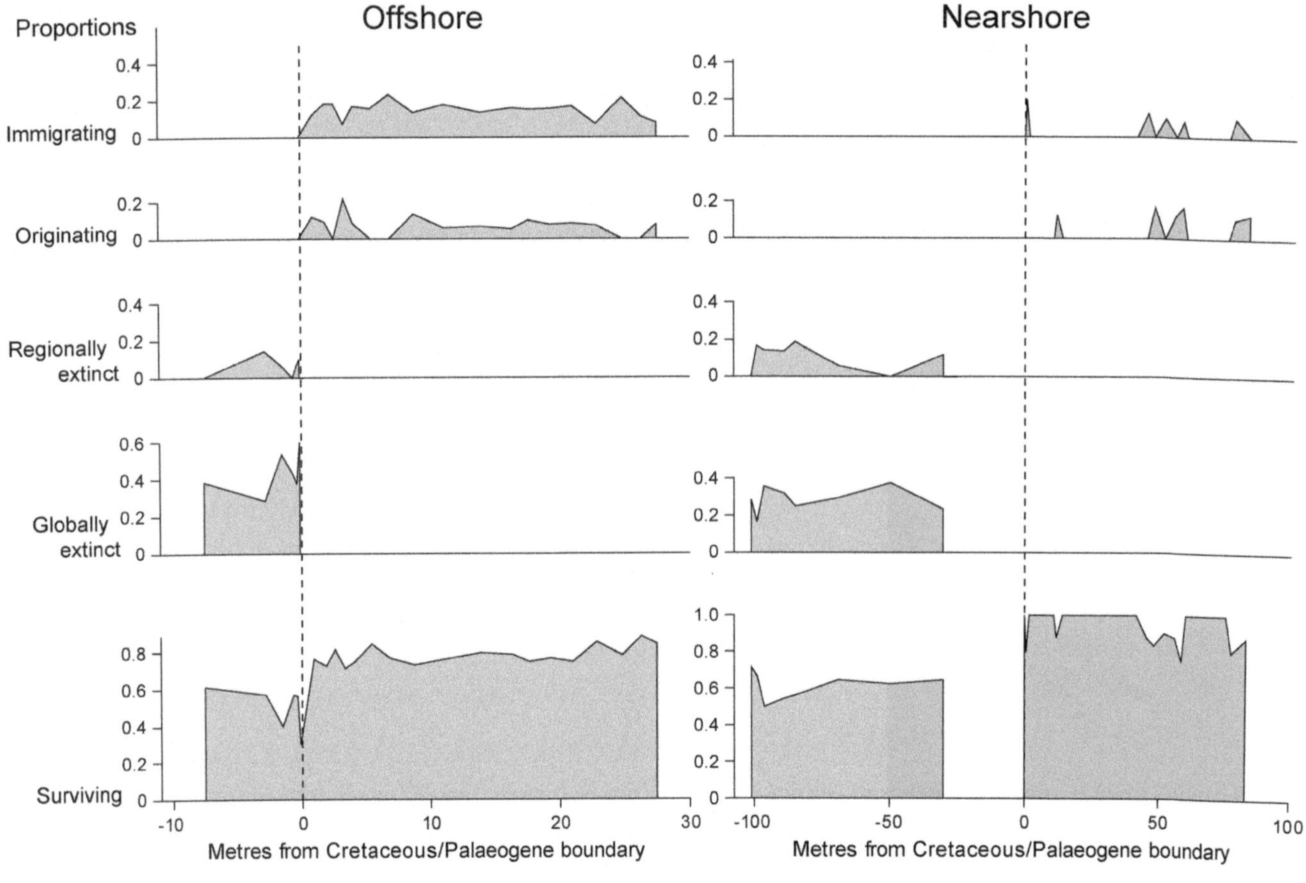

Figure 3. Proportional diversity of molluscan genera in five evolutionary/biogeographical categories across the Cretaceous-Palaeogene boundary. Proportions are shown for an offshore (panels on the left) and a nearshore boundary section (panels on the right) in Patagonia. Dashed vertical lines indicate position of the Cretaceous-Palaeogene boundary; values on x-axis refer to metres before (Cretaceous, negative values) or after the boundary. Nearshore, the uppermost *ca* 26 metres of the Cretaceous are characterised by low-diversity brackish-water assemblages, which are not included in the figure. Post-extinction faunas are dominated by survivors. Immigrating and originating genera play subordinate roles in the composition of Danian samples.

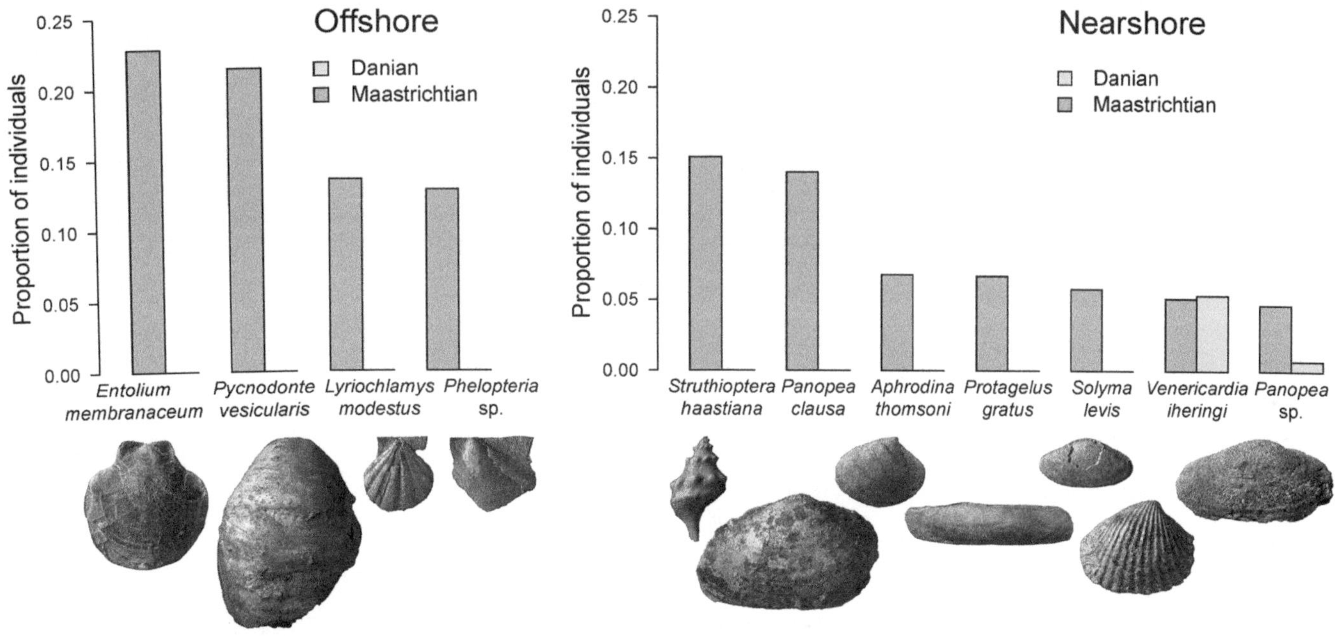

Figure 4. Proportional abundance of most common Maastrichtian molluscs contrasted with their abundances in the Danian. The displayed Maastrichtian species are those in the upper 90th percentile of the pooled samples of that time. Their Danian proportions are plotted alongside. The majority of the dominant Maastrichtian species did not survive into the Danian.

important mechanism for the delayed recovery in both onshore and offshore environments of Patagonia.

The extinction of dominants and the rise of rare species

In both studied environments the most common Maastrichtian species became extinct at the K-Pg boundary (Figure 4). This observation appears to be in contrast with ecological studies which suggest that high abundance promotes survival [36]. It can, however, be reconciled with the finding that post-Palaeozoic marine bivalves with high abundances exhibit extinction rates elevated over those of moderately common genera [37]. Also, results of a regional study of North American bivalves showed that

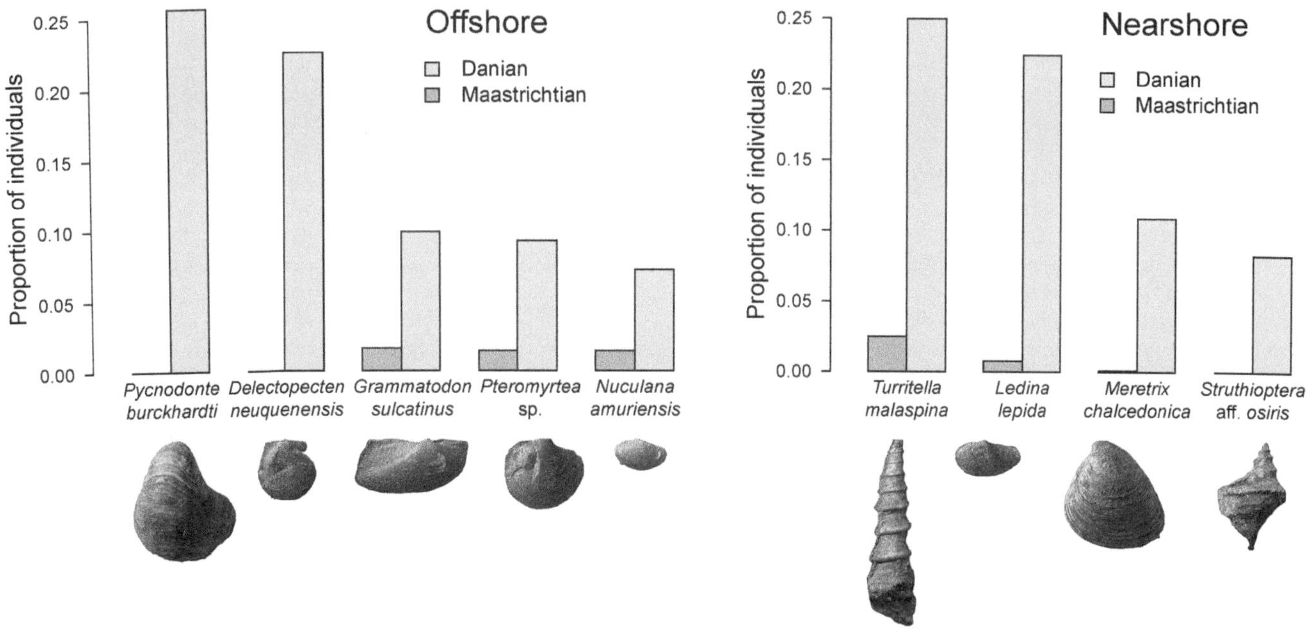

Figure 5. Proportional abundance of most common Danian molluscs contrasted with their abundances in the Maastrichtian. The displayed Danian species are those in the upper 90th percentile of the pooled samples of that time. Their Maastrichtian proportions are plotted alongside. Most of the dominant Danian species are survivors that were rare during the Maastrichtian. The two most abundant Danian offshore species are Patagonian endemics that newly originated in the early Danian.

abundant taxa were no more likely to survive the K-Pg mass extinction than rare ones [38].

Interestingly, survivors that dominate the Danian assemblages were rare during the Maastrichtian (Figure 5). We suggest that although being resistant to environmental stress, the new dominants were poor competitors before the boundary event. Once they were released from competition by the removal of dominants, the relative competitive ability of survivors increased. This suggests a particular type of incumbency, where numerically subordinate taxa arose to ecological dominance in the aftermath of a major environmental perturbation. Modelling of competitive interactions provides a potential mechanism for such reorganisations of ecosystems [39]: The loss of a competitively superior species releases otherwise unavailable resources, allowing the competitively inferior species to expand its realised niche. Thus, less abundant species may not only provide invasion resistance by their aggregate effects on resource utilization [40], but can be important by gaining dominance after environmental turmoil. Because distinct inversions in abundance across mass extinctions are known from other ecosystems [41] this may indicate a more general macroecological pattern.

The diversity-based conclusions stress the importance of survivors in post-extinction assemblages (Figure 3). This finding is generally confirmed by our abundance-based analyses with one important modification. In the offshore environment, the two top-ranking Danian species are newly evolved bivalves that originated in the early Danian of Patagonia (Figure 5). Their dominance indicates that newly originating post-extinction taxa can have high numbers of individuals even though overall numbers of originating taxa are low. There is a striking taxonomical, morphological, and ecological analogy between these two Danian species and the two most abundant Maastrichtian species in the offshore environment (Figures 4, 5). All four species are epifaunal suspension feeders and it seems that, after its extinction, the Maastrichtian pectinoid *Entolium membranaceum* was replaced by the Danian pectinoid *Delectopecten neuquenensis*, whereas the oyster *Pycnodonte vesicularis* was replaced by the congeneric *Pycnodonte burckhardti*. Although end-Cretaceous extinctions did not remove entire ecological guilds it is plausible that they created enough amount of open ecospace into which taxa with equivalent ecological traits could evolve.

The pace of recovery in sample-level diversity

The mean sample-level diversity remained unchanged across the K-Pg boundary in the studied Patagonian offshore environments at BJG and Opaso (Figure 1). The early Danian hiatus of *ca* 500 kyr at BJG [11] provides an upper limit for the duration necessary to restore sample-level diversity. The stratigraphically more complete section at Brazos River, east Texas, represents an environment similar to our offshore setting. Sample-level diversity at Brazos started to rebound *ca* 1 myr after the boundary [6] and full local diversity recovery to Cretaceous levels was attained after about 2 myr [42]. Thus, complete recovery of sample-level diversity to pre-extinction values took longer there than in Patagonia. Similarly, another study concluded that the local richness of offshore mollusc assemblages of the U.S. Gulf Coastal Plain recovered to Cretaceous values *ca* 2.7 myr after the event [43]. In conclusion, the time necessary to recover to pre-extinction diversity in mollusc-dominated offshore ecosystems varies and may take less than *ca* 0.5 myr according to our results from Patagonia and up to *ca* 2.7 myr according to data from the U.S. Gulf Coastal Plain. In contrast to offshore settings, Danian sample-level diversity did not recover to Maastrichtian levels in the Patagonian nearshore environments. We have shown from the data that there is a prolonged period of incomplete recovery in nearshore environments, although a lack of more detailed chronostratigraphic information prevents us from quantifying its duration more precisely.

Divergence of Maastrichtian and Danian diversity structure

Accumulation of total Danian diversity slowed down over time (Figure 2), seemingly compatible with a model of diversity-dependent niche-filling and a gradual approach to carrying capacity. These curves, however, should not be taken as accurate reflections of the rate at which post-extinction diversity builds up because they build on the assumption that (1) all genera found in any sample in the Danian are also present in all younger Danian samples, and (2) any genera not found in older samples represent true absences.

Geological time is critical for our interpretation that Figure 2 indicates systematic differences between Maastrichtian and Danian biodiversity. For example, the underlying (re-)population dynamics may be the same for each interval, but an apparent difference could be generated if Maastrichtian strata represented a much longer time span than Danian strata. Because the opposite is true in the offshore section (Danian samples encompass about twice as much time as Maastrichtian samples) any potential bias would operate only in the reverse direction. Interval lengths are less clear in the nearshore section but we assume roughly equal amounts of time (see above).

The reduced total diversity in the Danian and the longer-term stability of sample-level diversity across the boundary in offshore habitats are no contradiction. Jointly these patterns imply that, on average, Danian offshore assemblages retained, or rapidly regained, their Maastrichtian sample-level diversity despite being recruited from an overall smaller pool of Danian taxa. This indicates a shift in the partitioning of sample-level diversity and total diversity.

Erwin [44] pointed out that positive feedbacks may increase carrying capacity during biotic recovery. An example of a positive feedback in mollusc-dominated ecosystems may be the Cenozoic radiation of predatory gastropods and the resulting escalatory trends ([43] and references therein). If such effects were present in our area, they must have occurred subsequent to the studied time interval.

Nearshore–offshore differentials in recovery

In contrast to the offshore environment, the reduction in nearshore Danian sample-level diversity was pronounced and lasted over the entire studied interval. Differential evolutionary dynamics along onshore–offshore gradients – well-known in post-Palaeozoic benthic marine invertebrates, with preferential origination occurring onshore [45], [46] – is inadequate to account for the observed difference in our study. Not only would this mechanism favour the recovery of onshore diversity – which is just the opposite of the pattern we observe – but originations in general played only a minor role during recovery in our study areas. The nearshore–offshore difference might have arisen from unpredictable, stochastic processes, consistent with the contingent recovery model [47] which emphasises the importance of chance in the recovery of marine plankton from the end-Cretaceous extinction. A more mechanistic explanation focuses on the role of survivors as the most diverse group. We suggest that the recolonisation by survivors from nearby localities may have been facilitated in the spatially and temporally more homogeneous and continuous offshore environment. Under such conditions we expect that populations are well connected with each other. In ecology, high levels of connectivity within metapopulations are often considered to be beneficial for recovery after local depletion or extirpation [48]. In contrast, species inhabiting the more variable nearshore environments tend to have more patchy distributions. Their supposedly lower connectivity among populations may have hampered the recolonisation by survivors in nearshore settings.

Conclusions

We analysed so far underrepresented K-Pg boundary sections from the Southern Hemisphere, which contribute to a more complete understanding of recovery processes. Overall, our results corroborate and extend earlier work finding interregional and environmental variation in recovery dynamics in benthic [3], [7] and planktonic marine systems [28], [29]. Diversity of the post-extinction interval was controlled by taxa that survived the extinction event in Patagonia. We emphasise the importance of biotic interactions and conclude that survivors had negative effects on the potential origination of new taxa and the invasion of exotic genera through incumbency. Furthermore, survivors constituted the pool of taxa from which the majority of new dominants were recruited after the most abundant Maastrichtian species went extinct at the boundary. Such distinct shifts in the abundances of survivors are a feature these benthic marine ecosystems share with terrestrial plant communities across mass extinction boundaries [49]. Restoration of sample-level diversity of mollusc assemblages can be rapid ($<$ *ca* 500 kyr in offshore habitats). In contrast, total Danian diversity remained below that of the Maastrichtian, despite a marked Palaeogene increase in the global diversity of gastropods and bivalves [8].

Acknowledgments

We thank Dave Jablonski, Dave Lazarus, Jocelyn Sessa, Sonny Walton, and an anonymous reviewer for valuable discussions and constructive comments on the manuscript. We thank Francisco Medina, Roberto Scasso, Henning Scholz, and Sven Weidemeyer for assistance in the field, and Elke Siebert for support in figure preparation. This is Paleobiology Database publication number 200.

Author Contributions

Analyzed the data: WK MA. Wrote the paper: MA. Designed the study: MA. Interpreted results: MA WK. Prepared figures: WK MA. Improved final manuscript: WK.

References

1. McKinney ML (1997) Extinction vulnerability and selectivity: Combining ecological and paleontological views. Annu Rev Ecol Syst 28: 495–516.
2. Erwin DH (2001) Lessons from the past: Biotic recoveries from mass extinctions. Proc Natl Acad Sci USA 98: 5399–5403. (doi:10.1073/pnas.091092698).
3. Aberhan M, Weidemeyer S, Kiessling W, Scasso RA, Medina FA (2007) Faunal evidence for reduced productivity and uncoordinated recovery in Southern Hemisphere Cretaceous/Paleogene boundary sections. Geology 35: 227–230. (doi:10.1130/G23197A.1).
4. Solé RV, Saldaña J, Montoya JM, Erwin DH (2010) Simple model of recovery dynamics after mass extinction. J Theor Biol 267: 193–200. (doi:10.1016/j.jtbi.2010.08.015).
5. Schulte P, Alegret L, Arenillas I, Arz JA, Barton PJ, et al. (2010) The Chicxulub asteroid impact and mass extinction at the Cretaceous-Paleogene boundary. Science 327: 1214–1218. (doi:10.1126/science.1177265).
6. Hansen TA, Farrell BR, Upshaw B (1993) The first 2 million years after the Cretaceous-Tertiary boundary in east Texas: Rate and paleoecology of the molluscan recovery. Paleobiology 19: 251–265.
7. Jablonski D (1998) Geographic variation in the molluscan recovery from the end-Cretaceous extinction. Science 279: 1327–1330.
8. Alroy J (2010) The shifting balance of diversity among major marine animal groups. Science 329: 1191–1194. (doi:10.1126/science.1189910).
9. Scasso RA, Aberhan M, Ruiz L, Weidemeyer S, Medina FA, et al. (2012) Integrated bio- and lithofacies analysis of coarse-grained, tide-dominated deltaic environments across the Cretaceous/Paleogene boundary in Patagonia, Argentina. Cretaceous Research 36: 37–57. (doi:10.1016/j.cretres.2012.02.002).
10. Scasso RA, Concheyro A, Kiessling W, Aberhan M, Hecht L, et al. (2005) A tsunami deposit at the Cretaceous-Paleogene boundary in the Neuquén Basin of Argentina. Cretaceous Research 26: 283–297. (doi:10.1016/j.cretres.2004.12.003).
11. Keller G, Adatte T, Tantawy AA, Berner Z, Stinnesbeck W, et al. (2007) High stress late Maastrichtian – early Danian palaeoenvironment in the Neuquén Basin, Argentina. Cretaceous Research 28: 939–960. (doi:10.1016/j.cretres.2007.01.006).
12. Gradstein FM, Ogg JG, Schmitz MD, Ogg GM (2012) The geologic time scale 2012. Oxford: Elsevier. 1144 p.
13. Papú OH, Prámparo MB, Náñez C, Concheyro A (1999) Palinología y micropaleontología de la formación Jagüel (Maastrichtiano-Daniano), perfil Opaso, cuenca Neuquina, Argentina. Servicio Geológico Minero Argentino, Anales 33: 17–31.
14. Barreda VD, Cúneo NR, Wilf P, Currano ED, Scasso RA, et al. (2012) Cretaceous/Paleogene floral turnover in Patagonia: drop in diversity, low extinction, and a *Classopollis* spike. PLoS One 7: 1–8. (doi:10.1371/journal.pone.0052455).
15. Patzkowsky ME, Holland SM (2007) Diversity partitioning of a Late Ordovician marine biotic invasion: controls on diversity in regional ecosystems. Paleobiology 33: 295–309.
16. Kowalewski M, Kiessling W, Aberhan M, Fürsich FT, Scarponi D, et al. (2006) Ecological, taxonomic, and taphonomic components of the post-Paleozoic increase in sample-level species diversity of marine benthos. Paleobiology 32: 533–561.
17. Alroy J (2010) Fair sampling of taxonomic richness and unbiased estimation of origination and extinction rates. In: Alroy J, Hunt G, editors. Quantitative Methods in Paleobiology. The Paleontological Society Papers 16. 55–80.
18. Gotelli NJ, Colwell RK (2001) Quantifying biodiversity: procedures and pitfalls in the measurement and comparison of species richness. Ecol Lett 4: 379–391.
19. Sepkoski JJ (2002) A compendium of fossil marine animal genera. Bulletins of American Paleontology 363: 1–563.
20. Hansen T, Farrand RB, Montgomery HA, Billman HG, Blechschmidt G (1987) Sedimentology and extinction patterns across the Cretaceous-Tertiary boundary interval in east Texas. Cretaceous Research 8: 229–252. (doi:10.1016/0195-6671(87)90023-1).
21. Bryan JR, Jones DS (1989) Fabric of the Cretaceous–Tertiary marine macrofaunal transition at Braggs, Alabama. Palaeogeogr Palaeoclimatol Palaeoecol 69: 279–301. (doi:10.1016/0031-0182(89)90170-3).
22. Heinberg C (1999) Lower Danian bivalves, Stevns Klint, Denmark: Continuity across the K/T boundary. Palaeogeogr Palaeoclimatol Palaeoecol 154: 87–106. (doi:10.1016/S0031-0182(99)00088-7).
23. Jablonski D (2002) Survival without recovery after mass extinctions. Proc Natl Acad Sci USA 99: 8139–8144. (doi:10.1073/pnas.102163299).
24. Harries PJ, Kauffman EG, Hansen TA (1996) Models for biotic survival following mass extinction. In: Hart MB, editor. Biotic recovery from mass extinction events. London: The Geological Society. 41–60.
25. Erwin DH (1998) The end and the beginning: recoveries from mass extinctions. Trends Ecol Evol 13: 344–349.
26. Stilwell JD (2003) Patterns of biodiversity and faunal rebound following the K–T boundary extinction event in Austral Palaeocene molluscan faunas. Palaeogeogr Palaeoclimatol Palaeoecol 195: 319–356. (doi:10.1016/S0031-0182(03)00364-X).
27. Coxall HK, D'Hondt S, Zachos JC (2006) Pelagic evolution and environmental recovery after the Cretaceous-Paleogene mass extinction. Geology 34: 297–300. (doi:10.1130/G21702.1).
28. Jiang S, Bralower TJ, Patzkowsky ME, Kump LR, Schueth JD (2010) Geographic controls on nannoplankton extinction across the Cretaceous/Palaeogene boundary. Nat Geosci 3: 280–285. (doi:10.1038/ngeo775).
29. Hull PM, Norris RD (2011) Diverse patterns of ocean export productivity change across the Cretaceous-Paleogene boundary: New insights from biogenic barium. Paleoceanography 26: 1–10. (doi:10.1029/2010PA002082).
30. Valentine JW, Jablonski D, Krug AZ, Roy K (2008) Incumbency, diversity, and latitudinal gradients. Paleobiology 35: 169–178.
31. Algar AC, Mahler DL, Glor RE, Losos JB (2013) Niche incumbency, dispersal limitation and climate shape geographical distributions in a species-rich island adaptive radiation. Glob Ecol Biogeogr 22: 391–402. (doi:10.1111/geb.12003).

32. Sheehan PM (2008) Did incumbency play a role in maintaining boundaries between Late Ordovician brachiopod realms? Lethaia 41: 147–153.
33. Vermeij GJ, Roopnarine PD (2013) Reining in the Red Queen: the dynamics of adaptation and extinction reexamined. Paleobiology 39: 560–575. (doi:10.1666/13009).
34. Hallam A (1990) Biotic and abiotic factors in the evolution of early Mesozoic marine molluscs. In: Ross RM, Allmon WD, editors. Causes of evolution – a paleontological perspective. Chicago: The University of Chicago Press. 249–269.
35. Bambach RK, Knoll AH, Sepkoski JJ (2002) Anatomical and ecological constraints on Phanerozoic animal diversity in the marine realm. Proc Natl Acad Sci USA 99: 6854–6859. (doi:10.1073/pnas.092150999).
36. Gaston KJ, Blackburn TM, Greenwood JJD, Gregory RD, Quinn RM, et al. (2000) Abundance–occupancy relationships. J Appl Ecol 37 (Suppl. 1): 39–59.
37. Simpson C, Harnik PG (2009) Assessing the role of abundance in marine bivalve extinction over the post-Paleozoic. Paleobiology 35: 631–647.
38. Lockwood R (2003) Abundance not linked to survival across the end-Cretaceous mass extinction: Patterns in North American bivalves. Proc Natl Acad Sci USA 100: 2478–2482. (doi:10.1073/pnas.0535132100).
39. Roopnarine PD (2006) Extinction cascades and catastrophe in ancient food webs. Paleobiology 32: 1–19.
40. Lyons KG, Brigham CA, Traut BH, Schwartz MW (2005) Rare species and ecosystem functioning. Conserv Biol 19: 1019–1024. (doi:10.1111/j.1523-1739.2005.00106.x).
41. McElwain JC, Popa ME, Hesselbo SP, Haworth M, Surlyk F (2007) Macroecological responses of terrestrial vegetation to climatic and atmospheric change across the Triassic/Jurassic boundary in East Greenland. Paleobiology 33: 547–573.
42. Hansen TA, Kelley PH, Haasl DM (2004) Paleoecological patterns in molluscan extinctions and recoveries: comparison of the Cretaceous–Paleogene and Eocene–Oligocene extinctions in North America. Palaeogeogr Palaeoclimatol Palaeoecol 214: 233–242. (doi:10.1016/j.palaeo.2004.06.017).
43. Sessa JA, Bralower TJ, Patzkowsky ME, Handley JC, Ivany LC (2012) Environmental and biological controls on the diversity and ecology of Late Cretaceous through early Paleogene marine ecosystems in the U.S. Gulf Coastal Plain. Paleobiology 38: 218–239.
44. Erwin DH (2007) Increasing returns, ecological feedback and the Early Triassic recovery. Palaeoworld 16: 9–15. (doi:10.1016/j.palwor.2007.05.013).
45. Jablonski D (2005) Evolutionary innovations in the fossil record: the intersection of ecology, development, and macroevolution. J Exp Zool 304B: 504–519. (doi:10.1002/jez.b.21075).
46. Kiessling W, Aberhan M (2007) Environmental determinants of marine benthic biodiversity dynamics through Triassic-Jurassic times. Paleobiology 33: 414–434.
47. Hull PM, Norris RD, Bralower TJ, Schueth JD (2011) A role for chance in marine recovery from the end-Cretaceous extinction. Nat Geosci 4: 856–860. (doi:10.1038/ngeo1302).
48. Hughes TP, Carpenter S, Rockström J, Scheffer M, Walker B (2013) Multiscale regime shifts and planetary boundaries. Trends Ecol Evol 28: 389–395. (doi:10.1016/j.tree.2013.05.019).
49. McElwain JC, Punyasena SW (2007) Extinction events and the plant fossil record. Trends Ecol Evol 22: 548–557. (doi:10.1016/j.tree.2007.09.003).
50. del Río CJ (2004) Tertiary marine molluscan assemblages of eastern Patagonia (Argentina): a biostratigraphic analysis. J Paleontol 78: 1097–1122.
51. Lazo DG (2003) Taxonomy, facies relationships and palaeobiology of bakevelliid bivalves from the Lower Cretaceous of west-central Argentina. Cretaceous Research 24: 765–788. (doi:10.1016/j.cretres.2003.09.001).
52. Lazo DG (2007) Early Cretaceous bivalves of the Neuquén Basin, west-central Argentina: notes on taxonomy, palaeobiogeography and palaeoecology. Geol J 42: 127–142. (doi:10.1002/gj.1080).
53. Kauffman EG, Leanza HA (2004) A remarkable new genus of Mytilidae (Bivalvia) from the Lower Cretaceous of Southwestern Gondwanaland. J Paleontol 78: 1187–1191. (doi:10.1666/0022-3360(2004)078).
54. del Río CJ (2002) Molluscos del Terciario marino. Geología y Recursos Naturales de Santa Cruz 2: 1–22.
55. Zinsmeister WJ, Griffin M (1995) Late Cretaceous and Tertiary aporrhaid gastropods from the southern rim of the Pacific Ocean. J Paleontol 69: 692–702.
56. Chiesa JO, Camacho HH (2001) Invertebrados marinos Eocenos de la parte inferior de la Formacíon Río Foyel, Provincia de Río Negro, Argentina. Rev Esp Paleontol 16: 299–316.
57. Chiesa JO, Parma SG, Camacho HH (1995) Invertebrados marinos de la Formación El Chacay (Eoceno), provincia de Santa Cruz, Argentina. Sistemática y bioestratigrafía. Academia Nacional de Ciencias Exactas, Físicas y Naturales, Monografías 11: 17–68.
58. Casadío S, Griffin M, Parras A (2005) *Camptonectes* and *Plicatula* (Bivalvia, Pteriomorphia) from the Upper Maastrichtian of northern Patagonia: palaeobiogeographic implications. Cretaceous Research 26: 507–524. (doi:10.1016/j.cretres.2005.01.009).
59. Woods H (1917) The Cretaceous faunas of the north-eastern part of the South Island of New Zealand. New Zealand Geological Survey, Palaeontological Bulletin 4: 1–41.
60. Casadío S, Griffin M, Marenssi S, Net L, Parras A, et al. (2009) Paleontology and sedimentology of Middle Eocene rocks in Lago Argentino area, Santa Cruz Province, Argentina. Ameghiniana 46: 27–47.
61. Squires RL, Goedert JL (1994) New species of Eocene small to minute mollusks from the Crescent Formation, Black Hills, southwestern Washington. Veliger 37: 253–266.
62. del Río CJ, Concheyro A, Martínez SA (2011) The Maastrichtian – Danian at General Roca (Patagonia, Argentina): a reappraisal of the chronostratigraphy and biostratigraphy of a type locality. Neues Jahrb Geol Paläontol Abh 259: 129–156. (doi:10.1127/0077-7749/2011/0103).
63. Berggren WA, Kent DV, Swisher CC, Aubry M-P (1995) A revised Cenozoic geochronology and chronostratigraphy. In: Berggren WA, et al., editors. Geochronology, time scales and global stratigraphic correlation. SEPM Special Publications 54: 129–212.
64. Harrington HJ (1958) Geology of Kaitangata Coalfield. New Zealand Geological Survey Bulletin 59: 1–131.
65. Kafanov AI, Ogasawara K (2004) Neogene and Paleogene molluscan (Bivalvia) cenozones of the Russian Northeast. Science Reports of the Institute of Geoscience, University of Tsukuba, Section B 25: 9–45.
66. del Río CJ, Martínez SA, Stilwell JD, Concheyro A (2007) Paleontology of the Cerros Bayos section, Roca Formation (Danian), La Pampa Province, Argentina. Alcheringa 31: 241–269. (doi:10.1080/03115510701484713).
67. Waller TR, Marincovich L (1992) New species of *Camptochlamys* and *Chlamys* (Mollusca: Bivalvia: Pectinidae) from near the Cretaceous/Tertiary boundary at Ocean Point, North Slope, Alaska. J Paleontol 66: 215–227.
68. del Río CJ, Beu AG, Martínez SA (2008) The pectinoidean genera *Delectopecten* Stewart, 1930 and *Parvamussium* Sacco, 1897 in the Danian of Northern Patagonia, Argentina. Neues Jahrb Geol Paläontol Abh 249: 281–295.
69. Aguirre-Urreta MB (1989) The Cretaceous decapod Crustacea of Argentina and the Antarctic Pensinsula. Palaeontology 32: 499–552.
70. Darragh TA (1997) Gastropoda, Scaphopoda, Cephalopoda, and new Bivalvia of the Paleocene Pebble Point Formation, Victoria, Australia. Proceedings of the Royal Society of Victoria 109: 57–108.
71. Chavan A (1946) Nouvelles notes sur les *Jagonia* (Lamellibranches). Bulletin du Muséum National d'Histoire Naturelle Paris, sér. 2, 18: 345–347.
72. Lazo DG, Damborenea SE (2011) Barremian bivalves from the Huitrín Formation, west-central Argentina: taxonomy and paleoecology of a restricted marine association. J Paleontol 85: 721–749.
73. Stilwell JD (2003) Macropalaeontology of the *Trochocyathus-Trematotrochus* band (Paleocene/Eocene boundary), Dilwyn Formation, Otway Basin, Victoria. Alcheringa 27: 245–275. (doi:10.1080/03115510308619107).
74. Stilwell JD, Zinsmeister WJ, Oleinik AE (2004) Early Paleocene mollusks of Antarctica: systematics, paleoecology and paleobiogeographic significance. Bulletins of American Paleontology 367: 1–89.
75. Griffin M, Hünicken MA (1994) Late Cretaceous-Early Tertiary gastropods from southwestern Patagonia, Argentina. J Paleontol 68: 257–274.

2

Phylogenetic Distribution of Extant Richness Suggests Metamorphosis Is a Key Innovation Driving Diversification in Insects

James L. Rainford[1]*, Michael Hofreiter[1,2], David B. Nicholson[1,3,4], Peter J. Mayhew[1]

1 Department of Biology, University of York, York, United Kingdom, **2** Faculty of Mathematics and Natural Sciences, Institute for Biochemistry and Biology, University of Potsdam, Potsdam, Germany, **3** Department of Natural Sciences, National Museums Scotland, Edinburgh, United Kingdom, **4** Department of Earth Sciences, The Natural History Museum, London, United Kingdom

Abstract

Insects and their six-legged relatives (Hexapoda) comprise more than half of all described species and dominate terrestrial and freshwater ecosystems. Understanding the macroevolutionary processes generating this richness requires a historical perspective, but the fossil record of hexapods is patchy and incomplete. Dated molecular phylogenies provide an alternative perspective on divergence times and have been combined with birth-death models to infer patterns of diversification across a range of taxonomic groups. Here we generate a dated phylogeny of hexapod families, based on previously published sequence data and literature derived constraints, in order to identify the broad pattern of macroevolutionary changes responsible for the composition of the extant hexapod fauna. The most prominent increase in diversification identified is associated with the origin of complete metamorphosis, confirming this as a key innovation in promoting insect diversity. Subsequent reductions are recovered for several groups previously identified as having a higher fossil diversity during the Mesozoic. In addition, a number of recently derived taxa are found to have radiated following the development of flowering plant (angiosperm) floras during the mid-Cretaceous. These results reveal that the composition of the modern hexapod fauna is a product of a key developmental innovation, combined with multiple and varied evolutionary responses to environmental changes from the mid Cretaceous floral transition onward.

Editor: Axel Janke, BiK-F Biodiversity and Climate Research Center, Germany

Funding: The study was primarily funded by NERC (http://www.nerc.ac.uk) grant NE/J500197/1. Contributions were also made by the National Museums of Scotland and The Natural History Museum, London in funding for DN. The funders had no role in study design, data collection and analysis, decision to publish, or preparation of the manuscript.

Competing Interests: The authors have declared that no competing interests exist.

* Email: jlr516@york.ac.uk

Introduction

Hexapoda, including insects and their six-legged relatives, are the most species-rich animal clade in terrestrial ecosystems and collectively comprise over half of all described extant species [1,2]. Therefore understanding the origins of this exceptional richness is key to understanding the history of life on land and the assembly of terrestrial ecosystems [3]. In addition to their high overall species richness, insect groups are also remarkable for the degree of disparity in richness existing among the major sub-clades. For example the orders Zoraptera ("angel insects") and Coleoptera (beetles) differ in richness by four orders of magnitude (32 and 350,000 described extant species, respectively [2]). A key part of the discussion on these differences in extant richness relates to the hypothesized effects of potential key innovations that may have acted as drivers for hexapod richness [3]. Such proposed innovations include both major morphological developments including the origin of the insect body plan [2–4], flight [2–6], the capacity to fold the wings [7,8] and the origin of complete metamorphosis [2–4,9], and ecological opportunities or innovations, notably the evolution of flowering plants (angiosperms) [10–12] and parasitism [13].

Attempts to explicitly test these ideas within a phylogenetic framework have either been restricted to particular orders [14–16], thus omitting a wider context, or have ignored variation within orders [7,8]. Here we integrate these disparate approaches by producing a dated hypothesis of phylogenetic relationships across the hexapods that is near-complete at the family level, through the combination of previously published molecular sequence data and a set of literature derived constraints (see below and Supplementary materials). Our goal is therefore not to present a novel estimate of the hexapod phylogeny (see discussion below), but instead to focus on what current taxonomic, phylogenetic and paleontological evidence reveals about broad patterns of diversification within the group, and its relationship with key evolutionary innovations, environmental changes and mass extinctions [17–19].

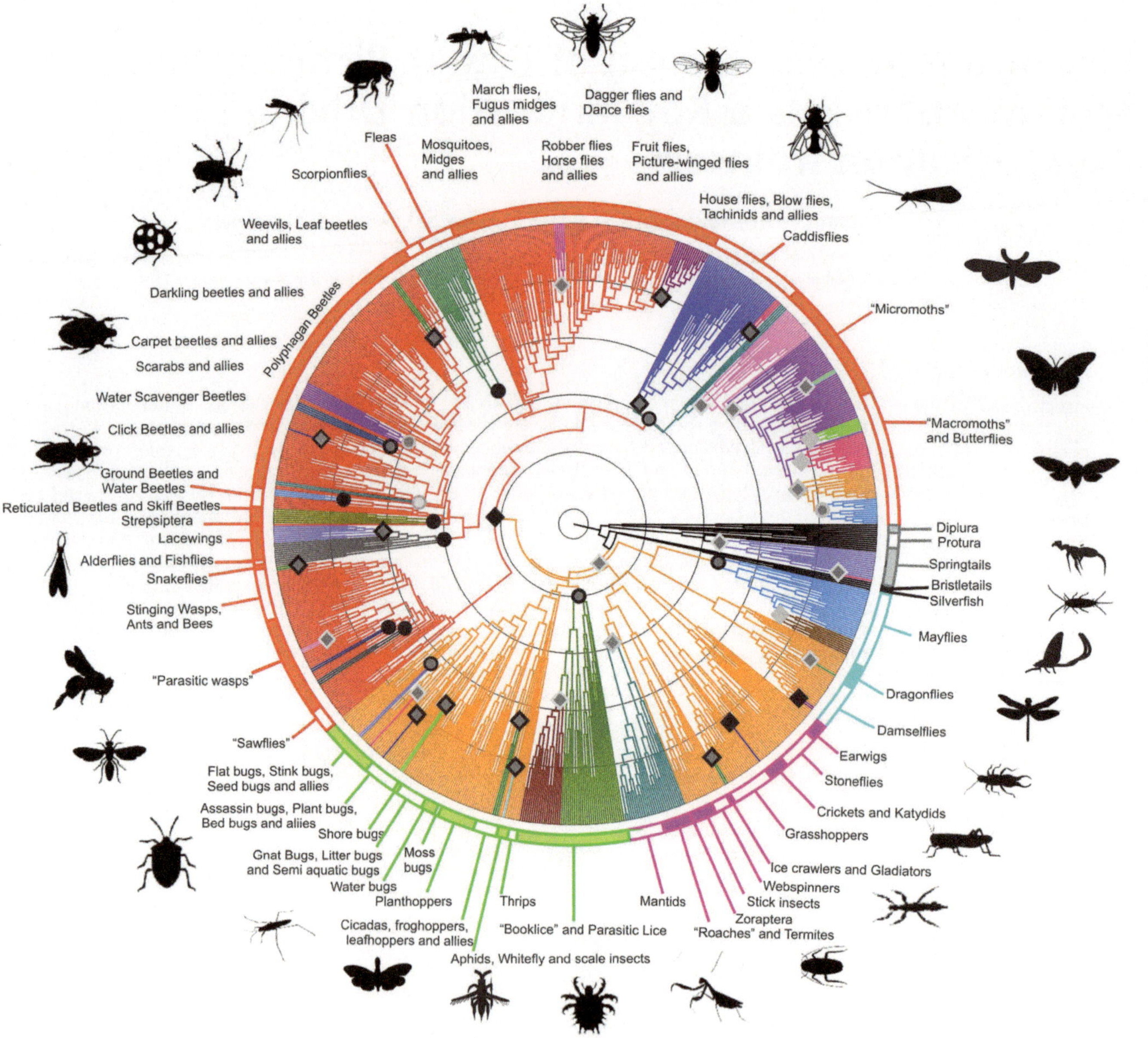

Figure 1. Dated phylogeny of extant hexapod families showing diversification rate shifts. The tree shown is from a maximum likelihood analysis of 8 genes, calibrated by 89 fossils. Membership of major clades is denoted by coloration of the ring (grey: Entognatha, black: basal insects, cyan: Palaeoptera, magenta: Polyneoptera, green: Paraneoptera, red: Holometabola). Changes in branch coloration denote diversification shifts identified using TurboMEDUSA (Table S3). Branch colors identify regions of the tree with the same underlying diversification model. Symbols at shifts denote a net upshift (diamond) or down shift (circle). Coloration of symbols reflects the robustness of the shift event across 500-scaled samples taken from the post-burin MCMC chain (black: shift recovered in >80% of samples, grey with black outline: recovery >50%, grey with pale outline: recovery >30%, pale grey: recovery<30%). Black rings are shown at 100 Ma increments from the present. See Supplementary materials for further details and discussion. See also Figures S1–S3, Tables S1–S4, and Datafiles S1, S2.

Results

The dated phylogeny used in this study contains 874 higher taxa of Hexapoda (Fig. 1). Taxa were variously resolved to family or superfamily level, such that the presented tree incorporates a total of 903 of the approximately 1100 recognized extant families, with taxonomy following that given by GenBank references up to August 2013 (see Supplementary materials for further discussion). The tree was reconstructed using a combination of eight widely sampled molecular markers and literature-derived constraints on certain widely recognized phylogenetic nodes ([20,21] see Supplementary materials for details). The tree topology was inferred using a partitioned RAxML (maximum likelihood) analysis [22,23]. This topology was dated using a relaxed molecular clock implemented in MrBayes 3.2 [24] and calibrated using 86 fossil dates taken from the recent palaeoentomological literature (Table S2).

Using our dated tree we estimated the crown divergence of Hexapoda, i.e. the divergence of true insects from Entognatha (basal hexapods including springtails) as occurring in the

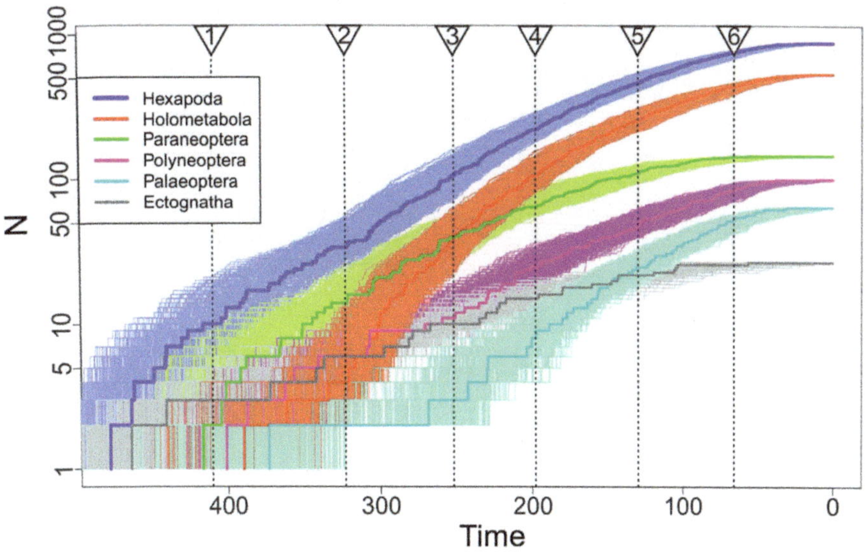

Figure 2. Lineage (y-axis; log scale) through time (x-axis; Ma) plot for the major groups of Hexapoda using the phylogeny in Fig. 1. Colors used identify the same clades as the ring in Fig. 1. Thick lines are calculated from the mean tree dates (Fig. 1). Shaded regions represent 500-scaled samples taken from the MCMC chain used in dating. Major events in the history of the group are denoted using dotted lines: 1. Oldest Hexapod fossil. 2. Oldest member of crown Pterygota (Polyneoptera). 3. Permo-Triassic mass extinction. 4. Origination of crown Angiosperms [44]. 5. Angiosperms become abundant in fossil record. 6. Cretaceous-Paleocene mass extinction.

Ordovician (mean estimate 474.4 Ma, 95% CI 439.6–502.9 Ma), which is consistent with other recent molecular clock estimates [25–27] (Fig S2). These estimates greatly exceed the age of the oldest securely placed hexapod fossils including the potential crown winged insect *Rhyniognatha hirsti* from the early Devonian [28]. Little is known regarding Devonian insect communities [2] and the nature of terrestrial communities at this early date remains poorly understood [29]. However, our results are in line with recent fossil evidence indicating an early (i.e. prior to the late Carboniferous) origin for major crown lineages, including the stem lineages of several orders of advanced Holometabola (insects that undergo complete metamorphosis) [30,31].

At higher taxonomic levels, lineage through time plots (Fig. 2) indicate a remarkable stability in divergence rate across all the major hexapod clades, with some suggestion of an elevated diversification rate in Holometabola during the late Permian corresponding to basal divergences within Coleoptera and Diptera (flies) [2,32]. Despite the conventional division between Paleozoic and post-Paleozoic insect faunas in paleoentomological research [2,33], our results reveal no evidence for changes in diversification rate around the time of the Permo-Triassic extinction event (P/T) (Fig. 2), suggesting that the radiation of extant groups was not strongly impacted by the loss of Paleozoic forms indicated by the fossil record [17,18]. A possible exception is an upshift in the diversity of Palaeoptera (dragonflies and mayflies) associated with the origin of crown members of the two orders, both of which undergo major taxonomic turnover during the P/T event [2,34] (Fig. 2).

Despite this apparent stability in the origination of higher taxa, the application of birth-death models [35,36] identifies two major transitions, characterized as shifts in the net diversification rate and turnover in the descendent clades, which together play a major role in defining the overall structure of hexapod diversification. These major shifts correspond to the origins of flight (Pterygota) and of complete metamorphosis (Holometabola) (Fig. 1, Table 1 and Table S3). Both in terms of the degree to which its inclusion improves the likelihood of diversification models (Table 1) and in its relative stability with respect to uncertainties in node age estimation (Fig. 1, Table S3), the upshift in diversification rate associated with the origin of complete metamorphosis represents the more strongly supported event. Previous studies proposing a link between complete metamorphosis and elevated diversification rates have been based on evidence in the fossil record [9,37], which for hexapods is highly incomplete [38]. In contrast, sister group comparisons, using earlier phylogenetic reconstructions [20,21], failed to recover a diversification shift associated with Holometabola [7,8]. However, likelihood ratio tests indicate that the birth-death models significantly favor this position over alternative proposals including Eumetabola (Holometabola plus its sister group) and Neoptera (insects able to fold their wings; Table 1). Earlier studies [8,39] have also provided some evidence supporting the role of flight in promoting hexapod diversification. Although our analysis supports this notion it also shows that the recovery of this shift is sensitive to uncertainties in divergence time estimates within the phylogeny rendering its overall role in hexapod diversification ambiguous (Fig. 1).

In addition to these broad patterns, diversification shift models identified a further forty-three clades on the tree potentially associated with shifts in diversification rates (Fig. 1, Table S3). These shifts vary in their intensity and robustness with respect to uncertainties in branch length and are distributed across the tree, with the majority occurring within the holometabolan radiation. Among the most robust and phylogenetically inclusive shifts are down-shifts impacting on known or suspected relict groups within the modern fauna. These include holometabolan groups such as Neuropterida (lacewings and their relatives), Mecoptera and Siphonaptera (scorpionflies and fleas [40]) and basal members of Coleoptera (beetles) and Lepidoptera (moths), as well as non-metamorphosing groups such as Ephemeroptera (mayflies) and Psocodea (booklice and parasitic lice) [41]. The fossil records for a number of these groups indicate a higher family richness during the Mesozoic, suggesting that their current representatives are surviving relics of taxa that were formerly more diverse [17,18], further supporting the results of the diversification shift models.

Table 1. Log likelihood and parameter estimates for models with a single shift in diversification rate.

Taxon 1	*lnL*	R root	R clade	Taxon 2	*lnL*	R root	R clade	ChiSquared	DF	p.value
Holometabola	−11299.630	0.00359686	0.0112969	**Uniform Model**	−11504.581	0.0087352	-	409.90	3	<0.001
Holometabola	−11299.630	0.00359686	0.0112969	**Obtectomera (Second Best Node)**	−11351.037	0.0073964	0.0264802	102.81	1	<0.001
Holometabola	−11299.630	0.00359686	0.0112969	**Paraneoptera**	−11433.738	0.0096620	0.0039822	268.22	1	<0.001
Holometabola	−11299.630	0.00359686	0.0112969	**Eumetabola**	−11393.178	0.0032761	0.0098645	187.10	1	<0.001
Holometabola	−11299.630	0.00359686	0.0112969	**Neoptera**	−11415.252	0.0020526	0.0093177	231.25	1	<0.001
Holometabola	−11299.630	0.00359686	0.0112969	**Pterygota**	−11474.292	0.0023812	0.0089912	349.32	1	<0.001
Holometabola	−11299.630	0.00359686	0.0112969	**Insecta**	−11481.919	0.0026486	0.0089487	364.58	1	<0.001
Holometabola	−11299.630	0.00359686	0.0112969	**Coleoptera**	−11495.184	0.0082814	0.0105405	391.11	1	<0.001
Holometabola	−11299.630	0.00359686	0.0112969	**Lepidoptera**	−11418.241	0.0075482	0.016791	237.22	1	<0.001
Holometabola	−11299.630	0.00359686	0.0112969	**Diptera**	−11486.288	0.0082253	0.0120813	373.32	1	<0.001
Holometabola	−11299.630	0.00359686	0.0112969	**Hymenoptera**	−11497.610	0.0085002	0.0111554	395.96	1	<0.001

Log likelihood tests performed to compare the optimal model, with shift placed on Holometabola, and various alternative models. Data shown are log likelihoods of the respective models estimated in laser using turnover estimates as estimated in the homogenous model in MEDUSA (i.e. that with no shifts). Net diversification rates are estimated for the partition including the root (R_{root}) and the descendants of the focal node (R_{clade}). Chi squared values, degrees of freedom and p values relate to results of likelihood ratio comparisons between the denoted models. See also Table S3.

In contrast with these relict groups, most of the shifts leading to a net increase in taxonomic richness are comparatively recent (Fig. 3) and are associated with restricted, but massively diverse lineages many of which are of large ecological significance in recent communities. Among the non-holometabolan insects these include large herbivorous radiations such as the katydids (Tettigoniidae), true and lubber grasshoppers (Acrididae and Romaleidae), aphids (Aphidoidea), leafhoppers and treehoppers (Membracoidea), as well as plant/lace bugs and stink bugs (Miridae/Tingidae and Pentatomidae). Also represented are predatory groups such as assassin bugs (Reduviidae) and certain families of dragonflies and damselflies (Odonata). The pattern of shifts within the Dictyoptera (which includes detritivorous roaches and termites as well as predatory mantids) [42] is unstable with respect to branch length, with the majority of samples failing to recover the small proposed shift encompassing the entirety of the group (Fig. S3, Table S4). These groups, with the exception of Dictyoptera, radiated during the mid to late Cretaceous, which may imply an association between these radiations and the restructuring of floral and faunal communities during this interval following the radiation of angiosperms [43,44].

Unsurprisingly, several upshifts in diversification within Holometabola also involve groups directly associated with the angiosperm radiation, with notable examples including leaf and longhorn beetles (Chrysomeloidea) [15,45] and advanced bees (Apidae and Megachilidae) [46]. Our results also strongly support an upshift encompassing the Calyptratae, which includes houseflies and the important parasitoid group Tachinidae [14]. The recovered pattern of diversification shifts in Lepidoptera is complex and highly sensitive to uncertainties in branch length estimation, reflecting the difficulties of accurately dating a group for which there is a lack of suitable calibration fossils [17,47], and which includes several regions of phylogenetic instability [48,49]. The pattern recovered from the mean estimates of node times indicates a nested model with an overall down-shift associated with the most basal moths followed by a series of up-shifts corresponding to the major clades Glossata (moths with a proboscis) and Ditrysia (moths with partitioned female reproductive tracts). The pattern of shifts within the advanced moths and butterflies is poorly resolved with a number of events showing limited robustness with respect to branch length variation. If shift recovery across multiple samples of the Markov Chain Monte Carlo used in dating is considered ([24], see Supplementary materials), several of these events are found to be collapsed into a single shift associated with the redefined Obtectomera [48,49] (Fig. S3, Table S4), which also corresponds to the shift associated with second largest improvement in overall model likelihood in single-shift models (Table 1, see Supplementary materials for further discussion). Comparable previous work on patterns of diversification within Diptera identified a series of nested shifts within the order that are not recovered in our study [14]. These differences can be attributed to the placement of radiations within a more inclusive phylogenetic context, i.e. within Holometabola in its entirety, resulting in greater estimated turnover within the group, as well as minor differences in taxonomic sampling and dating between analyses. Contrary to previous views, which have tended to emphasize the role of particular ecologies (notably phytophagy) [11,10] in determining patterns of hexapod richness, we find no evidence that the upshifted groups are associated with a particular set of life history traits. Instead, our results suggest diverse responses within the Mesozoic insect fauna to the ecological transition and novel opportunities provided by the Cretaceous angiosperm expansion [2].

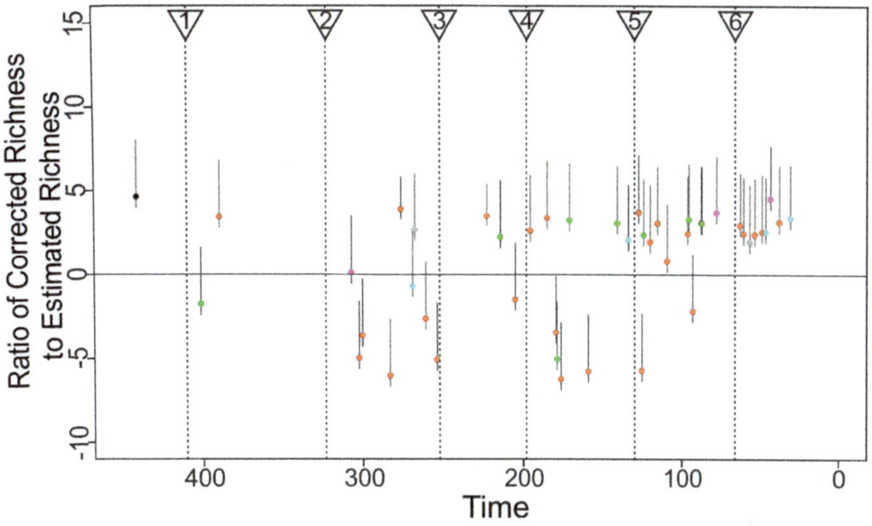

Figure 3. Change in species richness associated with shift events plotted through time. Values plotted show the ratios between the observed richness of the clade (after correction for nested shifts) and the mean estimated values of the richness of a clade of the appropriate age under the parental diversification model (see main text and Supplementary materials). Confidence intervals given are based on the change in richness associated with 95% CIs on the estimated outcomes of the stochastic diversification process. See also Table S3.

Discussion

Ultimate explanations of insect diversification can be classified into morphological key innovations and ecological interactions [3]. Our results highlight the importance of complete metamorphosis as the major key innovation underpinning the pattern of hexapod species richness. The mechanism by which complete metamorphosis promotes diversification is incompletely understood. However, previous workers have suggested that the ecological division of adult and juvenile life stages separated by a pupal stage in Holometabola may play a major role [3,4,9]. The adaptation to novel ecological niches likely played a role in promoting diversity within specific hexapod radiations, such as family-level or lower taxonomic levels, but there is no evidence here to support the idea that a single suite of ecological traits is generally associated with shifts in hexapod diversification. Instead, the patterns observed are consistent with distinct members of the community responding in a wide variety of ways to the ecological changes following the angiosperm radiation and continuing to the present day. However, we did find evidence that the radiation of angiosperms itself triggered a number of upshifts in diversification rate across both non-holometabolan and holometabolan groups, marking the evolution of angiosperms as a key ecological change in the evolutionary history of Hexapoda. It is important to note that our recognition of these patterns is dependent on the inferred phylogenetic topology, which contains some regions of considerable phylogenetic uncertainty (see Supplementary materials). However, it is unlikely that the major findings of our analysis – i.e. key roles of complete metamorphosis and angiosperm evolution as well as the failure to recover a distinct suite of ecological traits underlying a species group's phylogenetic richness – will change in the light of future improvements to the topology, dating, and extant species richness of the insect phylogenetic tree, which collectively will combine to further improve our understanding of the origins and diversification of this key component of terrestrial ecosystems.

Materials and Methods

The phylogeny of 874 terminal taxa, representing familial groups within Hexapoda was inferred based on eight widely sampled molecular markers including nuclear (CAD, Ef1α, PGD) and mitochondrial (COI, COII) protein coding sequences and 16S, 18S and 28S rRNA sequences. All included taxa are chimeric, i.e. the sequences used are assembled from multiple individuals and species, such as to reduce the problem of non-overlapping sampling within the source datasets and to maximize gene coverage for each of the sampled terminals. All gene partitions were aligned using MAFFT [50] with the exception of 18S and 28S rRNA, which were aligned using an automated profile alignment based on the structural reference database SILVA [51]. Conserved regions identified using the Gblocks protocol [52] and third codon positions for all the protein coding genes were excluded due to saturation. In total the concatenated sequence had a length of 7021bp and was 50.69% complete at the nucleotide level. Topological relationships within hexapods were inferred using a constrained maximum likelihood analysis in RAxML [22,23] implemented on the CIPRES web cluster [53]. Data was partitioned by nucleotide position and genome for protein coding sequences with the three ribosomal partitions each modeled independently. The tree topology was constrained in order to ensure comparability with other recent phylogenetic studies and to control the behavior of under-sampled and unstable taxa. The implemented constraints and further details relating to phylogenetic inference are described in Supplementary materials.

The fully resolved maximum likelihood topology estimated above was used as the basis for a relaxed independent gamma rates clock implemented in Mr. Bayes 3.2 [24]. Calibration was based on 86 fossils listed in Table S1, and implemented as hard minimum bounds on the ages of the relevant nodes with a hard maximum bound taken from a recent comparable molecular clock study [26]. Chains were run for 12 million generations with sampling conducted every 500 generations and a burn-in fraction of 50%. Further details are listed in Supplementary materials.

Estimates of extant richness for terminal taxa were compiled from recent encyclopedic sources (Supplementary materials). Models of clade diversification were implemented in R v 2.15.1 [54]. Topological shifts in diversification rate were identified using the stepwise algorithm MEDUSA [36] (package TurboMEDUSA [55]) on the dated consensus tree. Likelihood ratio tests to compare optimal placement of the first rate shift were implemented in the package Laser [56] using turnover estimates taken from MEDUSA. Estimates of richness of clades in the absence of rate shifts were calculated in the package Geiger [57,58]. The consistency of inferred shifts with respect to uncertainties in the node ages was accessed across 500 randomly sampled trees taken from the post-burnin phase of the dating chains. Further discussion of diversification analyses can be found in the Supplementary materials.

Supporting Information

Figure S1 Nodal support on the phylogeny. Nodes marked with circles are either constrained (red) or have high bootstrap support (light blue 50–80%, dark blue: over 80%).

Figure S2 Topology showing 95% confidence intervals on node ages (transparent blue bars). Black rings denote 100 Ma intervals from the present. Nodes denoted with red circles are involved in calibration (see Table S2 for details).

Figure S3 The fifty shifts with the highest rates of recovery in samples from the MCMC chain (Table S4) plotted together on the tree topology. Shifts are denoted as Fig. 1 with novel shifts not recovered on the mean tree denoted by red circles. Groupings on the ring and other information are as Fig.1.

Table S1 Estimates of extant species richness for terminal groups and GenBank accession numbers for sequences used in phylogenetic reconstruction.

Table S2 Fossil calibrations implemented in dating the tree topology. Calibrated nodes are plotted on Fig. S2. Where available, radiometric date estimates are referenced on the first occurrence of the deposit. Alternatively, the relevant stage termination is given based on ([1]-Supplementary references). References to cited studies in Supplementary materials.

Table S3 Parameter values and shifts in species richness associated with MEDUSA model shifts inferred across the mean topology. Shifts listed here are plotted on Fig. 1. Richness shifts are plotted on Fig. 3. See text and Supplementary materials for further discussion.

Table S4 The fifty most robustly recovered shifts inferred from 500 samples from the post-burnin Markov Chain Monte Carlo (MCMC). Shifts are plotted on Figure S3. Shifts without equivalents on the mean tree are highlighted in bold.

Datafile S1 Tree topology in Nexus format. Contains two tree files with a) an undated cladogram including bootstrap support for nodes and b) a dated topology including confidence intervals for the node ages and denotation of inferred shift events matching Fig. 1.

Datafile S2 Alignment and MrBayes instructions in Nexus format. Contains the implemented alignment after processing with the commands and priors used in setting the MrBayes dating run.

Text S1 Supplementary Experimental Procedures and Discussion. Contains further details of experimental procedures, discussion of topology and reliability of diversification shift estimates, and cited references for fossil calibrations and species richness estimates (Table S1, S2).

Acknowledgments

For advice on fossil calibrations we thank Dr Andrew Ross, Dr Olivier Béthoux and further colleagues in the paleoentomology community.

Author Contributions

Conceived and designed the experiments: JR MH PM. Performed the experiments: JR. Analyzed the data: JR. Contributed reagents/materials/analysis tools: JR DN. Wrote the paper: JR MH DN PM.

References

1. Gaston KJ (1991) The magnitude of global insect species richness. Conserv Biol 5: 283–296. doi:10.1111/j.1523-1739.1991.tb00140.x.
2. Grimaldi DA, Engel MS (2005) Evolution of the Insects. Cambridge University Press.
3. Mayhew PJ (2007) Why are there so many insect species? Perspectives from fossils and phylogenies. Biol Rev 82: 425–454. doi:10.1111/j.1469-185X.2007.00018.x.
4. Carpenter FM (1953) The Geological History and Evolution of Insects. Am Sci 41: 256–270.
5. Dudley R (2002) The biomechanics of insect flight: form, function, evolution. Princeton University Press. 500 p.
6. De Queiroz A (1998) Interpreting Sister-Group Tests of Key Innovation Hypotheses. Syst Biol 47: 710–718. doi:10.1080/106351598260699.
7. Davis RB, Baldauf SL, Mayhew PJ (2010) Many hexapod groups originated earlier and withstood extinction events better than previously realized: inferences from supertrees. Proc R Soc B Biol Sci 277: 1597–1606. doi:10.1098/rspb.2009.2299.
8. Mayhew PJ (2002) Shifts in Hexapod Diversification and what Haldane could have said. Proc R Soc B Biol Sci 269: 969–974.
9. Yang AS (2001) Modularity, evolvability, and adaptive radiations: a comparison of the hemi- and holometabolous insects. Evol Dev 3: 59–72. doi:10.1046/j.1525-142x.2001.003002059.x.
10. Mitter C, Farrell B, Wiegmann B (1988) The Phylogenetic Study of Adaptive Zones: Has Phytophagy Promoted Insect Diversification? Am Nat 132: 107–128.
11. Farrell BD (1998) "Inordinate fondness" explained: Why are there so many beetles? Science 281: 555–559. doi:10.1126/science.281.5376.555.
12. Nyman T (2010) To speciate, or not to speciate? Resource heterogeneity, the subjectivity of similarity, and the macroevolutionary consequences of niche-width shifts in plant-feeding insects. Biol Rev 85: 393–411. doi:10.1111/j.1469-185X.2009.00109.x.
13. Wiegmann BM, Mitter C, Farrell B (1993) Diversification of Carnivorous Parasitic Insects: Extraordinary Radiation or Specialized Dead End? Am Nat 142: 737–754.
14. Wiegmann BM, Trautwein MD, Winkler IS, Barr NB, Kim J-W, et al. (2011) Episodic radiations in the fly tree of life. Proc Natl Acad Sci 108: 5690–5695. doi:10.1073/pnas.1012675108.
15. Hunt T, Bergsten J, Levkanicova Z, Papadopoulou A, John OS, et al. (2007) A comprehensive phylogeny of beetles reveals the evolutionary origins of a superradiation. Science 318: 1913–1916. doi:10.1126/science.1146954.
16. Heikkilä M, Kaila L, Mutanen M, Peña C, Wahlberg N (2011) Cretaceous origin and repeated tertiary diversification of the redefined butterflies. Proc R Soc B Biol Sci 279: 1093–1099. doi:10.1098/rspb.2011.1430.
17. Labandeira CC, Sepkoski Jr JJ (1993) Insect diversity in the fossil record. Science 261: 310–315.

18. Labandeira CC (2005) The fossil record of insect extinction: new approaches and future directions. Am Entomol 51: 14–29.
19. Ross AJ, Jarzembowski EA, Brooks SJ (2000) The Cretaceous and Cenozoic record of insects (Hexapoda) with regard to global change. Biotic Response to Global Change: The Last 145 Million Years, S. J. Culver and P. Rawson Eds. Cambridge: Cambridge University Press. pp. 288–302.
20. Trautwein MD, Wiegmann BM, Beutel RG, Kjer KM, Yeates DK (2012) Advances in insect phylogeny at the dawn of the postgenomic era. Annu Rev Entomol 57: 449–468.
21. Yeates DK, Cameron SL, Trautwein M (2012) A view from the edge of the forest: recent progress in understanding the relationships of the insect orders. Aust J Entomol 51: 79–87. doi:10.1111/j.1440-6055.2012.00857.x.
22. Stamatakis A, Ludwig T, Meier H (2005) RAxML-III: A fast program for maximum likelihood-based inference of large phylogenetic trees. Bioinformatics 21: 456–463. doi:10.1093/bioinformatics/bti191.
23. Stamatakis A, Hoover P, Rougemont J (2008) A rapid bootstrap algorithm for the RAxML web servers. Syst Biol 57: 758–771.
24. Ronquist F, Teslenko M, van der Mark P, Ayres DL, Darling A, et al. (2012) MrBayes 3.2: Efficient Bayesian phylogenetic inference and model choice across a large model space. Syst Biol 61: 539–542. doi:10.1093/sysbio/sys029.
25. Rehm P, Borner J, Meusemann K, von Reumont BM, Simon S, et al. (2011) Dating the arthropod tree based on large-scale transcriptome data. Mol Phylogenet Evol 61: 880–887. doi:10.1016/j.ympev.2011.09.003.
26. Rota-Stabelli O, Daley AC, Pisani D (2013) Molecular timetrees reveal a Cambrian colonization of land and a new scenario for Ecdysozoan evolution. Curr Biol 23: 392–398. doi:10.1016/j.cub.2013.01.026.
27. Wheat C, Wahlberg N (2013) Phylogenomic insights into the Cambrian Explosion, the colonization of land and the evolution of flight in Arthropoda. Syst Biol 62: 93–109.
28. Engel MS, Grimaldi DA (2004) New light shed on the oldest insect. Nature 427: 627–630. doi:10.1038/nature02291.
29. Kenrick P, Wellman CH, Schneider H, Edgecombe GD (2012) A timeline for terrestrialization: consequences for the carbon cycle in the Palaeozoic. Philos Trans R Soc B Biol Sci 367: 519–536. doi:10.1098/rstb.2011.0271.
30. Nel A, Roques P, Nel P, Prokin AA, Bourgoin T, et al. (2013) The earliest known holometabolous insects. Nature 503: 257–261. doi:10.1038/nature12629.
31. Labandeira C (2011) Evidence for an Earliest Late Carboniferous Divergence Time and the Early Larval Ecology and Diversification of Major Holometabola Lineages. Entomol Am 117: 9–21.
32. Blagoderov V, Grimaldi DA, Fraser N (2007) How time flies for flies: Diverse Diptera from the Triassic of Virginia and early radiation of the order. Am Mus Novit 3572: 1–39.
33. Labandeira CC (1998) Early History of Arthropod and Vascular Plant Associations. Annu Rev Earth Planet Sci 26: 329–377. doi:10.1146/annurev.earth.26.1.329.
34. Davis RB, Nicholson DB, Saunders EL, Mayhew PJ (2011) Fossil gaps inferred from phylogenies alter the apparent nature of diversification in dragonflies and their relatives. BMC Evol Biol 11: 252. doi:10.1186/1471-2148-11-252.
35. Nee S (2006) Birth-death models in macroevolution. Annu Rev Ecol Evol Syst 37: 1–17. doi:10.1146/annurev.ecolsys.37.091305.110035.
36. Alfaro ME, Santini F, Brock C, Alamillo H, Dornburg A, et al. (2009) Nine exceptional radiations plus high turnover explain species diversity in jawed vertebrates. Proc Natl Acad Sci 106: 13410–13414. doi:10.1073/pnas.0811087106.
37. Nicholson DB, Ross AJ, Mayhew PJ (2014) Fossil evidence for key innovations in the evolution of insect diversity. Proc R Soc B Biol Sci: In press.
38. Wills MA (2001) How good is the fossil record of arthropods? An assessment using the stratigraphic congruence of cladograms. Geol J 36: 187–210. doi:10.1002/gj.882.
39. Mayhew PJ (2003) A tale of two analyses: estimating the consequences of shifts in hexapod diversification. Biol J Linn Soc 80: 23–36. doi:10.1046/j.1095-8312.2003.00217.x.
40. Whiting MF (2002) Mecoptera is paraphyletic: multiple genes and phylogeny of Mecoptera and Siphonaptera. Zool Scr 31: 93–104. doi:10.1046/j.0300-3256.2001.00095.x.
41. Yoshizawa K, Johnson KP (2010) How stable is the "Polyphyly of Lice" hypothesis (Insecta: Psocodea)?: A comparison of phylogenetic signal in multiple genes. Mol Phylogenet Evol 55: 939–951. doi:10.1016/j.ympev.2010.02.026.
42. Davis RB, Baldauf SL, Mayhew PJ (2009) Eusociality and the success of the termites: insights from a supertree of dictyopteran families. J Evol Biol 22: 1750–1761. doi:10.1111/j.1420-9101.2009.01789.x.
43. Fiz-Palacios O, Schneider H, Heinrichs J, Savolainen V (2011) Diversification of land plants: insights from a family-level phylogenetic analysis. BMC Evol Biol 11: 341.
44. Clarke JT, Warnock RCM, Donoghue PCJ (2011) Establishing a time-scale for plant evolution. New Phytol 192: 266–301. doi:10.1111/j.1469-8137.2011.03794.x.
45. Gómez-Zurita J, Hunt T, Kopliku F, Vogler AP (2007) Recalibrated Tree of Leaf Beetles (Chrysomelidae) Indicates Independent Diversification of Angiosperms and Their Insect Herbivores. PLoS ONE 2: e360. doi:10.1371/journal.pone.0000360.
46. Davis RB, Baldauf SL, Mayhew PJ (2010) The origins of species richness in the Hymenoptera: insights from a family-level supertree. BMC Evol Biol 10: 109. doi:10.1186/1471-2148-10-109.
47. Sohn J-C, Labandeira CC, Davis DR, Mitter C (2012) An annotated catalog of fossil and subfossil Lepidoptera (Insecta: Holometabola) of the world. Zootaxa 3286: 1–132.
48. Regier JC, Mitter C, Zwick A, Bazinet AL, Cummings MP, et al. (2013) A Large-Scale, Higher-Level, Molecular Phylogenetic Study of the Insect Order Lepidoptera (Moths and Butterflies). PLoS ONE 8: e58568. doi:10.1371/journal.pone.0058568.
49. Cho S, Zwick A, Regier JC, Mitter C, Cummings MP, et al. (2011) Can Deliberately Incomplete Gene Sample Augmentation Improve a Phylogeny Estimate for the Advanced Moths and Butterflies (Hexapoda: Lepidoptera)? Syst Biol 60: 782–796. doi:10.1093/sysbio/syr079.
50. Katoh K, Misawa K, Kuma K, Miyata T (2002) MAFFT: a novel method for rapid multiple sequence alignment based on fast Fourier transform. Nucleic Acids Res 30: 3059–3066. doi:10.1093/nar/gkf436.
51. Pruesse E, Quast C, Knittel K, Fuchs BM, Ludwig W, et al. (2007) SILVA: a comprehensive online resource for quality checked and aligned ribosomal RNA sequence data compatible with ARB. Nucleic Acids Res 35: 7188–7196. doi:10.1093/nar/gkm864.
52. Castresana J (2000) Selection of Conserved Blocks from Multiple Alignments for Their Use in Phylogenetic Analysis. Mol Biol Evol 17: 540–552.
53. Miller MA, Pfeiffer W, Schwartz T (2010) Creating the CIPRES Science Gateway for inference of large phylogenetic trees. Proceedings of the Gateway Computing Environments Workshop. New Orleans: GCE. pp. 1–8.
54. R Developement Core Team (2011) R: A Language and Environment for Statistical Computing. Available: http://www.R-project.org.
55. Brown JW, FitzJohn RG, Alfaro ME, Harmon LJ (2012) MEDUSA: Modeling Evolutionary Diversification Using Stepwise AIC,. Available: http://www.webpages.uidaho.edu/~lukeh/software/software.html. Accessed 24 February 2013.
56. Rabosky DL (2007) LASER: A Maximum Likelihood Toolkit for Detecting Temporal Shifts in Diversification Rates From Molecular Phylogenies. Evol Bioinforma Online 2: 247–250.
57. Magallon S, Sanderson MJ (2001) Absolute Diversification Rates in Angiosperm Clades. Evolution 55: 1762–1780. doi:10.1111/j.0014-3820.2001.tb00826.x.
58. Harmon LJ, Weir JT, Brock CD, Glor RE, Challenger W (2008) GEIGER: investigating evolutionary radiations. Bioinformatics 24: 129–131. doi:10.1093/bioinformatics/btm538.

3

Laurentian Great Lakes Phytoplankton and Their Water Quality Characteristics, Including a Diatom-Based Model for Paleoreconstruction of Phosphorus

Euan D. Reavie[1]*, Adam J. Heathcote[2], Victoria L. Shaw Chraïbi[3]

1 Center for Water and the Environment, Natural Resources Research Institute, University of Minnesota Duluth, Duluth, Minnesota, United States of America, 2 Groupe de Recherche Interuniversitaire en Limnologie, Département des Sciences Biologiques, Université du Québec à Montréal, Montréal, Québec, Canada, 3 Department of Earth and Atmospheric Sciences, University of Nebraska-Lincoln, Lincoln, Nebraska, United States of America

Abstract

Recent shifts in water quality and food web characteristics driven by anthropogenic impacts on the Laurentian Great Lakes warranted an examination of pelagic primary producers as tracers of environmental change. The distributions of the 263 common phytoplankton taxa were related to water quality variables to determine taxon-specific responses that may be useful in indicator models. A detailed checklist of taxa and their environmental optima are provided. Multivariate analyses indicated a strong relationship between total phosphorus (TP) and patterns in the diatom assemblages across the Great Lakes. Of the 118 common diatom taxa, 90 (76%) had a directional response along the TP gradient. We further evaluated a diatom-based transfer function for TP based on the weighted-average abundance of taxa, assuming unimodal distributions along the TP gradient. The r^2 between observed and inferred TP in the training dataset was 0.79. Substantial spatial and environmental autocorrelation within the training set of samples justified the need for further model validation. A randomization procedure indicated that the actual transfer function consistently performed better than functions based on reshuffled environmental data. Further, TP was minimally confounded by other environmental variables, as indicated by the relatively large amount of unique variance in the diatoms explained by TP. We demonstrated the effectiveness of the transfer function by hindcasting TP concentrations using fossil diatom assemblages in a Lake Superior sediment core. Passive, multivariate analysis of the fossil samples against the training set indicated that phosphorus was a strong determinant of historical diatom assemblages, verifying that the transfer function was suited to reconstruct past TP in Lake Superior. Collectively, these results showed that phytoplankton coefficients for water quality can be robust indicators of Great Lakes pelagic condition. The diatom-based transfer function can be used in lake management when retrospective data are needed for tracking long-term degradation, remediation and trajectories.

Editor: Francois G. Schmitt, CNRS, France

Funding: This research was largely supported by grants to E. Reavie from the U.S. Environmental Protection Agency under Cooperative Agreements GL-00E23101 and GL-00E00790-2. This document has not been subjected to the Agency's required peer and policy review and therefore does not necessarily reflect the view of the Agency, and no official endorsement should be inferred. Sedimentary assemblages from Lake Superior were collected and assessed under research sponsored by the Minnesota Sea Grant College Program supported by the NOAA office of Sea Grant, United States Department of Commerce, under grant No. R/CC-01-10. The funders had no role in study design, data collection and analysis, decision to publish, or preparation of the manuscript.

Competing Interests: The authors have declared that no competing interests exist.

* Email: ereavie@d.umn.edu

Introduction

Recent trends from monitoring data demonstrate rapid changes in the Laurentian Great Lakes system associated with anthropogenic drivers. Most notably, changes in the water quality [1], phytoplankton (unpublished data), zooplankton [2] and benthos [3] collectively confirm or suggest detriments associated with invasive species, nutrient imbalances and climate. While algal abundances have been inordinately low in some regions (e.g., Lake Huron; unpublished data), new eutrophication problems have also manifested as substantial algal blooms (e.g., [4]). Such quantitative assessments of environmental problems have been valuable to support lake management [5] because they provide information on the abundance of various organisms, traits that are critical to understanding and managing food webs. However, taxon-specific evaluations of pelagic indicators in the Great Lakes are lacking. In addition to these more intuitive evaluations of organism numbers and biomass, taxon-specific ecological data can provide additional tools to inform management [6].

Developing effective indicators of ecological condition requires that they be calibrated to identify their responses to important environmental stressors [7]. The main goals of calibration are to identify environmental characteristics of potential indicator taxa so that they may be subsequently used to infer condition. Assemblages of algae, which are physiologically subject to water quality, have the potential to provide time-integrated inferences of limnological conditions. Such bioindicators are particularly needed to monitor the impacts of human activities that are increasing nutrient supplies to water bodies, introducing non-native species, and changing climate. Great Lakes coastal algae, particularly diatoms, have been shown to provide a more temporally integrated assessment of water quality conditions than discrete water quality measurements [8], but similar indicators have not been developed for the pelagic ecosystem. Diatoms are particularly advantageous as indicators in paleolimnological studies because their diagnostic cell walls (frustules) remain archived in lake sedimentary records [9]. Hence, the fossil diatom assemblages may be used to reconstruct historical degradation and recovery and estimate future trajectories of condition.

The characteristics of the Great Lakes phytoplankton community as indicators of the prevailing eutrophication that was noted in the 1960s and 1970s was well described by Stoermer [10]. At that time little taxonomic treatment of the phytoplankton had been accomplished, so Stoermer advocated the indicator potential that would result from better taxonomic assessment of the flora and species autecology. Significant collections of phytoplankton have occurred since that time, such as the United States Environmental Protection Agency's (USEPA) biological monitoring program that has been active since 1983 [11], among other monitoring data (e.g., [12]). We aimed to provide the tools to develop species- and assemblage-based indicators from Great Lakes phytoplankton data.

Modern datasets (also known as training sets) provide the basis for developing indicator transfer functions by relating contemporary assemblages with their corresponding environmental measurements [13]. Algal assemblages, in particular, are proven robust indicators of stressors such as nutrients ([14],[15]), water clarity [16] and acidification [17], as well as a suite of other water quality problems in freshwater ecosystems [18]. Training sets using diatoms have been developed for lakes, rivers and the coastlines of the Great Lakes [8]. Algae are known to have definable optima along gradients of environmental conditions. Species tend to be taxonomically distinct and abundant, and they respond rapidly to changing conditions. Hence, by using training set data to calibrate the environmental characteristics of algae, researchers can use changes in community composition to classify and quantify long-term environmental changes that result from anthropogenic activities. This is particularly needed because few pelagic monitoring data were collected prior to the 1970s; existing data provide a spatially and temporally incomplete picture of environmental conditions in the lakes. Species-environmental coefficients should allow reconstructions of past conditions based on sedimentary assemblages.

A transfer function is derived by relating taxa assemblages (usually diatoms) in a training set of samples (in this case, pelagic samples collected throughout the Great Lakes) to an environmental variable of interest [19]. The transfer function consists of taxa coefficients (environmental optima and, optionally, tolerances) that can be used to infer quantitative information about the variable of interest based on the abundance of each taxon in a sample assemblage. These assemblages are usually characterized from recently accumulated surface sediments [20]; to our knowledge, a transfer function based on diatom phytoplankton samples has not been attempted.

For the current assessment, modeling approaches were applied to characterize the environmental coefficients of the Great Lakes phytoplankton taxa. A checklist of the common taxa and their ecological indicator values for a suite of water quality variables is presented. To illustrate one way these coefficients may be used, a diatom-based transfer function was developed for paleolimnological applications. Transfer function evaluation and testing typically involve the comparison of algal-inferred water quality to measured water quality to evaluate function robustness, which is usually characterized by a coefficient of determination (r^2) and a prediction error. It is well-known that adjacent phytoplankton samples (e.g. 5–100 km apart) from the Great Lakes tend to have similar species assemblages and environmental conditions, and this lack of independence among sites can violate statistical assumptions. Such autocorrelation may result in misleading estimates of transfer function performance [21], and with new analyses revealing previously undocumented weaknesses with diatom-based nutrient models [22], it is imperative that transfer functions undergo thorough testing. So, we evaluated how (1) autocorrelation, (2) relationships between taxon abundance and phosphorus and (3) redundancies among environmental variables affected our training set and its predictive power. We further tested the model by applying it to fossil diatom assemblages from Lake Superior to determine whether pelagic diatom-inferred phosphorus serves as an effective paleoecological tool in the Great Lakes.

Materials and Methods

We evaluated phytoplankton data collected as part of the USEPA's biological monitoring program for the Great Lakes. The standard operating procedure for phytoplankton collection and analysis is described in detail in the published procedures [11], but abbreviated details were as follows. The EPA data were based on twice-annual synoptic sampling ("spring" = typically the month of April, "summer" = typically the month of August) from standard stations throughout the Great Lakes basin. Our analyses focused on phytoplankton and water quality samples collected from 2007 through 2011, a total of 717 unique sampling events. No specific permissions were required for these locations or activities; namely, sample locations in the pelagic Great Lakes. Field studies did not involve endangered or protected species.

Whole water samples were collected from the rosette sampler on-board the Research Vessel Lake Guardian. Phytoplankton samples were composites of water sampled at discrete depths from the euphotic zone of the water column. For isothermal spring samples, the sample integrated equal volumes of water from 1, 5, 10 and 20 m. In shallower locations in Lake Erie the 20-m sample was replaced by an above-bottom collection. If the total depth was less than 15 m, equal volumes were integrated from surface, mid-depth and above-bottom samples. For the stratified (summer) water column, equal volumes were taken from 1 m, 5 m, 10 m and the lower epilimnion and integrated. If the epilimnion was very shallow, equal volumes were integrated from a maximum of four and a minimum of two sampling depths. Samples were split and analyzed separately for the whole phytoplankton assemblage (i.e., "soft" algae) and diatoms. Analysis of soft algae used the quantitative Utermöhl method of counting preserved specimens in a settling chamber on an inverted microscope [23]). During soft algal analyses, diatoms containing cytoplasm were identified as centric or pennate forms. The second split sample was digested in nitric acid and subsequently in peroxide to isolate the diatom valves which were then plated on slides and counted using oil immersion (1000× or higher) to identify taxa. All counting included measurements of cell dimensions so that algal biovolumes could be calculated. Ultimately, analyses afforded detailed taxonomic resolution, and data were available in quantitative data formats (cell density [cells/ml] and biovolume [μm^3/ml]).

A 30-cm sediment core from eastern Lake Superior was used as a test subject for a diatom-based phosphorus transfer function. Details of sediment coring and sample preparation are provided by Shaw Chraïbi et al. [24]. The stratigraphic record of diatom assemblages dating back to the early 1700s was the subject of testing in the present assessment.

Algal associations with 11 environmental variables (Table 1), collected simultaneously with phytoplankton, were explored. Additional variables were considered but were redundant (specific conductivity with chloride, beam attenuation with turbidity) or poorly represented in the dataset (dissolved oxygen, irradiance, total nitrogen). Geophysical variables (e.g., depth, latitude, longitude) were not used in multivariate analyses in order to better describe relationships with water quality, although the importance of "lake" as a nominal variable was explored in terms of species specificity. Collection and analysis of environmental data is described in detail in the USEPA (2010) standard operating procedures. Three additional molar ratio variables (N:P, N:Si, P:Si) were calculated from other variables in the list so that they may be used in the evaluation of algal responses across nutrient ratio gradients. Because of substantial correlation among these variables (Figure 1), multivariate approaches were deemed necessary to determine species-environmental relationships.

All analyses were performed using the R statistical software [25] and associated packages. All scripts and datasets are available from the authors on request. Phytoplankton and water quality data are archived in the Great Lakes Environmental Database (http://www.epa.gov/greatlakes/monitoring/data_proj/glenda), and the paleoecological data are provided electronically in Shaw Chraïbi [26].

Ordination

Multivariate ordinations were performed using the R package "vegan" [27]. An initial principal components analysis (PCA) was used to characterize the major environmental gradients in the water quality data. All environmental variables were scaled to zero mean and unit variance. The first two principal components (axes), which are a linear combination of the greatest variation in environmental condition across sites, were used as new environmental variables (Figure 1) in the exploration of algal species autecologies.

For diatom-environmental analyses, redundancy analysis (RDA), the constrained form of PCA, was performed to condense the complex dataset into summary variables (axes) that capture the majority of environmental variation (Juggins and Birks [28]). An RDA with a Monte Carlo permutation test was used to identify the environmental variables with significant ($P<0.05$) relationships with the diatom data. The subset of significant variables was then selected to generate a new RDA to identify species-environmental relationships and orthogonal gradients that capture variation in the diatom data. Explained variation from this RDA was compared to a partial RDA that partitioned variation in the diatom data by characterizing the total and unique contributions

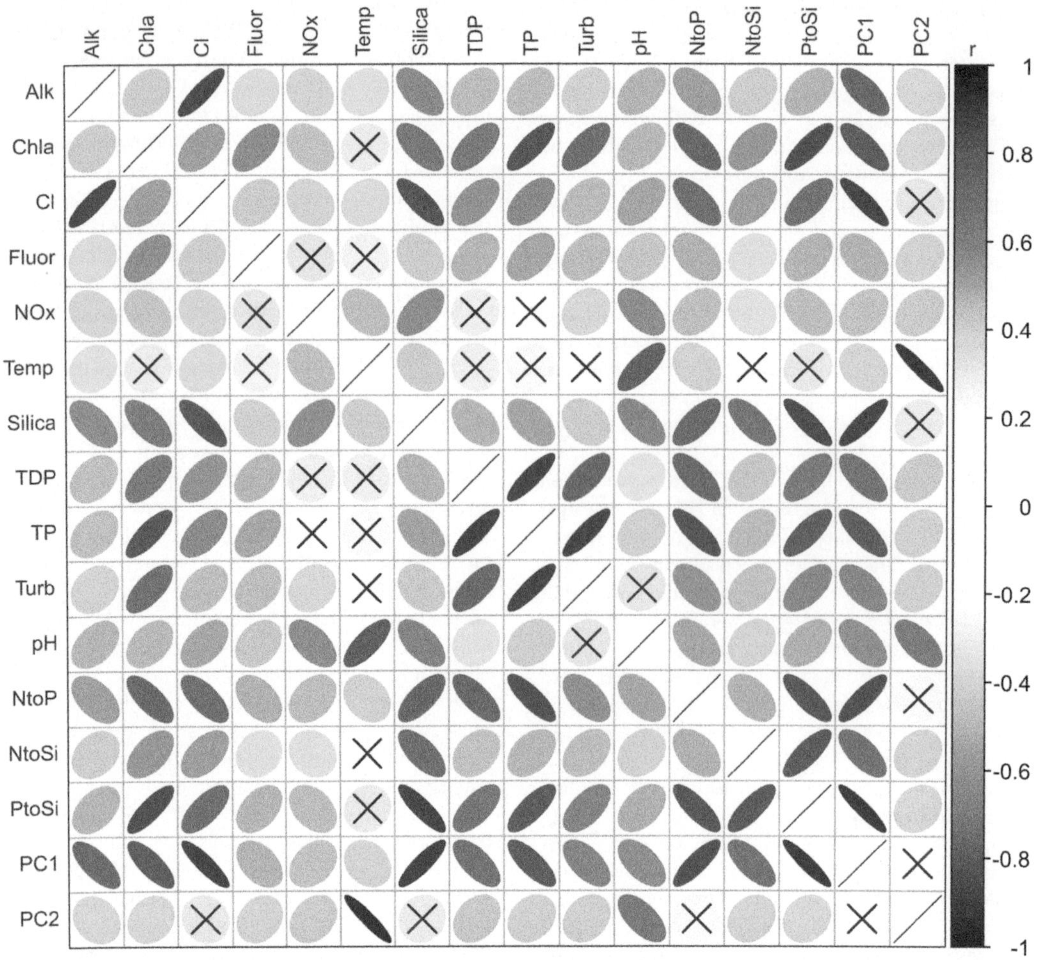

Figure 1. Correlation matrix of Great Lakes water quality variables. Correlation matrix of Great Lakes water quality data, using Pearson product-moment correlation coefficients. Ellipses summarize positive or negative correlations, with narrower, darker ellipses indicating stronger correlations. Variable pairs containing an × were not significant ($P=0.05$ with Bonferroni correction for multiple comparisons). All variables were transformed as necessary to minimize skew and approximate or achieve normality (Table 1). Scores from the first two axes of a principal components analysis are included to indicate important variables for these gradients.

Table 1. The environmental variables used in this study.

Variable	Abbreviation	Transformation	Minimum	5th percentile	Median	Mean	95th percentile	Maximum
Alkalinity (mg/L)	Alk	sqrt	41.90	43.00	86.00	78.40	109.00	124.00
Chlorophyll a (µg/L)	Chla	log	0.00	0.44	1.03	2.54	10.12	35.39
Chloride (mg/L)	Cl	sqrt	1.20	1.29	11.57	10.39	22.69	29.27
Fluorescence (µg/L)	Fluor	log	0.00	0.11	1.39	2.41	7.53	50.96
Nitrates + Nitrites (mg/L)	NOx	sqrt	0.00	0.08	0.32	0.30	0.43	1.63
Temperature (°C)	Temp	sqrt	0.11	1.02	7.19	11.23	24.99	26.81
Silica (mg/L)	Silica	-	0.00	0.02	0.82	0.69	1.11	1.63
Dissolved Phosphorus (µg/L)	TDP	log	0.45	0.85	1.90	2.75	6.40	69.70
Total Phosphorus (µg/L)	TP	log	0.72	1.54	2.96	6.26	18.08	226.78
Turbidity (NTU)	Turb	log	0.11	0.16	0.30	1.64	5.23	178.00
pH	pH	-	7.58	7.89	8.31	8.33	8.80	9.20
N:P (molar ratio)	NtoP	sqrt	0.00	0.01	0.10	0.10	0.22	0.37
N:Si (molar ratio)	NtoSi	log	0.00	0.29	0.37	1.49	5.74	120.95
P:Si (molar ratio)	PtoSi	log	0.98	1.58	3.37	84.48	414.08	3702.76
Principal Component 1	PC1	-	−0.91	−0.68	0.12	0.00	0.46	0.58
Principal Component 2	PC2	-	−0.63	−0.52	0.04	0.00	0.50	2.29

Provided are untransformed ranges of the variables and the transformations applied to remove skew. A principal components analysis was performed to derive the summary axes (PCs).

that can be attributed to TP. This was further applied to the variables alkalinity (Alk), chloride (Cl) and nitrates + nitrites (NOx) to examine similar characteristics for variables that may be confounded with TP (e.g., Alk, NOx) or have strong independent controls on diatom autecology (e.g, Cl).

Species-environmental relationships

Algae-based inference models typically assume the taxa respond unimodally across the environmental gradient of interest [29]; however, based on many possible factors (e.g., narrow gradients, multivariate controls, inadequate sampling or unique ecological distributions) taxa may vary in their responses, or have no apparent response. It is important to characterize the goodness of fit of species to the environmental variable, as well as to consider linear, Gaussian or skewed responses along gradients in order to validate the assumption that diatom species responses actually reflect the environmental variable of interest.

Responses of algae taxa along environmental gradients were evaluated using linear, Gaussian and weighted-average (WA) assumptions. First, environmental variables were transformed as necessary to minimize skew and achieve normality in their frequency distribution, if possible. Taxa that were observed in more than five samples were evaluated. Responses were evaluated using four measures of algal abundance: cell densities (cells/ml), biovolume (μm^3/ml), relative cell density and relative biovolume.

Using the "stats" (version 3.0.0) package [25] the function "lm" was used to generate a linear model, then the function "cor.test" was used to determine linear model relationships between taxon abundance and each environmental variable. Significance ($P = 0.05$) was tested based on Pearson's product moment correlation coefficient.

Using the "gam" (version 1.08) package [30], the "gam" function was used to fit Gaussian model relationships between taxon abundance and each environmental variable. Based on a comparison of residuals, a paired *t*-test ($P = 0.05$) was applied to determine whether a Gaussian fit was better than a linear fit. The point on the environmental gradient where a Gaussian fit was maximized was considered a taxon's optimum. The environmental weighted-average optimum for a taxon was calculated based on the average of environmental values where the taxon occurred, weighted by taxon abundance. Taxon optima based on Gaussian and WA calculations were rescaled to a range of 0 to 10, which was based on the range of the transformed (if applied) environmental variable. This rescaling resulted in more user-friendly coefficients (i.e. "environmental indicator values"), but coefficients based on actual variable units are available from the authors.

To support the utility of taxa included in the diatom-based transfer function, generalized additive models (GAMs [31]) were used to test the significance of the response of each diatom taxon to the measured TP gradient using the R package "mgcv" [32]. The additive models test several response forms (linear, unimodal, skewed) across the selected gradient. We further tested the significance of the unique explanatory power of TP once the effect of alkalinity (another important water quality variable in the Great lakes) was removed. GAMs were tested on the relative density data for taxa with 10 or more occurrences in samples and significance of fitted models was assessed using 199 Monte Carlo permutations of the deviance explained by TP for each species response GAM. (Although a less stringent criterion for occurrence [five samples] was used in the development of indicator values, here we use 10 samples to ensure sufficient abundance to identify responses along the TP gradient.) Data were evaluated based on relative density and relative biovolume as these are the data types that may be used in diatom-based paleoecological studies.

Diatom-based transfer functions

Transfer functions for Great Lakes total phosphorus (TP) were derived using the diatom components of the algal assemblages. As such a function is most likely to be used for down-core analyses in paleolimnological assessments [29], the soft algae that are less likely to preserve in the sedimentary record were not used. Transfer functions were developed using weighted averaging (WA) regression with inverse de-shrinking with log-normal taxa transformation ("rioja" package [33]). Diatom-inferred estimates of TP (DI-TP) for each sample were calculated by taking the optimum of each taxon to that variable, weighting it by its abundance in that sample, and calculating the average of the combined weighted taxa optima ([34], [35]). The apparent strengths of the transfer functions were evaluated by calculating the squared correlation coefficient (r^2) and the root mean square error of prediction (RMSEP) between measured TP and transfer function estimates of those values for all samples. The predictive error was estimated by jackknifing [36] which corrects for under-estimation of root-mean squared error due to predicting values only from samples that were included in the model. As above, variations of these functions were tested using two possible representations of the algal data, relative cell density and relative biovolume. Developing such functions based on absolute abundances was not deemed useful for future applications because species abundances in sedimentary assemblages (e.g., cells per g dry sediment) cannot be assumed to be comparable to prevailing phytoplankton concentrations (e.g., cells per mL in the water column).

Transfer function testing

We evaluated redundancy in the training set using the modern analog technique (MAT; [37]) with five analogs. Instead of using the weighted abundance of taxa, a reconstructed value was based on the five most taxonomically similar samples in the training set. The effect of autocorrelation on transfer function performance was evaluated using the "rne" function in the R package "palaeoSig" [38]. The rne function compares the performance (r^2) of the transfer function during cross-validation as sites are deleted (1) randomly, (2) that are geographically close to the test site, and (3) that are environmentally similar to the test site [21]. Reducing the number of sites will worsen the performance as the number of potential analogues decreases [39]. In the case of autocorrelation, deleting the nearest sites would remove the best analogues, and should be more detrimental to performance than random deletion. Such degradation would also be expected when sites in close environmental proximity (e.g., with similar TP) are selectively deleted. If performance deteriorates more by deletion of close sites than by deletion of environmental neighbors, the performance loss relative to the random deletion must in part be due to spatial autocorrelation between diatom assemblages.

We further checked the diatom-TP transfer function performance against several randomized simulations using multiple iterations of the "multi.mat" function in palaeoSig [21]. This method simulates the environmental variable across sites using the same autocorrelation structure as the measured data and recalculates performance (r^2). Creating simulated environmental data requires many geographical considerations, including development of an empirical variogram to determine the spatial structure of TP in the Great Lakes, fitting a theoretical variogram model, and a simulation to generate a spatially structured random variable with the same spatial structure as the original data. We employed the gstat package [40] in R to support these calculations, and the method is detailed by Telford and Birks [21]. A transfer function is considered to have statistically significant predictive power ($p<0.05$) if the real value of r^2 for the transfer function exceeds 95% of the simulations.

Application to paleolimnology

Reconstructed estimates of past TP concentrations were obtained from sedimentary samples collected as part of a recent paleolimnological analysis of eastern Lake Superior (Shaw Chraïbi et al. [24]). Despite the training set samples representing seasonal snapshots of modern diatom assemblages as opposed to sedimentary assemblages containing year-round integrations of settling diatoms, paleo-assemblages had several good modern analogues in the training set, thus providing preliminary evidence that the training set may be applicable (Shaw Chraïbi [26]). Diatom-inferred TP from the uppermost sediment samples matched well with contemporary TP monitoring data (Shaw Chraïbi et al. [24]). Downcore TP estimates were calculated using WA with inverse de-shrinking (Juggins and Birks [28]). To track the historical trajectory in relation to historical diatom conditions, sedimentary assemblages were projected on the RDA containing the suite of training set diatom data and significant environmental variables.

We then calculated the ratio of two values: (1) λ_R, the variance in sedimentary diatom assemblages captured by the first axis of a RDA constrained to DI-TP; (2) λ_P, the variance explained by the first axis of an unconstrained PCA of the diatom data. The λ_R/λ_P ratio expresses the proportion of variation explained by DI-TP as a fraction of the maximum explainable variance in the sedimentary samples. We also calculated the correlation coefficient for past DI-TP versus corresponding axis 1 sample scores from the unconstrained PCA of the sedimentary diatom assemblages. We would expect both of these values to be high (i.e. close to a proportion of 1.0) if downcore assemblages were strongly related to changes in TP [22].

Results and Discussion

Taxon-specific results

263 taxa were sufficiently abundant (i.e. in more than five samples) to be considered for evaluation along environmental gradients. The common taxa comprised centric diatoms (42 taxa), pennate diatoms (103 taxa), chrysophytes (37 taxa), green algae (45 taxa), cryptomonads (8 taxa), blue-green algae (17 taxa), euglenoids (2 taxa), dinoflagellates (6 taxa) and some unknown entities (3 taxonomic categories). Some were unknown species or genera not identifiable to the species level (a genus followed by "spp.") and others were members of known divisions but with few diagnostic characteristics (e.g., "unidentifiable chrysophyte ovoid"). Although these less specific taxonomic categories are difficult to understand ecologically due to the likelihood of multiple species comprising a category, we present them as possible inference tools because several of them had significant relationships with environmental variables.

Table 2 presents environmental characteristics for select taxa relative to nine variables, as well as their lake and seasonal specificity, based on observed algal biovolumes and relative cell densities in phytoplankton samples. It is beyond the scope of this article to characterize the numerous species-specific mechanisms for these environmental relationships, but some of these taxa are recognized as important components of the Great Lakes phytoplankton community. The full suite of species coefficients are provided in supplementary appendices that detail species coefficients based on cell densities (Table S1), relative cell density (Table S2) biovolume (Table S3) and relative biovolume (Table S4). These tables further characterize within-lake conditions and may serve as a set of coefficients suited to further indicator development and monitoring assessments.

Several taxa have high (e.g., *Aulacoseira distans*) or low (e.g., *Cyclotella michiganiana*) optima for TP. However, it is interesting that all of the significant linear species relationships with TP are positive (Table 2, Table S3). A similar negative effect occurred for silica. This appears to be due to the dominating effect of summer in Lake Erie, a time when TP is high, silica is low, and several algal taxa are highly abundant, thus driving positive slopes in the biovolume-TP relationship. A comparative examination of relative densities (Table S2) indicates a greater diversity in species responses to TP because relative abundances were more evenly distributed across lakes and environmental gradients.

Based on weighted abundance across all samples, the centric diatom *Aulacoseira islandica* is the most abundant alga in the Great Lakes phytoplankton ([41]). However, it has a very specific spring bloom period in Lake Erie, with very low occurrence across the other lakes. Its distribution reflects high chlorophyll *a* and fluorescence, which is not surprising as it comprises most of Lake Erie's spring algal bloom. *A. islandica* has a significantly negative relationship with silica, which seemingly counters its very high silica requirement in the formation of its heavily silicified cell walls. This may imply that by the time spring sampling occurs in April of each year diatoms have exhausted the dissolved silica in the water that is available to incorporate into their cell walls.

While structurally similar to *A. islandica*, *Aulacoseira granulata* is a distinctly summer diatom in Lake Erie. *A. granulata* has an apparent tolerance for warmer, late-summer conditions when it coexists with green and blue-green algae [42] that account for high chlorophyll concentrations. *Fragilaria crotonensis* is a pennate diatom that also occurs in the summer in Lake Erie, with lower abundances in Ontario and Michigan. It has high optima for pH and temperature and appears to be associated with a low N:P ratio.

The cyanophyte *Microcystis aeruginosa* is a well-known coastal bloom species in Lake Erie [43], and it has been observed abundantly in recent offshore phytoplankton samples [42]. Our assessment confirms its fairly strict occurrence in the summer in Lake Erie, as well as its occurrence at times of high chlorophyll *a*, temperature and pH, as well as low nitrates and N:P.

Three cryptomonad algae were abundant: *Cryptomonas rostratiformis*, *C. reflexa* and *C. erosa*. These species occurred year-round in all of the lakes. While seemingly cosmopolitan, there were clear distinctions within this genus, such as *C. reflexa* being tolerant of high chloride and temperature while *C. rostratiformis* occurred under conditions of low chloride and temperature.

The diatom-based model

Under the full diatom-inferred TP model, optimal model parameters included use of WA with tolerance down-weighting, which offered minimal improvement. The model was developed using log-transformation of the TP data and it was cross-validated using the jackknifing procedure [35]. There were no transformations of the taxonomic data or removal of taxa or samples as these procedures did not improve apparent model performance. We tested removal of benthic and rare taxa (i.e., occurring in fewer than 5 samples) but such considerations also degraded model performance. Comparisons between observed TP and DI-TP resulted in a jackknifed r^2 of 0.79 and RMSEP of 0.38 (log-TP units) using relative density data and an r^2 of 0.77 and RMSEP of 0.36 using relative biovolumes (Table 3).

Based on GAM analyses of relative density data for each common diatom taxon, 90 out of 118 taxa (76%) taxa had a

Table 2. Environmental indicator values for select algae taxa in the Great Lakes.

	Season		Lake					Chlorophyll a			Chloride			Fluorescence			Nitrates + Nitrites			Temperature			Silica			Total Phosphorus			pH			N:P		
Taxon Full Name	**Spring**	**Summer**	**Superior**	**Michigan**	**Huron**	**Erie**	**Ontario**	**Lin**	**Gau**	**WA**	**Lin**	**Gau**	**WA**	**Lin**	**Gau**	**WA**	**Lin**	**Gau**	**WA**	**Lin**	**Gau**	**WA**	**Lin**	**Gau**	**WA**	**Lin**	**Gau**	**WA**	**Lin**	**Gau**	**WA**	**Lin**	**Gau**	**WA**
Aulacoseira distans (Ehr.) Simonsen	9	1	1	7	1	1	0	+	3*	2	+	5*	5		7	2		10	3	+	3*	3	–	10*	6	+	10*	2		3	3	–	3	3
Aulacoseira granulata (Ehr.) Simonsen	0	10	0	0	0	10	0	+	10*	8		6	6	+	8	6		0*	1	+	10	10		3	4		7	5	+	10	8		0	0
Aulacoseira islandica (O. Mull.) Simonsen	10	0	0	0	0	10	0	+	10*	7	+	7*	7	+	10*	5	–	0*	1		3	2	–	0*	0	+	5*	5	+	6*	5	–	0*	0
Cyclotella michiganiana Skv.	0	10	0	7	2	0	0		0	2	+	5	5		0	1	–	2*	2	+	9*	8	–	4*	4		0	1	+	6*	6	–	3	3
Fragilaria crotonensis Kitton	0	10	0	1	0	8	2	+	6	5	+	7*	7	+	6	4	–	0*	2	+	10*	9	–	1*	2	+	4	3	+	10*	7	–	1*	1
Cryptomonas erosa Ehr.	3	7	1	1	0	6	2	+	6	4	+	7*	6	+	4	3	–	0*	2	+	10*	7	–	0	3	+	5	3	+	10*	6	–	0*	1
Cryptomonas reflexa Skuja	2	8	1	1	0	6	2	+	6	4	+	6*	6	+	4*	3	–	0*	2	+	10*	8	–	4	3	+	5	3	+	10*	6	–	0*	1
Cryptomonas rostratiformis Skuja	3	7	4	0	0	2	4		6	3	+	10*	4		1*	2	–	1	2	+	8*	6	–	0*	4		3	2	+	0*	5		10	3
Microcystis aeruginosa (Kütz.) emend. Elenkin	0	10	0	0	0	10	0		10	9		6*	6	+	8	6		0*	1	+	10*	10		3*	4	+	7*	6		10	9		0*	0

Select algae taxa in the Great Lakes and their environmental indicator values based on biovolume distributions across the environmental gradients. Negative (–) and positive (+) signs indicate significant linear relationships ($P < 0.05$) based on the Pearson product moment correlation. Gaussian optima (Gau) display an asterisk (*) if fitted residuals were significantly smaller (t-test; $P<0.05$) than residuals from the linear fit. WA indicates optima based on weighted-averaging. "Season" and "Lake" coefficients reflect seasonal and spatial occurrence of the taxa. Indicator values from 0 through 10 respectively represent whether a given taxon had its greatest modeled abundance at a low or high region of a variable's gradient. Indicator numbers for seasons or lakes represent the relative proportion that each taxon achieved in 2007–2011, again scaled from 0 to 10.

Table 3. Results of DI-TP model testing using diatom data based on relative cell densities and biovolumes.

	Density	Biovolume
Model $r^2_{jackknife}$	0.79	0.77
Model RMSEP (log-TP units)	0.38	0.36
Number of taxa with significant GAM relationship to TP (out of 118)	90	89
Number of taxa with significant relationship to TP after factoring out Alk (out of 118)	65	67
Percentile of actual model in simulated remapping of training set	99.8%	99.8%
RDA 1 eigenvalue	16.7%	20.0%
RDA 2 eigenvalue	6.4%	7.4%
Total variance explained by TP	6.8%	9.1%
Unique variance explained by TP	4.8%	6.0%
Downcore DI-TP versus PCA 1 r	0.98	0.68

significant response across the TP gradient. After removing the effects of alkalinity, 65 out of 118 (55%) had a significant response to TP, indicating that 25 of the taxa with significant relationships with TP had a signal that was confounded with alkalinity. Comparable results using relative biovolumes revealed 89 taxa with significant responses to phosphorus, 67 after accounting for alkalinity. Basing the model on diatom densities or biovolumes appeared to have little effect on model performance. It is more typical in paleolimnological studies to use count data without the additional benefit of cell measurements to calculate biovolumes, so subsequent figures focus on the use of relative diatom valve densities.

In a check of the effects of sample removal in the training set, random removal of samples had a small effect on apparent model performance (Figure 2). For instance, the observed-inferred r^2 decreased from 0.77 for the full model to 0.66 after removal of 80% of samples. However, removal of spatially and environmentally similar sites had a more substantial impact on performance, indicating autocorrelation among sites. Removal of sites that were geographically close to the test site in cross-validation resulted in a sudden drop in the r^2, as low as 0.13 after sites within 300 km were removed. This result is not surprising given the known uniqueness of the phytoplankton assemblages in each lake [44], so removal of training set samples in the same lake as the test site undoubtedly had a substantial effect on assemblage analogs. A similarly precipitous drop in performance occurred due to removal of sites that were environmentally similar to the test site, but model degradation was not noticeably different from that due to spatial removal. The relative importance of geographical and environmental neighbors shows how important adjacent sites are for the performance of the transfer function. As explained by Telford and Birks [21], the occurrence of autocorrelation indicates potential problems in a transfer function, and so the modern analog technique (MAT [37]) of inference is not recommended for this model. However, the model may still have predictive power under weighted-averaging scenarios. Based on models derived by random re-mapping of the environmental data over the species map, the actual model performed better 99.8% of the time (Table 3; Figure S1), indicating that the diatom-TP WA transfer function is statistically significant, suggesting that TP can be reconstructed from this training set.

An RDA of the diatom assemblages constrained by the 11 significant environmental variables indicated that these variables account for a major part (16.7% on the first axis; Table 3, Figure 3) of the underlying gradients in the diatom relative density data. Axis 1 of an unconstrained PCA captured 20.6% of the variation in the diatom data, indicating that the significant environmental variables accounted for a large portion of the underlying species gradients. Environmental data accounted for a larger portion of the variance in diatom relative biovolume data (20.0%; Table 3). TP (and the related TDP) were strongly correlated to axis 1, the main gradient of species turnover. There is also a clear alkalinity/chloride gradient that diagonally represents axes 1 and 2. Axis 2 captures a smaller proportion of variance (6.4%), although it is closely correlated to the NOx gradient. These proportions of explained variance are typical and informative in multivariate analyses of biological abundance data that tend to be noisy (Gauch [45]).

Figure 4 illustrates variance partitioning of the diatom relative density data by selected environmental variables. The total and unique effects of TP were 6.8% and 4.8% respectively (Table 3), a 29% reduction in explained variance. Based on relative biovolumes the total and unique effects were 9.1% and 6.0%, indicating a closer linkage between TP and diatom biovolumes than densities. That the unique TP effect is lower than total is due to redundancies, such as correlation to variables like chlorophyll *a* and turbidity (Figure 1). While it is difficult to confirm the relative importance of these variance partitioning values, the unique variance explained by TP surpasses that observed for TP training sets developed in Europe [46] and Minnesota [47], [48]; 3.9% and 2.5% respectively, as presented by Juggins et al. [22]. Of the selected variables, TP independently explained the most variation in the assemblage data, in contrast to alkalinity and chloride which had more than 70% reduction from total to uniquely explained variance. NOx explained the least variation but was minimally confounded by other variables. For TP to be confounded with other correlated variables is not surprising, but these results suggest that changes in the diatom assemblages across the Great Lakes are meaningfully related to TP.

Downcore testing of the diatom-based model

Figure 5 presents the common diatom taxa (>5% in any sample) in a sediment core from eastern Lake Superior (unpublished data). The stratigraphy is based on relative diatom density, including the DI-TP results which indicate a temporary enrichment event. This inference was largely driven by a short-term abundance of *A. islandica* which temporarily replaces lower-nutrient *Cyclotella* and *Discostella* species through much of the 20th century. Passively plotting the Lake Superior fossil samples on the RDA traces the path of historical changes in the diatom

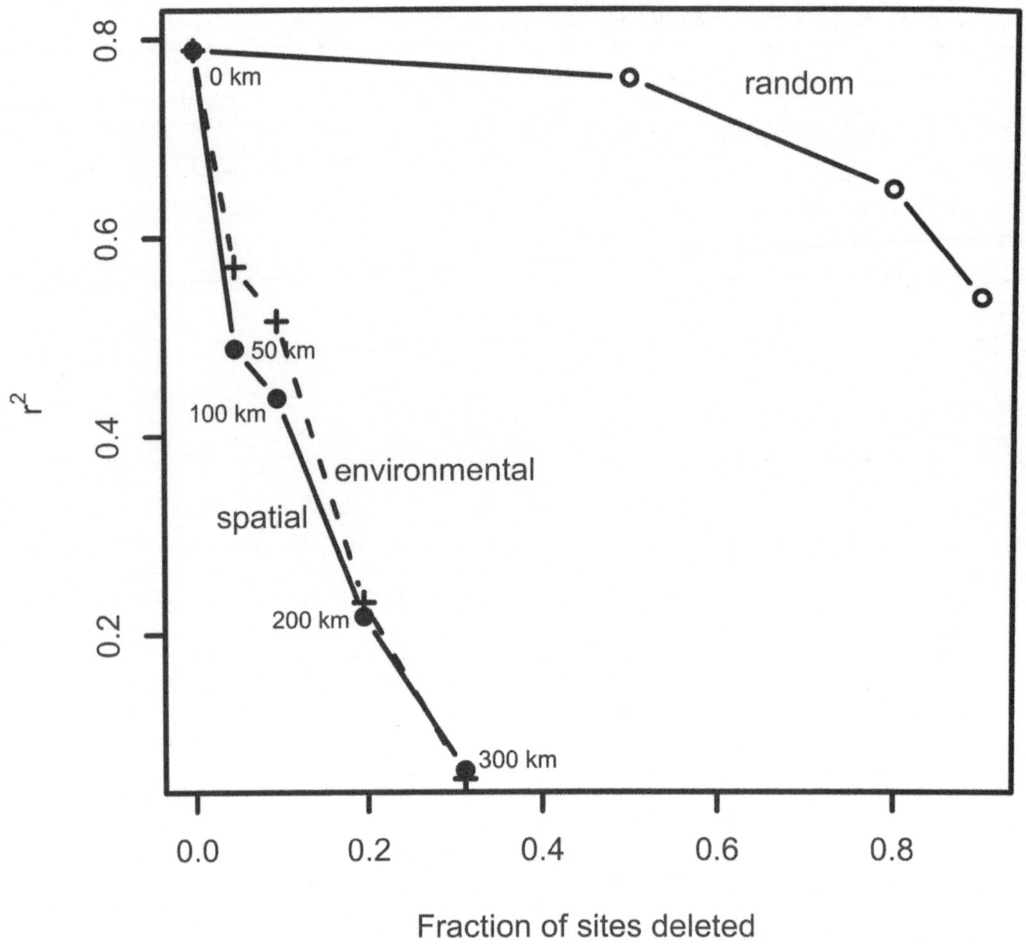

Figure 2. Effect of deleting sites on diatom-TP model performance. The effect on transfer function r^2 of deleting sites at random (open circles), from the geographical neighborhood of the test site (filled circles), and those that are environmentally similar (crosses) during cross-validation of the Great Lakes diatom training set. The radius of each distance is labelled.

assemblages (Figure 3). The long-term trend indicates a temporary movement of the scores in the direction of the TP vector from 1922 through 1970, a period that reflects nutrient enrichment in Lake Superior in association with the increased abundance of *Aulacoseira islandica* (unpublished data; [49]). Subsequent to 1970 the sample scores revert to locations close to pre-Euro-American samples, indicating TP reductions as the diatom assemblage shifted to a greater dominance by *Cyclotella* (unpublished data).

There was a high correlation ($r = 0.98$; $N = 26$) between DI-TP and the main gradient in the Lake Superior sedimentary diatom data (Figure S2), indicating TP was a strong driver of Superior's assemblages. The fraction of variance explained by DI-TP, represented as a fraction of the maximum explainable variance, was also high ($\lambda_R/\lambda_P = 0.97$). These results suggest that phosphorus has been an important determinant of diatom assemblages in Lake Superior, and that the TP transfer function was well-suited to reconstructing the nutrient history of the lake.

General Discussion

We provide a suite of environmental coefficients for the dominant phytoplankton taxa in the Laurentian Great Lakes. We hope that these data increase understanding of taxon-specific ecology, and expect these data will be used to develop inference models such as the diatom-based model tested herein. We endorse using these species coefficient tables (Table 2, Tables S1, S2, S3, S4) as checklists for future indicator studies. For instance, Kelly and Whitton [50] provided a checklist of trophic coefficients for diatom specimens, calibrated to nutrient and organic pollution variables, from rivers in the United Kingdom. Their coefficients have been used repeatedly in ecological assessments intended to manage river ecosystems [51]. Model-specific coefficients (linear, Gaussian, WA) and their statistical significance may also be used selectively in development of an indicator.

We recommend some cautions in using these data to infer environmental conditions. The coefficients are taxon-environmental relationships based on simultaneous sampling of algae and water quality. A phytoplankton assemblage may be a product of water quality conditions prior to sampling. For instance, despite the low silica optimum for *Aulacoseira islandica*, we know this taxon requires high silica concentrations for cell wall development [52], and so the Lake Erie spring diatom assemblages are undoubtedly a product of high silica concentrations prior to spring sampling. So, while high densities of *A. islandica* may infer low dissolved silica, the taxon may not represent a prevailing condition of low silica. We expect coefficients based on water quality variables that were not limiting at the time of sampling (e.g., spring nutrients [53]) are more reliable.

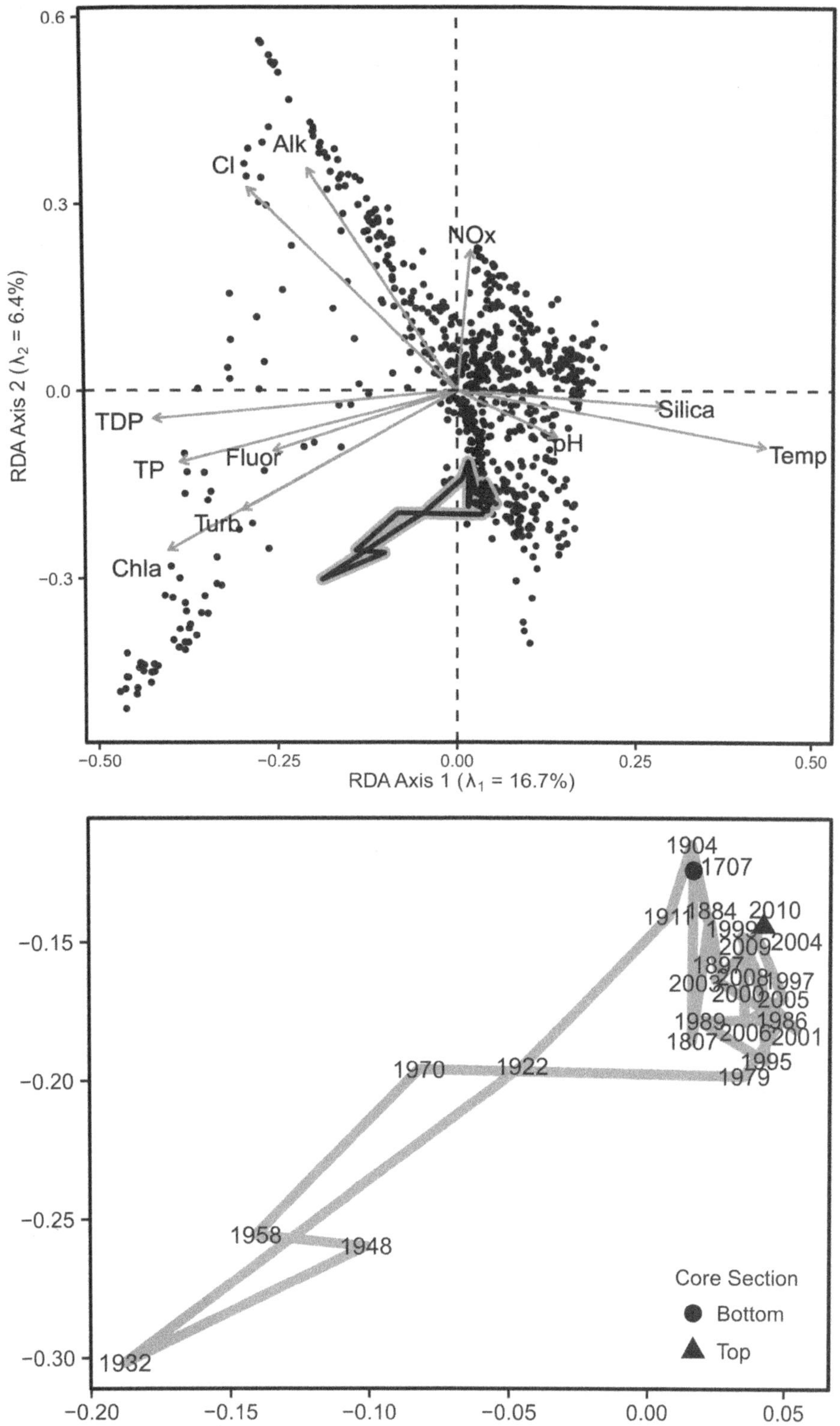

Figure 3. Redundancy analysis of the diatom training set constrained to significant water quality variables. Eigenvalues indicate variance explained by each axis. Passive plotting of the Lake Superior sedimentary assemblages is indicated by the black line set in a thicker gray line. The lower panel is a zoomed-in plot of the passive ordination of Lake Superior sedimentary diatom assemblages. The core top (2010) and bottom (~1707) are indicated and the date of each interval (based on ^{210}Pb dating) is provided for each interval.

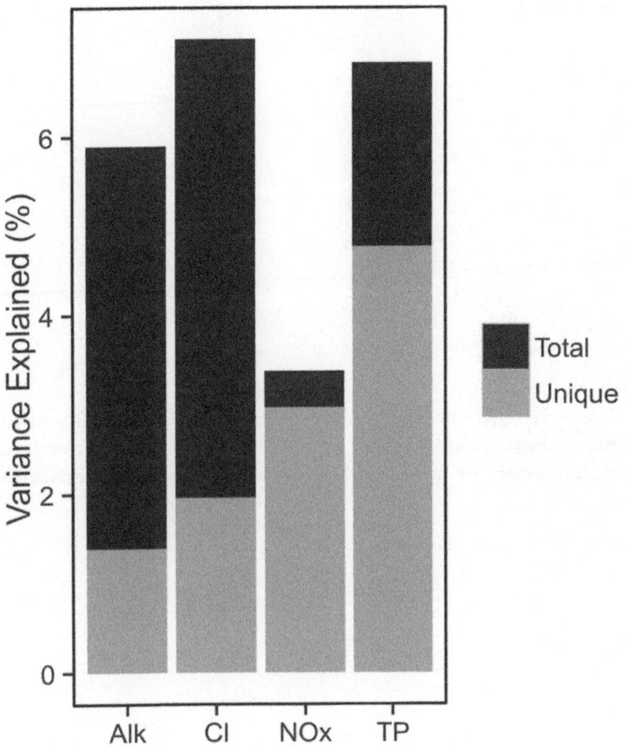

Figure 4. Histograms show percentage of variation in data explained by four water quality gradients. Total height indicates the total variation explained while gray height indicates the unique fraction of variance captured by that variable.

The provided transfer function, a diatom-based total phosphorus model, shows promise for future applications in the Great Lakes. Although internal testing of the model showed substantial spatial and environmental autocorrelation among sample locations, which is to be expected in large lakes that are likely homogeneous across vast areas, further testing indicates that phosphorus is an important driver of diatom assemblages in the Great Lakes, and that paleolimnological studies may benefit from this indicator model.

Unlike the snapshot approach of phytoplankton sampling, diatom assemblages retrieved from sediment cores reflect year-round integration of valves. For our Lake Superior example we confirm good analogues between fossil and training set assemblages, as well as a credible long-term TP reconstruction. Past assemblages, which included relatively high abundances of *A. islandica* during the mid-20th century, differ greatly from the phytoplankton in Lake Superior today, so the TP reconstruction relied on taxon coefficients that were largely derived from other lakes that presently contain abundant *A. islandica* (e.g., Lake Erie). It is unknown whether we may assume similar performance from cores collected elsewhere in the Great Lakes system, so we recommend thorough downcore validation of the model as we have done here, and as others recommend [22], before making conclusions about paleoecological nutrient trends. Also, the reconstructed range of TP for Lake Superior was on the low end of the range for the Great Lakes as a whole, so future testing should consider whether reconstructions in more nutrient-rich lakes are possible.

Lake Superior has encountered substantial diatom assemblage reorganization in the last 200 years, and as speculated by others [49] these changes were largely driven by anthropogenic nutrients (i.e. the rise and fall of pollutant flux to the lake). A sedimentary record containing similarly strong historical revisions in assem-

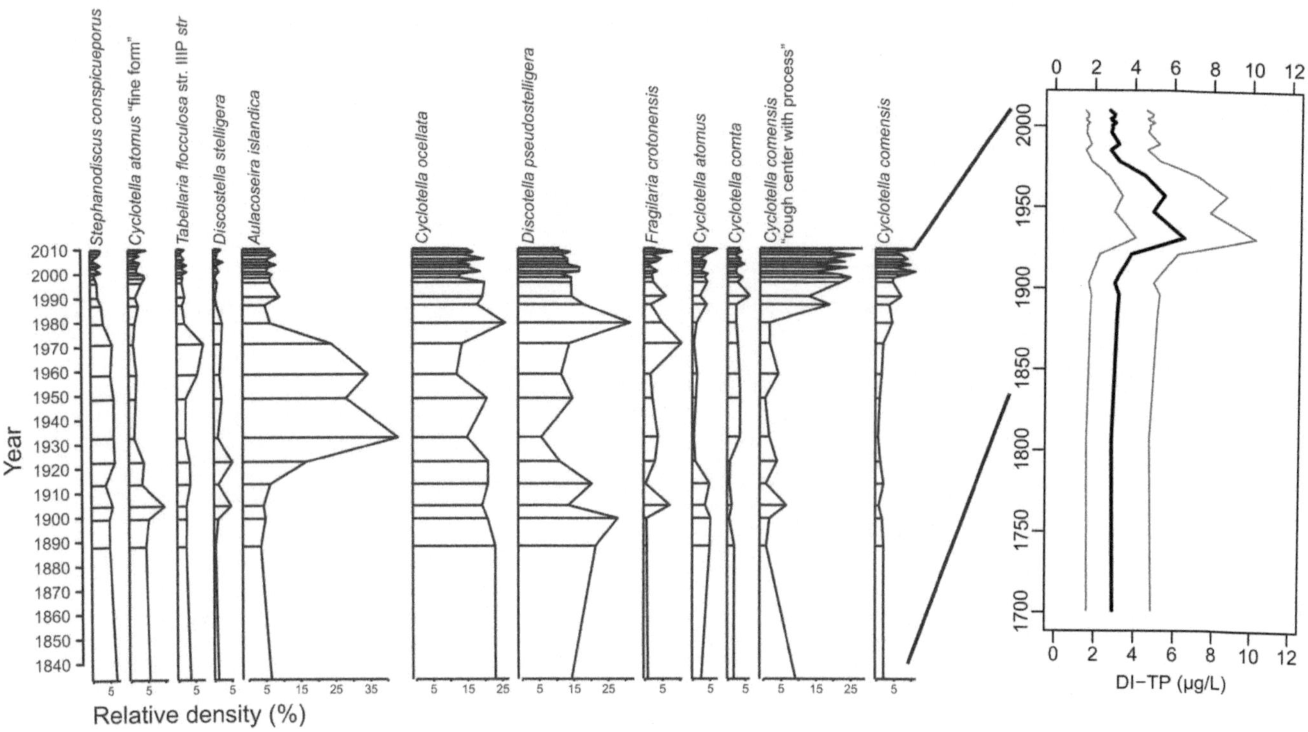

Figure 5. Dominant diatom species (>5% relative abundance) for the eastern core of Lake Superior. A plot of corresponding DI-TP for the fossil assemblages is shown on the right. Black line indicates inferred TP and grey lines indicate the range of model error (RMSEP).

blages should have a strong assemblage-DI relationship as we observed for TP in Lake Superior, otherwise an investigator must explore other variables as the cause of historical shifts. However, monotonous stratigraphic records of diatoms will have narrow ecological breadth, and a probable narrow range of DI-TP. Such cases may result in a weak correlation between assemblage patterns and DI data, so other means would be necessary to confirm the importance of TP to the sedimentary record.

In conclusion, the taxon-specific autecological information provided herein may be used to inform future observations of these species within the Great Lakes and elsewhere. More importantly, Great Lakes phytoplankton assemblages can make a contribution to indicator studies as suggested by Stoermer [10]. We present one of many indicator possibilities; the transfer function presented here appears to be suitable for downcore reconstructions. Using similar rule-based models with clear validation protocols should lead to refined, defensible environmental predictions and reconstructions.

Supporting Information

Figure S1 Performance of 1000 transfer functions with simulated environmental variables based on re-mapping of the TP dataset. 998 simulated transfer functions had poorer performance than the actual model r^2 based on the observed-inferred TP relationship, as indicated by the dotted line.

Figure S2 Relationship between Lake Superior downcore DI-TP and PCA axis 1 for diatom data.

Table S1 Environmental indicator values for the common algae taxa (>5%) in the Great Lakes based on distribution of cell densities across the environmental gradients.

Table S2 Environmental indicator values for the common algae taxa (>5%) in the Great Lakes based on distribution of relative cell densities across the environmental gradients.

Table S3 Environmental indicator values for the common algae taxa (>5%) in the Great Lakes based on biovolume distribution across the environmental gradients.

Table S4 Environmental indicator values for the common algae taxa (>5%) in the Great Lakes based on distribution of relative biovolume across the environmental gradients.

Acknowledgments

M. Agbeti supported assessments of the recent GLNPO phytoplankton samples. The U.S. Government is authorized to reproduce and distribute reprints for government purposes, notwithstanding any copyright notation that may appear hereon.

Author Contributions

Conceived and designed the experiments: EDR AJH VLSC. Performed the experiments: EDR AJH VLSC. Analyzed the data: EDR AJH VLSC. Contributed reagents/materials/analysis tools: EDR. Wrote the paper: EDR AJH.

References

1. Chapra SC, Dove A, Warren GJ (2012) Long-term trends of Great Lakes major ion chemistry. J Great Lakes Res 38:550–560.
2. Vanderploeg HA, Pothoven SA, Fahnenstiel GL, Cavaletto JF, Liebig JR, et al. (2012) Seasonal zooplankton dynamics in Lake Michigan: Disentangling impacts of resource limitation, ecosystem engineering, and predation during a critical ecosystem transition. J Great Lakes Res 38: 336–352.
3. Barbiero RP, Schmude K, Lesht BM, Riseng CM, Warren GJ, et al. (2011) Trends in Diporeia populations across the Laurentian Great Lakes, 1997–2009. J Great Lakes Res 37: 9–17.
4. Michalak AM, Anderson EJ, Beletsky D, Boland S, Bosch NS, et al. (2013) Record-setting algal bloom in Lake Erie caused by agricultural and meteorological trends consistent with expected future conditions. Proc Nat Acad Sci 110: 6448–6452.
5. McCormick PV, Cairns J Jr (1994) Algae as indicators of environmental change. J Appl Phycol 6: 509–526.
6. Holt EA, Miller SW (2011) Bioindicators: Using organisms to measure environmental impacts. Nat Educ Knowl 3: 8.
7. Brazner JC, Danz NP, Niemi GJ, Regal RR, Trebitz AS, et al. (2007) Evaluation of geographic, geomorphic and human influences on Great Lakes wetland indicators: a multi-assemblage approach. Ecol Indic 7: 610–635.
8. Reavie ED, Axler RP, Sgro GV, Danz NP, Kingston JC, et al. (2006) Diatom-based weighted-averaging transfer functions for Great Lakes coastal water quality: relationships to watershed characteristics. J Great Lakes Res 32: 321–347.
9. Dixit SS, Smol JP, Kingston JC, Charles DF (1992) Diatoms: Powerful indicators of environmental change. Environ Sci Technol 26: 22–33.
10. Stoermer EF (1978) Phytoplankton assemblages as indicators of water quality in the Laurentian Great Lakes. Trans Am Microsc Soc 97: 2–16.
11. U.S. Environmental Protection Agency (2010) Sampling and Analytical Procedures for GLNPO's Open Lake Water Quality Survey of the Great Lakes. Great Lakes National Program Office, U.S. Environmental Protection Agency, document EPA 905-R-05-001, Chicago, IL.
12. Munawar M, Munawar IF (1982) Phycological studies in Lakes Ontario, Erie, Huron, and Superior. Can J Bot 60: 1837–1858.
13. Imbrie J, Kipp NG (1971) A new micropaleontological method for quantitative paleoceanography based on microfossils. Dev Mar Geol 1:523–563.
14. Tibby J (2004) Development of a diatom-based model for inferring total phosphorus in southeastern Australian water storages. J Paleolim 31: 23–36.
15. Meriläinen JJ, Hynynen J, Palomäki A, Mäntykoski K, Witick A (2003) Environmental history of an urban lake: A palaeolimnological study of Lake Jyväsjärvi, Finland. J Paleolim 30:387–406.
16. Dixit SS, Smol JP (1994) Diatoms as indicators in the Environmental Monitoring and Assessment Program–Surface Waters. Environ Monit Assess 31: 275–306.
17. Siver PA, Ricard R, Goodwin R, Giblin AE (2003) Estimating historical in–lake alkalinity generation from sulfate reduction and its relationship to lake chemistry as inferred from algal microfossils. J Paleolim 29: 179–197.
18. Smol JP (2002) Pollution of Lakes and Rivers: A Paleoenvironmental Perspective. London, UK: Arnold Publishers.
19. Charles DF (1990) A checklist for describing and documenting diatom and chrysophyte calibration data sets and equations for inferring water chemistry. J Paleolim 3: 175–178.
20. Last WM, Smol JP, Birks HJ, editors (2001) Tracking environmental change using lake sediments, vol 3: Terrestrial, algal, and siliceous indicators. Springer.
21. Telford RJ, Birks HJB (2009) Evaluation of transfer functions in spatially structured environments. Quat Sci Rev 28: 1309–1316.
22. Juggins S, Anderson NJ, Hobbs JMR, Heathcote AJ (2013) Reconstructing epilimnetic total phosphorus using diatoms: statistical and ecological constraints. J Paleolim 49: 373–390.
23. Utermöhl H (1958) Zur Vervollkommnung der quantitativen Phytoplankton-Methodik. Mitt Int Verein Limnol 9:1–38.
24. Shaw Chraïbi VL, Kireta AR, Reavie ED, Brown TN, Cai M (2014) A Paleolimnological Assessment of Human Impacts on Lake Superior. J Great Lakes Res (in press).
25. R Core Team (2013) R: A language and environment for statistical computing. R Foundation for Statistical Computing, Vienna, Austria. URL http://www.R-project.org/
26. Shaw Chraïbi VL (2013) A 250-year assessment of human impacts on Lake Superior: an updated paleolimnological perspective. M.Sc. Thesis, University of Minnesota Duluth. Available: http://conservancy.umn.edu/bitstream/11299/157356/1/Chraibi_umn_0130M_13654.pdf. Accessed 2014 June18.
27. Oksanen J, Blanchet FG, Kindt R, Legendre P, Minchin PR, et al. (2013) Vegan: Community ecology package. R package version 2.0-9. http://CRAN.R-project.org/package=vegan
28. Juggins S, Birks HJB (2012) Quantitative environmental reconstructions from biological data. In: Birks HJB, Lotter AF, Juggins S, Smol JP, editors. Tracking environmental change using lake sediments. Springer Netherlands. pp 431–494.

29. Battarbee RW, Jones VJ, Flower RJ, Cameron NG, Bennion H, et al. (2001) Diatoms. In Smol JP, Birks HJB, Last WM, editors. Tracking environmental change using lake sediments. vol 3: Terrestrial, algal, and siliceous indicators. Dordrecht Kluwer Academic Publishers. pp. 155–202.
30. Hastie TJ (2013) gam: Generalized additive models. R package version 1.08. http://CRAN.R-project.org/package = gam
31. Yee TW, Mitchell ND (1991) Generalized additive models in plant ecology. J Veg Sci 2: 587–602.
32. Wood SN (2006) Generalized additive models: an introduction with R. Chapman Hall/CRC, Boca Raton.
33. Juggins S (2012) rioja: Analysis of quaternary science data, R package version (0.8–5). (http://cran.r-project.org/package = rioja).
34. Birks HJB, Line JM, Juggins S, ter Braak CJF (1994) WACALIB version 3.3 – a computer program to reconstruct environmental variables from fossil assemblages by weighted averaging and to derive sample–specific errors of prediction. J Paleolim 10: 147–152.
35. Juggins S, Birks HJB (2012) Quantitative environmental reconstructions from biological data. In: Birks HJB, Lotter AF, Juggins S, Smol JP, editors. Tracking environmental change using lake sediments: Data handling and statistical techniques. Dordrecht: Springer. pp.431–494.
36. Efron B, Gong G (1983) A leisurely look at the bootstrap, the jackknife, and cross-validation. Am Stat 37: 36–48.
37. Overpeck JT, Webb T, Prentice IC (1985) Quantitative interpretation of fossil pollen spectra: Dissimilarity coefficients and the method of modern analogs. Quat Res 23: 87–108.
38. Telford RJ (2013) palaeoSig: Significance tests of quantitative palaeoenvironmental reconstructions. R package version 1.1-2. http://cran.r-project.org/web/packages/palaeoSig
39. Reavie ED, Juggins S (2011) Exploration of sample size and diatom-based indicator performance in three North American phosphorus training sets. Aquat Ecol 45:529–538.
40. Pebesma EJ (2004) Multivariable geostatics in S: the gstat package. Comput Geosci 30:683–691.
41. Reavie ED, Barbiero RP, Allinger LE, Warren GJ (2014) Phytoplankton trends in the Laurentian Great Lakes: 2001–2011. J Great Lakes Res (in press).
42. Allinger LE, Reavie ED (2013) The ecological history of Lake Erie as recorded by the phytoplankton community. J Great Lakes Res 39: 365–382.
43. Millie DF, Fahnenstiel GL, Dyble Bressie J, Pigg RJ, Rediske RR, et al. (2009) Late-summer phytoplankton in western Lake Erie (Laurentian Great Lakes): bloom distributions, toxicity, and environmental influences. Aquat Ecol 43: 915–934.
44. Barbiero RP, Tuchman ML (2001) Results from the US EPA's biological open water surveillance program of the Laurentian Great Lakes: I. Introduction and phytoplankton results. J Great Lakes Res 27: 134–154.
45. Gauch HG (1982) Multivariate analysis in community ecology. Cambridge: Cambridge University Press. 298 p.
46. Bennion H, Juggins S, Anderson NJ (1996) Predicting epilimnetic phosphorus concentrations using an improved diatom-based transfer function and its application to lake eutrophication management. Environ Sci Technol 30: 2004–2007.
47. Ramstack JM, Fritz SC, Engstrom DR, Heiskary SA (2003) The application of a diatom-based transfer function to evaluate regional water-quality trends in Minnesota since 1970. J Paleolim 29: 79–94.
48. Reavie ED, Robbins JA, Stoermer EF, Douglas MS V, Morehead NR, et al. (2005) Paleolimnology of a fluvial lake downstream of Lake Superior and the industrialized region of Sault Saint Marie. Can J Fish Aquat Sci 62: 2586–2608.
49. Stoermer EF, Kociolek CL, Schelske CL, Conley DJ (1985) Siliceous microfossil succession in the recent history of Lake Superior. Proc Acad Nat Sci Phila 137: 106–118.
50. Kelly MG, Whitton BA (1995) The trophic diatom index: a new index for monitoring eutrophication in rivers. J Appl Phyc 7: 433–444.
51. Szczepocka E, Szulc B (2009) The use of benthic diatoms in estimating water quality of variously polluted rivers. Oceanol Hydrobiol Stud 38: 17–26.
52. Barbiero RP, Carrick HJ, Volerman JB, Tuchman ML (2000) Factors affecting temporal and spatial distribution of diatoms in Lake Michigan. Verh Int Verein Limnol 27: 1788–1794.
53. Lean DRS, Abbott AP, Charlton MN, Rao SS (1983) Seasonal phosphate demand for Lake Erie plankton. J Great Lakes Res 9: 83–91.

4

Exploratory Hydrocarbon Drilling Impacts to Arctic Lake Ecosystems

Joshua R. Thienpont[1], Steven V. Kokelj[2], Jennifer B. Korosi[1,3], Elisa S. Cheng[1], Cyndy Desjardins[3], Linda E. Kimpe[3], Jules M. Blais[3], Michael F.J. Pisaric[4,5], John P. Smol[1]*

1 Paleoecological Environmental Assessment and Research Lab (PEARL), Department of Biology, Queen's University, Kingston, Ontario, Canada, 2 Northwest Territories Geoscience Office, Government of the Northwest Territories, Yellowknife, Northwest Territories, Canada, 3 Department of Biology, University of Ottawa, Ottawa, Ontario, Canada, 4 Department of Geography and Environmental Studies, Carleton University, Ottawa, Ontario, Canada, 5 Department of Geography, Brock University, St. Catharines, Ontario, Canada

Abstract

Recent attention regarding the impacts of oil and gas development and exploitation has focused on the unintentional release of hydrocarbons into the environment, whilst the potential negative effects of other possible avenues of environmental contamination are less well documented. In the hydrocarbon-rich and ecologically sensitive Mackenzie Delta region (NT, Canada), saline wastes associated with hydrocarbon exploration have typically been disposed of in drilling sumps (i.e., large pits excavated into the permafrost) that were believed to be a permanent containment solution. However, failure of permafrost as a waste containment medium may cause impacts to lakes in this sensitive environment. Here, we examine the effects of degrading drilling sumps on water quality by combining paleolimnological approaches with the analysis of an extensive present-day water chemistry dataset. This dataset includes lakes believed to have been impacted by saline drilling fluids leaching from drilling sumps, lakes with no visible disturbances, and lakes impacted by significant, naturally occurring permafrost thaw in the form of retrogressive thaw slumps. We show that lakes impacted by compromised drilling sumps have significantly elevated lakewater conductivity levels compared to control sites. Chloride levels are particularly elevated in sump-impacted lakes relative to all other lakes included in the survey. Paleolimnological analyses showed that invertebrate assemblages appear to have responded to the leaching of drilling wastes by a discernible increase in a taxon known to be tolerant of elevated conductivity coincident with the timing of sump construction. This suggests construction and abandonment techniques at, or soon after, sump establishment may result in impacts to downstream aquatic ecosystems. With hydrocarbon development in the north predicted to expand in the coming decades, the use of sumps must be examined in light of the threat of accelerated permafrost thaw, and the potential for these industrial wastes to impact sensitive Arctic ecosystems.

Editor: Stephen J. Johnson, University of Kansas, United States of America

Funding: This work was funded by the Natural Sciences and Engineering Research Council (NSERC) of Canada through Discovery grants to MFJP, JMB and JPS, and an NSERC Northern Supplement to MFJP. The Polar Continental Shelf Program (PCSP) also provided logistical support to MFJP. The cumulative impact monitoring program (CIMP) provided support for collection of water chemistry results. The funders had no role in study design, data collection and analysis, decision to publish, or preparation of the manuscript.

Competing Interests: The authors have declared that no competing interests exist.

* E-mail: smolj@queensu.ca

Introduction

The Mackenzie Delta in Canada's western Arctic is underlain by significant discovered and predicted reserves of hydrocarbons [1], but is also amongst the most rapidly warming regions globally [2]. Activities associated with the exploitation of these hydrocarbon resources, including enhanced exploration, as well as infrastructure development through the extraction, production and transmission of hydrocarbons to market, constitute an additional stressor to the freshwater ecosystems of the region. The Mackenzie Delta region is ecologically important, as identified by the establishment of the Kendall Island Migratory Bird Sanctuary in 1961, as well as being culturally significant for local indigenous communities [3]. Much recent attention has focused on oil and gas activities increasing the delivery of toxic polycyclic aromatic hydrocarbons (PAHs) into the environment [4,5]; however the potential effects of industrial activities on aquatic ecosystems are widespread, and PAH contamination is just one example of the environmental consequences of oil and gas exploration and development.

Hydrocarbon exploration has been occurring in the Mackenzie Delta region (Fig. 1a) since the 1960s, and was particularly intense during the 1970–80s and around 2000 [6]. In the Canadian Arctic, wastes associated with the drilling of exploratory onshore hydrocarbon wells have been typically disposed of using in-ground sumps (Fig. 1b, c; [7]). These large excavations into the permafrost, usually located next to the exploratory well, are intended to act as a permanent containment location for the wastes produced during exploratory well development, including mud and rock cuttings, and drilling fluids. Drilling fluids are made up of, surfactants and detergents, as well as large quantities of highly concentrated saline solutions (primarily potassium chloride, with concentrations up to 100 g L^{-1}), which are used as a freezing point depressant during winter drilling operations [8–10].

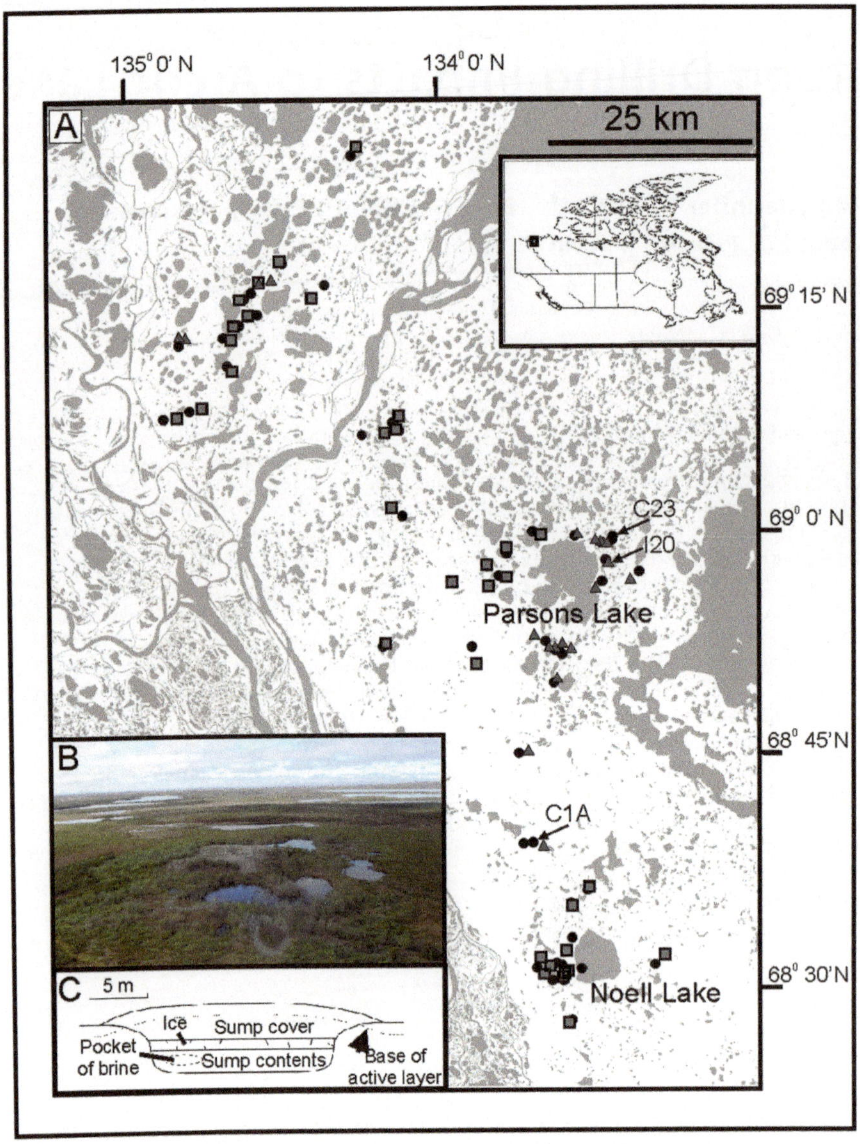

Figure 1. Map of study area, schematic of typical drilling sump and image of a degrading sump. A) The location of the 101 study lakes in the Mackenzie Delta uplands (Northwest Territories, Canada) (triangles – drilling-sump lakes; squares – thaw-slump lakes; circles – control lakes). Inset shows the region in the context of Canada. B) Image of a degrading drilling sump from the Mackenzie Delta uplands, near Parsons Lake, exhibiting significant surface and perimeter ponding. C) Generalized schematic of a drilling mud sump. A large pit is excavated into the permafrost and filled with the drilling wastes and fluids. These drilling fluids are then allowed to partially or completely freeze, and backfilled with the excavated material. The assumption is that the material will be permanently contained in the permafrost. Redrawn from [8].

Historically, a typical three-kilometre deep well, required approximately 40,000 m^3 of drilling fluid alone [7], though this volume has been reduced in more recent operations due to an improvement in recycling technologies.

More than 150 drilling sumps have been constructed in the Mackenzie Delta region since the mid-1960s [6]. As most exploration activity occurs in the winter (this practice has been required by law since 1986 in an attempt to minimize the environmental impact of drilling activities), these fluids are meant to freeze *in situ*, and then capped with the material excavated from the sump. It was assumed the drilling muds are permanently encapsulated within the surrounding permafrost, and that drilling sumps represent a permanent containment mechanism for materials associated with hydrocarbon exploration [7].

Pronounced climate warming in the western Canadian Arctic has serious implications for sump containment, as increasing air temperatures are increasing tall shrub cover and causing permafrost temperatures to rise [3]. Recent studies of sump integrity have observed increased conductivity beyond the extent of the sump at 74% of the sites studied [6], and as many as one third of these sites are exhibiting surface ponding, suggesting significant thaw of the sump contents has occurred [9]. Vegetation communities growing on sump caps were found to be distinct from surrounding, undisturbed areas, related to factors including drainage, active-layer depth and soil salt concentrations [10]. Increased shrub growth on sump covers enhances snow accumulation and accelerates warming of the sump and lease areas [11]. No assessment of the impacts of saline drilling wastes on nearby freshwater ecosystems exists, although the Mackenzie Delta region and Tuktoyaktuk Coastlands are amongst the most lake-rich environments in the Arctic. This is particularly important, given that many of the compounds present in drilling muds (e.g. KCl, caustic soda, barite) are known to be toxic to freshwater organisms at the concentrations common in drilling muds [7,12,13], and because the majority of the drilling sumps constructed are in the catchments of lakes in this water-rich landscape. Understanding the effects of drilling sump failure is essential to evaluate the relative impact of environmental stressors on freshwater in this region.

In this study, we use a combination of contemporary limnological sampling and inferences of past conditions using material preserved in lake sediments (i.e. paleolimnology) to assess the impacts of drilling fluids on the freshwater ecosystems of the Mackenzie Delta uplands region. We compare contemporary water chemistry measurements of lakes with drilling sumps in their catchments to lakes undergoing another major stressor, intense permafrost degradation in the form of retrogressive thaw slumping, as well as undisturbed control lakes with no evidence of localized disturbance. Retrogressive thaw slumps are a common form of thermokarst, which currently occur on the shoreline of approximately 10% of the lakes in the Mackenzie Delta uplands, and are increasing in size and growth rate as a result of recent warming [14]. Thaw slumping is known to result in major changes to contemporary lakewater chemistry [15,16] and biology [17]. Our previous diatom-based paleolimnological study showed that the main biological response to permafrost thaw in the lakes of the region was related to changes in aquatic habitat availability and water clarity[18]. In this study we expand on our previous sediment-based research in order to understand the impact of hydrocarbon exploration on the lakes of this sensitive region. We used a large present-day limnological dataset of 101 sites to compare lakes with drilling sumps in their catchments to permafrost thaw-affected lakes which are known to be highly-disturbed aquatic systems in the region. Due to the fact that drilling fluids are saline, we hypothesized that if migration of drilling fluids to receiving surface waters has occurred, we should record elevated levels of major ions and conductivity in impacted lakes. Also, due to a lack of long-term biomonitoring data, a variety of sedimentary proxies were analyzed in sediment cores from three lakes in order to put the timing and nature of any limnological change in a historical context. We further hypothesized that if saline-rich wastes from drilling sumps are impacting lakes, shifts in the assemblages of sentinel biological indicators should be detected.

Materials and Methods

One hundred and one lakes (20 drilling sump, 34 permafrost thaw slump, and 47 control lakes) in the Mackenzie Delta uplands (NT, Canada) were sampled for their present-day limnological conditions in the summers of 2005 or 2007. Thirteen measured physical and chemical variables observed above detection limits in the majority of the sites were selected and normalized. Principal components analysis (PCA) was conducted using the vegan package [19] for R [20]. Analysis of similarity (ANOSIM; [21]) and similarity percentages (SIMPER) were conducted using PRIMER v.6 in order to assess the relationships between the three *a priori* assigned groups, and to determine the variables that contributed to any dissimilarity.

Sediment cores were obtained from lakes I20 (68.9742°N, 133.5253°W; sump-impacted), C23 (68.9978°N, 133.5129°W; control), and C1A (68.6589°N, 133.7602°W; control) (all names unofficial) in August 2009 or July 2010 and sectioned at 0.5 cm resolution using our standard, high-resolution paleolimnological techniques [18]. Lake I20 is located downslope of a drilling sump that shows cover subsidence indicative of permafrost degradation. Lakes C23 and C1A are located nearby in similar terrain to the impacted lake I20, but with no drilling sumps in their respective catchments, and are thus classified as control lakes. Sediment age determination was conducted using ^{210}Pb and ^{137}Cs radiometric techniques (Fig. S1) [22], following which sedimentary subfossil indicators were isolated and analyzed using standard methods (diatoms: [23]; cladocerans: [24]). Relative percentage diagrams were generated using Tilia v.1.7.16 [25]. Constrained incremental sums of squares (CONISS) cluster analyses were conducted in order to identify biostratigraphic zones of change [26], with the broken stick model used to determine the number of significant zones [27]. Overall lake primary production was estimated by inferring sedimentary chlorophyll *a* concentrations via visual reflectance spectroscopy [28]. PAHs were extracted from wet sediments using accelerated solvent extraction (ASE, Dionex). Approval for all field-based research and the issuing of relevant scientific research licenses was conducted through the Aurora Research Institute (Inuvik, NT). No protected areas or protected species were sampled as part of this research. A more detailed description of the methods used in this study is presented in Text S1.

Results and Discussion

The three *a priori*-defined groups (drilling sump, permafrost thaw slump and control lakes) were found to be significantly different based on their present-day water chemistry (ANOSIM, Global $R = 0.307$, $p = 0.001$), and lakes with drilling sumps in their catchments had significantly higher concentrations of chloride (Cl^-) than either the control lakes or lakes impacted by permafrost thaw slumps (ANOVA, $df = 2, 98$, $F = 7.91$, $p < 0.001$; Tukey's HSD post-hoc test) (Fig. 2). The glaciogenic sediments of the Mackenzie Delta uplands are primarily derived from shale bedrock, and as such are naturally high in sulphate, but relatively low in chloride [29,30]. As a result of permafrost thaw-slump development, the influx of terrestrially-derived materials to lakes results in a significant difference in SO_4^{-2} concentration, but no significant difference in Cl^- concentrations (Fig. S2, Table S1). Significantly greater Cl^- concentrations in the drilling-sump impacted lakes (Fig. 2), therefore, suggests a source other than the underlying geological material, such as the wastes from drilling activity, which in most cases are brine-based (e.g. KCl) [7]. Based on similarity percentages (SIMPER), Cl^- concentrations contributed 19% of the difference between the drilling-sump and control lakes (more than any other variable), and 13% of the difference between the drilling-sump and thaw-slump lakes (Table S2). Complete comparisons of the measured physical and chemical variables are presented in Text S1. These results indicate that not only is chloride elevated in drilling-sump impacted lakes, but it is also the single most important environmental variable distinguishing sump-impacted lakes from the undisturbed control sites in the region.

Leaching of drilling fluids is clearly the most plausible explanation for the trends in the present-day chemical conditions observed in the drilling-sump impacted lakes in this dataset. Lakes impacted by drilling sumps also exhibited significantly higher concentrations of Ca^{2+}, Na^+, and specific conductivity compared to the control sites, though significantly less than lakes with thaw slumps (Fig. S1). Drilling fluids are the most likely source of these ions to impacted lakes, as materials such as caustic soda (NaOH) are common components [12]. In addition, as drilling fluids

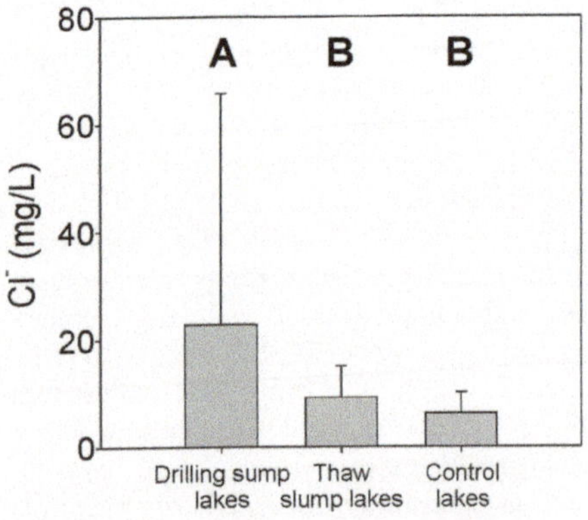

Figure 2. Mean chloride concentration in the 101 lake dataset separated into three *a priori* defined groups. Vertical error bars representing the standard deviation are included. Letters A and B indicate drilling-sump impacted lakes have significantly higher chloride levels than either thaw-slump affected lakes or control lakes, which are not statistically different (Tukey HSD post-hoc test, following ANOVA run on normalized environmental data)

migrate through the active layer downslope to lake ecosystems, other ions present in soils would be translocated to the lake. This impact is exacerbated by the fact that ion-rich permafrost on top of drilling sumps and in adjacent lease areas is thawing, as active layer depth is increasing due to enhanced shrub growth [10], liberating base cations, and contributing to elevated ionic concentrations in the nearby water bodies.

Principal components analysis (PCA) was used to characterize the water chemistry variation in the 101 study lakes, taking into account the combined importance of all measured environmental variables. The first PCA axis primarily represents a response to ionic strength (Fig. 3), which is not surprising given the importance of these variables in accounting for the differences among groups. Lakes impacted by thaw slumps separate in the PCA plot from control sites primarily based on these variables [16], while the drilling-sump lakes are more widely distributed within the ordination space (Fig. 3). Some drilling-sump lakes, such as I15, I32B, I23A and I17, exhibit environmental conditions similar to sites impacted by large, active permafrost thaw slumps (Fig. 3). Drilling-sump impacted Lake I15 is chemically most similar to thaw-slump affected Lake 10B, which has an active thaw slump impacting over 25% of the lake's catchment [15], despite the lack of any natural geomorphic disturbance in Lake I15's catchment. Other drilling-sump impacted lakes, such as I1 or I12B, are chemically more similar to the control lakes, and thus the sumps in their catchments are likely providing better containment of drilling fluids (Fig. 3). Concentrations of major ions and conductivity

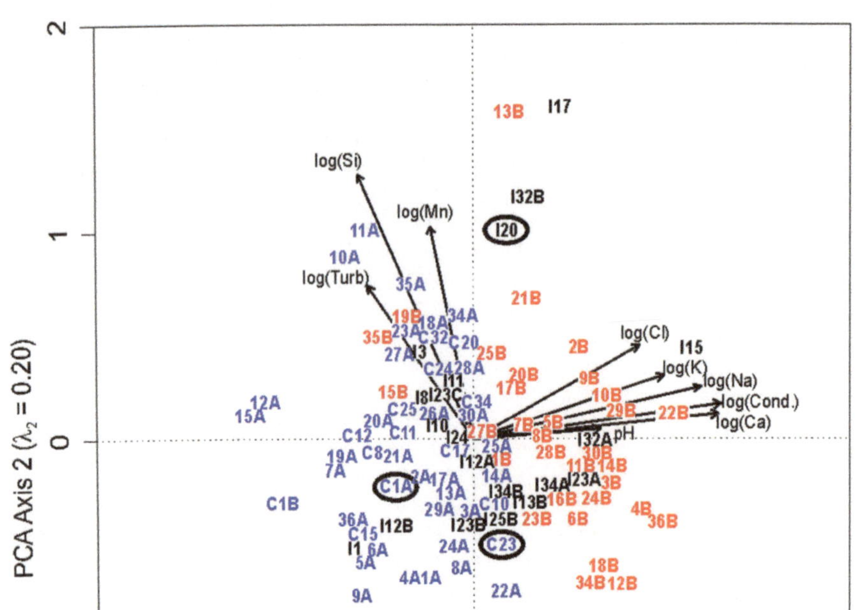

Figure 3. Principal components analysis (PCA) ordination of select chemical variables in the 101 lake dataset. Black labels represent the position of the drilling-sump lakes in the ordination space (n = 20). Red labels represent lakes impacted by retrogressive thaw slumping (n = 34). Blue labels represent control lake ordination results (n = 47). Variables were all normalized using log transformation. Arrows represent importance of given chemical variables in structuring the distribution of sites. Ordination results of study lakes I20, C23, and C1A, from which sediment cores were analyzed, are circled. Si = reactive silica.

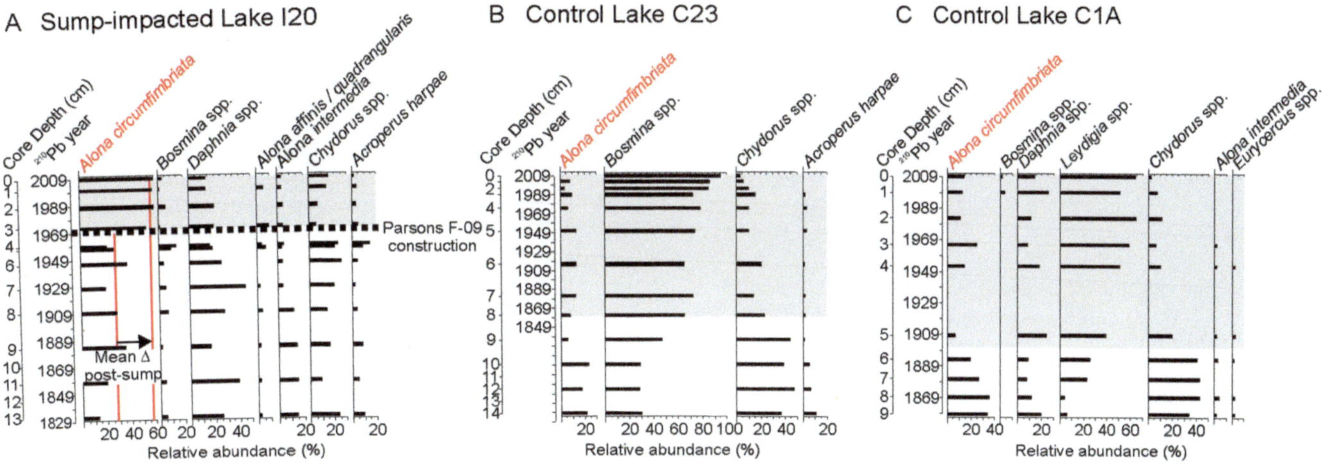

Figure 4. Stratigraphic profile of the most common cladoceran taxa from the three study lakes. Relative abundance diagrams from: lakes A) I20, impacted by drilling sump degradation; and control lakes B) C23; and C) C1A. Species assemblages (x axes) are scaled by relative abundance. Down-core sedimentary profiles (y axes) are scaled by date, based on ^{210}Pb radiometric dating techniques, with the depth in the sediment core included as a secondary axis. For all three lakes, two biostratigraphic zones were identified (constrained incremental sum of squares cluster analysis with the broken stick model) and are plotted with the background colour of one zone in grey the other white. The known timing of construction of the compromised drilling sump near Lake I20 (industry ID: Parsons F-09) is included as a horizontal line. The vertical red lines represent the pre- and post-sump construction mean of the relative abundance of the cladoceran *Alona circumfimbriata*.

appear to be a useful tracer for identifying the influx of materials from deteriorating drilling sumps into nearby lake ecosystems. The variability in the response of drilling-sump impacted lakes is expected, given the range of sump conditions currently observed in the region, with some sumps experiencing significant deterioration while other sites exhibit reasonable to good cover integrity [9]. Exploratory hydrocarbon drilling appears to have resulted in impacts to the freshwater ecosystems of the Mackenzie Delta uplands, in some cases more severe than large-scale, natural,

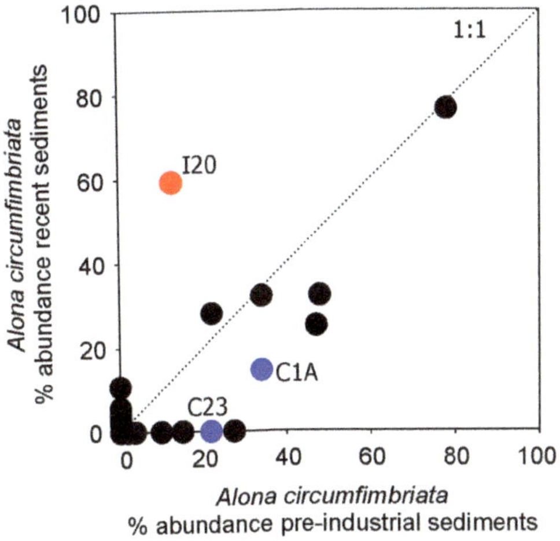

Figure 5. Change in the relative abundance of *Alona circumfimbriata* between present-day and pre-industrial sediment samples. Based on a comparison of the top of the sediment core (representing present-day conditions) with a bottom sample (pre-industrial) from 50 lakes in the Canadian sub-Arctic. Black circles represent 47 lakes from the only regional cladoceran survey in the Canadian sub-Arctic, spanning present-day treeline in the Northwest Territories (data recalculated from [34]). Blue circles represent the control lakes (C23, C1A) from this study. The red circle represents drilling sump-impacted Lake I20. The 1:1 line is also shown.

localized geomorphic disturbances, such as retrogressive thaw slump development, which represent a major stressor to these ecosystems.

Due to the lack of long-term biological monitoring data, paleolimnological approaches were used to assess if sump leakage has resulted in any biological changes in sentinel indicator species [31]. Multiple sedimentary proxies were analyzed in radiometrically dated sediment cores from three lakes in the Mackenzie Delta uplands: two sites near Parsons Lake, a location that has undergone significant exploratory hydrocarbon drilling over the last 50 years [6]; and one lake to the south (Fig. 1). Lake I20, impacted by a drilling sump in its catchment, shows elevated ionic water concentrations compared to the majority of the control lakes (Fig. 3), suggesting that drilling fluids may be impacting this lake. PCA (Fig. 3) revealed that some control sites, such as Lake C23, exhibited ion concentrations similar to those of some drilling sump-impacted lakes. Because these lakes have no known history of localized disturbance, we hypothesized that the elevated conductivity levels in C23 are natural, and sedimentary indicators in lake sediments should exhibit no change related to sump development. Lake C1A has conductivity levels more typical of other control lakes (Fig. 3). The analysis of I20 and two control lakes that differ in present-day conductivity will allow any changes in the drilling sump-impacted site to be placed in the context of natural chemical variability in the region.

Concentrations of polycyclic aromatic hydrocarbons did not increase following the construction of the drilling sump near Lake I20 (Text S1, Fig. S3). This suggests that, unlike the impact of large-scale extraction operations such as the Alberta oilsands [4,5], impacts from exploratory drilling activities in this region of the Arctic do not appear related to contamination by hydrocarbons themselves. This is not surprising, given that most drilling in the delta region utilizes brine-based drilling muds (compared to oil-based), and thus sump contents proportionately contain very little hydrocarbons. Instead, any environmental impacts of these activities are likely to be related to chemicals present in the drilling fluids, notably salts.

Many species of Cladocera (Crustacea, Branchiopda) are considered poor osmoregulators, and thus we hypothesized that changes in ionic strength following drilling-sump failure would result in a shift in the species assemblage. In addition there is evidence that potassium chloride in the concentrations found in drilling fluids produces both lethal and sub-lethal effects on Cladocera in laboratory experiments [13]. Cladocera are a key component of aquatic foodwebs, and thus understanding their response to this environmental stressor is essential.

In Lake I20, an abrupt increase in the relative abundance of *Alona circumfimbriata* subfossils, a taxon which is known to be relatively saline tolerant [32,33], occurred near the time of sump development in ~1972 (Fig. 4). In a survey of Cladocera in sub-Arctic lakes spanning treeline in the NWT, this taxon was most common in the highest conductivity sites [34]. While the overall magnitude of this assemblage change is relatively subtle, the timing and rapid nature of this change occurring at the time of (or very soon after) the construction of the sump suggests sump disturbance for at least this site may be primarily related to construction and abandonment practices. No comparable changes in the cladoceran assemblage of either control lake were observed coincident with, or subsequent to, the construction of the sumps near to (but not in the catchment of) those sites, despite the natural variability in ion-related water chemistry (Fig. 3). Instead, the primary changes that do occur in the control lakes are gradual and more indicative of regional warming, which has been extensive in this region and has been inferred to be impacting the cladoceran assemblages in another lake in the Mackenzie Delta [35]. The increase in *A. circumfimbriata* in I20 is unique when analyzed in the context of the only other regional dataset available for Cladocera in the Canadian sub-Arctic [34] (Fig. 5). Sedimentary diatom assemblages recorded no changes inferred to be as a result of drilling-sump containment loss, and instead are responding to climate warming which has been significant in this region (Text S1, Fig. S4). The lack of diatom response to these still relatively subtle changes in lakewater chemistry is not surprising. In lakes impacted by large permafrost thaw slumps, diatom assemblage shifts were inferred to be in response to modifications in aquatic habitat, and not chemical changes [18].

As the sub-Arctic and Arctic continue to warm rapidly, and permafrost temperatures increase, the likelihood of waste containment issues in drilling mud sumps will increase, potentially increasing impacts on aquatic ecosystems. Possible climate-induced increases in evaporation, as have been observed in other high-latitude regions [36,37], could also lead to reductions in lakewater level, and further concentration of salts. This will be exacerbated by increased ionic flux due to warming and the thaw of containing permafrost. It is therefore possible that the impacts inferred from the comparison of present-day chemistry and long-term sediment records of these lakes in the Mackenzie Delta represent as yet a small component of the potential impact of drilling sumps on lake ecosystems. This is significant due to the fact that, to date, at some sites, the impacts on contemporary limnology have been as large as those associated with spectacular and conspicuous permafrost degradation. Drilling sumps represent only one example of the multiple stressors impacting Arctic lakes, which when combined with other natural and anthropogenic stressors may result in cumulative impacts to aquatic systems. If the contents of drilling sumps leach into the aquatic ecosystem due to loss of containment following permafrost thaw, the quantity and type of materials leaching into nearby lakes may increase and could result in severe, deleterious impacts on aquatic life. As the scale of hydrocarbon exploration and development in the Arctic increases in the coming decades, the potential for widespread contamination of freshwater lakes could result in cascading effects on both aquatic and terrestrial ecosystems, as well as the local indigenous communities which rely on them. In conjunction with work on permafrost conditions at drilling mud sumps [11] our data demonstrate that permafrost has not performed well as a waste containment medium, and thus if total containment of drilling wastes is a primary disposal objective, then alternative methods for waste disposal must be explored.

Supporting Information

Figure S1 Radioisotopic activity and CRS-model derived age for select intervals from the three analyzed sediment cores.

Figure S2 Boxplots of select environmental variables exhibiting significant differences between *a priori* defined groups. SU – drilling sump lakes; CO – control lakes; SL – thaw slump lakes. In each plot, letters indicate significantly different groups (calculated using a Tukey HSD post-hoc test, following ANOVA run on normalized environmental data).

Figure S3 Profiles of PAH concentrations and inferred primary production in sediment cores from three study lakes. Visual reflectance spectroscopically-inferred chlorophyll *a* (VRS chl *a*: circles) and summed 12 priority polycyclic aromatic hydrocarbon (PAH) concentrations (squares) for study lakes A) I20, B) C23 and C) C1A. The limit of detection for VRS-chl *a*, which includes its diagenetic products, is 0.01 mg/g dry weight, and thus the entire C1A profile is below detection.

Figure S4 Stratigraphic profile of the most common diatom taxa from three study lakes. Relative abundance diagrams from lakes A) I20, impacted by drilling sump failure; and control lakes B) C23 and C) C1A. Down-core sedimentary profiles (y axes) are scaled by date, based on ^{210}Pb radiometric dating techniques, with the depth in the sediment core included as a secondary axis. For lakes I20 and C23, two biostratigraphic zones were identified (constrained incremental sum of squares cluster analysis with the broken stick model) and are plotted with the background colour of one zone in grey the other white. For Lake C1A, no significant biostratigraphic zones were identified. The known timing of construction of the leaching drilling sump near Lake I20 (industry ID: Parsons F-09) is included as a horizontal line.

Table S1 Summary statistics of water chemistry data. Sites were sampled in the summers of 2005 and 2007, with samples taken at approximately one metre water depth.

Table S2 Results of similarity percentages (SIMPER) analysis.

Text S1 Supplementary methods, hydrocarbon, diatom and inferred primary production results and discussion.

Acknowledgments

The collection and compilation of water chemistry data from the Mackenzie Delta region was facilitated by the Cumulative Impact Monitoring Program, AANDC.

Author Contributions

Conceived and designed the experiments: JRT SVK JBK JMB MFJP JPS. Performed the experiments: JRT SVK JBK ESC CD LEK. Analyzed the data: JRT SVK JBK ESC CD LEK JMB. Wrote the paper: JRT SVK JBK ESC CD JMB MFJP JPS.

References

1. Dixon J, Morrell GR, Dietrich JR (1994) Part 1: Basin analysis. In: Dixon J, editor. Petroleum Resources of the Mackenzie Delta and Beaufort Sea. Bulletin 474Ottawa: Geological Survey of Canada. pp. 1–37.
2. ACIA (2005) Arctic Climate Impact Assessment. New York: Cambridge University Press.1042 p.
3. Burn CR, Kokelj SV (2009) The environment and permafrost of the Mackenzie Delta area. Permafrost Periglac Process 20:83–105.
4. Kelly EN, Short JW, Schindler DW, Hodson PV, Ma M, et al. (2009) Oil sands development contributes polycyclic aromatic compounds to the Athabasca River and its tributaries. Proc Natl Acad Sci USA 106:22346–22351.
5. Kurek J, Kirk JL, Muir DCG, Wang X, Evans MS, et al. (2013) Legacy of a half century of Athabasca oil sands development recorded by lake ecosystems. Proc Natl Acad Sci USA 110:1761–1766.
6. Kanigan J, Kokelj SV (2010) Review of current research on drilling-mud sumps in permafrost terrain, Mackenzie Delta region, NWT, Canada. In: GEO2010: 63rd Canadian Geotechnical Conference & 6th Canadian Permafrost Conference. Richmond: Canadian Geotechnical Society. pp. 1473–1479.
7. French H (1980) Terrain, land use and waste drilling fluid disposal problems, Arctic Canada. Arctic 33:794–806.
8. Dyke LD (2001) Contaminant migration through the permafrost active layer, Mackenzie Delta area, Northwest Territories, Canada. Polar Rec 37:215–228.
9. Jenkins R, Kanigan J, Kokelj SV (2008) Factors contributing to the long-term integrity of drilling-mud sump caps in permafrost terrain, Mackenzie Delta Region, Northwest Territories, Canada. In: Kane DL, Hinkel KM, editors. Proceedings of the Ninth International Conference on Permafrost. Fairbanks: University of Alaska Fairbanks Press. pp 833–843.
10. Johnstone JF, Kokelj SV (2008) Environmental conditions and vegetation recovery at abandoned-drilling mud sumps in the Mackenzie Delta region, NWT, Canada. Arctic 61:199–211.
11. Kokelj SV, Riseborough D, Coutts R, Kanigan JCN (2010) Permafrost and terrain conditions at northern drilling-mud sumps: Impacts of vegetation and climate change and the management implications. Cold Reg Sci Technol 64:46–56.
12. Falk MR, Lawrence MJ (1973) Acute toxicity of petrochemical drilling fluid components and wastes to fish. Ottawa: Fisheries and Marine Services, Environment Canada.108 p.
13. Utz L, Bohrer M (2001) Acute and chronic toxicity of potassium chloride (KCl) and potassium acetate ($KC_2H_3O_2$) to *Daphnia similis* and *Ceriodaphnia dubia* (Crustacea; Cladocera). Bull Environ Contam Toxicol 66:379–385.
14. Lantz TC, Kokelj SV (2008) Increasing rates of retrogressive thaw slump activity in the Mackenzie Delta region, N.W.T., Canada. Geophys Res Lett 35:L06502, doi:10.1029/2007GL032433.
15. Kokelj SV, Jenkins RE, Milburn D, Burn CR, Snow N (2005) The influence of thermokarst disturbance on the water quality of small upland lakes, Mackenzie Delta region, Northwest Territories, Canada. Permafrost Periglac Process 16:343–353.
16. Kokelj SV, Zajdlik B, Thompson MS (2009) The impacts of thawing permafrost on the chemistry of lakes across the subarctic boreal tundra transition, Mackenzie Delta region, Canada. Permafrost Periglac Process 20:185–200.
17. Mesquita PS, Wrona FJ, Prowse TD (2010) Effects of retrogressive permafrost thaw slumping on sediment chemistry and submerged macrophytes in Arctic tundra lakes. Freshw Biol 55:2347–2358.
18. Thienpont JR, Rühland KM, Pisaric MFJ, Kokelj SV, Kimpe LE, et al. (2013) Biological responses to permafrost thaw slumping in Canadian Arctic lakes. Freshw Biol 58:337–353.
19. Oksanen J, Blanchet FG, Kindt R, Legendre P, Minchin PR, et al. (2010) Vegan: Community Ecology Package. R package version 1. 17–4.
20. R Development Core Team (2010) R: A language and environment for statistical computing. Vienna: R foundation for Statistical Computing.
21. Clarke K (1993) Nonparametric multivariate analyses of changes in community structure. Aust J Ecol 18:117–143.
22. Appleby PG (2001) Chronostratigraphic techniques in recent sediments. In: Last WM, Smol JP, editors. Tracking Environmental Changes Using Lake Sediments. Volume 1: Basin Analysis, Coring, and Chronological Techniques. Dordrecht: Kluwer Academic Publishers. pp. 171–203.
23. Battarbee RW, Jones VJ, Flower RJ, Cameron NG, Bennion H, et al. (2001) Diatoms. In: Smol JP, Last WM, editors. Tracking Environmental Changes Using Lake Sediments. Volume 3: Terrestrial, Algal, and Siliceous Indicators. Dordrecht: Kluwer Academic Publishers. pp 155–202.
24. Korosi JB, Smol JP (2012) An illustrated guide to the identification of cladoceran subfossils from lake sediments in northeastern North America: part 1—the Daphniidae, Leptodoridae, Bosminidae, Polyphemidae, Holopedidae, Sididae, and Macrothricidae. J Paleolimnol 48:571–586.
25. Grimm EC (2011) Tilia v1.6 computer program. Springfield: Illinois State Museum, Research and Collections Center.
26. Grimm EC (1987) CONISS – a FORTRAN-77 program for stratigraphically constrained cluster-analysis by the method of incremental sum of squares. Comput Geosci 13:13–35.
27. Bennett KD (1996) Determination of the number of zones in a biostratigraphical sequence. New Phytol 132:155–170.
28. Michelutti N, Blais JM, Cumming BF, Paterson AM, Rühland KM, et al. (2010) Do spectrally-inferred determinations of chlorophyll *a* reflect trends in lake trophic status? J Paleolimnol 43:205–217.
29. Rampton VN (1988) Quaternary geology of the Tuktoyaktuk Coastlands, Northwest Territories. Geological Survey of Canada Memoir 423.Ottawa: Energy Mines and Resources Canada. 98 p.
30. Kokelj SV, Burn CR (2005) Geochemistry of the active layers and near-surface permafrost, Mackenzie delta region, Northwest Territories, Canada. Can J Earth Sci 42:37–48.
31. Smol JP (2008) Pollution of lakes and rivers: A paleoenvironmental perspective, 2nd ed. Oxford: Blackwell Publishing.383 p.
32. Bos DG, Cumming BF, Smol JP (1999) Cladocera and Anostraca from the Interior Plateau of British Columbia Canada, as paleolimnological indicators of salinity and lake level. Hydrobiologia 392:129–141.
33. Chengalath R (1982) A faunistic and ecological survey of littoral Cladocera of Canada. Can J Zool 60:2668–2682.
34. Sweetman JN, Rühland KM, Smol JP (2010) Environmental and spatial factors influencing the distribution of cladocerans in lakes across the central Canadian Arctic treeline region. J Limnol 69: 76–87.
35. Deasley K, Korosi JB, Thienpont JR, Kokelj SV, Pisaric MFJ, et al. (2012) Investigating the response of Cladocera to a major saltwater intrusion event in an Arctic lake from the outer Mackenzie Delta (NT, Canada). J Paleolimnol 48:287–296.
36. Smol JP, Douglas MSV (2007) From controversy to consensus: making the case for recent climatic change in the Arctic using lake sediments. Front Ecol Environ 5:466–474.
37. Smol JP, Douglas MSV (2007). Crossing the final ecological threshold in high Arctic ponds. Proc Natl Acad Sci USA 104:12395–12397.

5

Broad-Scale Patterns of Late Jurassic Dinosaur Paleoecology

Christopher R. Noto[1]*, Ari Grossman[2]

1 Department of Biomedical Sciences, Grand Valley State University, Allendale, Michigan, United States of America, 2 Department of Anatomy, Midwestern University, Glendale, Arizona, United States of America

Abstract

Background: There have been numerous studies on dinosaur biogeographic distribution patterns. However, these distribution data have not yet been applied to ecological questions. Ecological studies of dinosaurs have tended to focus on reconstructing individual taxa, usually through comparisons to modern analogs. Fewer studies have sought to determine if the ecological structure of fossil assemblages is preserved and, if so, how dinosaur communities varied. Climate is a major component driving differences between communities. If the ecological structure of a fossil locality is preserved, we expect that dinosaur assemblages from similar environments will share a similar ecological structure.

Methodology/Principal Findings: This study applies Ecological Structure Analysis (ESA) to a dataset of 100+ dinosaur taxa arranged into twelve composite fossil assemblages from around the world. Each assemblage was assigned a climate zone (biome) based on its location. Dinosaur taxa were placed into ecomorphological categories. The proportion of each category creates an ecological profile for the assemblage, which were compared using cluster and principal components analyses. Assemblages grouped according to biome, with most coming from arid or semi-arid/seasonal climates. Differences between assemblages are tied to the proportion of large high-browsing vs. small ground-foraging herbivores, which separates arid from semi-arid and moister environments, respectively. However, the effects of historical, taphonomic, and other environmental factors are still evident.

Conclusions/Significance: This study is the first to show that the general ecological structure of Late Jurassic dinosaur assemblages is preserved at large scales and can be assessed quantitatively. Despite a broad similarity of climatic conditions, a degree of ecological variation is observed between assemblages, from arid to moist. Taxonomic differences between Asia and the other regions demonstrate at least one case of ecosystem convergence. The proportion of different ecomorphs, which reflects the prevailing climatic and environmental conditions present during fossil deposition, may therefore be used to differentiate Late Jurassic dinosaur fossil assemblages. This method is broadly applicable to different taxa and times, allowing one to address questions of evolutionary, biogeographic, and climatic importance.

Editor: Andy Hector, University of Zurich, Switzerland

Funding: The authors have no support or funding to report.

Competing Interests: The authors have declared that no competing interests exist.

* E-mail: notoc@gvsu.edu

Introduction

Over the past twenty years new fossil discoveries, novel technologies, and a proliferation of analytical techniques have greatly increased our knowledge of dinosaur morphology, phylogeny, and behavior. These data, when combined with our growing knowledge of dinosaur biogeography, not only make it possible to address complex questions about changing dinosaur distributions, but also broad-scale ecological questions about the nature of dinosaur-dominated communities.

Advances in computer software enabled the creation of large, accessible databases recording data from an ever-increasing number of localities as new discoveries are made [1–3]. Coupled with geographic information system (GIS) technology, this allows us to examine patterns of dinosaur distribution at broad–regional to global–scales. There have been numerous studies of this type on dinosaur biogeography, focusing on vicariance, dispersal, and extinction patterns [4–10], which have helped identify possible areas of endemism, directions of dispersal, and even test the validity of different paleogeographic reconstructions. However, these data and methods have not yet been applied to ecological questions.

Historically, the ecology and behavior of dinosaurs has been reconstructed with reference to modern analogs, living animals that contain similar physical attributes or are closely related. Depending on the researcher, the modern analogs used to describe dinosaur ecology and behaviors have ranged from crocodilians and birds to mammals [11–21]. These models for dinosaur ecology and behavior are valuable starting points, providing a necessary conceptual framework from which to evaluate the unusual morphology of these extinct creatures. In many ways such thinking has been highly informative, as it is impossible to explain the biology of extinct taxa without first studying how living organisms operate. On the other hand it may lead to erroneous or unrealistic reconstructions based on constraints imposed by the organismal model used in the analogy, especially when those comparisons exist in the absence of quantitative data. Also, the focus on reconstructing the ecology of particular dinosaur taxa has hampered exploration of dinosaur-dominated ecosystems as a whole. While understanding the ecology of individual taxa is important, this information is insufficient outside of a broader ecosystem context.

An excellent example of a whole-ecosystem study is that of Foster [22], who carried out a detailed ecological analysis of the Late Jurassic Morrison Formation in the western United States. Such studies are few due to the immense effort involved in bringing together the multiple lines of data necessary to carry out such a comprehensive analysis. Therefore, detailed work of this nature is currently lacking for the large number of dinosaur-fossil bearing formations around the world. To gain a broad-scale view of dinosaur-dominated ecosystems a different approach is needed.

An ecological approach that reconstructs general habitat characteristics and ecological diversity among dinosaur faunas (as interpreted from fossil assemblages) in different regions of the globe is an important tool for understanding some of the many forces shaping dinosaur distribution patterns. In addition, deciphering changes in the ecological structure of dinosaur communities over time is a powerful tool for elucidating the role of global changes in climate, continental arrangement, and land area in shaping dinosaur biogeography, ecology, and evolution. Various dynamics, such as Milankovich cycles (which impact solar energy distribution across the Earth's surface), or plate tectonics (which largely determine the location, size, and geology of terrestrial landmasses), affect climate patterns and determine the abiotic conditions to which ecosystems are exposed [23]. Changes in these processes over time are likely to be reflected in community structure and can be recorded in the fossil record [24–26].

This study utilizes dinosaur taxa from several Late Jurassic (161–145 Mya) fossil localities from around the world. The Late Jurassic is notable for its extremely warm and equable climate, which was dominated by a monsoonal circulation pattern [27–30]. These extreme conditions played a large role in the distribution and diversity of Late Jurassic biota [31]. Furthermore, throughout the Jurassic many dinosaur clades diversified [32], leading to the evolution of many extreme morphologies characteristic of these groups. While often speculated upon, the ecological role of these adaptations remains poorly understood, particularly how these adaptations were integrated to form stable, operational communities. How were dinosaur-dominated communities structured and what role did climatic conditions play? It is well known that biological communities evolving under similar environmental conditions often contain convergent adaptations [33–36]. We therefore predict that dinosaur fossil assemblages falling under similar climatic conditions will exhibit convergent community structure irrespective of their individual phylogenic or biogeographic history.

Materials and Methods

Ecological Structure Analysis

The reconstruction of dinosaurs as living animals has greatly benefited from the increased emphasis on biomechanics and functional morphology as applied to fossil organisms. Use of biomechanical principals and a better understanding of the relationships between function and form substantially improve our ability to generate hypotheses about the behavior of extinct taxa [37–43]. These studies have led to many new insights about dinosaur paleoecology.

Traditionally, community analysis (as represented by fossil assemblages) uses indices of species richness [44] and/or taxonomic diversity indices (e.g., Simpson's Diversity Index). Species richness only analyzes the abundance of taxa at a site, and becomes extremely problematic when we lack tight control over the rate(s) of fossil accumulation at a site. In many cases, dinosaur fossil assemblages are a time-averaged collection representing a prolonged period of accumulation. Taxonomic diversity indices are more often used to determine the relative age of sites [45,46] but may sometimes be used to make inferences about habitat as well [47]. However, on its own taxonomy is not sufficient to determine ecology. Different taxa may converge onto similar ecological niches, or alternatively closely related taxa may be ecologically diverse, especially when looking across broad spans of time. Thus, neither of these methods is sufficient for reconstructing the ecosystem represented at a fossil site.

Here we employ Ecological Structure Analysis (ESA), which uses functional morphology, to produce ecological profiles for different Late Jurassic localities. This method allows us to compare dinosaur assemblages, with varying taxonomic profiles and diversity, from a large number of localities. In applying ESA, taxa are classified using ecological criteria. Therefore, it is important to note that this is not a "taxon-free" method but merely a different way of classifying taxa. Reed [48–50] demonstrated that modern mammal communities located in different habitats differ significantly in the percentages of taxa found in trophic, locomotor, and body size categories (using both Kruskal-Wallis and the Mann-Whitney U tests), while those in similar habitats are more alike. She also demonstrated that the same ecological categories would correctly classify sites with different environments (using discriminant function analysis). Thus, it is possible to compare fossil mammal sites with modern ones and classify fossil localities according to habitat differences.

For this study, ESA was applied in the following way. A database of fossil localities and taxon lists was compiled from the Paleobiology Database [51] and Weishampel et al. [52]. Each locality was plotted on a paleogeographic map (150 mya reconstruction, Mollweide projection) using the mapping function available on the Paleobiology Database. Geographically close localities or those forming natural clusters were grouped together to form composite assemblages (Figure 1). This was done to provide the necessary sample size for analysis. The twelve assemblages include (Table 1): six from North America (M1–M6), two from Europe (E1–E2), one from Africa (A1), one from South America (S1) and two from Asia (C1–C2). Complete data are available in Table S1.

For each assemblage genera were counted only once and higher taxa (i.e., generically indeterminate remains) were counted only if the group was not represented by generic remains, unless published reports provided a reasonable case that the remains represent distinct taxa. Isolated teeth were used only if they were assigned to a taxon for which a specific size range is known. Each taxon was classified using three separate ecological categories: Body Size, Locomotor Mode, and Trophic Mode, themselves divided into multiple classes (Table 2). Classification within each category was determined using morphological correlates as determined from the fossil itself, published reconstructions (if available), and closely related taxa (especially if the taxa are fragmentary). While other categories may also be useful, we prefer these particular three because osteological and dental evidence is potentially available in the fossil record for all taxa studied. For each assemblage the percentage of taxa in each class was calculated, thus generating an ecological profile. It is important to note that this analysis deals with paleoecological diversity and assumes a degree of correspondence between the diversity and relative abundance of ecomorphs in an assemblage. A detailed description of all ecological categories and classes is given below.

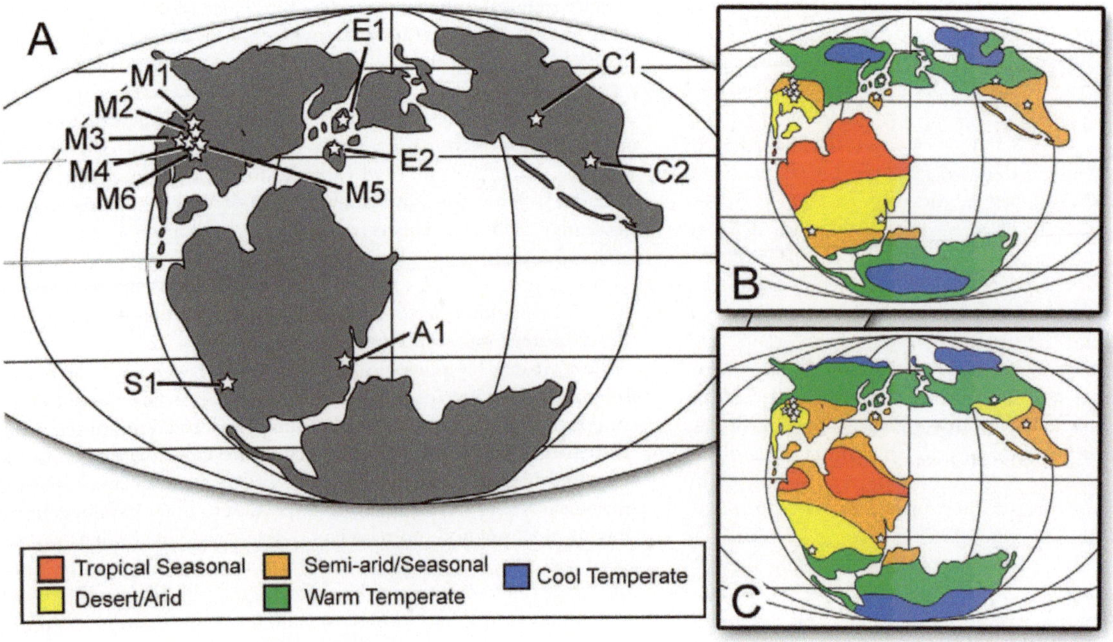

Figure 1. Late Jurassic paleogeographic map and reconstructed biomes. A) Positions of fossil assemblages (stars): A1 = Africa, C1–C2 = China, E1–E2 = Europe, M1–M6 = North America, and S1 = South America. Base map adapted from [27]. B) Biome reconstruction based on Rees et al. [27] model. C) Biome reconstruction based on Sellwood and Valdes [30] model.

Body Size. Body size is perhaps the single most important ecological character. It not only influences lifestyle and behavior in an individual species, but is also responsible for driving ecosystem dynamics at higher levels of community organization. This occurs through the scaling of metabolic rate with body size, which influences population dynamics and species diversity, ultimately determining the flow of energy between trophic levels [53–57].

For individuals and species, body size can determine such ecologically meaningful characters as techniques of predator avoidance, the type of substrate an animal may utilize during locomotion [58], and the size of its home and day range [59]. In addition, body size generally affects overall mass and body proportions due to the mechanical constraints inherent in biological materials [60]. The forces that bodies of different size generate affect performance and therefore strategies in feeding, locomotion, and reproduction [61,62]. In dinosaurs and other extinct organisms lacking modern analogs, many of these traits will remain difficult to determine directly from available fossil

Table 1. Location, biome assignments(s), and formations constituting each assemblage.

Assem.	Biome R[1]	Biome SV[2]	Region	Countries/States	Formation(s)
E1	2	1	Europe	England, France	Sables de Glos, Argiles d'Octeville, Marnes de Bléville, Kimmeridge Clay, Calcareous Grit, Corallian Oolite, Oxford Clay, Portland Stone
E2	2	2	Europe	Portugal, Spain	Villar del Arzobis, Alcobaça, Guimarota, Sobral, Amoreira-Porto Novo, Bombarral, Freixial, Louriñha
M1	2	3	North America	Wyoming, Montana, South Dakota	Morrison
M2	2	3	North America	Wyoming	Morrison
M3	3	3	North America	Colorado, Utah	Morrison
M4	3	3	North America	Colorado, Utah	Morrison
M5	3	3	North America	Colorado	Morrison
M6	3	3	North America	Colorado, New Mexico, Oklahoma	Morrison
C1	1	1	Asia	China (Xinjiang)	Shishugou, Kalazha
C2	2	2	Asia	China (Sichuan)	Shangshaximiao (Upper Shaximiao)
S1	2	3	South America	Chile, Argentina	Toqui, Cañadón Cálcereo
A1	2	3	Africa	Tanzania	Tendaguru

Biome Key: 1 = temperate, 2 = semi-arid/seasonally wet, 3 = desert/arid.
[1]Climate model of Rees et al. [27].
[2]Climate model of Sellwood and Valdes [30].

Table 2. Description of ecological categories and constituent classes used in this study.

Ecological Category[1]	Characteristics	
Trophic Mode		**Code**
Carnivore	Eats meat	C
Herbivore	High browsing: above 5 meters	HH
	Intermediate browsing: up to 5 meters	HI
	Low browsing up to 2 meters	HL
	Ground foraging: up to 1 meter	HG
Locomotor Mode		
Terrestrial Quadruped	Moves quadrupedally on the ground	TQ
Terrestrial Biped	Moves bipedally on the ground	TB
Facultative Biped	Capable of quadru- and bipedal motion	TF
Terrestrial/Arboreal Biped	Moves and feeds on the ground and in trees	TA
Body Size[2]		
Tiny	<1 kg	T
Very Small	1–10 kg	VS
Small	10–100 kg	S
Medium	100–1000 kg	M
Large	1000–10,000 kg	L
Very Large	>10,000 kg	VL

[1]Ecological categories after [49].
[2]Mass calculations from [63].

evidence. Use of body size estimates therefore can provide a first-order approximation of many important ecological characters when comparing taxa and assemblages.

Dinosaur body mass can be determined through estimate only. Each method has its own strengths and weaknesses, but discussion of these is outside the scope of this paper. More important is to provide a relative ranking of taxa even if absolute values are inaccurate. For continuity, we used body mass estimates based on Seebacher [63], which provides a wide list of taxa. For taxa not found on this list, we used the mass estimation equations given in the paper. We used only adult representatives where possible for mass estimation. Six size classes were used: Tiny (<1 kg), Very Small (1–10 kg), Small (10–100 kg), Medium (100–1000 kg), Large (1000–10 000 kg), and Very Large (>10 000 kg). Size classes are broad enough that expected body size variation among adults and estimate uncertainty is included for most taxa.

Trophic Mode. Trophic mode refers to the diet and food processing strategies of an animal, the two primary categories being carnivore and herbivore. Further divisions depend on our knowledge of the morphology and behavior of the group in question. Morphology of the teeth is most useful in determining what type of food the animal processed during its life. While it is true that the type of prey consumed and the manner in which it is captured varies considerably among carnivores, these specializations are often much harder to decipher from available fossil material. Nevertheless, among non-mammalian carnivores, teeth tend to be elongate, sharp, and pointed; sometimes laterally compressed with a blade-like edge. In addition, many non-mammalian carnivores have long, laterally compressed skulls and possess sharp, recurved claws that aid in prey capture, dispatch, and processing. In contrast, non-mammalian herbivores tend to possess shorter, blunter teeth with grinding, slicing, or shearing surfaces. Their skulls tend to be broad and short, and the limbs lack the kinds of claws typically seen in carnivores.

Due to the unknown nature of plant preference in herbivorous dinosaurs, we assign herbivores to categories of browse height rather than plant type. We assign only an upper browsing limit based on characters such as neck posture and limb length. Ground level forms the lower browsing limit. These classes are (from lowest to highest feeding height): Ground Forager (<1 m), Low Browser (<2 m), Intermediate Browser (<5 m), and High Browser (>5 m). The neck posture of sauropods remains contentious given their extreme and unusual morphology [38,42]. However, general differences in vertical feeding position between sauropod taxa are supported by morphological, biomechanical, and tooth wear analysis [38,42,64,65].

Locomotor Mode. We divide locomotor mode into two major categories, quadruped and biped. All locomotor modes used in this analysis are terrestrial in nature. Terrestrial quadrupeds are distinguished by having forelimbs that are closer in length to the hindlimbs and a robust radius locked with the ulna in a pronated position, increasing the stability of the front leg for weight-bearing during locomotion. In contrast, the forelimbs of bipeds are relatively shorter than the hind limbs. In addition, since the forelimbs are not necessary for weight-bearing the elbow joint is more mobile and the forelimb is not permanently locked in a pronated position. An additional division, facultative biped, was used for those taxa with intermediate axial and limb morphologies, such as shorter forelimbs capable of weight-bearing, suggesting that these taxa were capable of both bipedal and quadrupedal locomotion (e.g., some ornithopods) [66–68]. Some groups of dinosaurs evolved a quadrupedal stance secondarily, having descended from bipedal ancestors; however, these patterns in limb proportion and structure generally still hold.

The division Arboreal Biped was introduced because the locomotor behavior of some small maniraptorans and early birds (i.e, *Archaeopteryx*) is still debated [69–73]. For this analysis, we expect arboreal bipeds to have structures adapted for a combination of climbing, flight (gliding or active), and terrestrial locomotion, indicating they spent some amount of time both in trees and on the ground.

We acknowledge the great morphological variability in dinosaurs within the biped and quadruped categories, signaling important differences in locomotor strategy. More realistically, the locomotor strategies of taxa fall along a continuum, with discrete categories serving to delineate, rather than fully describe, how organisms move [74]. Again, further biomechanical work is necessary to determine how different limb and axial morphologies affected locomotor performance in dinosaurs. This work is currently underway and should lead to greater refinement of locomotor categories in the future [43,75–78].

Comparison with Late Jurassic Climate

Biomes were used over individual locality-level environmental reconstructions to observe how well independent climate data reflect ecological differences between assemblages. Biomes are characterized by different groups of plants, which help form the basic structure of every environment and are related to the individual climatic tolerance of each species. This structure plays an important role in directing the subsequent evolution of constituent species inhabiting these areas. Therefore, biomes provide an estimate of the general environmental and climatic conditions that prevailed over the regions where different fossil localities formed. Furthermore, like ecological categories, biomes can be extended and applied to other time periods because they in part rely on morphological categories for assigning plant taxa [27,79].

Each assemblage was assigned a biome using reconstructions from Rees et al. [27] or Sellwood and Valdes [30]. The former is based primarily on the distribution and morphology of fossil plant taxa and the distribution of climate sensitive sediments (e.g., evaporites and coals), similar to methods used in delineating modern biomes. The latter is based on a general circulation model (GCM), delineating each biome from model-predicted temperature and precipitation patterns (Table 1). While generally similar, each reconstruction differs in their interpretation of certain dinosaur habitats and should therefore be considered when comparing the assemblages.

Statistical Analyses

In order to compare ecological profiles, data were first arcsin transformed prior to analysis to approximate normality [80]. Similarity between assemblages was assessed with cluster analysis, utilizing the unweighted pair-group average (UPGMA) algorithm and Euclidean similarity measure. The stability of each node was assessed with a bootstrap test of 1000 replicates. In order to explore which ecological classes are most responsible for separating assemblages, a principal components analysis (PCA) was performed utilizing a covariance matrix. The number of significant components was determined using the eigenvalue bootstrapping method described in Peres-Neto et al. [81], in which we used the minimum recommended number of 1000 replicates. All analyses were run with PAST v 1.85 [82].

Results

Cluster Analysis

Each assemblage exhibited a distinct ecological profile (Table 3). Cluster analysis separated the assemblages into several groups (Figure 2). S1 was found distinct from all other assemblages, while C1 was the next most distinct. The remaining ten assemblages are split into two groups, with the North American assemblages (M1–M6) split evenly between them. A1 and C2 are associated with one of these groups, while the other contains both E1 and E2. However, support for most nodes does not surpass 50%, except for those linking M1, M5, and M6 (73%) and E2, M2, M3, and M4 (52%). The relationship of C2 to either of the large nodes appears especially uncertain given its very low support (5%). Due to its lack of relationship with any other assemblage, S1 was excluded from further analysis.

Principal Components Analysis

Two significant principal components were recovered, which together account for 64.6% of the variance in the data (Figure 3). PC 1 accounts for 47.8% and PC 2 accounts for 16.8% of the variance, respectively. Positive values of PC 1 are associated with Ground Foragers, Small and Very Small sizes, and Bipeds, while negative values are associated with High Browsers, Very Large taxa, and Quadrupeds (Table 4). Positive values of PC 2 indicate a larger proportion of Very Small taxa while negative values suggest a greater proportion of Small taxa and Carnivores. The assemblages do not group tightly together but instead appear to fall along the continuum of PC 1, with members of the two major groups noted in the cluster analysis falling to either side of the origin. In order of increasing PC 1 score: M6, M1, M5, C2, A1, E1, E2, M4, M2, M3, and C1. Along PC 2 the greatest outliers are C2, E1, and A1. With the exception of assemblages M2, M3, and M4, each assemblage generally falls along PC 1 according to its biome. Assemblages with greater aridity have negative values, semi-arid assemblages remain near the origin, and assemblages representing moister environments are more positive.

Discussion

Relationship with Climate Reconstructions

Based on the correlations between ecological categories and principal components, general characteristics are ascribed to the biota and climatic conditions of each assemblage based on its position in the component plot. PC 1 represents climatic differences between the assemblages: negative scores indicate aridity and positive scores more temperate conditions. PC 2 is discussed below.

Table 3. Percentages of taxa within ecological categories used in this study.

		Trophic Mode					Locomotor Mode				Body Size					
Assem.	Total Taxa	C	HH	HI	HL	HG	TQ	TB	TF	TA	T	VS	S	M	L	VL
E1	12	25.0	16.7	8.3	41.7	8.3	41.7	41.7	16.7	0.0	0.0	8.3	8.3	33.3	25.0	25.0
E2	24	37.5	12.5	25.0	20.8	4.2	50.0	41.7	4.2	4.2	4.2	12.5	8.3	16.7	29.2	29.2
M1	14	28.6	14.3	35.7	21.4	0.0	57.1	35.7	7.1	0.0	0.0	0.0	7.1	21.4	28.6	42.9
M2	22	36.4	9.1	27.3	18.2	9.1	45.5	50.0	4.5	0.0	0.0	9.1	22.7	18.2	13.6	36.4
M3	14	42.9	14.3	14.3	14.3	14.3	35.7	57.1	7.1	0.0	7.1	7.1	21.4	21.4	14.3	28.6
M4	20	40.0	10.0	20.0	25.0	5.0	45.0	50.0	5.0	0.0	5.0	5.0	15.0	25.0	20.0	30.0
M5	14	35.7	14.3	21.4	28.6	0.0	50.0	42.9	7.1	0.0	0.0	0.0	7.1	28.6	28.6	35.7
M6	10	30.0	20.0	30.0	20.0	0.0	60.0	30.0	10.0	0.0	0.0	0.0	0.0	20.0	30.0	50.0
C1	10	30.0	0.0	20.0	20.0	30.0	40.0	60.0	0.0	0.0	0.0	10.0	30.0	10.0	30.0	20.0
C2	11	27.3	9.1	27.3	27.3	9.1	63.6	36.4	0.0	0.0	0.0	9.1	0.0	27.3	27.3	36.4
S1	6	33.3	33.3	16.7	16.7	0.0	66.7	33.3	0.0	0.0	0.0	0.0	16.7	0.0	33.3	50.0
A1	11	45.5	9.1	27.3	18.2	0.0	45.5	54.5	0.0	0.0	0.0	0.0	18.2	36.4	18.2	27.3

Trophic Mode: C = carnivore, HH = high browser, HI = intermediate browser, HL = low browser, HG = ground foraging. Locomotor Mode: TQ = quadruped, TB = biped, TF = facultative biped, TA = arboreal biped. Body Size: VS = very small, S = small, M = medium, L = large, VL = very large. See text and Table 1 for explanation.

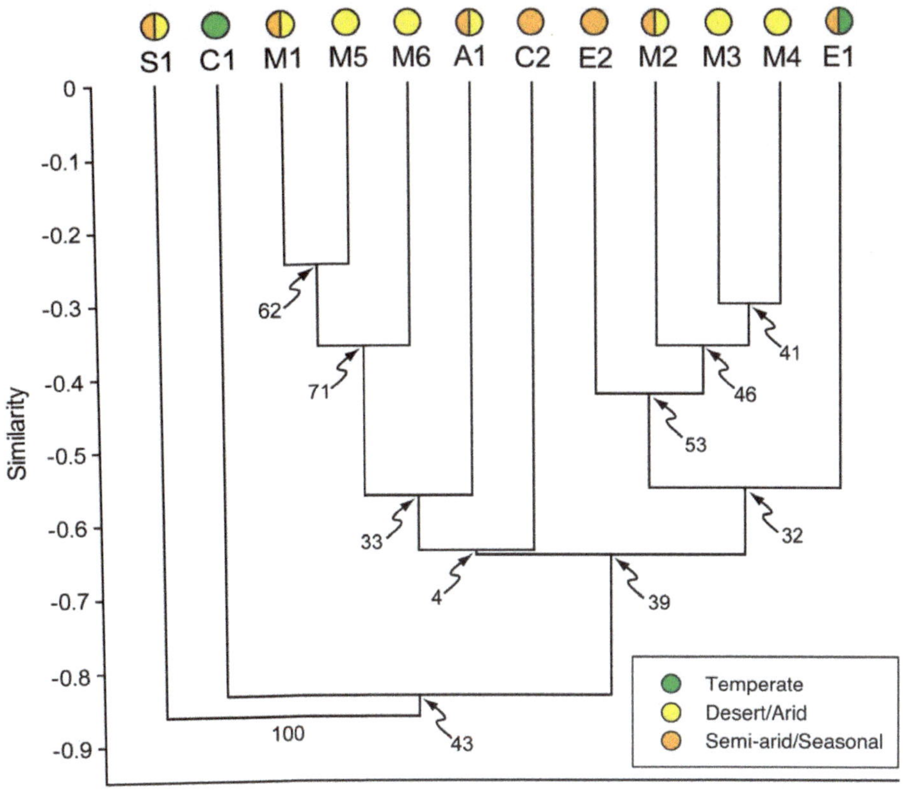

Figure 2. Results of cluster analysis (UPGMA with Euclidean similarity) for twelve Late Jurassic dinosaur fossil assemblages. Numbers indicate support for node (bootstrap: N = 1000). Labels refer to the fossil assemblage (see Table 1). Symbols denote the climate conditions assigned from the biome map. Two colors are shown if the biome assignment differs between the climate models.

Ultimately, assemblages did not group together by biome as expected. This discrepancy is related to 1) the different climate reconstructions used and 2) small-scale environmental variation between the localities comprising each assemblage. The biome assignments for each assemblage from Rees et al. [27] and Sellwood and Valdes [30] largely agree. Key differences occur in regions that are transitional zones between adjacent biomes. The actual boundaries between biomes are gradational and much less distinct and likely migrated in response to long-term shifts in climate patterns. This mainly affects the assessment of assemblages M1, M2, E1and A1, which are alternatively assigned semi-arid or desert/arid conditions (Table 1). In many of the assemblages, published environmental reconstructions differ from their assigned biome and may help interpret the results. These discrepancies point to major differences in the way climate models and ecological data reconstruct environmental conditions. Therefore, the model-assigned designations may not conform to the modern conception of these environments and instead relate to relative differences specific to the equable climate of the Late Jurassic [Foster, pers. comm.]. The conditions assessed for each assemblage are briefly discussed below.

Europe (E1 and E2). Throughout the Late Jurassic western Europe became progressively more arid with a strongly seasonal, Mediterranean-type climate [28,83–86]. E1 likely represents a drier environment, making it more similar to E2, and in accordance with the Rees et al. [27] biome model. This is supported by both the cluster analysis and PCA results, which place E1 and E2 close to the middle of the continuum. Both assemblages are therefore considered to have had a semi-arid climate.

North America (M1–M6). The Morrison Fm. is reconstructed as a seasonally dry, savannah-like environment; much moister than both models, which indicate greater aridity. In addition these assemblages do not consistently group together. The spatial extent and topography of the Morrison depositional basin sheds some light on this pattern. Recent work suggests two spatial gradients exist in the Morrison Fm.: a south to north, arid to temperate climate gradient and an east to west precipitation/drainage gradient [87,88]. These conditions created a greater proportion of lakes and wetlands towards the center of the depositional basin [89]. This work supports the observed division between Morrison Fm. assemblages, with M3 as a possible exception. M3 is expected to group with "drier" Morrison assemblages due to its southwestern position, although this relationship may not be resolvable with the current arrangement of assemblages and/or using dinosaurs only. Therefore, assemblages M2, M3 (tentatively), and M4 are considered to have been semi-arid or more seasonally wet, while M1, M5, and M6 were likely more arid or strongly seasonal.

Asia (C1 and C2). Climate reconstructions for many formations in China indicate semi-arid and seasonal conditions [90–92], however Hallam [28,84,86] reconstructs eastern Eurasia as being moister than the west. Each biome model places C1 as a temperate assemblage and C2 as seasonally semi-arid. Both analyses find C1 is quite different from the other assemblages, which could be indicative of a temperate climate, although this is unlikely. C1 occurs close to the boundary making its biome assignment tenuous at best. Sedimentary indicators of seasonality further refute the temperate nature of this assemblage [91,92]. Tectonic uplift throughout the Jurassic increased seasonality throughout the region, which was not accounted for in either biome model. Therefore C1 was likely semi-arid, but perhaps less seasonal or

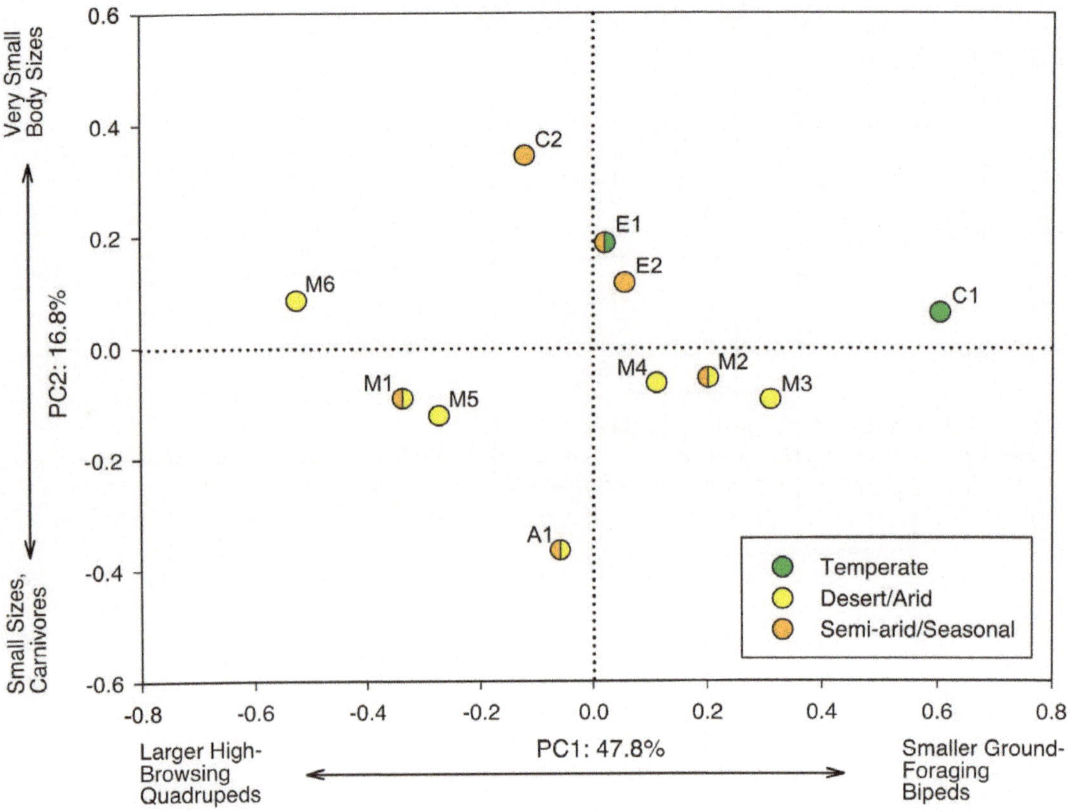

Figure 3. Results of Principal Components Analysis on twelve Late Jurassic dinosaur fossil assemblages. Symbols denote the climate conditions assigned from the biome map. Two colors are shown if the biome assignment differs between the climate models. PC 1 corresponds to differences in the proportion of herbivore body sizes and feeding strategies, while PC 2 follows other environmental, taphonomic, and historical differences between the assemblages.

experienced more intense moist periods than other areas. C2, on the other hand, lies between the two major groupings making it semi-arid and more seasonal than C1.

Table 4. Variable correlation values for the two major principal component axes.

Variable	Code	PC Axis 1	PC Axis 2
Carnivore	C	0.239	−0.7485
High Browser	HH	−0.7404	−0.01778
Intermediate Browser	HI	−0.4834	−0.183
Low Browser	HL	−0.2177	0.4712
Ground Forager	HG	0.8879	0.4218
Quadruped	TQ	−0.7707	0.3876
Biped	TB	0.864	−0.467
Facultative Biped	TF	−0.3563	0.07475
Arboreal Biped	TA	0.05912	0.2034
Tiny	T	0.3478	−0.07956
Very Small	VS	0.7331	0.5727
Small	S	0.8023	−0.545
Medium	M	−0.4491	−0.2363
Large	L	−0.3843	0.4739
Very Large	VL	−0.8327	0.06462

Africa (A1). The Tendaguru Fm. has been reconstructed with a semi-arid climate with coastal influences that maintained somewhat higher moisture levels than seen inland [93], supporting the biome model of Sellwood and Valdes [30]. The intermediate position of A1 in both analyses agrees with the assignment of a semi-arid climate to this assemblage.

South America (S1). The region surrounding S1 may have shared a similar coastal semi-arid climate [94,95, Rauhut, pers. comm.], but the number of dinosaur fossils from this area remains too sparse to allow a full comparison here. We will have to wait until further dinosaur fossils from the area are described.

Assemblage-Level Patterns

Results suggest that climatic and ecological factors played an important role in the distribution of Late Jurassic dinosaurs. In mammals, ecological preferences are often shared at the genus level or above [96], and the same was likely true of dinosaurs. The climatic and ecological similarity found here between North America, Europe, and southern Africa supports the biogeographic connections between these regions [5,97,98]. C2 shares relatively few taxa in common with the other semi-arid assemblages [5,98], yet its general ecological similarity points to an ecosystem that evolved convergently under the same climatic conditions.

Generally, the faunas of the Morrison Fm. have been treated as a single unit, however it appears they were more ecologically diverse than previously thought, despite the relatively uniform distribution of dinosaur taxa [99,100]. The ecological subdivisions present within the Morrison Fm. may also be related to its long history of intense study, which led to hundreds of described localities. This number is unmatched by the other assemblages. Future work may find further ecological subdivisions as more fossil localities are added in other regions.

C1, as the northernmost assemblage represented here, remains taxonomically and ecologically distinct. C1 may be like the central Morrison Fm. (M2–M4), maintaining greater moisture levels despite a seasonally semi-arid climate, or it may represent an entirely different semi-arid fauna that has yet to be encountered elsewhere.

PC2 appears related to assemblage-specific differences that contributed to the low bootstrap support in the cluster analysis. The wide distribution of scores among assemblages suggests PC2 cannot be directly attributed environmental variation. Instead, PC2 most likely represents some mixture of environmental, taphonomic, and sampling effects. For example, A1, E1, and C2 have high PC2 scores and also have a small number of taxa. Alternatively, this may be some relative measure of error in the spatial and/or temporal grouping of localities in an assemblage. Unfortunately the current database does not allow for more complex analysis, but the connection of PC2 to the ecology of these assemblages is certainly worthy of further inquiry.

Relationship with Dinosaur Paleoecology

The distribution of ecological categories varies accordingly with the proposed climatic differences between assemblages (Figure 4). Assemblages from more arid conditions (M1, M5, M6) tend to have a greater proportion of large-bodied taxa and lack anything smaller than 10 kg. High- and intermediate-browsing herbivorous sauropods and large carnivorous theropods such as *Allosaurus* dominate these assemblages. Smaller carnivores and ground-foraging herbivores are rare or absent. This pattern may reflect a lack of low-lying vegetation for a period of the season that restricted the numbers of smaller dinosaurs. Larger dinosaurs, on the other hand, were better able to cope with lower resource density and quality [101–104]. Assemblages representing semiarid or seasonally wet conditions usually contain representatives from each size class, though with no discernable decrease in the proportion of larger size classes. Ground Foragers are present as a larger portion of the herbivore population and High Browsers are less prevalent, indicating that more ground cover was likely available capable of supporting a greater diversity of herbivores. With increasingly moist conditions, assemblages exhibit greater proportions of Ground Foragers and Low- and Intermediate Browsers. In the extreme case of C1 High Browsers are conspicuously absent, reflecting perhaps a lack of suitable habitat or resources.

The environmental dependence of certain ecological categories demonstrates the importance of habitat structure in driving the relative abundance of dinosaur herbivore guilds. As observed in modern ecosystems, this is due to differences in habitat structure and resource availability. In the PCA results none of the herbivore classes are strongly positively correlated with each other. When the proportion of one herbivore class is high, the remaining classes are typically lower, reflecting the varying ecological roles each class plays in different environments or their differing habitat requirements. A similar trend is found among the herbivorous dinosaurs of the Morrison Formation [22], although this pattern occurs across different depositional environments instead of biomes.

Arid climates typically host sparse, open environments more suited for the largest animals. Small herbivores may suffer not only from a lack of food, but suitable cover from predators. Increasing moisture levels lead to more abundant growth, which inhibits the largest herbivores through changes to habitat structure and/or resource distribution, while becoming more favorable for smaller herbivores. The overall proportion of carnivores appears largely independent from environmental conditions although the absence of smaller predators in arid environments may be related to the lack of appropriately-sized prey and/or successful competitive exclusion by larger predators.

Limitations of the Current Study

Despite the encouraging results, it is important to note some caveats. First, not all formations and regions have been equally explored. Historically, excavations have been conducted in North America and Europe more extensively than elsewhere. Spectacular discoveries from China in recent decades show that major strides in the number of vertebrate localities from a region can be accomplished rapidly. The attention now afforded to Africa and South America will hopefully yield similar results in the future.

The role of taphonomic factors presents the greatest uncertainty because its ultimate effect may depend on the scale of observation. Environment-specific taphonomic filtering may drive taxonomic variation between individual localities; aggregating the localities into assemblages minimizes this effect. However, at larger spatial scales climatic processes, especially the onset of arid conditions, affect vertebrate preservation [25,31,91,105]. Under such condi-

Figure 4. Schematic representation of variation in herbivore types and inferred habitat structure along PC 1. Relatively arid conditions occur to the left (negative values), where very large, high-browsing herbivores dominate among sparser foliage. Assemblages with semi-arid/seasonal conditions are towards the center (low values), which includes a greater diversity of feeding modes, including high, intermediate, and low browsers among increased ground cover. To the right are more moist conditions (positive values), where smaller, ground-foraging herbivores are more prevalent within a more densely vegetated environment. Carnivorous theropods appear largely independent of this pattern. Green = high browser, orange = intermediate browser, blue = low browser, red = ground forager.

tions small size classes should be underrepresented in the fossil sample, potentially biasing the ecological profiles towards larger size classes with increasing aridity. The addition of vertebrate microfossil localities to an assemblage may help overcome this problem because these sites better reflect the community structure of the surrounding landscape, including both large and small taxa [106]. Dinosaur microfossil sites are found in the Morrison (M1–M4) and Camadas de Guimarota (E2) Formations, but are absent elsewhere [22,97,100]. Dinosaur microfossil remains usually consist of small theropod teeth, only some of which are assignable to useable taxa (see Materials and Methods and Table S1). The relationship between some assemblages change with removal of tooth taxa, but the climatic associations noted above remain more or less the same, indicating such localities are not necessary to assign climatic conditions but are useful is resolving ecological relationships among assemblages. Nevertheless, the role of microfossil localities and taphonomic filtering requires further scrutiny.

In addition to addressing taphonomic biases, it was necessary to group localities in order to achieve a minimum sample size for analysis. The relationships of assemblages with fewer than 10 taxa were found to be unresolvable, as in the case of S1. A small number of taxa skews the content of the categories since class data are calculated as a proportion of the total number of taxa and 10 is suggested here as a minimum sample for this type of study. While necessary, grouping spatially and stratigraphically distinct localities in this way increases the likelihood of including taxa and environments that never coexisted in life. Each assemblage therefore represents a coarse average of ecological conditions. Multiple studies have found overall taxonomic stability of the Morrison fauna through time [22,99,100]. The Tendaguru fauna of Africa was also similarly stable through the Late Jurassic [93]. The majority of formations included here lack such detailed biostratigraphic study. Paleoecological studies across a broad spectrum of scales, environments, and taxonomic groups have found that many past ecosystems maintained a stable structure over timescales of 100 ky to 2 my or more, despite major taxonomic turnovers or climatic events [107–113]. In this case, resolution of small-scale ecological differences are lost that may lead to interpretations of individual fossil localities disagreeing with the general results presented here.

A continuing challenge that requires more attention regards the assignment of ecological classes to dinosaurian taxa. In many cases the ecological niche of a dinosaur is still assessed using qualitative comparisons with living forms. Even with complex biomechanical models, very different interpretations of ecologically relevant morphological and behavioral reconstructions continue to arouse debate [114,115]. This problem becomes more acute in taxa based on incomplete and fragmentary remains. Altering one or more classes for a single taxon has little effect, though more than these can alter the results. As always, new fossil finds may lead us to radically revise our view on the ecology of certain taxa. If multiple interpretations do exist separate analyses should be run using all combinations of interpretations to assess their effect on the results.

Conclusions

Our study demonstrates that ESA is a useful tool for quantifying ecological differences between Late Jurassic dinosaur assemblages. The grouping of climatically similar assemblages supports the preservation of ecological structure at large scales and helped assess the accuracy of two different paleoclimate models. Ecological similarities are most likely related to differences in habitat structure due to variation in moisture availability; most important among these are the relative proportions of herbivore and body size classes in an assemblage. Not only can these proportions be used as additional climate indicators, but also may provide evidence of ecosystem convergence when comparing taxonomically distinct dinosaur localities. Although most Late Jurassic fossil localities come from relatively arid or semi-arid environments [31], there is a surprising amount of ecological variation that warrants further study.

A more comprehensive understanding of broad-scale ecological patterns is important in understanding the effect of climate patterns on dinosaur ecology and evolution. Perhaps more importantly, this study provides a framework for studying the long-term evolutionary dynamics of terrestrial communities related to climate change, major adaptive radiations, or evolutionary events (e.g., angiosperm evolution) and whether these events had a significant impact on the subsequent structure of vertebrate communities. These data can then be used to test hypotheses related to community formation and ecosystem function. A great deal of effort has been spent quantifying global taxonomic diversity levels through the Phanerozoic [116,117], however these estimates tell us little about the ecological factors responsible for producing that diversity, nor how it was distributed on the surface [118].

Supporting Information

Table S1 List of Late Jurassic dinosaur taxa and their assigned ecological categories used in this analysis, organized by assemblage. Diet: C = carnivore, HH = high browser, HI = intermediate browser, HL = low browser, HG = ground foraging. Locomotion: TQ = quadruped, TB = biped, TF = facultative biped, TA = arboreal biped. Body Mass: VS = very small, S = small, M = medium, L = large, VL = very large. See text and Table 1 for explanation. "PBDB Collection #" refers to the collection number of the specimen in the Paleobiology Database. Formations given in the table do not necessarily reflect all units in which the taxon is found. Data come from [51] and [52].

Acknowledgments

Paleocoordinates for each locality were made available through the Paleobiology Database (http://paleodb.org) utilizing the paleogeographic reconstructions of Chris Scotese (University of Texas, Arlington). Oliver Rauhut (Bayerische Staatssammlung für Paläontologie und Geologie) provided important information on the Cañadón Calcáreo Formation. Thanks go to John Foster (Museum of Western Colorado) and John Fleagle, (Stony Brook University), whose insightful comments helped significantly improve earlier versions of this manuscript. Catherine Forster (George Washington University), Mike Bell (Stony Brook University), Louis Jacobs (Southern Methodist University), Chris Scotese, and John Foster provided important discussion and encouragement. The final version of this manuscript was significantly improved by the feedback of three anonymous reviewers.

Author Contributions

Conceived and designed the experiments: CRN. Performed the experiments: CRN. Analyzed the data: CRN AG. Wrote the paper: CRN AG.

References

1. Schiermeier Q (2003) Palaeobiology: Setting the record straight. Nature 424: 482–483.
2. Rees PM, Noto CR (2005) A new online database of dinosaur distributions. Journal of Vertebrate Paleontology 25: 103A.

3. Alroy J (2003) Global databases will yield reliable measures of global biodiversity. Paleobiology 29: 26–29.
4. Russell DA (1993) The role of Central Asia in dinosaurian biogeography. Canadian Journal of Earth Sciences 30: 2002–2012.
5. Holtz TR, Chapman RE, Lamanna MC (2004) Mesozoic biogeography of Dinosauria. In: Weishampel DB, Dodson P, Osmolska H, eds. The Dinosauria, second edition. Berkeley: University of California Press. pp 627–642.
6. Forster CA (1999) Gondwanan dinosaur evolution and biogeographic analysis. Journal of African Earth Sciences 28: 169–185.
7. Upchurch P, Hunn CA, Norman DB (2002) An analysis of dinosaurian biogeography: evidence for the existence of vicariance and dispersal patterns caused by geological events. Proceedings of the Royal Society of London Series B-Biological Sciences 269: 613–621.
8. Sereno PC (1999) Dinosaurian biogeography: Vicariance, dispersal, and regional extinction. In: Tomida Y, Rich TH, Vickers-Rich P, eds. Proceedings of the Second Gondwanan Dinosaur Symposium. Tokyo: National Science Museum Monographs 15. pp 249–257.
9. McCarthy D (2005) Biogeographical and geological evidence for a smaller, completely-enclosed Pacific Basin in the Late Cretaceous. Journal of Biogeography 32: 2161–2177.
10. Barrett PM, Hasegawa Y, Manabe M, Isaji S, Matsuoka H (2002) Sauropod dinosaurs from the Lower Cretaceous of Eastern Asia: Taxonomic and biogeographical implications. Palaeontology 45: 1197–1217.
11. Van Valkenburgh B, Molnar RE (2002) Dinosaurian and mammalian predators compared. Paleobiology 28: 527–543.
12. Carrano MT, Janis CM, Sepkoski JJ (1999) Hadrosaurs as ungulate parallels: Lost lifestyles and deficient data. Acta Palaeontologica Polonica 44: 237–261.
13. Currie PJ (1989) Long distance dinosaurs. Natural History 6: 60–65.
14. Bakker RT (1972) Anatomical and ecological evidence of endothermy in dinosaurs. Nature 238: 81–&.
15. Hopson JA (1975) The evolution of cranial display structures in hadrosaurian dinosaurs. Paleobiology 1: 21–43.
16. Seebacher F, Grigg GC, Beard LA (1999) Crocodiles as dinosaurs: Behavioural thermoregulation in very large ectotherms leads to high and stable body temperatures. Journal of Experimental Biology 202: 77–86.
17. Bailey JB (1997) Neural spine elongation in dinosaurs: Sailbacks or buffalo-backs? Journal of Paleontology 71: 1124–1146.
18. Geist NR, Jones TD (1996) Juvenile skeletal structure and the reproductive habits of dinosaurs. Science 272: 712–714.
19. Abler WL (1992) The serrated teeth of tyrannosaurid dinosaurs, and biting structures in other animals. Paleobiology 18: 161–183.
20. Farlow JO, Dodson P (1975) Behavioral significance of frill and horn morphology in ceratopsian dinosaurs. Evolution 29: 353–361.
21. Farlow JO, Dodson P, Chinsamy A (1995) Dinosaur biology. Annual Review of Ecology and Systematics 26: 445–471.
22. Foster JR (2003) Paleoecological analysis of the vertebrate fauna of the Morrison Formation (Upper Jurassic), Rocky Mountain Region, U.S.A. Bulletin of the New Mexico Museum of Natural History and Science 23: 1–95.
23. Barnosky AD, Hadly EA, Maurer BA, Christie MI (2001) Temperate terrestrial vertebrate faunas in North and South America: Interplay of ecology, evolution, and geography with biodiversity. Conservation Biology 15: 658–674.
24. Graham RW, Lundelius EL, Graham MA, Schroeder EK, Toomey RS, et al. (1996) Spatial response of mammals to late Quaternary environmental fluctuations. Science 272: 1601–1606.
25. Smith RMH, Swart R (2002) Changing fluvial environments and vertebrate taphonomy in response to climatic drying in a mid-Triassic rift valley fill: The Omingonde Formation (Karoo Supergroup) of Central Namibia. Palaios 17: 249–267.
26. Wendler J, Graefe K-U, Willems H (2002) Reconstruction of mid-Cenomanian orbitally forced palaeoenvironmental changes based on calcareous dinoflagellate cysts. Palaeogeography, Palaeoclimatology, Palaeoecology 179: 19–41.
27. Rees PM, Ziegler AM, Valdes PJ (2000) Jurassic phytogeography and climates: new data and model comparisons. In: Huber BT, MacLeod KG, Wing SL, eds. Warm Climates in Earth History. Cambridge, England: University of Cambridge Press. pp 297–318.
28. Hallam A (1993) Jurassic climates as inferred from the sedimentary and fossil record. Philosophical Transactions of the Royal Society of London Series B: Biological Sciences 341: 287–296.
29. Hallam A (1998) The determination of Jurassic environments using palaeoecological methods. Bulletin De La Societe Geologique De France 169: 681–687.
30. Sellwood BW, Valdes PJ (2008) Jurassic climates. Proceedings of the Geologists' Association 119: 5–17.
31. Rees PM, Noto CR, Parrish JM, Parrish JT (2004) Late Jurassic climates, vegetation, and dinosaur distributions. Journal of Geology 112: 643–654.
32. Fastovsky DE, Huang Y, Hsu J, Martin-McNaughton J, Sheehan PM, et al. (2004) Shape of Mesozoic dinosaur richness. Geology 32: 877–880.
33. Ricklefs RE (2006) Evolutionary diversification and the origin of the diversity-environment relationship. Ecology 87: 3–13.
34. Harmon LJ, Kolbe JJ, Cheverud JM, Losos JB (2005) Convergence and the multidimensional niche. Evolution 59: 409–421.
35. Melville J, Harmon LJ, Losos JB (2006) Intercontinental community convergence of ecology and morphology in desert lizards. Proceedings of the Royal Society B: Biological Sciences 273: 557–563.
36. Lamouroux N, Poff NL, Angermeier PL (2002) Intercontinental convergence of stream fish community traits along geomorphic and hydraulic gradients. Ecology 83: 1792–1807.
37. Plotnick RE, Baumiller TK (2000) Invention by evolution: Functional analysis in paleobiology. Paleobiology 26: 305–323.
38. Taylor M, Wedel M, Naish D (2009) Head and neck posture in sauropod dinosaurs inferred from extant animals. Acta Palaeontologica Polonica 54: 213–220.
39. Rayfield EJ (2005) Aspects of comparative cranial mechanics in the theropod dinosaurs *Coelophysis*, *Allosaurus* and *Tyrannosaurus*. Zoological Journal of the Linnean Society 144: 309–316.
40. Manning PL, Payne D, Pennicott J, Barrett PM, Ennos RA (2006) Dinosaur killer claws or climbing crampons? Biology Letters 2: 110–112.
41. Hutchinson JR, Gatesy SM (2006) Dinosaur locomotion: Beyond the bones. Nature 440: 292–294.
42. Stevens KA, Parrish JM (2005) Neck posture, dentition, and feeding strategies in Jurassic sauropod dinosaurs. In: Tidwell V, Carpenter K, eds. Thunder-Lizards: The Sauropodomorph Dinosaurs. Bloomington: Indiana University Press. pp 212–232.
43. Bonnan MF (2004) Morphometric analysis of humerus and femur shape in Morrison sauropods: Implications for functional morphology and paleobiology. Paleobiology 30: 444–470.
44. Magurran AE (1988) Ecological diversity and its measurement. Princeton: Princeton University Press. 179 p.
45. Miller ER (1999) Faunal correlation of Wadi Moghara, Egypt: Implications for the age of *Prohylobates tandyi*. Journal of Human Evolution 36: 519–533.
46. Nakaya H (1994) Faunal change of Late Miocene Africa and Eurasia: Mammalian fauna from the Namurungule Formation, Samburu Hills, Northern Kenya. African Study Monographs Supplementary Issue. pp 1–112.
47. Andrews P (1996) Palaeoecology and hominoid paleoenvironments. Biological Reviews of the Cambridge Philosophical Society 71: 257–300.
48. Reed KE (1996) The paleoecology of Makapansgat and other African Plio-Pleistocene hominid localities [PhD]. Stony Brook: Stony Brook University. 624 p.
49. Reed KE (1998) Using large mammal communities to examine ecological and taxonomic structure and predict vegetation in extant and extinct assemblages. Paleobiology 24: 384–408.
50. Reed KE (2002) The use of paleocommunity and taphonomic studies in reconstructing primate behavior. In: Plavcan JM, ed. Reconstructing Behavior in the Primate Fossil Record. New York: Kluwer Academic/Plenum Publishers. pp 217–259.
51. The Paleobiology Database (2010) http://paleodb.org. Accessed March 15, 2010.
52. Weishampel DB, Le Loeuff J, Coria RA, et al. (2004) Dinosaur distribution. In: Weishampel DB, Dodson P, Osmolska H, eds. The Dinosauria, second edition. Berkeley: University of California Press. pp 517–606.
53. White CR, Phillips NF, Seymour RS (2006) The scaling and temperature dependence of vertebrate metabolism. Biology Letters 2: 125–127.
54. Loeuille N, Loreau M (2006) Evolution of body size in food webs: Does the energetic equivalence rule hold? Ecology Letters 9: 171–178.
55. Marquet PA, Quinones RA, Abades S, Labra F, Tognelli M, et al. (2005) Scaling and power-laws in ecological systems. Journal of Experimental Biology 208: 1749–1769.
56. Brown JH, Gillooly JF, Allen AP, Savage VM, West GB (2004) Toward a metabolic theory of ecology. Ecology 85: 1771–1789.
57. Greenwood JJD, Gregory RD, Harris S, Morris PA, Yalden DW (1996) Relations between abundance, body size and species number in British birds and mammals. Philosophical Transactions of the Royal Society of London Series B-Biological Sciences 351: 265–278.
58. Fleagle JG, Mittermeier RA (1980) Locomotor behavior, body size, and comparative ecology of seven Surinam monkeys. American Journal of Physical Anthropology 52: 301–314.
59. Milton K, May ML (1976) Body weight, diet, and home range area in primates. Nature 259: 459–462.
60. Christiansen P (2002) Mass allometry of the appendicular skeleton in terrestrial mammals. Journal of Morphology 251: 195–209.
61. Iossa G, Soulsbury CD, Baker PJ, Harris S (2008) Body mass, territory size, and life-history tactics in a socially monogamous canid, the red fox *Vulpes vulpes*. Journal of Mammalogy 89: 1481–1490.
62. Fa JE, Purvis AP (1997) Body size, diet and population density in Afrotropical forest mammals: A comparison with Neotropical species. The Journal of Animal Ecology 66: 98–112.
63. Seebacher F (2001) A new method to calculate allometric length-mass relationships of dinosaurs. Journal of Vertebrate Paleontology 21: 51–60.
64. Upchurch P, Barrett PM (2000) The evolution of sauropod feeding mechanisms. In: Sues H-D, ed. Evolution of herbivory in terrestrial vertebrates: Perspectives from the fossil record. Cambridge: Cambridge University Press. pp 79–122.
65. Fiorillo AR (1998) Dental microwear patterns from the sauropod dinosaurs *Camarasaurus* and *Diplodocus*: Evidence for resource partitioning in the Late Jurassic of North America. Historical Biology 13: 1–16.
66. Wright JL (1999) Ichnological evidence for the use of the forelimb in iguanodontid locomotion. Cretaceous Fossil Vertebrates. pp 209–219.

67. Christian A, Preuschoft H (1996) Deducing the body posture of extinct large vertebrates from the shape of the vertebral column. Palaeontology 39: 801–812.
68. Heinrich RE, Ruff CB, Weishampel DB (1993) Femoral ontogeny and locomotor biomechanics of *Dryosaurus lettowvorbecki* (Dinosauria, Iguanodontia). Zoological Journal of the Linnean Society 108: 179–196.
69. Sanz JL, Chiappe LM, Perez-Moreno BP, Buscalioni AD, Moratalla JJ, et al. (1996) An Early Cretaceous bird from Spain and its implications for the evolution of avian flight. Nature 382: 442–445.
70. Alonso PD, Milner AC, Ketcham RA, Cookson MJ, Rowe TB (2004) The avian nature of the brain and inner ear of Archaeopteryx. Nature 430: 666–669.
71. Long CA, Zhang GP, George TF, Long CF (2003) Physical theory, origin of flight, and a synthesis proposed for birds. Journal of Theoretical Biology 224: 9–26.
72. Thulborn RA (2003) Wind-assisted flight of *Archaeopteryx*. Neues Jahrbuch für Geologie und Palaontologie-Abhandlungen 229: 61–74.
73. Speakman JR (1993) Flight Capabilities in *Archaeopteryx*. Evolution 47: 336–340.
74. Carrano MT (1999) What, if anything, is a cursor? Categories versus continua for determining locomotor habit in mammals and dinosaurs. Journal of Zoology 247: 29–42.
75. Bonnan MF (2005) Pes anatomy in sauropod dinosaurs: Implications for functional morphology, evolution, and phylogeny. In: Tidwell V, Carpenter K, eds. Thunder-Lizards: The Sauropodomorph Dinosaurs. Bloomington: Indiana University Press. pp 346–380.
76. Alexander RM (2006) Dinosaur biomechanics. Proceedings of the Royal Society B-Biological Sciences 273: 1849–1855.
77. Hutchinson JR (2006) The evolution of locomotion in archosaurs. Comptes Rendus Palevol 5: 519–530.
78. Sellers WI, Manning PL (2007) Estimating dinosaur maximum running speeds using evolutionary robotics. Proceedings of the Royal Society B: Biological Sciences 274: 2711–2716.
79. Willis KJ, McElwain JC (2002) The Evolution of Plants. Oxford: Oxford University Press. 392 p.
80. Sokal RR, Rohlf FJ (1995) Biometry. New York: W.H. Freeman and Company. 887 p.
81. Peres-Neto PR, Jackson DA, Somers KM (2003) Giving meaningful interpretation to ordination axes: Assessing loading significance in principal component analysis. Ecology 84: 2347–2363.
82. Hammer Ø, Harper D, Ryan P (2001) PAST: Paleontological statistics software package for education and data analysis. Palaeontologia Electronica 4: 9.
83. Abbink O, Targarona J, Brinkhuis H, Visscher H (2001) Late Jurassic to earliest Cretaceous palaeoclimatic evolution of the southern North Sea. Global and Planetary Change 30: 231–256.
84. Hallam A (1984) Continental humid and arid zones during the Jurassic and Cretaceous. Palaeogeography, Palaeoclimatology, Palaeoecology 47: 195–223.
85. Ziegler AM, Parrish JM, Jiping Y, Gyllenhaal ED, Rowley DB, et al. (1993) Early Mesozoic phytogeography and climate. Philosophical Transactions: Biological Sciences 341: 297–305.
86. Hallam A (1985) A review of Mesozoic climates. Journal of the Geological Society 142: 433–445.
87. Demko TM, Currie BS, Nicoll KA (2004) Regional paleoclimatic and stratigraphic implications of paleosols and fluvial/overbank architecture in the Morrison Formation (Upper Jurassic), Western Interior, USA. Sedimentary Geology 167: 115–135.
88. Turner CE, Peterson F (2004) Reconstruction of the Upper Jurassic Morrison Formation extinct ecosystem–a synthesis. Sedimentary Geology 167: 309–355.
89. Dunagan SP, Turner CE (2004) Regional paleohydrologic and paleoclimatic settings of wetland/lacustrine depositional systems in the Morrison Formation (Upper Jurassic), Western Interior, USA. Sedimentary Geology 167: 269–296.
90. Lucas SG (2001) Chinese fossil vertebrates. New York: Columbia University Press. 375 p.
91. Eberth DA, Brinkman DB, Chen P-J, Yuan F-T, Wu S-Z, et al. (2001) Sequence stratigraphy, paleoclimate patterns, and vertebrate fossil preservation in Jurassic–Cretaceous strata of the Junggar Basin, Xinjiang Autonomous Region, People's Republic of China. Canadian Journal of Earth Sciences 38: 1627–1644.
92. Tütken T, Pfretzschner HU, Vennemann TW, Sun G, Wang YD (2004) Paleobiology and skeletochronology of Jurassic dinosaurs: Implications from the histology and oxygen isotope compositions of bones. Palaeogeography, Palaeoclimatology, Palaeoecology 206: 217–238.
93. Aberhan M, Bussert R, Heinrich W-D, Schrank E, Schultka S, et al. (2002) Palaeoecology and depositional environments of the Tendaguru Beds (Late Jurassic to Early Cretaceous, Tanzania). Mitteilungen aus dem Museum für Naturkunde in Berlin Geowissenschaftliche Reihe 5: 19–44.
94. Salgado L, De La Cruz R, Suárez M, Fernández M, Gasparini Z, et al. (2008) First Late Jurassic dinosaur bones from Chile. Journal of Vertebrate Paleontology 28: 529–534.
95. Soto M, Perea D (2008) A Ceratosaurid (Dinosauria, Theropoda) from the Late Jurassic–Early Cretaceous of Uruguay. Journal of Vertebrate Paleontology 28: 439–444.
96. Hadly EA, Spaeth PA, Li C (2009) Niche conservatism above the species level. Proceedings of the National Academy of Sciences 106: 19707–19714.
97. Antunes MT, Mateus O (2003) Dinosaurs of Portugal. Comptes Rendus Palevol 2: 77–95.
98. Mateus O (2006) Late Jurassic dinosaurs from the Morrison Formation (USA), the Lourinha and Alcobaça Formations (Portugal), and the Tendaguru Beds (Tanzania): A Comparison. New Mexico Museum of Natural History and Science Bulletin 36.
99. Turner CE, Peterson F (1999) Biostratigraphy of dinosaurs in the Upper Jurassic Morrison Formation of the Western Interior, U.S.A. In: Gillette DD, ed. Vertebrate Paleontology in Utah. Salt Lake City: Miscellaneous Publication - Utah Geological Survey. pp 77–114.
100. Dodson P, Behrensmeyer AK, Bakker RT, McIntosh JS (1980) Taphonomy and paleoecology of the dinosaur beds of the Morrison Formation. Paleobiology 6: 208–232.
101. Engelmann GF, Chure DJ, Fiorillo AR (2004) The implications of a dry climate for the paleoecology of the fauna of the Upper Jurassic Morrison Formation. Sedimentary Geology 167: 297–308.
102. Farlow JO (1987) Speculations about the diet and digestive physiology of herbivorous dinosaurs. Paleobiology 13: 60–72.
103. Midgley JJ, Midgley G, Bond WJ (2002) Why were dinosaurs so large? A food quality hypothesis. Evolutionary Ecology Research 4: 1093–1095.
104. Hummel J, Gee CT, Südekum K-H, Sander PM, Nogge G, et al. (2008) In vitro digestibility of fern and gymnosperm foliage: Implications for sauropod feeding ecology and diet selection. Proceedings of the Royal Society B: Biological Sciences 275: 1015–1021.
105. Garcia AJV, da Rosa AAS, Goldberg K (2005) Paleoenvironmental and paleoclimatic control on early diagenetic processes and fossil record in Cretaceous continental sandstones of Brazil. Journal of South American Earth Sciences 19: 243–258.
106. Rogers RR, Brady ME (2010) Origins of microfossil bonebeds: Insights from the Upper Cretaceous Judith River Formation of north-central Montana. Paleobiology 36: 80–112.
107. Kidwell SM, Flessa KW (1995) The quality of the fossil record: Populations, species, and communities. Annual Review of Ecology and Systematics 26: 269–299.
108. Jablonski D, Sepkoski JJ (1996) Paleobiology, community ecology, and scales of ecological pattern. Ecology 77: 1367–1378.
109. Morris PJ, Ivany LC, Schopf KM, Brett CE (1995) The challenge of paleoecological stasis: Reassessing sources of evolutionary stability. Proceedings of the National Academy of Sciences 92: 11269–11273.
110. Van Valkenburgh B (1995) Tracking ecology over geological time—Evolution within guilds of vertebrates. Trends in Ecology & Evolution 10: 71–76.
111. Jackson JBC, Erwin DH (2006) What can we learn about ecology and evolution from the fossil record? Trends in Ecology & Evolution 21: 322–328.
112. McGill BJ, Hadly EA, Maurer BA (2005) Community inertia of Quaternary small mammal assemblages in North America. Proceedings of the National Academy of Sciences 102: 16701–16706.
113. DiMichele WA, Behrensmeyer AK, Olszewski TD, Labandeira CC, Pandolfi JM, et al. (2004) Long-term stasis in ecological assemblages: Evidence from the fossil record. Annual Review of Ecology, Evolution and Systematics 35: 285–322.
114. Antón M, Sánchez IM, Salesa MJ, et al. (2003) The muscle-powered bite of *Allosaurus* (Dinosauria; Theropoda): An interpretation of cranio-dental morphology. Estudios Geológicos 59.
115. Rayfield EJ, Norman DB, Horner CC, Horner JR, Smith PM, et al. (2001) Cranial design and function in a large theropod dinosaur. Nature 409: 1033–1037.
116. Raup DM (1972) Taxonomic diversity during Phanerozoic. Science 177: 1065–1068.
117. Benton MJ (1985) Patterns in the diversification of Mesozoic nonmarine tetrapods and problems in historical diversity analysis. Special Papers in Palaeontology 33: 185–202.
118. Vermeij GJ, Leighton LR (2003) Does global diversity mean anything? Paleobiology 29: 3–7.

6

Dynamics of Green Sahara Periods and Their Role in Hominin Evolution

Juan C. Larrasoaña[1,2]*, Andrew P. Roberts[2], Eelco J. Rohling[2,3]

1 Instituto Geológico y Minero de España, Unidad de Zaragoza, Zaragoza, Spain, **2** Research School of Earth Sciences, The Australian National University, Canberra, Australia, **3** School of Ocean and Earth Science, University of Southampton, National Oceanography Centre, Southampton, United Kingdom

Abstract

Astronomically forced insolation changes have driven monsoon dynamics and recurrent humid episodes in North Africa, resulting in green Sahara Periods (GSPs) with savannah expansion throughout most of the desert. Despite their potential for expanding the area of prime hominin habitats and favouring out-of-Africa dispersals, GSPs have not been incorporated into the narrative of hominin evolution due to poor knowledge of their timing, dynamics and landscape composition at evolutionary timescales. We present a compilation of continental and marine paleoenvironmental records from within and around North Africa, which enables identification of over 230 GSPs within the last 8 million years. By combining the main climatological determinants of woody cover in tropical Africa with paleoenvironmental and paleoclimatic data for representative (Holocene and Eemian) GSPs, we estimate precipitation regimes and habitat distributions during GSPs. Their chronology is consistent with the ages of Saharan archeological and fossil hominin sites. Each GSP took 2–3 kyr to develop, peaked over 4–8 kyr, biogeographically connected the African tropics to African and Eurasian mid latitudes, and ended within 2–3 kyr, which resulted in rapid habitat fragmentation. We argue that the well-dated succession of GSPs presented here may have played an important role in migration and evolution of hominins.

Editor: Michael D. Petraglia, University of Oxford, United Kingdom

Funding: This work was supported by projects NE/E01531X/1 (RESET) of the UK Natural Environment Research Council, PR2011-0480 of the Spanish Ministry of Education, and AQUIFER: Hidrología Subterránea (E72) of the Aragon Government. The funders had no role in study design, data collection and analysis, decision to publish, or preparation of the manuscript.

Competing Interests: The authors have declared that no competing interests exist.

* E-mail: jc.larra@igme.es

Introduction

Debate about the role of climate variability in hominin evolution, through its impact on landscapes, has largely neglected North Africa because of the present-day hyper-arid nature of the Sahara and the lack of a fossil record comparable to that of East and South Africa [1,2]. However, growing evidence indicates that wetland-spotted savannah landscapes spread throughout the Sahara during past periods of enhanced monsoonal precipitation back to the late Miocene [3–40], thereby enabling its occupation by hominins [7,9,11–14,19,24,31,41–50]. These so-called "green Sahara" periods (GSPs) are important for paleoanthropology because most East African hominin sites back to the late Miocene have been related to savannah ecosystems dominated by either open [51] or mosaics of open and closed (often gallery forest) landscapes [52,53] with permanent freshwater sources [52–54]. The Sahara has a massive size compared with the rest of tropical Africa, and its current hyperaridity effectively blocks the gateway to Eurasia. Understanding the timing, extent and duration of GSPs is therefore crucial for a biogeographical approach in studies of hominin evolution to complement a meagre hominin fossil record that represents only a small fraction of their potential biogeographic range [55]. Such recognition has been hindered by: 1) lack of knowledge of the timing, extent and duration of GSPs beyond two of the most recent (Holocene and Eemian) humid periods; and 2) uncertainties about ecological aspects of past Saharan savannah landscapes that are required to assess biome habitability by hominins (except for the Holocene GSP).

Although landscape variability can be established from continental sediments scattered throughout the Sahara, such records provide only a fragmentary view that becomes scarcer and more difficult to date with increasing age. Here we combine continental and marine records of paleoenvironmental variability at different latitudinal bands to provide insights into the timing and duration of GSPs at timescales associated with hominin evolution. By considering the main climatological determinants of woody cover in tropical Africa [56], we translate paleoprecipitation estimates for well-known, representative GSPs (Holocene and Eemian) into specific ecosystems within the savannah biome inhabited by hominins [51]. We combine this information with estimates of hominin dispersal rates [57] and life history data [58] to put biogeographical constraints on hominin population dynamics and evolution in tropical Africa back to 8 million years ago (Ma).

Materials and Methods

North African Climate and Vegetation

Climatic conditions in North Africa are dictated by the position of the continent around the equator and by regional factors such as topography [4,59] (Figures 1, 2). North Africa has a narrow fringe with Mediterranean temperate climates in which westerly low-pressure systems (Figure 1) bring mean annual precipitation (MAP) in excess of 250 mm/yr, mainly during boreal winter and

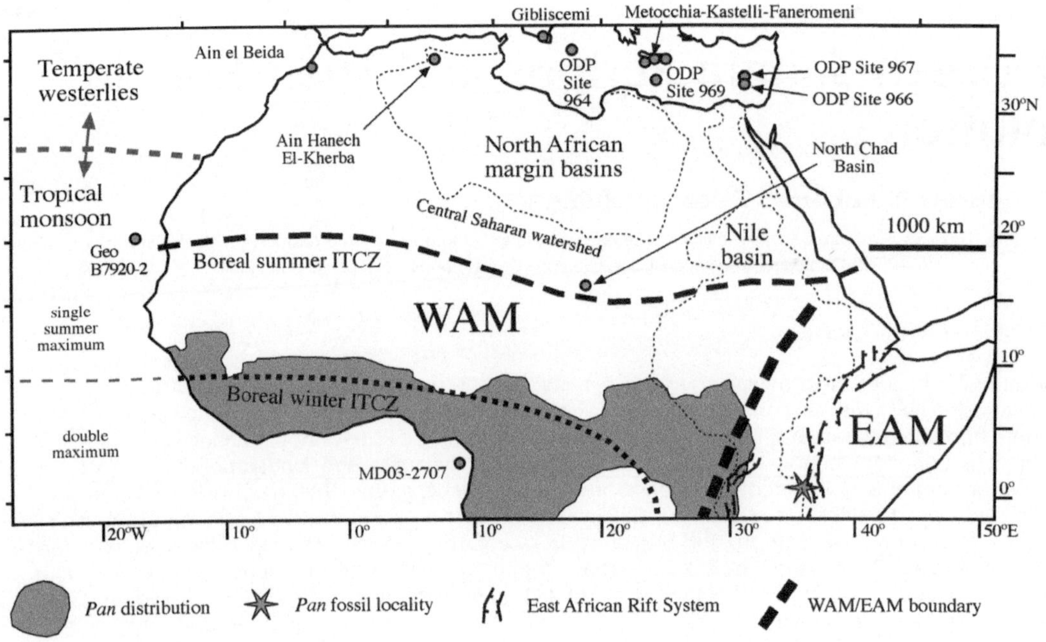

Figure 1. Map of North Africa with relevant meteorological and physiographic features. Meteorological features are after ref. [4], the location of normal faults in the East African Rift System (EARS) is after ref. [88], and the extents of the Nile and North African margin basins are after refs. [8,78–80]. The map includes the present-day distribution (grey shading) and occurrence of fossil remains (star) of chimpanzees (*Pan*) [69], and the locations of records mentioned in the text. WAM: West African monsoon; EAM: East African monsoon; ITCZ: Inter-tropical convergence zone.

spring (Figure 2). Although this precipitation may reach to latitudes of 25°N in extreme events [4], it normally rapidly decreases to less than 100 mm/yr at distances of 100–200 km from the coast.

Tropical Africa has a monsoonal climate in which precipitation is governed by seasonal changes in the differential heating over land and the neighbouring oceans [4,59] (Figure 1). Equatorial regions receive the most precipitation (>2000 mm/yr) due to their year-round location within the reaches of the boreal summer and winter meteorological equator (the Intertropical Convergence Zone), which drives precipitation with two main wet seasons governed by the West African Monsoon (WAM) (Figures 1, 2). Northward migration of the WAM front into North Africa during boreal summer (Figure 1) drives a pronounced progressive decrease in monsoonal rainfall from ~1500 mm/yr in near-equatorial regions down to ~100 mm/yr at around 18°N (Figure 2). Today, WAM rains reach latitudes of up to 25°N during exceptional events [4]. The high topography associated with the East African Rift System (EARS) and a different moisture source, the Indian Ocean, drive the so-called East African Monsoon (EAM) [4,59] (Figure 1). The broad zonal precipitation pattern observed in Africa north of the equator (Figure 2) is disrupted in East Africa by a rain-shadow effect, which causes reduced precipitation throughout the region, including equatorial areas (500–1500 mm/yr) (Figure 2).

Present-day vegetation in tropical North Africa is influenced by the prevailing climate [60] (Figure 2). Evergreen rainforest occupies the wettest regions around the equator and near the Atlantic coast in West Africa (FOR in Figure 2). The driest part of North Africa (the Sahara Desert) is largely devoid of vegetation with the exception of wadis, oases and mountain areas, where there is sparse vegetation (DES in Figure 2). Between the rainforest and dry desert lies the savannah, which includes three main belts that reflect the gradient of northward-decreasing rainfall. These belts include: 1) a narrow fringe on the periphery of the rainforest that extends from the Atlantic coast to the great lakes region in East Africa and that consists of a mosaic of semi-evergreen forest, woodlands and secondary grasslands (FMT in Figure 2); 2) a grassland, bushland and thicket fringe along the southern margin of the Sahara throughout much of North and East Africa (GBT in Figure 2); and 3) an intermediate woodland belt that lies between the two other ecosystems (WOO in Figure 2). Superimposed on this pattern are the montane forests, bushlands and shrublands that occupy mountain regions of East and North Africa.

Chronology of Saharan Pleistocene-Holocene Wetter Conditions

We draw on a recent compilation of lacustrine, palustrine and fluvial sediments that are indicative of wetter Holocene conditions throughout the Sahara south of 28°N [6]. The dataset includes 1237 radiocarbon- and luminescence-dated sediments, which have been grouped in three latitudinal bands (13–18°N, 18–23°N and 23–28°N) to examine spatial variations in wetter conditions during the Holocene GSP in response to monsoon dynamics.

For earlier GSPs, we have compiled 136 ages for sediments indicative of wetter conditions throughout the Sahara south of 28°N (Algeria, Chad, Egypt, Libya, Mali, Mauritania and Sudan) [11–31]. Of these sediments, 116, 16 and 3 have been dated using uranium-series, luminescence, and electron spin resonance methods, respectively. An additional age derives from $^{40}Ar/^{39}Ar$ dating of meteorite impact glass embedded in lacustrine sediments in the Dakhleh paleolake (Egypt) [32]. We have excluded 15 additional ages for which only minimum estimates (of up to >450 kyr) were provided [12,13,17,19,21,22]. The main difference between our data set and those recently published by Smith (ref. [9]) and Drake et al. (ref. [10]) is that ours excludes sediments located north of 28°N to isolate signals due to monsoon dynamics.

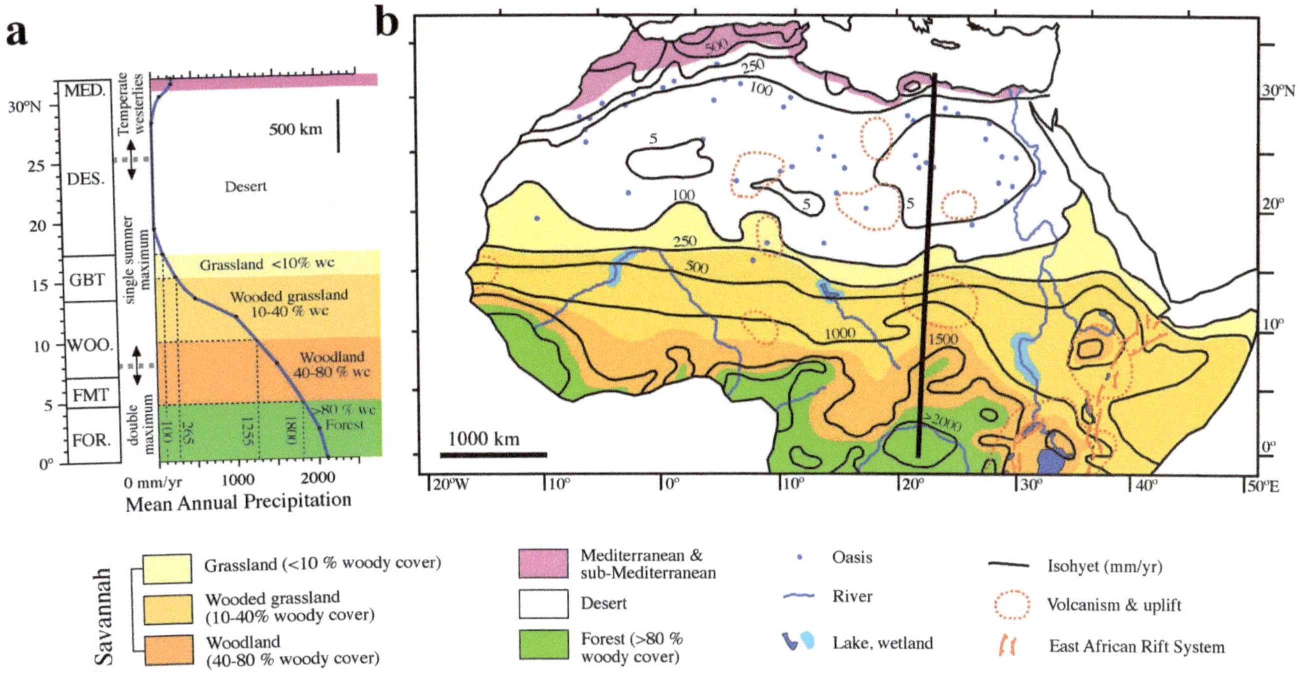

Figure 2. Present day meteorology and vegetation in North Africa. (a) Latitudinal distribution of present-day vegetation belts (MED.: Mediterranean and sub-Mediterranean; DES.: desert; GBT: grassland, bushland and thicket; WOO: woodland; FMT: forest mosaics and transitions; FOR.: rain forest) [60], and mean annual precipitation (MAP) and meteorological elements [3,4,59] projected onto a cross-section along the eastern Sahara (dark N-S line in b). (b) Map of the main physiographic and tectonic [88,97] elements. Coloured vegetation belts are based on a structural classification of vegetation (indicating percentages of woody cover: %wc) using MAP values and main climatological determinants on tropical African biomes [56].

We combine the terrestrial data with three marine middle Pleistocene to Holocene climate records that are widely recognized as indicative of paleoclimate conditions at different latitudinal bands (Figures 1, 3): 1) the sea surface salinity (SSS) record of core MD03-2707, which responds to riverine runoff in the Niger and Sanaga river catchments, and integrates WAM signals between 5° and 15°N [61]; 2) the humidity index from core GeoB7920-2, which responds to changes in eolian and fluvial contributions from the Sahel/Sahara boundary region, and integrates WAM signals between 18° and 22°N [62]; and 3) the sapropel record of sites 964, 966, 967 and 969 of the Ocean Drilling Program Leg 160, which is representative of sapropel deposition throughout the eastern Mediterranean Sea [63]. The later responds to a combination of enhanced Nile and other riverine discharge along the wider African-Mediterranean margin, which integrates WAM signals north of the central Saharan watershed at ~21°N [63–66]. We select these records because they provide specific information about monsoonal precipitation at different locations around North Africa that can be directly compared with their terrestrial counterparts.

Biogeography of Hominins in Tropical Africa

Recently, hominin habitats have been characterized by translating $\delta^{13}C$ data from paleosol carbonates into a structural definition (in terms of woody cover) of the different ecosystems with which fossil hominins are associated [51] (Table 1). Accordingly, nearly all (>99%) East African hominins are associated with three ecosystems within the savannah biome [51], which is defined as a mixed tree-grass system characterized by a discontinuous tree canopy (up to 80% of woody cover) in a continuous C_4 grass layer [67]. Soil carbonates mainly form when evaporation exceeds precipitation, therefore this method may provide a view of the amount of tree cover associated with hominin sites that is somewhat biased toward drier habitats [68]. However, qualitative multiproxy paleoenvironmental reconstructions imply a recurrent association of hominin sites with open landscapes or with landscape mosaics including grasslands, woodlands and forests typically associated with permanent freshwater lakes, rivers and springs [52–54]. Regardless of whether this association is genuine or the result of temporal or spatial averaging due to taphonomic processes [52,53], it clearly illustrates a link between hominin sites and dynamic savannah landscapes with variable woody cover that includes riparian forests along freshwater sources. This independent support for the quantitative reconstructions of Cerling et al. (ref. [51]) enables identification of hominin habitats back to at least 6.5 Ma despite variable background temperature and atmospheric CO_2 conditions, which also affect vegetation.

Despite the well-established link between vegetation and precipitation in tropical Africa [60] (Figure 2), the main climatological determinants of woody cover in tropical Africa have only recently become established, based on a comparison of daily precipitation and woody cover records derived from satellite data [56]. A least-squares regression between F_c (fraction of woody cover), P_w (mean wet season precipitation), and α_w (average wet season storm intensity, defined by normalizing P_w to the duration of the wet season) provides a robust statistical relationship ($r^2 = 0.65$) between F_c and the quantity and intensity of rainfall according to [56]:

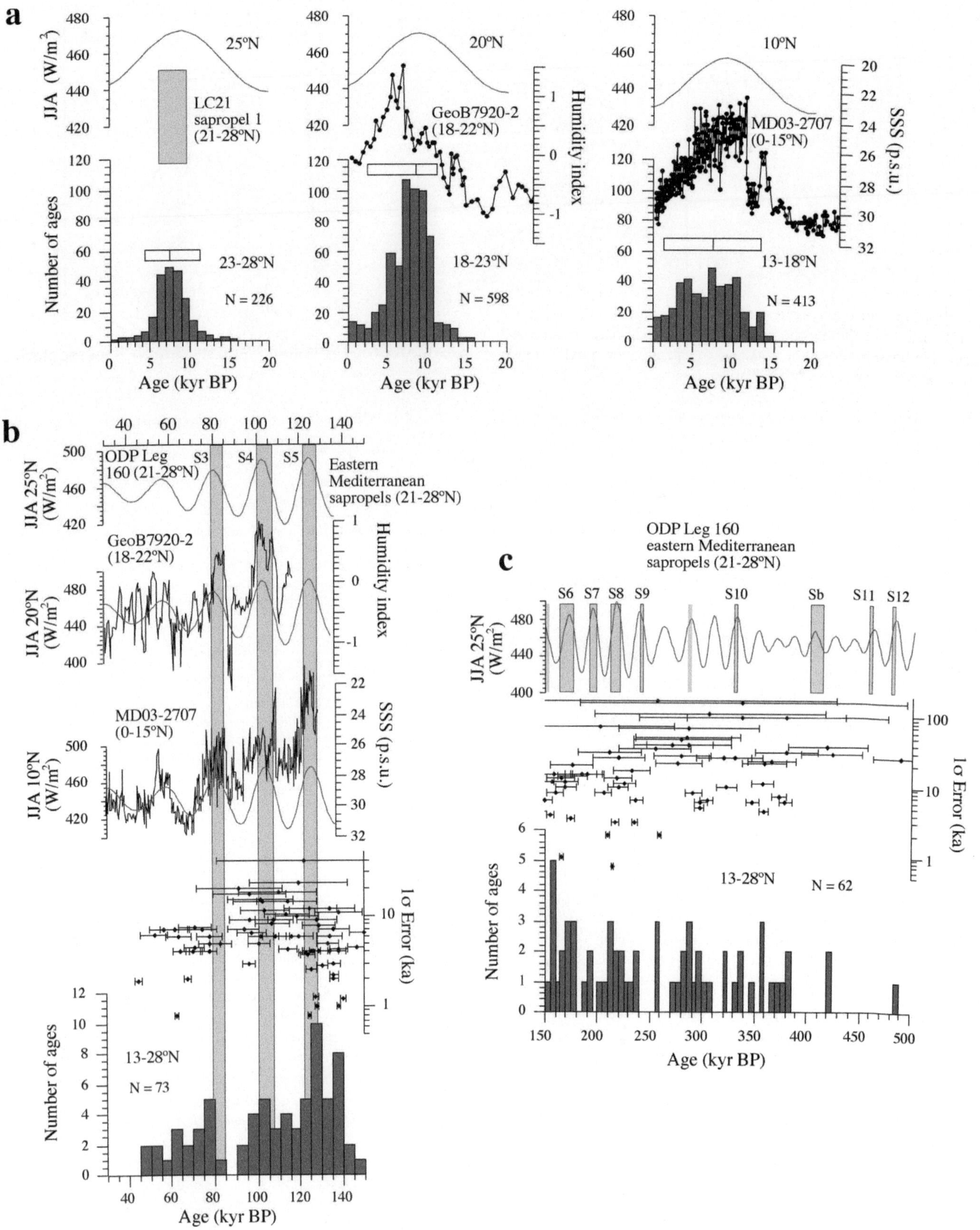

Figure 3. Late Pleistocene-Holocene climate variability in North Africa. (a) Histograms of ages of continental sediments indicative of wetter conditions at different latitudinal bands across North Africa between 0 and 16 kyr BP [6], plotted with marine records that reveal North African climate variability at comparable latitudes (sea surface salinity (SSS) from core MD03-2707 [61], humidity index from core GeoB7920-2 [62] and

sapropel record from ODP Leg 160 [63,64] and summer insolation curves (June, July, August; JJA) at 10°, 20° and 25°N [107]. Bars indicate the 5 and 95 percentiles around the median of the histograms. (b) Histogram of ages of continental sediments indicative of wetter conditions throughout the Sahara between 30 and 150 kyr BP, with SSS from core MD03-2707 [61], humidity index from core GeoB7920-2 [62] and the sapropel record from core ODP Leg 160 [63] and their relevant summer insolation curves [107]. The chronology of sapropels is after ref. [65]. (c) Histogram of ages of continental sediments indicative of wetter conditions throughout the Sahara between 150 and 500 kyr BP, plotted with the eastern Mediterranean sapropel record from ODP Leg 160 [63] and summer insolation at 25°N [107]. Unlabelled sapropels correspond to those erased by post-depositional oxidation [63].

$$Fc = 0.054Pw - 0.66\alpha w - 0.0017Pw\alpha w$$

Across tropical Africa, around 79% of the annual rainfall arrives in the wet season, which has an average duration of 172 days [56]. Considering these average estimates for P_w and α_w, this equation can be used to estimate the percentage of woody cover from MAP data. We have inverted the equation to calculate the MAP value at which the fraction of woody cover attains certain values. By setting the fractions of woody cover to 10, 40 and 80%, we can estimate the isohyets that delineate the different structurally-defined savannah ecosystems with which nearly all East African fossil hominin sites are associated [51] (Table 1). Based on the present-day coincidence of the northern boundary of grasslands with the 100 mm/yr isohyet, and given that this boundary is not defined in terms of woody cover, this MAP value has been used to delineate the boundary between savannah and desert. Since the equation does not capture the full range of rainfall variability for the highest fractions of woody cover [56], we set the limit between savannah and rainforest at the northernmost rainforest boundary at ~1800 mm/yr of precipitation. Good coherency between structural types of savannahs determined from MAP values and the vegetation types of White (ref. [60]) (left and middle panels in Figure 2) validates our approach.

Next, we consider the combined distribution of different savannah habitats throughout tropical Africa as the biogeographic range of hominins. The need for such an approach is illustrated by the distribution of extant and fossil chimpanzees and gorillas. While the only secure report of fossil chimpanzee remains to date comes from middle Pleistocene successions at Tugen Hills (Kenya) in the EARS, the present-day (before historical habitat disruption) distribution of chimpanzees spans a broad belt across equatorial Africa (between 10°S and 15°N) that includes rainforest, woodland and dry savannah, but that does not extend east into the EARS [69] (Figure 1). The case for gorillas is even more marked, with no secure identification of fossil remains so far. Failing to recognise the biogeography of chimpanzees and gorillas with respect to their relevant biomes, and relying only on their fossil remains, would inevitably lead to erroneous inferences concerning their past distribution, habitats and population dynamics. The same holds for hominins, so that ignoring past habitats suitable for their occupation would result in an incomplete environmental backdrop to frame their evolution. We attribute the scant occurrence of fossil hominins in North Africa as due to geological conditions unfavourable for creating sediment accommodation space (as opposed to the EARS and some South African karst systems [2]) and to sampling bias (historical focus on East and South Africa [55]).

Distribution of Hominin Habitats during Representative GSPs

Paleoprecipitation estimates for the Holocene GSP (~11 to 6 ka) based on multiple geological, archeozoological and archeobotanical data indicate a MAP of around 100 mm/yr at the core of the northeastern Sahara (between 23° and 29°N) and of 450 mm/yr at around 18°N [41]. Further south, pollen data from lakes Oyo (~19°N) and Malha (~15°N) suggest MAP values of ≥400 mm/yr [33] and 700 mm/yr [34], respectively. Paleohydrological modelling of the West Nubian (18°–19°N) and Chad (10°–18°N, centered at 14°N) megalakes points to MAP values of 500–900 mm/yr [35] and 600–650 mm/yr [36], respectively. An increase in MAP of 250 mm/yr is indicated in present-day rainforest areas of equatorial West Africa (~5°N) on the basis of pollen data from Lake Barombi Mbo [70]. These estimates have been projected into a N-S section across the eastern Sahara and into equatorial Africa (Figure 4a). We then fitted a profile of these estimates along the same cross-section used for present-day precipitation variations in Figure 2. From this new profile we then established the positions of the 250, 500, 1000, 1500 and >2000 mm/yr isohyets (Figure 4a). Given the lack of error estimates in most paleoprecipitation reconstructions, we have drawn paleoprecipitation throughout North Africa based on a broad zonal pattern with a curvature in East Africa and taking into account the influence of major topographic features. This approach also affects the 265, 1255 and 1800 mm/yr isohyets that delineate the grassland/wooded grassland, wooded grassland/woodland, and woodland/forest boundaries, respectively.

Paleoprecipitation estimates for the Eemian are available from paleohydrological modelling of the Dakhla (~26°N) and Chad megalakes and from fossil vertebrate assemblages from Bir Tarfawi (22°N) (Figure 4b). Paleohydrological modelling of the Dakhla and Chad megalakes suggests MAP values of 410–670 mm/yr [16] and ≥730 mm/yr [39], respectively. Fossil faunas from Bir

Table 1. Reconstructed woody cover and estimates of hominin sites associated with different tropical ecosystems (after ref. [51]), along with our calculated mean annual precipitation (MAP) values for such ecosystems.

Biome	Ecosystem	Woody cover (%)	Hominin sites (%)	Inferred MAP (mm/yr)
	Grassland	<10	10	100–265
Savannah	Wooded grassland	10–40	60	265–1255
	Woodland	40–80	29	1255–1800
Tropical forest	Forest	>80	<1	>1800

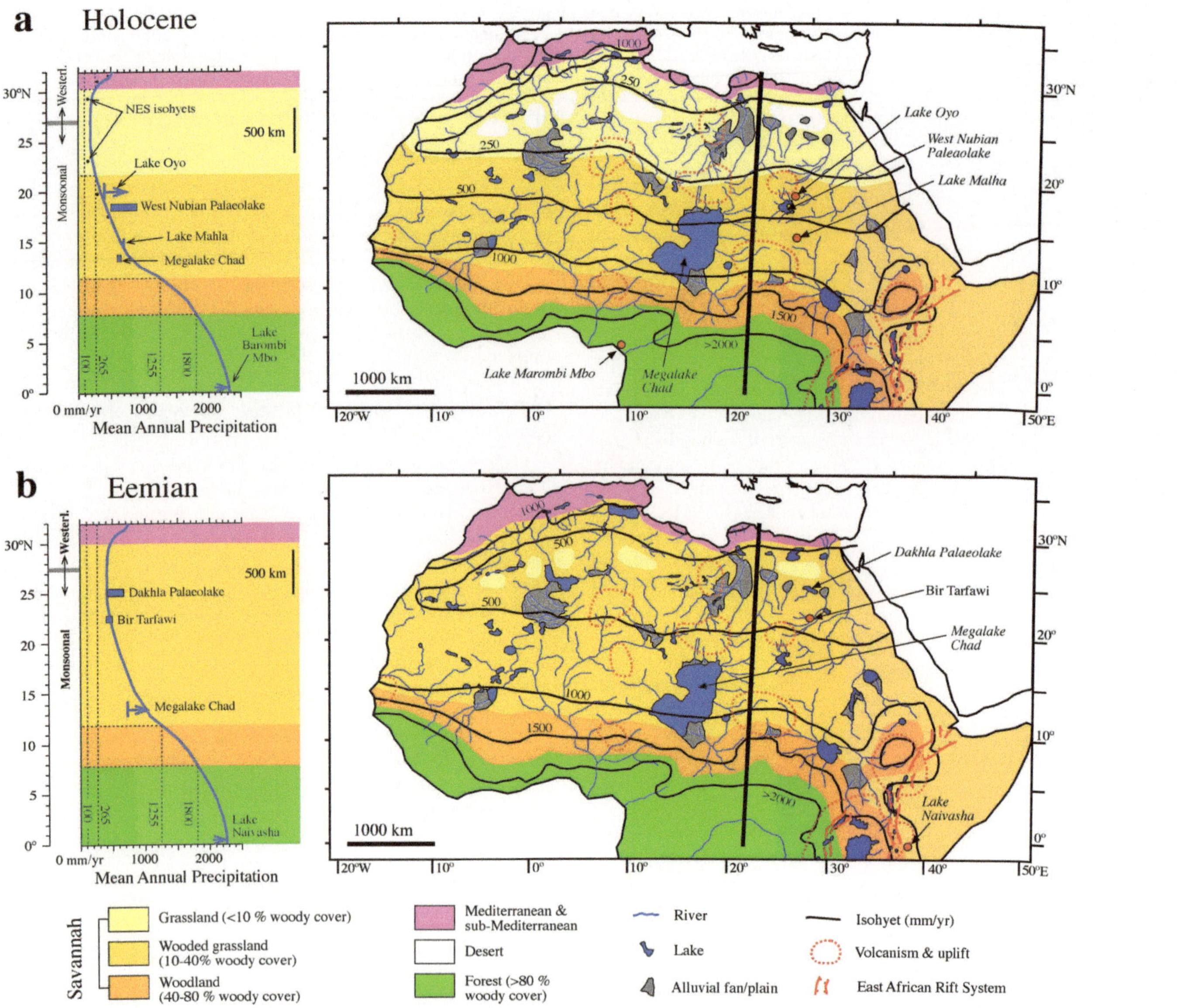

Figure 4. Reconstruction of North African vegetation during past green Sahara periods. (a) Estimated and reconstructed MAP for the Holocene GSP (6–10 kyr BP) projected onto a cross-section along the eastern Sahara (left panel) and map view of reconstructed MAP, vegetation and physiographic elements [7,8,11,45] (right panel). (b) Estimated and reconstructed MAP for the Eemian GSP (122–128 kyr BP) projected onto a cross-section along the eastern Sahara (left panel) and map view of reconstructed MAP, vegetation and physiographic elements [14,15,44,45] (right panel).

Tarfawi indicate a MAP value of ≥500 mm/yr [40]. We are unaware of Eemian paleoprecipitation reconstructions from present-day equatorial West African rainforest areas. We have, therefore, considered a MAP increase of 270 mm/yr reported from paleohydrological modelling of Lake Naivasha [71] (Figure 4b) to be representative of Eemian precipitation changes in equatorial Africa. We then followed the same procedure used for the Holocene to reconstruct MAPs and ecosystem distribution throughout North Africa during the Eemian GSP.

Results and Discussion

Middle Pleistocene to Holocene Climate Variability in North Africa

In the southern Sahara (13–18°N), ages of latest Pleistocene to Holocene sediments indicative of humid conditions cluster around 7.5 ka, with a wide distribution (90% of ages are between 1.5 and 13.5 ka) that matches the broad interval of increased monsoonal precipitation between 0° and 15°N [61] and encompasses the local summer insolation peak (Figure 3a). In the central Sahara (18–23°N), the distribution of ages is narrower around 8.5 ka (90% of ages are between 2.5 and 11.5 ka), in correspondence with a shorter period of insolation-driven increased monsoonal precipitation at around 20°N [62]. Further north (23–28°N), ages cluster more tightly around 7.5 ka (90% of ages are between 4.5 and 11.5 ka) and broadly coincide with deposition of an organic-rich layer in the eastern Mediterranean Sea known as sapropel S1, and which reflects monsoon runoff flooding from North Africa into the eastern Mediterranean Sea [64,65]. These data indicate a progressive (2–3 kyr) northward expansion and southward retraction of monsoonal precipitation at the onset and termination of the GSP, respectively. Monsoon-driven wetter conditions throughout the Sahara peaked over a ~4 kyr period that is broadly marked by deposition of sapropel S1. Both the mid-point of sapropel S1 and clusters of sediments indicative of wet conditions in the Sahara lag by 1.5–3 kyr relative to the insolation peak, which has been attributed to groundwater recharge preceding development of large lakes at the onset of the GSP [6] and arrival of runoff to the Mediterranean Sea.

Between 30 and 150 ka, scarcity of dated terrestrial sediments prevents separation of ages within different latitudinal bands. Due to potential prolonged lake and spring activity related to groundwater discharge and to the errors (typically >5 kyr) associated with most published ages, ages of sediments indicative of wetter conditions are scattered (Figure 3b) [9,10]. Field evidence indicates that these sediments are grouped into genetically related units that are typically separated by eolian deflational surfaces [8,11,12,13,15,17,24]. This points to prevailing hyperarid conditions punctuated by short wetter spells. With these caveats in mind, we identify three prominent clusters of ages at around 80, 105 and 125 that we interpret as genuine periods of widespread wetter conditions in the Sahara in agreement with other studies [9,10]. These three clusters coincide, when considering chronological errors, with distinctive peaks in the SSS record of core MD03-2707 [61], the humidity index of core GeoB7920-2 [62], sapropels S3 to S5 in the eastern Mediterranean [63,65], and local summer insolation maxima (Figure 3b). These data corroborate the causal link between monsoon dynamics and green Sahara conditions, which peaked throughout the Sahara during deposition of eastern Mediterranean sapropels. The decreasing duration of wet periods from south to north, as portrayed by the marine records (Figure 3b), further highlights the time-transgressive onset and demise of GSPs. Smaller clusters of ages between 40 and 60 ka broadly coincide with wetter conditions in core MD03-2707 [61] and other records that indicate enhanced precipitation at Africa around 10°N [72]. These ages also coincide with slightly wetter conditions in core GeoB7920-2 [62], but not with any eastern Mediterranean sapropel (Figure 3b). This demonstrates the occurrence of other periods during which monsoonal precipitation was enhanced, but which lacked the power to drive expansion of savannah landscapes beyond the central Sahara. The low amplitude of insolation peaks at around this time (Figure 3b) suggests a link with cold sea surface temperatures in the North Atlantic, which exerted a dominant role on North African humidity changes at times of weaker monsoon activity [10,47,61,72]. The interval between 150 and 500 ka is marked by fewer continental sediment ages, which become progressively scarcer back in time (Figure 3c) due to protracted erosion of older sediments [9,10]. The errors associated with the ages of these sediments are larger (>10 kyr), which prevents correlation of clusters of ages to deposition of eastern Mediterranean sapropels (Figure 2c). Despite this, ages cluster with a ~20 kyr frequency that is indicative of control by insolation-driven enhanced monsoon rainfall [9,12,23]. Based on the lesson learned from the much better documented late Pleistocene-Holocene period, we interpret that short-lived middle Pleistocene wet periods in the Sahara were also concurrent with eastern Mediterranean sapropel deposition during GSPs.

Timing of green Sahara periods since 8 Ma

Our compilation of middle Pleistocene to Holocene paleoenvironmental records (Figure 3) indicates that marine records provide excellent documentation of circum-Saharan monsoon variability, which complements a continental record that becomes more scarce, fragmentary and difficult to date at older times. Prior to 130 ka, the sapropel record provides the only continuous and precisely (astronomically) dated archive with the potential for identifying peak savannah expansions at timescales associated with hominin evolution.

Sapropel formation has been traditionally attributed to increased Nile runoff during summer insolation maxima [73]. The main sources for Nile runoff are the highlands of Uganda and Ethiopia in East Africa (Figure 1); some authors attribute increased Nile runoff periods to rainfall driven mainly by the WAM [73], while others relate them to EAM dynamics [74]. However, $\delta^{18}O$ data from sapropels [75,76] demonstrate that the lightest values are typically found off Libya rather than off the Nile mouth. Significant northward penetration of the WAM summer front beyond the central Saharan watershed (~21°N) (Figure 1) must have occurred simultaneously with enhanced Nile runoff to allow direct drainage of low-$\delta^{18}O$ monsoonal precipitation into the eastern Mediterranean [77] (Figure 1). Such massive drainage was channelled toward the gulfs of Sirte and Gabes through now-extinct river systems identified along the North African margin [8,78–80] (Figure 1), as demonstrated with neodymium isotope ratios [44]. Penetration of the summer WAM front as far as ~25°N during GSPs, which would have enabled substantially enhanced monsoon runoff into the eastern Mediterranean, is evidenced by widespread formation of lake and fluvial systems between 20° and 28°N that are disconnected from more southerly (monsoonal) and northerly (Mediterranean) hydrological systems [4,6,8–12,14–20,44,45,78–80]. This conclusion is further supported by modelling results that highlight the positive feedbacks between lake formation and northward expansion of monsoon rainfall [38]. Enhanced WAM (as opposed to EAM) precipitation during GSPs also provides an explanation for enhanced pedogenic activity [81] and dampening of dust production [66] throughout the Sahara during periods of sapropel formation. Such a prominent role of the WAM as the main driver of GSPs explains the accumulation of sapropels back to 10 Ma, under the same scenario of enhanced monsoonal discharge during local summer insolation maxima [63,82,83], well before the earliest clear evidence for a river connecting the East African tropical regions with the Mediterranean Sea across Egypt sometime between the latest Miocene (ca 6 Ma [79]) and the early Pliocene (4–5 Ma [84]).

Based on the mechanistic link established here between sapropels and GSPs, we take eastern Mediterranean sapropels [63,85,86] as markers of over 230 GSPs back to 8 Ma (Figure 5f,g). *In situ* evidence for recurrent GSPs over millions of years is provided, despite its fragmentary nature, by the lacustrine, peri-lacustrine and eolian sedimentary record of the North Chad Basin, which demonstrates an alternation of humid and dry conditions since ~8 Ma [48] (Figure 5c). The only gap in sapropel sedimentation is observed between ca 5.3 and 6 Ma, due to disruption of normal marine sedimentation during the Messinian Salinity Crisis (MSC) (Figure 5f). Given the link between GSPs and insolation maxima, we infer that at least the most prominent local summer insolation maxima also led to GSPs between 5.3 and 6.6 Ma. This inference is supported by observation of an undisturbed pattern of North African climate variability across the MSC [87].

Overall, we infer that WAM variability led to recurrent GSPs back to 8 Ma (Figure 5g), with GSPs forming 400-kyr and 100-kyr clusters that attest to the impact of eccentricity modulation of precession (and hence of insolation) on monsoon variability [63,82,83,85,86] (Figure 5a,e,f). A less regular sapropel pattern after 0.9 Ma [63,66] suggests that GSPs were somewhat less frequent after development of large polar ice caps during the Mid-Pleistocene Transition (Figure 5a,f,g). The notion of eccentricity

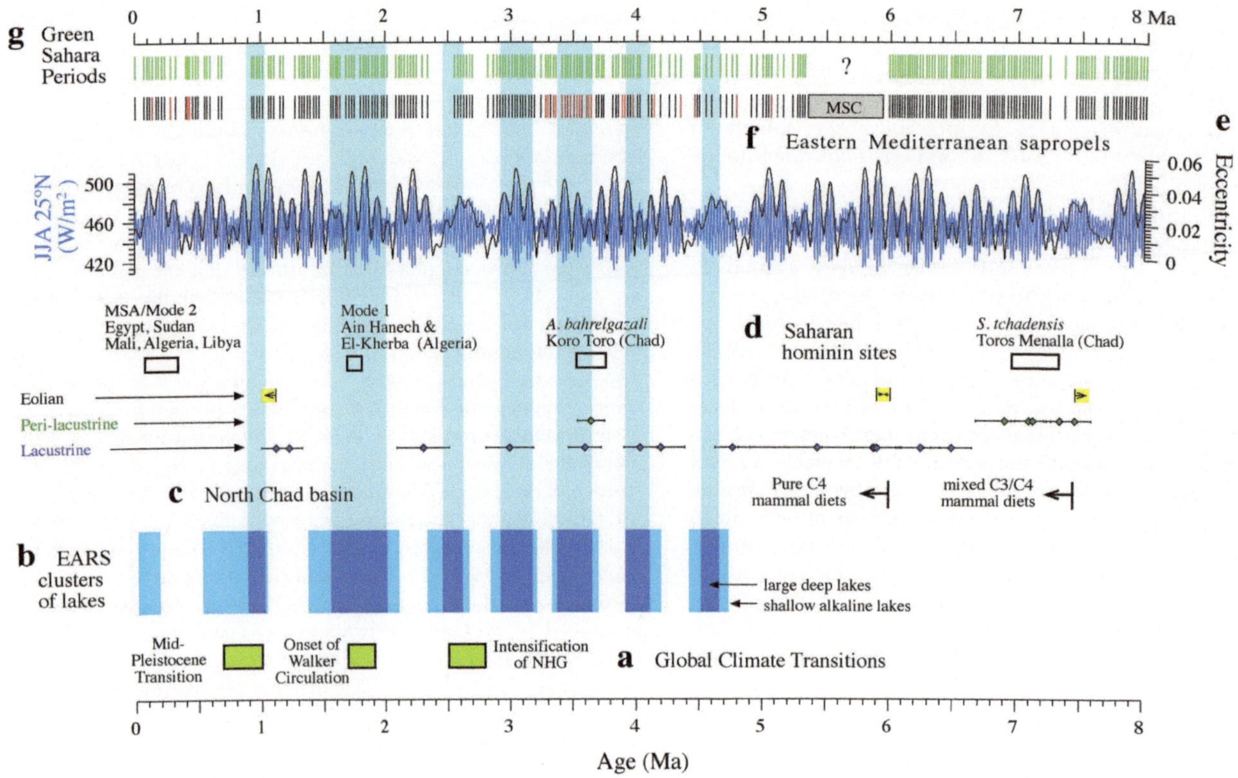

Figure 5. Saharan climate and hominin occupation 8 Ma to present. (a) Chronology of global climate transitions [88] (NHG: northern hemisphere glaciation). (b) Chronology of East Africa Rift System (EARS) lake periods [88]. Formation of large lakes was prevented by the lack of full-graben basin morphologies before 5 Ma. Vertical shaded bars indicate periods with large, deep, freshwater lakes. (c) Chronology of lacustrine, peri-lacustrine and eolian sediments in the North Chad Basin [48] plotted with earliest evidence for mixed C_3/C_4 and pure C_4 mammal diets in the basin [96]. (d) Chronology of Saharan hominin occupation sites [11,12,14,47,48,50]. (e) Variations in summer insolation (June, July, August; JJA) at 25°N and eccentricity back to 8 Ma [107]. (f) Composite eastern Mediterranean sapropel record from ODP Leg 160 sites [63] and land sections at Gibliscemi, Metochia, Kastelli and Faneromeni [85,86] (red bars indicate oxidized sapropels). (g) Green Sahara periods back to 8 Ma inferred from the eastern Mediterranean sapropel record.

modulation of precession as the main driver of WAM variability provides strong support for a similar influence on the EAM, given that phases of expanded lakes in the EARS coincided with 400 kyr eccentricity maxima [2,88,89] (Figure 5b) during which moisture availability was driven by insolation changes at precessional timescales [2,90]. Simultaneous response of WAM and EAM systems to insolation forcing has been demonstrated based on the synchronicity of periods of higher water availability in East Africa and of eastern Mediterranean sapropel deposition at least at 0.1–0.2 Ma [91] and 1.85–1.95 Ma [92].

Landscape Composition and Hominin Occupation during Green Sahara Periods

We translate Holocene precipitation estimates [33–36,41,70] into percentages of woody cover in tropical savannahs to infer the presence of wooded grasslands (10–40% woody cover) between 10° and 20°N and grasslands (woody cover <10%) north of ~22°N (Figure 4a). Although more paleoprecipitation estimates are needed to provide a detailed picture of the distribution of different habitats, the proposed distribution is consistent with both paleoenvironmental evidence across North Africa [4,5,37,41–43] and climate modelling data [38]. Enhanced precipitation led to development of interconnected rivers, lakes, playas and terminal alluvial fans throughout the Sahara [3–7,11,35,41–43,45,80] (Figure 4a). This drainage network enabled expansion of Sudanian and Sahelian vegetation into the desert as gallery forests [5] that coexisted with grasslands [42] to create a biodiverse ecosystem with no present analogue [5]. Paleontological and archeological data indicate widespread trans-Saharan dispersals of aquatic and savannah faunas along interconnected drainage systems and grasslands [45]. Field [41] and modelling data [38] suggest precipitation <100 mm/yr in some areas north of ~28°N, especially in the northeast Sahara. This points to continued desert conditions in isolated regions, separated by major rivers and lake systems. Paleoclimatic and modelling data also indicate intensification and southward displacement of mid-latitude cyclones down to ~30°N during GSPs [9,10,81,93], which brought increased rainfall to the Mediterranean rainfall belt of the North African margin [38,41]. This must have resulted in slight southward expansion of Mediterranean vegetation and in the narrowing of isolated arid regions between ~28° and ~30°N (Figure 4a).

The scarcer paleoprecipitation estimates available for the Eemian GSP (122–128 kyr BP) [16,39,40,71] point to markedly increased precipitation across North Africa compared to the Holocene GSP (Figure 4a, b), as is evident also from quantifications of monsoonal runoff into the eastern Mediterranean [44,77] and from modelling data [39]. Eemian GSP paleoprecipitation points to wooded grassland occurrences (10–40% woody cover) across the Sahara, which is supported by the occurrence of tropical humid plants (*Ficus*, *Celtis*, and ferns) and subtropical vertebrate fauna as far north as 23–26°N in the Western Desert [14,40]. Given these overall wetter conditions, it is likely that the discontinuous arid belt between the wooded grasslands to the south and Mediterranean vegetation to the north was narrower and less arid compared to the Holocene. Although paleohydrological information on the Eemian GSP is not as abundant as for the Holocene GSP, evidence for megalakes and fluvial activity throughout the Sahara is compelling [9–16,44–46,80] so that reconstruction guided by the Holocene GSP is tenable [45] (Figure 4b). Field data indicate a marked size increase of older Saharan paleolakes, which attest to a significant progressive precipitation decrease during peak GSPs since 500 ka [8,9,12,15]. This is supported by pollen records off northwest Africa [94], and indicates that the Eemian GSP can be taken as representative of other middle Pleistocene GSPs.

Development of fluvial networks and megalakes throughout the Sahara during Messinian, Pliocene and lower Pleistocene times [8,78,80] provides support for recurrent GSPs since the late Miocene. Paleoprecipitation estimates are not available for such older GSPs, but their paleolakes were considerably larger than for middle Pleistocene to Holocene GSPs [8]. Although the complex interplay of drainage networks with volcanic, tectonic and geomorphic features may have influenced lake development and size [8], the extension of older large paleolakes to as far north as 28°N suggests that conditions were wetter during older peak GSPs [8,80]. This is consistent with the more regular sapropel pattern before 0.9 Ma [63,66], which suggests that WAM penetration over the Sahara was favoured before development of large polar ice caps around the Mid-Pleistocene Transition. Regardless, the continued occurrence of desert plant pollen in marine sediments off northwest Africa since at least the middle Pleistocene [94], coupled with the persistence of arid adapted mammals in the Sahara back to ~5 Ma [95], suggests that the discontinuous arid belt between the subtropical savannah and the Mediterranean eco-region never disappeared completely. It therefore appears that rivers emanating from the central Saharan watershed must have played a key role in enabling dispersals across the Sahara during GSPs, especially during relatively drier GSPs such as the Holocene.

Widespread archeological evidence indicates human occupation of the entire Sahara during the Holocene GSP [7,11,41–43,45]. Earlier and Middle Stone Age remains are far less abundant than Neolithic ones, but their association with fluvial, lacustrine, and spring deposits compellingly links human occupation of the Sahara with middle and late Pleistocene GSPs [9,12–14,46,47], especially during the Eemian [9,11–14,44,45,47]. Although archeological remains scattered throughout the Sahara provide evidence of human occupation before 0.5 Ma [49], well dated sites are restricted to the Earlier Stone Age archeological remains in Ain Hanech and El-Kherba in Algeria (1.7–1.8 Ma) [50] and to fossil hominin remains in Koro Toro and Toros Menalla (North Chad Basin), which date to 3.5–3.7 Ma and 6.9–7.3 Ma, respectively [48] (Figure 5d). All of these sites have been linked either to open landscapes that included permanent freshwater lakes and streams (Ain Hanech and El-Kherba) [50], or to mosaics of open to gallery forest landscapes with nearby permanent lakes that evolved from and to desert conditions (Koro Toro and Toros Menalla) [53,48]. Paleoenvironmental reconstructions of these sites indicate GSP-like conditions that were much wetter than present-day arid climates. Evidence from these sites therefore strongly support a case for recurrent hominin occupation of the Sahara during GSPs since the late Miocene. Within chronological uncertainties, these hominin occupations during GSPs coincided with large East African lake phases and 400-kyr eccentricity maxima (Figure 5).

Mammal teeth carbonate $\delta^{13}C$ data from the North Chad Basin demonstrate the occurrence of C_4 grasses by 7.45 Ma [96], which implies the episodic presence of true savannah [67] in the Sahara since at least that time. Pollen data indicate uniquely biodiverse savannah landscapes [5] during GSPs. Saharan grasslands and wooded grasslands with gallery forests along interconnected drainage systems during GSPs back to 8 Ma therefore provide an ecological context to hominin occupations that is comparable to the landscapes associated with most East African hominin sites back to the late Miocene [50–53]. Central areas of the Sahara were also affected by local tectonic uplift and volcanism during the late Neogene and Quaternary [97], which would have contributed to creation of dynamic landscapes favourable for hominin occupation as they do in East and South Africa [97]. From a biogeographic perspective, supported by periods with incontestable evidence for hominin presence in the Sahara [7,9,11–14,41–50], it appears that the multiple GSPs since 8 Ma represented massive episodic expansions of prime hominin habitats into North Africa.

Implications for Hominin Evolution

Our compilation of marine and continental paleoenvironmental data (Figure 3) indicates a time-trangressive onset of GSPs from south to north over periods of a few thousand years. Estimated dispersal rates of hominins and other large mammals are rapid, of the order of 13–21.5 km/yr [57]. Even if Saharan reoccupation by hominins and large mammals during GSPs was limited by the lower dispersal rate of tropical plants (0.5 km/yr [5]), then colonization throughout North Africa would still have been achieved within 2–3 millennia of initial savannah expansion. Values of ~15 years for the age at first birth [58] assumed for hominins, along with observed durations of peak green Sahara conditions of 4–8 millennia (based on typical sapropel durations [64,65,82,98]), suggest that hundreds of generations of hominins could have thrived in North Africa during each GSP. Subsequent GSP terminations would have caused major population collapses and displacements, and habitat fragmentation with associated potential for isolating small residual populations. These conditions would be conducive to genetic drift and strong selection pressures, possibly to the point of speciation or extinction.

The fossil record attests to frequent connections between East and Northwest African mammal faunas since the late Miocene, with the faunas of the two regions reaching maximum affinity in the middle Pleistocene [99]. Archeological and fossil records also indicate recurrent hominin occupation of the Sahara since the late Miocene [7,9,11–14,41–50] (Figure 5d). Our corroboration of astronomically-controlled monsoon variability as the main driver of changes in tropical and subtropical African vegetation [2,47,88,89] points to recurrent GSPs as the underlying mechanism that enabled frequent connection between East and Northwest African mammal faunas and recurrent hominin occupation of the Sahara. The fossil record also suggests frequent bidirectional exchanges of large mammal faunas between Africa and Asia since the late Miocene, including hominins [57,100]. Furthermore, genetic evidence points to recurrent gene flow between African and Asian hominins (modulated by isolation due to distance) after their first out-of-Africa dispersal [101]. In the absence of land bridges across the southern Red Sea after the Late Miocene [102] and of unequivocal evidence for navigation skills before *H. sapiens* [103], GSPs appear to provide (in addition to seafaring along the Red Sea coast) a critical control mechanism for hominin migrations between Africa and Asia through the Levantine corridor after enabling their first out-of-Africa dispersal. This control may be ancient. Recent re-evaluation of the initial out-of-Africa dispersal of the genus *Homo* and its subsequent colonization of Asia, thought to date to around 1.85 Ma [50,57], suggests that this dispersal may be considerably older, probably between 1.9 and 2.4 Ma [55,104]. It has been also suggested that *Homo* may have speciated in Asia from an earlier African immigrant hominin, before eventually dispersing throughout Eurasia and back again into Africa [104]. The frequent GSPs that we document prior to 1.85 Ma would have facilitated such earlier, bidirectional dispersals [104] that we tentatively link with GSPs around the eccentricity maxima at 2.1–2.3 Ma (Figure 5e,g). Our interpretation of older GSPs supports the idea that biological evolution or social organization factors, rather than climate [57], may have been limiting to earlier first out-of-Africa dispersal of hominins.

The last four GSPs (at 6–10, 77–81, 102–108 and 122–128 kyr BP) coincided with wetter conditions across North [61,62,63,65] and East Africa [91], the Arabian Peninsula [105] and the Levant [106], due to similar regional climatic responses to enhanced insolation maxima. Over longer timescales, 400-kyr eccentricity changes appear to have caused clusters of GSPs that coincided with large East African lake phases (Figure 5). GSPs, therefore, facilitated critical biogeographic connections within a widespread region with enhanced water availability that stretched from the African tropics to the Eurasian mid latitudes. In addition, North African savannahs during GSPs were larger ($\sim 14\times10^6$ km^2) than East and South African savannahs combined ($\sim 6\times10^6$ km^2), so that GSPs represented at least an episodic threefold extension of prime hominin habitats, followed by rapid habitat fragmentation. Overall, Saharan savannahs likely enabled modulation of hominin population dynamics in tropical Africa and controlled accessibility of the Levantine gateway to Asia. Thus, we infer that the recurrent GSPs that we document over the past 8 Myr may have had a large, but underappreciated, potential for driving hominin speciation, extinction, adaptation, and migration.

Acknowledgments

We are grateful to Robin Dennell and three anonymous reviewers whose constructive comments greatly improved the manuscript.

Author Contributions

Conceived and designed the experiments: JCL. Performed the experiments: JCL. Analyzed the data: JCL APR EJR. Contributed reagents/materials/analysis tools: JCL APR EJR. Wrote the paper: JCL APR EJR.

References

1. Balter M (2011) Was North Africa the launch pad for modern human migrations? Science 331: 20–23.
2. Maslin MA, Christensen B (2007) Tectonics, orbital forcing, global climate, and human evolution in Africa: introduction to the African paleoclimate special volume. J Hum Evol 53: 443–464.
3. Petit-Maire N, Guo Z (1997) Holocene palaeoprecipitation over the present-day Sahara desert: implications for the future. Episodes 20: 232–234.
4. Gasse F (2000) Hydrological changes in the African tropics since the Last Glacial Maximum. Quat Sci Rev 19: 189–211.
5. Watrin J, Lézine AM, Hély C, Cour P, Ballouche A, et al. (2009) Plant migration and plant communities at the time of the "green Sahara". C R Geoscience 341: 656–670.
6. Lézine A, Hély C, Grenier C, Braconnot P, Krinner G (2011) Sahara and Sahel vulnerability to climate changes, lessons from Holocene hydrological data. Quat Sci Rev 30: 3001–3012.
7. Pachur HJ, Hoelzmann P (2000) Late Quaternary paleoecology and paleoclimates of the eastern Sahara. J Afr Earth Sci 30: 929–939.
8. Drake NA, El-Hawat AS, Turner P, Armitage SJ, Salem MJ, et al. (2008) Palaeohydrology of the Fazzan Basin and surrounding regions: the last 7 million years. Palaeogeogr Palaeoclimatol Palaeoecol 263: 131–145.
9. Smith JR (2012) Spatial and temporal variations in the nature of Pleistocene pluvial phase environments across North Africa. In: Hublin JJ, McPherron SP, eds. Modern Origins: A North African Perspective. Springer. 35–47.
10. Drake NA, Breeze P, Parker A (2013) Palaeoclimate in the Saharan and Arabian Deserts during the Middle Palaeolithic and the potential for hominin dispersals. Quat Int 300: 48–61.
11. Petit-Maire N (1986) Palaeoclimates in the Sahara of Mali. A multidisciplinary study. Episodes 9: 7–16.
12. Szabo BJ, Haynes Jr CV, Maxwell TA (1995) Ages of Quaternary pluvial episodes determined by uranium-series and radiocarbon dating of lacustrine deposits of the eastern Sahara. Palaeogeogr Palaeoclimatol Palaeoecol 113: 227–242.
13. Smith JR, Giegengack R, Schwarcz HP, McDonald MMA, Kleindienst MR, et al. (2004) A reconstruction of Quaternary pluvial environments and human occupations using stratigraphy and geochronology of fossil-spring tufas, Kharga Oasis, Egypt. Geoarchaeology 19: 407–439.
14. Smith JR, Hawkins AL, Asmerom Y, Polyac V, Giegengack R (2007) New age constraints on the Middle Stone Age occupations of Kharga Oasis, Western Desert, Egypt. J Hum Evol 52: 690–701.
15. Geyh MA, Thiedig F (2008) The Middle Pleistocene Al Mahrúqah Formation in the Murzuq Basin, northern Sahara, Libya: evidence for orbitally-forced humid episodes during the last 500,000 years. Palaeogeogr Palaeoclimatol Palaeoecol 257: 1–21.
16. Kieniewicz JM, Smith JR (2009) Paleoenvironmental reconstruction and water balance of a mid-Pleistocene pluvial lake, Dakhleh Oasis, Egypt. Geol Soc Am Bull 121: 1154–1171.
17. Armitage SJ, Drake NA, Stokes S, El-Hawat A, Salem MJ, et al. (2007) Multiple phases of North African humidity recorded in lacustrine sediments from the Fazzan Basin, Libyan Sahara. Quat Geochron 2: 181–186.
18. Causse C, Conrad G, Fontes J, Gasse F, Gibert E, et al. (1988) The last Pleistocene humid phase in the Western Sahara, dated to about 80–100 ka. C R Acad Sci Paris 306: 1459–1464.
19. Wendorf F, Schild R, Close AE, Schwarcz HP, Miller GH, et al. (1994) A chronology for the middle and late Pleistocene wet episodes in the Eastern Sahara. In: Bar-Yosef O, Kra R, eds. Egypt During the Last Interglacial: the Middle Paleolithic of Bir Tarwafi and Bir Sahara East. New York: Plenum Press. 147–168.
20. Petit Maire N, Reyss JL, Fabre J (1994) A last interglacial palaeolake in the hyperarid Sahara of Mali, 23°N. C R Acad Sci Paris 319: 805–809.
21. Crombie MK, Arvidson RE, Sturchio NC, Alfy ZE, Zeid KA (1997) Age and isotopic constraints on Pleistocene pluvial episodes in the Western Desert, Egypt. Palaeogeogr Palaeoclimatol Palaeoecol 130: 337–355.
22. Sultan M, Sturchio N, Hassan FA, Hamdan MAR, Mahmood AM, et al. (1997) Precipitation source inferred from stable isotopic composition of Pleistocene groundwater and carbonate deposits in the Western Desert of Egypt. Quat Res 48: 29–37.
23. Osmond JK, Dabous AA (2004) Timing and intensity of groundwater movement during Egyptian Sahara pluvial periods by U-series analysis of secondary U in ores and carbonates. Quat Res 61: 85–94.
24. Kleindienst MR, Schwarcz HP, Nicoll K, Churcher CS, Frizano J, et al. (2008) Water in the desert: first report on uranium-series dating of Caton-Thompson's and Gardner's "classic" Pleistocene sequence at Refuf Pass, Kharga Oasis. In: Wiseman MF, ed. Oasis Papers II: Proceedings of the Second Dakhleh Oasis Project Research Seminar. Oxford: Oxbow Books. 25–54.
25. Blackwell BAB, Skinner AR, Mashriqi F, Deely AE, Long RA, et al. (2012) Challenges in constraining pluvial events and hominin activity: examples of ESR dating molluscs from the Western Desert, Egypt. Quat Geochron 10: 430–435.
26. Brookes I (1993) Geomorphology and Quaternary geology of the Dakhla Oasis region, Egypt. Quat Sci Rev 12: 529–552.
27. Brookes I (2010) Spatially variable sedimentary responses to orbitally driven pluvial climate during Marine Oxygen Isotope Stage 5.1, Dakhla Oasis, Egypt. Quat Res 74: 252–264.
28. Hamdam MAE (2000) Quaternary travertines of Wadis Abu Had-Dib Area, eastern desert, Egypt: palaeoenvironment through field, sedimentology, age and isotopic study. Sedimentology of Egypt 8: 49–62.
29. Lézine A, Casanova J (1991) Correlated oceanic and continental records demonstrate past climate and hydrology of North Africa (0–140 ka). Geology 19: 307–310.
30. Ouda B, Zouari K, Ben Ouezdou H, Chkir N, Causse C (1998) Nouvelles donnees paleoenvironnementales pour le Quaternaire recent en Tunisie centrale (basin de Maknassy). C R Acad Sci Paris 326: 855–861.
31. Wendorf F, Schild R, Close AE (1993) Summary and conclusions. In: Wendorf F, Schild R, Close AE, eds. Egypt During the Last Interglacial, the Middle Paleolithic of Bir Tarwafi and Bir Sahara East. New York: Plenum Press. 552–573.
32. Osinski GR, Schwarcz HP, Smith JR, Kleindienst MR, Haldemann AFC, et al. (2007) Evidence for a ~200-100 ka meteorite impact in the Western Desert of Egypt. Earth Planet Sci Lett 253: 378–388.
33. Ritchie JC (1994) Holocene pollen spectra from Oyo, northwestern Sudan: problems of interpretation in an hyper-arid environment. Holocene 4: 9–15.
34. Mees F, Nijs R, Dumont H (1991) Holocene evolution of the crater lake at Malha, northwestern Sudan. J Paleolimnol 5: 227–253.

35. Hoelzmann P, Kruse HJ, Rottinger F (2000) Precipitation estimates for the eastern Saharan palaeomonsoon based on a water balance model of the West Nubian Palaeolake Basin. Global Planet Change 26: 105–120.
36. Kutzbach JE (1980) Estimates of past climate at Paleolake Chad, North Africa, based on a hydrological and energy-balance model. Quat Int 14: 210–223.
37. Kröpelin S, Verschuren D, Lézine AM, Eggermont H, Cocquyt C, et al. (2008) Climate-driven ecosystem succession in the Sahara: the past 6000 years. Science 320: 765–768.
38. Krinner G, Lézine AM, Braconnot P, Sepulchre P, Ramstein G, et al. (2012) A reassessment of lake and wetland feedbacks on the North African Holocene climate. Geophys Res Lett 39: L07701, doi:10.1029/2012GL050992.
39. Prell WL, Kutzbach JE (1987) Monsoon variability over the past 150,000 years. J Geophys Res 92: 8411–8425.
40. Kowalski K, Bochenski Z, Mlynarski M, Rzebik-Kowalska B, Szyndlar Z, et al. (1989) A last interglacial fauna from the eastern Sahara. Quat Res 32: 335–341.
41. Kuper R, Kröpelin S (2006) Climate-controlled Holocene occupation in the Sahara: motors of Africa's evolution. Science 313: 803–807.
42. Nicoll K (2004) Recent environmental change and prehistoric human activity in Egypt and northern Sudan. Quat Sci Rev 23: 561–580.
43. Sereno PC, Garcea EAA, Jousse H, Stojanowski CM, Saliège JF, et al. (2008) Lakeside cemeteries in the Sahara: 5000 years of Holocene population and environmental change. PLoS ONE 3: e2995.
44. Osborne AH, Vance D, Rohling EJ, Barton N, Rogerson M, et al. (2008) A humid corridor across the Sahara for the migration of early modern humans out of Africa 120,000 years ago. Proc Natl Acad Sci USA 105: 16444–16447.
45. Drake NA, Blench RM, Armitage SJ, Bristow CS, White KH (2011) Ancient watercourses and biogeography of the Sahara explain the peopling of the desert. Proc Natl Acad Sci USA 108: 458–462.
46. McHugh PM, Breed CS, Schaber GS, McCauley JF, Szabo BJ (1988) Acheulian sites along the "radar rivers", southern Egyptian Sahara. J Field Archaeol 15: 361–379.
47. Whiting-Blome M, Cohen AS, Tyron CA, Brooks A, Russell J (2012) The environmental context for the origins of modern human diversity: a synthesis of regional variability in African climate 150,000–30,000 years ago. J Hum Evol 62: 563–592.
48. Lebatard AE, Bourlès DL, Braucher R, Arnold M, Duringer P, et al. (2010) Application of the authigenic $^{10}Be/^{9}Be$ dating method to continental sediments: reconstruction of the Mio-Pleistocene sedimentary sequence in the early hominid fossiliferous areas of the northern Chad Basin. Earth Planet Sci Lett 297: 57–70.
49. Lahr MM (2010) Saharan corridors and their role in the evolutionary geography of "Out of Africa I". In: Fleagle JG, et al. eds. Springer. 27–46.
50. Carrión JS, Rose J, Stringer C (2011) Early human evolution in the western Palaearctic: ecological scenarios. Quat Sci Rev 30: 1281–1295.
51. Cerling TE, Wynn JG, Andanje SA, Bird MI, Kimutai Korir D, et al. (2011) Woody cover and hominin environments in the past 6 million years. Nature 476: 51–56.
52. Wood B, Strait D (2004) Patterns of resource use in early *Homo* and *Paranthropus*. J Hum Evol 46: 119–162.
53. Elton S (2008) The environmental context of human evolutionary history in Eurasia and Africa. J Anat 212: 377–393.
54. Ashley GM, Barboni D, Dominguez-Rodrigo M, Bunn HT, Mabulla AZP, et al. (2010) Paleoenvironmental and paleoecological reconstruction of a freshwater oasis in savannah grassland at FLK North, Olduvai Gorge, Tanzania. Quat Int 74: 333–343.
55. Dennell R, Roebroeks W (2005) An Asian perspective on early human dispersal from Africa. Nature 438: 1099–1104.
56. Good SG, Caylor KK (2011) Climatological determinants of woody cover in Africa. Proc Natl Acad Sci USA 108: 4902–4907.
57. Van der Made J (2011) Biogeography and climatic change as a context to human dispersal out of Africa and within Eurasia. Quat Sci Rev 30: 1353–1367.
58. Robson SL, Wood B (2008) Hominin life history: reconstruction and evolution. J Anat 212: 394–425.
59. Gasse F, Chalié F, Vincens A, Williams MAJ, Williamson D (2008) Climatic patterns in equatorial and southern Africa from 30,000 to 10,000 years ago reconstructed from terrestrial and near-shore proxy data. Quat Sci Rev 27: 2316–2340.
60. White F (1983) The Vegetation of Africa. Paris: UNESCO. 356 p.
61. Weldeab S, Lea DW, Schneider RR, Andersen N (2007) 155,000 years of West African monsoon and thermal ocean evolution. Science 316: 1303–1307.
62. Tjallingii R, Claussen M, Stuut JBW, Fohlmeister J, Jahn A, et al. (2008) Coherent high- and low-latitude control of the northwest African hydrological balance. Nat Geoscience 1: 670–675.
63. Emeis KC, Sakamoto T, Wehausen R, Brumsack HJ (2000) The sapropel record of the eastern Mediterranean Sea - results of Ocean Drilling Program Leg 160. Palaeogeogr Palaeoclimatol Palaeoecol 158: 371–395.
64. Rohling EJ, Mayewski PA, Abu-Zied RH, Casford JSL, Hayes A (2002) Holocene atmosphere-ocean interactions: records from Greenland and the Aegean Sea. Clim Dyn 18: 587–593.
65. Grant KM, Rohling EJ, Bar-Matthews MB, Ayalon A, Medina-Elizalde M, et al. (2012) Rapid coupling between ice volume and polar temperature over the past 150,000 years. Nature 491: 744–747.
66. Larrasoaña JC, Roberts AP, Rohling EJ, Winklhofer M, Wehausen R (2003) Three million years of monsoon variability over the northern Sahara. Clim Dyn 21: 689–698.
67. Ratnam J, Bond WJ, Fensham RJ, Hoffmann WA, Archibald S, et al. (2011) When is a "forest" a savannah, and why does it matter? Global Ecol Biogeogr 20: 653–660.
68. Feibel CS (2011) Shades of the savannah. Nature 476: 39–40.
69. McBrearty S, Jablonski NG (2005) First fossil chimpanzee. Nature 437: 105–108.
70. Lebamba J, Vincens A, Maley J (2010) Pollen, biomes, forest successions and climate at Lake Barombi Mbo (Cameroon) during the last ca. 33000 cal yr BP – a numerical approach. Clim Past Discuss 6: 2703–2740.
71. Bergner AGN, Trauth MH, Bookhagen B (2003) Paleoprecipitation estimates for the Lake Naivasha basin (Kenya) during the last 175 k.y. using a lake-balance model. Global Planet Change 36: 117–136.
72. Castañeda IS, Mulitza S, Schefuβ E, Lopes dos Santos R, Sinninghe Damsté JS, et al. (2009) Wet phases in the Sahara/Sahel region and human dispersal patterns in North Africa. Proc Natl Acad Sci USA 106: 20159–20163.
73. Rossignol-Strick M (1983) African monsoons, an immediate climate response to orbital insolation. Nature 304: 46–49.
74. Caley T, Malaizé B, Revel M, Ducassou E, Wainer K, et al. (2011) Orbital timing of the Indian, East Asian and African boreal monsoons and the concept of a "global monsoon". Quat Sci Rev 30: 3705–3715.
75. Rohling EJ, Cane TR, Cooke S, Sprovieri M, Bouloubassi I, et al. (2002) African monsoon variability during the previous interglacial maximum. Earth Planet Sci Lett 202: 61–75.
76. Emeis KC, Schulz H, Struck U, Rossignol-Strick M, Erlenkeuser H, et al. (2003) Eastern Mediterranean surface water temperatures and $\delta^{18}O$ composition during deposition of sapropels in the late Quaternary. Paleoceanography 18: 1005, doi:10.1029/2000PA000617.
77. Rohling EJ, Sprovieri M, Cane TR, Casford JSL, Cooke S, et al. (2004) Reconstructing past planktic foraminiferal habitats using stable isotope data: a case history for Mediterranean sapropel S5. Mar Micropal 50: 89–123.
78. Griffin DL (2002). Aridity and humidity: two aspects of the late Miocene climate of North Africa and the Mediterranean. Palaeogeogr Palaeoclimatol Palaeoecol 182: 65–91.
79. Issawi B, McCauley JF (1992) The Cenozoic rivers of Egypt: the Nile problem. In: Adams B, Friedman R, eds. The Followers of Horus. Oxford: Oxbow Press. 1–18.
80. Paillou P, Tooth S, Lopez S (2012) The Kufrah paleodrainage system in Libya: a past connection to the Mediterranean Sea? C R Geoscience 344: 406–414.
81. Zhao Y, Colin C, Liu Z, Paterne M, Siani G, et al. (2012) Reconstructing precipitation changes in northeastern Africa during the Quaternary by clay mineralogical and geochemical investigations of Nile deep-sea fan sediments. Quat Sci Rev 57: 58–70.
82. Wehausen R, Brumsack HJ (2000) Chemical cycles in Pliocene sapropel-bearing and sapropel-barren eastern Mediterranean sediments. Palaeogeogr Palaeoclimatol Palaeoecol 158: 325–352.
83. Schenau SJ, Antonarakou A, Hilgen FJ, Lourens LJ, Nijenhuis IA, et al. (1999) Organic-rich layers in the Metochia section (Gavdos, Greece): evidence for a single mechanism of sapropel formation during the past 10 My. Mar Geol 153: 117–135.
84. Macgregor DS (2012) The development of the Nile drainage system: integration of onshore and offshore evidence. Petrol Geosci 18: 417–341.
85. Hilgen FJ, Krijgsman W, Langereis CG, Lourens JL, Santarelli A, et al. (1995) Extending the astronomical (polarity) time scale into the Miocene. Earth Planet Sci Lett 136: 495–510.
86. Hüsing SK, Kuiper KF, Link W, Hilgen FJ, Krijgsman W (2009) The upper Tortonian-lower Messinian at Monte dei Corvi (Northern Apennines, Italy): completing a Mediterranean reference section for the Tortonian Stage. Earth Planet Sci Lett 282: 140–157.
87. Van der Laan E, Hilgen FJ, Lourens JL, de Kaenel E, Gaboardi S, et al. (2012) Astronomical forcing of Northwest African climate and glacial history during the late Messinian. Palaeogeogr Palaeoclimatol Palaeoecol 313–314: 107–126.
88. Maslin MA, Trauth MH (2009) Plio-Pleistocene East African pulsed climate variability and its influence on early human evolution. In. Grine F, Fleagle JG, Leakey RE, eds. The First Humans – Origin and Early Evolution of the Genus *Homo*. Springer. 151–158.
89. Trauth MH, Larrasoaña JC, Mudelsee M (2009) Trends, rhythms and events in Plio-Pleistocene African climate. Quat Sci Rev 28: 399–411.
90. Magill CR, Ashley GM, Freeman KH (2013) Water, plants and early human habitats in eastern Africa. Proc Natl Acad Sci USA 105: 1175–1180.
91. Brown FH, McDougall I, Fleagle JG (2012) Correlation of the KHS Tuff of the Kibish Formation to volcanic ash layers at other sites, and the age of early *Homo sapiens* (Omo I and Omo II). J Hum Evol 63: 577–585.
92. Joordens JCA, Vonhof HB, Feibel CS, Lourens JL, Dupont-Nivet G, et al. (2011) An astronomically-tuned climate framework for hominins in the Turkana Basin. Earth Planet Sci Lett 307: 1–18.
93. Arz HW, Lamy F, Pätzold J, Müller PJ, Prins M (2003) Mediterranean moisture source for an Early-Holocene humid period in the northern Red Sea. Science 300: 118–121.
94. Hooghiemstra H, Lézine AM, Leroy SAG, Dupont L, Marret F (2006) Late Quaternary palynology in marine sediments: a synthesis of the understanding of pollen distribution patterns in the NW African setting. Quat Int 148: 29–44.

95. Ndiaye A, Bâ K, AniskinV, Benazzou T, Chevret P, et al. (2012) Evolutionary systematics and biogeography of endemic gerbils (Rodentia, Muridae) from Morocco: an integrative approach. Zool Scrip 41: 11–28.
96. Ségalen L, Lee-Thorp JA, Cerling TE (2007) Timing of C_4 grass expansion across sub-Saharan Africa. J Hum Evol 53: 549–559.
97. Bailey GN, Reynolds SC, King GCP (2011) Landscapes of human evolution: models and methods of tectonic geomorphology and the reconstruction of hominin landscapes. J Hum Evol 60: 257–280.
98. Negri A, Morigi C, Giunta S (2003) Are productivity and stratification important to sapropel deposition? Microfossil evidence from late Pliocene insolation cycle 180 at Vrica, Calabria. Palaeogeogr Palaeoclimatol Palaeoecol 190: 243–255.
99. Geraads D (2010) Biogeographic relationships of Pliocene and Pleistocene North-western African mammals. Quat Int 212: 159–168.
100. O'Regan HJ, Turner A, Bishop LC, Elton S, Lamb AL (2011) Hominins without fellow travellers? First appearance and inferred dispersals of Afro-Eurasian large-mammals in the Plio-Pleistocene. Quat Sci Rev 30: 1343–1352.
101. Templeton AR (2005) Haplotype trees and modern human origins. Yearb Phys Anthropol 48: 33–59.
102. Fernandes CA, Rohling EJ, Siddall M (2006) Absence of post-Miocene Red Sea land bridges. J Biogeogr 33: 961–966.
103. Derricourt R (2005) Getting "Out of Africa": sea crossings, land crossings and culture in the Hominin migrations. J World Prehist 19: 119–132.
104. Agustí J, Lordkipanidze D (2011) How "African" was the early human dispersal out of Africa? Quat Sci Rev 30: 1338–1342.
105. Groucutt HS, Petraglia MD (2012) The prehistory of the Arabian Peninsula: deserts, dispersals, and demography. Evol Anthropol 21: 113–125.
106. Vaks A, Bar-Matthews M, Ayalon A, Matthews A, Halicz L, et al. (2007) Desert speleothems reveal climatic window for African exodus of modern humans. Geology 35: 831–834.
107. Laskar J, Gastineau M, Joutel F, Robutel P, Levrard B et al. (2004) A long-term numerical solution for the insolation quantities of the Earth. Astron Astrophys 428: 261–285.

7

Local Ecological Knowledge and Scientific Data Reveal Overexploitation by Multigear Artisanal Fisheries in the Southwestern Atlantic

Mariana G. Bender[1], Gustavo R. Machado[2], Paulo José de Azevedo Silva[3], Sergio R. Floeter[1], Cassiano Monteiro-Netto[2], Osmar J. Luiz[4], Carlos E. L. Ferreira[2]*

1 Departamento de Ecologia e Zoologia, Universidade Federal de Santa Catarina, Florianópolis, SC, Brazil, **2** Departamento de Biologia Marinha, Universidade Federal Fluminense, Niteroi, RJ, Brazil, **3** Fundação Instituto de Pesca, Arraial do Cabo, RJ, Brazil, **4** Department of Biological Sciences, Macquarie University, Sydney, NSW, Australia

Abstract

In the last decades, a number of studies based on historical records revealed the diversity loss in the oceans and human-induced changes to marine ecosystems. These studies have improved our understanding of the human impacts in the oceans. They also drew attention to the shifting baseline syndrome and the importance of assessing appropriate sources of data in order to build the most reliable environmental baseline. Here we amassed information from artisanal fishermen's local ecological knowledge, fisheries landing data and underwater visual census to assess the decline of fish species in Southeastern Brazil. Interviews with 214 fishermen from line, beach seine and spearfishing revealed a sharp decline in abundance of the bluefish *Pomatomus saltatrix*, the groupers *Epinephelus marginatus, Mycteroperca acutirostris, M. bonaci* and *M. microlepis*, and large parrotfishes in the past six decades. Fisheries landing data from a 16-year period support the decline of bluefish as pointed by fishermen's local knowledge, while underwater visual census campaigns show reductions in groupers' abundance and a sharp population decline of the Brazilian endemic parrotfish *Scarus trispinosus*. Despite the marked decline of these fisheries, younger and less experienced fishermen recognized fewer species as overexploited and fishing sites as depleted than older and more experienced fishermen, indicating the occurrence of the shifting baseline syndrome. Here we show both the decline of multigear fisheries catches – combining anecdotal and scientific data – as well as changes in environmental perceptions over generations of fishermen. Managing ocean resources requires looking into the past, and into traditional knowledge, bringing historical baselines to the present and improving public awareness.

Editor: Konstantinos I. Stergiou, Aristotle University of Thessaloniki, Greece

Funding: Conselho Nacional de Desenvolvimento Científico e Tecnológico (CNPq - www.cnpq.org) provided individual grants to CELF, SRF and CMN. CNPq, Fundação do Amparo à Pesquisa do Estado do Rio de Janeiro (FAPERJ - www.faperj.br) and EcoHub (www.ecohub.com.br) supported field sampling. The funders had no role in study design, data collection and analysis, decision to publish, or preparation of the manuscript.

Competing Interests: The authors have declared that no competing interests exist.

* Email: carlosferreira@id.uff.br

Introduction

Fishing is the most ancient form of exploitation of coastal resources, preceding all other human disturbances to marine ecosystems, such as pollution, eutrophication, habitat loss, disease outbreaks, human induced climate change and species invasions [1–4]. Fishing has caused the worldwide depletion of large predatory fishes, including shark populations [5,6], and driven several species across different ecosystems to ecological extinction [1]. Moreover, fishing is reported to impact fish populations since prehistoric times [2,7–10], affecting species diversity and size [11,12]. Today, fishing remains as the major source of impact upon marine and coastal environments, contributing to global biodiversity loss [13–15]. The continuity of unsustainable historical fishing has left few truly pristine ecosystems in the marine realm [1,5]. In the Southwestern Atlantic, patterns of marine resources' overexploitation are similar to what have been reported worldwide [5,14,16], with widespread population declines and collapsed stocks [17–19]. It has been estimated that 23% of all Brazilian marine fish stocks are fully exploited and 33% are overexploited, including species from low trophic levels [20–22]. Overfishing has been changing the density and the size structure of reef fish top predators occurring in Brazil [17,23], consequently elevating their threatened status [24,25].

Environmental changes and lack of baselines for pristine marine ecosystems have profound implications in our perceptions of what is a natural environment [4,26]. The compromised ability of people in perceiving environmental modifications and past ecological conditions is called the shifting baseline syndrome (SBS) [27]. This phenomenon was first noted among fisheries scientists, who perceived as a 'natural' baseline for stock size and composition the condition they observed at the start of their careers, while failing to incorporate past and historical data [27–29]. As generations change, environmental baselines become increasingly shifted, misinforming fisheries management. In order to adjust shifted environmental perceptions, old travel diaries, naturalists' observations, historical data and fishermen' anecdotes are important sources of information [28–32], as well archaeological and paleontological data [1,4]. Baselines constructed from historical data are critical to better gauge and interpret long-term changes and to set appropriate targets for management and restoration [1,4,33].

The Local Ecological Knowledge (LEK) held by traditional fishing communities is considered an important tool for the assessment and management of tropical fisheries [29,30,37]. Local Ecological Knowledge is a set of perceptions and experiences of traditional communities regarding its surrounding natural environment [34,35], this knowledge being handed-down through generations by cultural transmission [36]. This knowledge includes perceptions of fishing resources – fisheries' composition and abundance patterns, as well as fish species biological and ecological aspects [38,39] – all of which are important to conservation and management strategies [40]. Because humans have modified natural systems, local perceptions of the status of species and ecosystem resources are unlikely to remain constant over time [41]. Thus, the SBS could influence the validity of LEK derived data, whose use for species and ecosystems assessments is becoming more common [42,43]. However, specific conditions are required for SBS to occur, which include a combination of environmental events and observer perceptions [44]. Here we combine multiple sources of information to understand the status of local fisheries of Arraial do Cabo, Rio de Janeiro state, Southeastern Brazil (Fig. 1). Specifically, we aim to (i) investigate the occurrence of SBS among fishermen communities of Arraial do Cabo; (ii) utilize LEK to obtain past estimates of abundance and population trends for exploited fish species; and (iii) demonstrate the importance of anecdotal data to adjust perceptions regarding marine resources in the subtropical coast of Brazil.

Methods

Study site

Arraial do Cabo (Rio de Janeiro state, Brazil) constitutes a traditional fishing village with approximately 20,000 residents and 1,340 active fishermen. The area was declared a Marine Extractive Reserve (MER) in 1997 – the first of its kind in Brazil – a collaboratively managed marine protected area where only local fishermen are allowed to exploit resources. In Arraial do Cabo, while there are some fishing regulations, enforcement is poor or non-existent, resulting in a typical overexploitation scenario [17]. Among fishing gears, hook and line, gillnet, beach seine and spearfishing are the most used. The beach seine fishery is the most traditional practice, using large canoes and seine nets that encircle passing schools of bluefish (*Pomatomus saltatrix*), mullets (Mugilidae), jacks and trevallies (Carangidae), and tunas (Scombridae). Spearfishing commenced *c.* 50 years ago in Brazil, being today widespread along the coast. The region is formed by an isthmus and two islands dominated by rocky shores and sand beaches. Coastal morphology associated to prevailing winds trigger small-scale upwelling events, favoring a rich marine environment (Fig. 1).

Data collection

Ethical considerations. The Marine Biology Graduate Program at Universidade Federal Fluminense (PPGBM-UFF), Niterói, Rio de Janeiro state, Brazil, approved the data collection of our study that comprised interviews with local fishermen communities. Prior to each interview, fishermen were informed on research purpose and we obtained verbal consent from participants. This methodology was chosen because some interviewees would lack reading and writing skills. Once the participant agreed on being interviewed, we recorded participant consent by writing the participant's name and interview date on questionnaire sheets. This procedure was considered adequate by the Graduate Program review board. We also informed participants that all information provided in interviews would be anonymized.

Interviews. To assess fishermen perceptions regarding the status of overexploited fish species, we conducted individual interviews with 214 fishermen – from August 2007 to August 2008 – in Arraial do Cabo, Southeastern Brazilian coast (Fig.1). Interviews with fishermen included three different gears: hook and line, beach seine and spearfishing. As line and beach seine are traditional fishing gears in the region and have been practiced for centuries, this enabled a large number of experienced fishermen to be surveyed. Making possible the unveiling of a rich ecological

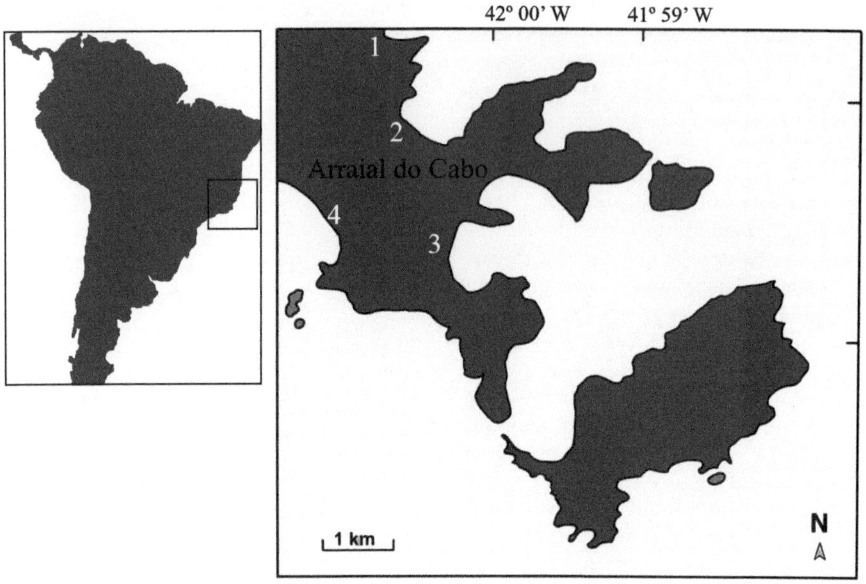

Figure 1. Locations of interview sites at Arraial do Cabo.

knowledge related to these gears and a robust historical baseline for local fisheries. Respondents were categorized into three groups according to fishing experience. For line fishing and beach seine fishing the categories were: beginners (<16 years of practice, n = 45), intermediate (16–30 years of practice, n = 50) and experienced (>30 years of practice, n = 43). However, since spearfishing is a recent fishing practice when compared to beach seine and line fishing, experience categories for spearfishermen were adapted to a smaller period: beginners (<8 years of practice, n = 25), intermediate (8–15 years of practice, n = 28) and experienced (>15 years of practice, n = 23).

The survey started a few weeks following arrival to the fishing village. Fishermen were randomly approached at the beach – as they were encountered either going fishing or making repairs on nets and fishing gears – and inquired on the availability to participate on the research. Following this approach, fishermen in different groups (experience category per fishing gear) were selected through snowball sampling [37,45], where one or more peers indicate the respondent. This method was applied so that we could achieve the target sample size in each experience category and fishing gear (line, beach seine and spearfishing). This method also led us to interview some elder and retired informants at their homes. Respondents were always interviewed separately in order to avoid interfering from others. Our questionnaire constitued of structured and open-ended questions [29] (see File S1 for questionnaire).

The bluefish, *Pomatomus saltatrix* was chosen among various species based on its importance as a resource in the region [45,47]. We asked fishermen about the best bluefish catch they ever remembered landing and the year in which this catch was made [47]. For the spearfishing practice interviews, we selected four Epinephelidae (groupers) species [48] that are important targets in the region (*Mycteroperca acutirostris*, *M. bonaci*, *M. microlepis* and *Epinephelus marginatus*) [17]. To explore the possible fishing impacts at lower trophic levels, we included in questionnaires four endemic large herbivorous-detritivorous species of the subfamily Scarinae (parrotfishes), namely, *Scarus trispinosus, Sparisoma axillare, Sparisoma frondosum* and *Sparisoma amplum.* These fishes became targets for spearfishing in the last 20 years [17,20]. For each species we asked divers about the best catch they remembered spearing (number of fishes/individuals), the largest individual they ever caught (in kilograms), and the year in which these catches were made [29]. When fishermen were asked about the largest fishes they had ever caught, the majority of respondents provided information relative to species weight (in kilograms). However, when fishermen referred to fish body length, species length–weight relationships were used to convert to weight [49]. Finally, we assessed shifts in environmental perceptions among generations of fishermen. We inquired each respondent to cite the number of overexploited species and sites in the region, as well as the number of species that were formerly discarded but are now fishing targets. Given the richness of common names for Brazilian reef fish [46], we used photographs to clarify species identifications during the interviews [37].

Bluefish landing data. We have included fisheries landing data in our analysis in order to improve the overall understanding of the status of fish stocks in the region. Bluefish (*Pomatomus saltatrix*) catch data (total catch – kg and effort – fishing hours) was recorded daily and pooled over the year during 16 years (1992 to 2008) at the local landing point (Marina dos Pescadores, Porto do Forno) in Arraial do Cabo.

Reef fish abundance. The abundance of groupers (Epinephelidae) and parrotfishes (Labridae, Scarinae) was assessed through an underwater visual census (UVC) program performed in Arraial do Cabo region across different years (1992, 2012). The program is led by one of the authors (C.E.L.F.), and includes UVC and other sampling techniques applied to studies/monitoring of the reef community at Arraial do Cabo. Fish density was estimated using 20×2 m strip transects [50]. In each transect, all fishes within the sample area were identified, counted and categorized into five size classes. Data across all sampled sites were pooled together to show temporal changes for target species in Arraial do Cabo region.

Data analysis

Interviews. From the data obtained through interviews with local fishermen we measured the decline in fish species catches, as well as the differences in the set of environmental perceptions across fishermen generations. With the purpose of measuring the decline in fish species abundance and body-size patterns across the years, we applied regressions and plotted (i) the best day catch reported by each respondent, versus the year in which the catch was made; and (ii) the larger individual ever caught by the respondent (kg) versus the year in which a given species was caught. We tested distinct regression types (exponential, polynomial and linear regressions) and r^2 values were compared to assess the quality of the regression fit to data distribution. Fish catches measured for time intervals are presented in the results section as mean ± standard deviation.

To investigate the potential differences in respondents perceptions, we tested differences in the mean number of overexploited species, overexploited sites, and the mean number of target species (formerly discarded) mentioned by fishermen in different categories (beginner, intermediate, experienced) through one-way ANOVA ($\alpha = 0.05$; Student-Newman-Keuls post-hoc test) [51]. All data was tested for normality (Kolmogorov-Smirnov test) and transformed when necessary (square root or log transformed) before conducting analysis.

Bluefish landing data. In order to measure *P. saltatrix* catch trends across the years (1992 to 2008) from landing data we have applied regression analysis, plotting the catch per unit effort (tons/hour) throughout the years. Again, we tested for distinct regression types (exponential, polynomial and linear regressions) and r^2 values were compared to assess the quality of the regression fit to data distribution.

Reef fish abundance. In order to examine the population decline trends within targeted reef fish families (groupers and parrotfishes) in Arraial do Cabo we tested the Spearman's rank correlation between the mean density (of groupers and parrotfishes) per transect and year of sampling. Only the most abundant and large sized parrotfish species (*Scarus trispinosus, Sparisoma axillare, S. frondosum, S.amplum*) were used for analysis [52].

Results

Multiple sources reveal fish species declines

The pattern for bluefish catches across the years reveals significant reductions in *P. saltatrix* stocks in Arraial do Cabo region over time (Fig. 2a). Captures (mean ±standard deviation) of 1,200 kg (±186) were considered a good catch during the 1960s among line fishermen. This number decreased to an average of 970 kg (±260) in the 1980s. Today, 370 kg (±220) of bluefish is considered a great catch (Fig. 2a). Results reveal that in fifty years (1960–2010) the amount of bluefish caught on a good day catch was reduced by approximately 70%.

Experienced beach seine fishermen also reported great bluefish catches in Arraial do Cabo decades ago, and approximately 80% decline in the past six decades, a pattern similar to that identified

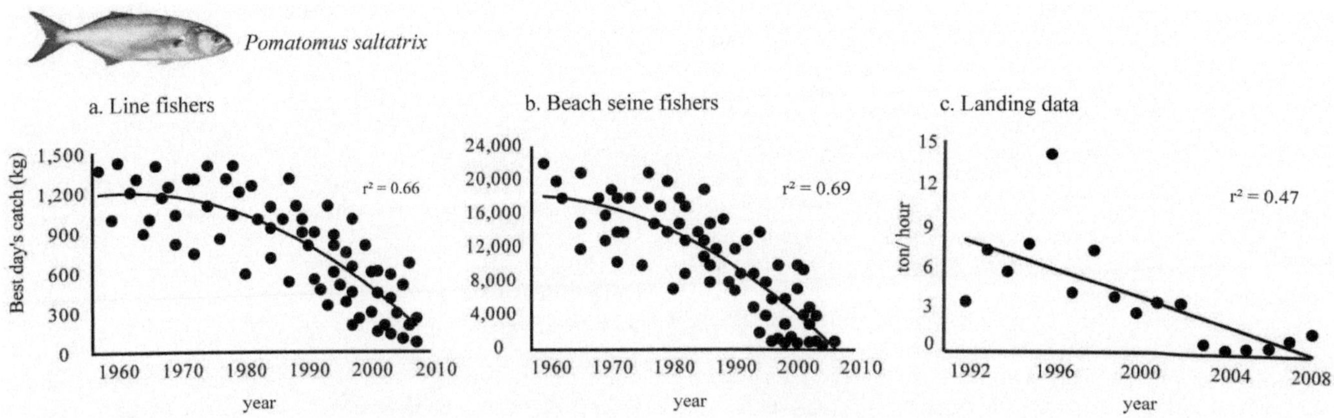

Figure 2. Bluefish *(Pomatomus saltatrix)* catches according to line fishermen (a) and beach seine fishermen (b), both are second order polynomial regressions; and catch per unit effort ($t \times h^{-1}$) (c) from 1992 to 2008 in Arraial do Cabo, Brazil.

by line fishermen (Fig. 2b). During the 1960s, large catches were in the order of 18,000 kg (±3,517) on average. In the 1980s, this number has fallen to13,570 kg (±3,753), and today *P. saltatrix* catches greater than 3,700 kg (±3,340) are rare. Moreover, landing data from *P. saltatrix* catches in the region reveal the same pattern found by accessing the knowledge of local fishermen (Fig. 2c), with a declining trend in bluefish fisheries along 16 years (1992–2008). CPUE varied between 14.2 tons/hour in 1996 and 0.3 tons/hour in 2004, with an annual average of 4 tons hour (Fig. 2c).

Fish species exploited by spearfishing also exhibit declining trends. There was a decrease in the abundance of the assessed Epinephelidae species (Fig. 3). The relationships considering the largest catch and year again indicate that large individuals of *E. marginatus*, *M. bonaci*, *M. acutirostris* and *M. microlepis* have declined in the region, and that the big ones were frequently caught not so far back, in the 1970s (Fig. 3). For the endemic parrotfishes *Sparisoma axillare*, *S. frondosum* and *S. amplum* there were no significant differences neither in the abundance nor in the size of individuals caught along the years (Fig. S1 in File S1). However, for *Scarus trispinosus*, we have identified reductions on the greatest individuals captured (kg) and on the best catches made (Fig. 3).

Data from visual census campaigns conducted in the region from 1992 to 2012 – revealed a declining trend in the local densities of *Scarus trispinosus*, in fact, individuals of the species are not recorded on transects conducted in the region since 1995 (Fig. 4). The decline in *S. trispinosus* is followed by an overall decline of Scarinae (Labridae) species. For Epinephelidae species, the situation is similar, with decreasing densities in local reef fish communities from 1997 onwards (Fig. 4).

Shifting baselines across multigear fisheries

Out of 214 interviewed fishermen, ~83% (n = 177) recognized reductions in the abundance of exploited fish species in the region of Arraial do Cabo. Less experienced fishermen are less aware of overexploitation, indicating the occurrence of the SBS among local fishermen. Significant age-related differences were evident in assessed fishing communities in terms of respondent perception regarding fisheries declines. Experienced line, beach seine and spearfishermen mentioned a greater number of overexploited species when compared to less experienced ones (Table 1).

We detected significant experience-related differences in terms of respondent perception of shifts in target species (Table 1). Not only for line and beach seine fishermen (line fishing: experienced = 6.1, intermediate = 3.1 and beginners = 1.6; beach seine: experienced = 5.5, intermediate = 3.4 and beginners = 1.4), but also for spearfishing there were differences between the mean number of species cited by each category (spearfishing: experienced = 7.4, intermediate = 5.1 and beginners = 1.8) (see Table 1). For instance, surgeonfishes (e.g. *Acanthurus* spp.) were identified as targets exclusively by beginner spearfishermen. The mean number of sites mentioned as depleted by different fishermen categories also suggests the occurrence of the SBS. The more experienced and older fishermen referred to a greater number of overexploited sites when compared to younger, less experienced fishermen. For instance, experienced line fishermen cited approximately 11 sites, while intermediate and beginners referred to 4 and 2 sites, respectively (see Table 1).

Discussion

From local to global: the impacts of changes in fishermen's perceptions

Since Daniel Pauly reported the global phenomenon that he termed 'the shifting baseline syndrome' (SBS) [27], much supporting evidence have been reported across marine [29,53], freshwater [41,54] and terrestrial ecosystems [43,55]. Fishermen gather their environmental references based on personal experiences, which may lead to perception differences among generations. In Arraial do Cabo, experienced line and beach seine fishermen often mentioned the old days when top predators such as the sharks *Carcharhinus milberti*, *Squatina argentina*, *Isurus oxyrinchus* and the shovelnose ray *Rhinobatos percellens* were abundant in their catches, being witnesses to the decline of those species. Even though most of the older beach seine fishermen (73%) recognized the shark *Carcharhinus maculipinnis* as overexploited, beginners believe that the species is naturally rare in the region. Also, contrary to older fishermen, young spearfishermen did not mention the snappers *Lutjanus analis, Lutjanus cyanopterus*, and the groupers *Mycteroperca microlepis* and *Epinephelus morio* as depleted. The decline of these fish species forced fishermen to change their target species. Those species that were not important to fisheries in the past – or were only utilized as bait – turned out to be today's main targets. This replacement process is occurring slowly and gradually over the years, making it more difficult for future generations of fishermen to perceive these changes. Experienced and older fishermen have seen the effects of

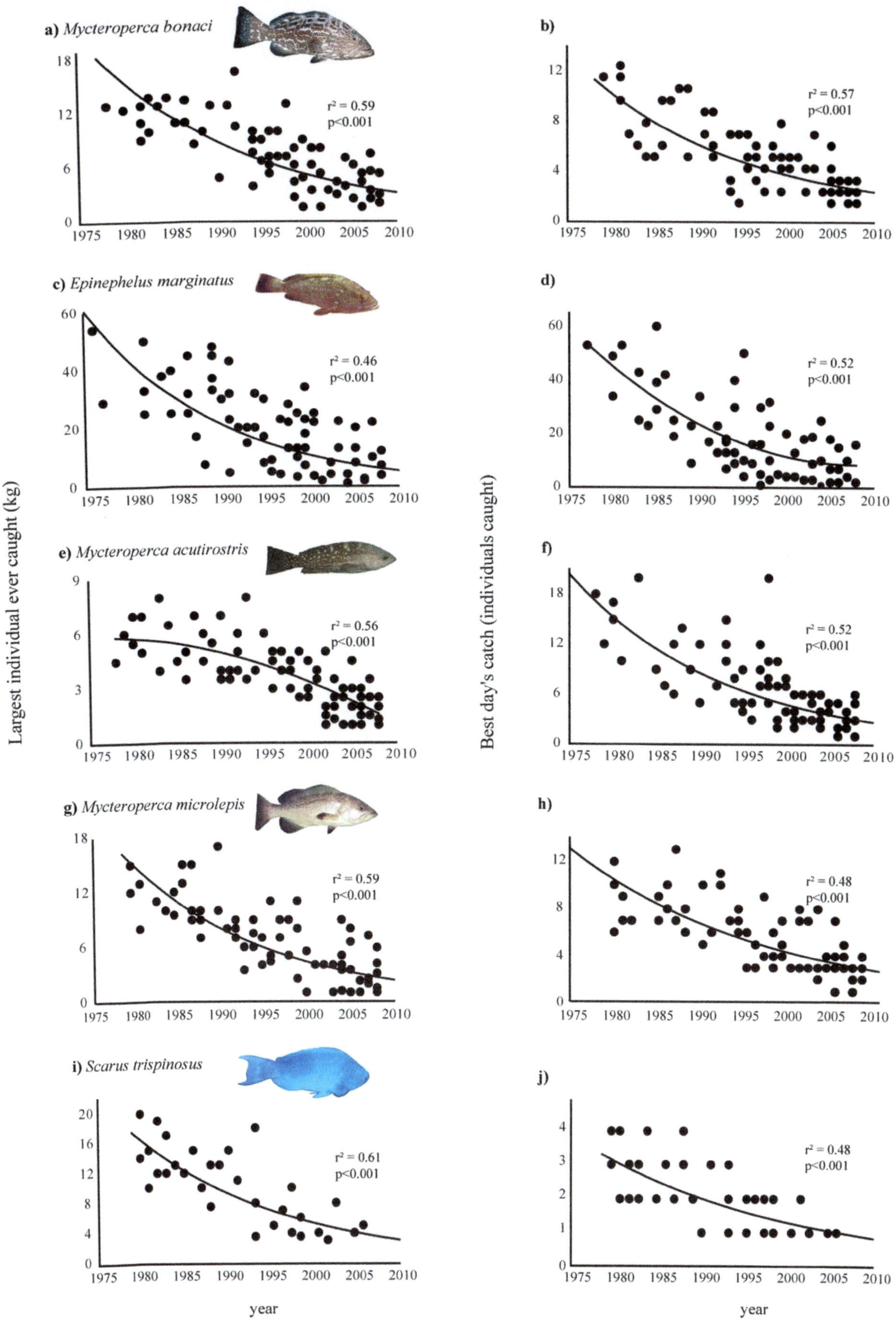
a) Mycteroperca bonaci
r² = 0.59
p<0.001
b)
r² = 0.57
p<0.001
c) Epinephelus marginatus
r² = 0.46
p<0.001
d)
r² = 0.52
p<0.001
e) Mycteroperca acutirostris
r² = 0.56
p<0.001
f)
r² = 0.52
p<0.001
g) Mycteroperca microlepis
r² = 0.59
p<0.001
h)
r² = 0.48
p<0.001
i) Scarus trispinosus
r² = 0.61
p<0.001
j)
r² = 0.48
p<0.001
Largest individual ever caught (kg)
Best day's catch (individuals caught)
year
year
1975 1980 1985 1990 1995 2000 2005 2010

Figure 3. Largest individual (in kilograms) and greatest catch (number of individuals caught) that spearfishermen remembered landing. a) and b) *Mycteroperca bonaci*, second order polynomial regression; c) and d) *Epinephelus marginatus*, exponential regression; e) and f) *Mycteroperca acutirostris*, second order polynomial regression; g) and h) *Mycteroperca microlepis*, second order polynomial regression; i) and j) *Scarus trispinosus*, exponential regression.

overexploitation upon marine ecosystems for many more years, whereas young fishermen begun fishing when stocks were either reduced or on the path for depletion. This led young fisherman to assume that sharks and large groupers were naturally rare in Arraial do Cabo, and that overexploited fishing grounds were still very productive. Thus perceptions are subjected to the effects of time and possibly compromised knowledge transmission between community members [30].

Baseline shifts regarding the past conditions of the world's oceans might lead to a collective amnesia, making people more tolerant to population declines and biodiversity losses [56]. Despite the fact that fishermen are usually in contact with nature on a daily-basis, exhibiting greater awareness of environmental changes, baselines shift rapidly even among them [29,41]. In Arraial do Cabo, sixty years of fisheries exploitation and impacts upon marine ecosystems have affected the environmental baselines of three consecutive generations. Such changes in the perceptions of fishermen regarding the status of marine ecosystems are of special concern given the importance of LEK to adjust and improve our understanding of the oceans' past conditions.

A possible short-term solution to slow the reference loss process is to recreate old conditions of marine life through historical data [4] such as films, photos [57], booklets and reports [31,50] of early explorers. Such kind of data can provide important detail on former species size and catch composition [30], potentially improving our knowledge on the effects of overfishing. Recreating past environmental baselines is a unique opportunity to assess the historical changes that ecosystems have undergone, as well as for improving social awareness on the status of biodiversity. Recent studies have pointed that underestimating empirical data is a mistake [27,29,45,56,58,59], especially in places like Arraial do Cabo, where written records on species' former size and abundance are rare.

When assessing fishing impacts, few studies make such parallel use of LEK and species population status, mostly because LEK is considered anecdotal, non-scientifically sampled, and of limited application among scientists and managers [40,60]. Our work is among the first to combine landing, underwater visual census and LEK data to assess the temporal patterns of exploited fish populations [58,61]. Combining data from different sources renders more robust and reliable assessments, which are especially important in times where decision makers require scientific certainty. More importantly, local management and conservation efforts will certainly gain more commitment from fishermen when their own experience is taken into account on the establishment of management strategies.

Fishing collapse in the Southwestern Atlantic

The combined methodologies utilized to assess multigear fisheries reveal the decline of different fish species in the region of Arraial do Cabo. The decline of the bluefish (*P. saltatrix*) fishery revealed through information provided by fishermen was supported by evidence from landing data. The declining trend for Epinephelidae species (*E. marginatus, Mycteroperca bonaci, M. acutirostris* and *M. microlepis*) in the subtropical reefs of Arraial do Cabo has been demonstrated through information provided by fishermen, and through data from underwater visual census campaigns conducted in the region. Another study for the region underscores recent declines in the abundance of groupers using the underwater visual census technique [17]. Not only the abundance of fish is diminishing, but also the larger individuals are consistently more difficult to capture and sight in the field [17].

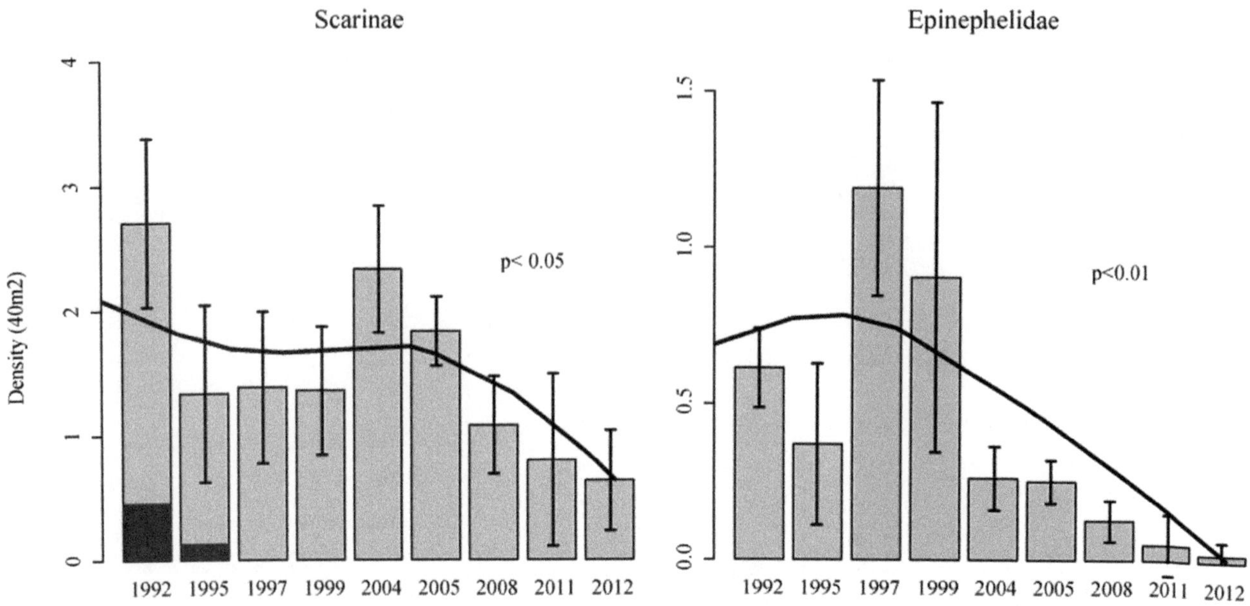

Figure 4. Mean densities per 40 m^2 of Scarinae and Epinephelidae species in Arraial do Cabo from underwater visual census. In the Scarinae plot, black portion of bars correspond to the contribution of *Scarus trispinosus* to densities.

Table 1. Mean number of overexploited fish species, new target species and depleted sites mentioned by fishermen from different fishing gears and according to experience categories.

Fishing modality	Fishers experience category (n)	Overexploited species (mean ±SD)	ANOVA*	New target species (mean ±SD)	ANOVA*	Depleted sites (mean ±SD)	ANOVA*
Line	Beginners (22)	2.0±1.0	$f=139.3$	1.6±1.1	$f=35.7$	2.0±1.0	$f=135$
	Intermediate (25)	5.5±1.8	$p<0.001$	3.1±1.5	$p<0.001$	4.5±1.9	$p<0.001$
	Experienced (21)	10.5±1.7		6.0±1.0		11.3±2.3	
Beach seine	Beginners (23)	2.3±1.5	$f=76.1$	1.4±1.5	$f=60.8$	2.4±2.1	$f=63.6$
	Intermediate (26)	6.0±2.2	$p<0.001$	3.4±1.1	$p<0.001$	5.5±2.0	$p<0.001$
	Experienced (21)	9.0±1.5		5.5 ± 1.9		10.5±3.1	
Spearfishing	Beginners (25)	1.4±0.9	$f=187.7$	1.8±0.7	$f=107.2$	1.8±1.4	$f=133.5$
	Intermediate (28)	5.8±1.5	$p<0.001$	5.1±1.7	$p<0.001$	6.3±1.7	$p<0.001$
	Experienced (23)	8.9±1.7		7.4±1.9		11.4±2.9	

*f and p-values of each ANOVA (one-way) test.
All values associated to different age categories were significantly different according to ANOVA (one-way) and Student-Newman-Keuls comparison test ($p<0.05$).

Overfishing is affecting the size of fish individuals, removing the largest ones at an unsustainable rate, and preventing the smallest ones from reaching large sizes. This can seriously compromise fish population rebound since larger and older females produce greater quantity and better quality offspring when compared to younger fish [62]. Most of these grouper species are important fishing targets elsewhere in the Brazilian coast, and the overfished state of populations has been demonstrated by different studies [17,20]. Moreover, they are highly vulnerable to overexploitation given its critical combination of biological attributes [25].

As populations of top predators collapse, other large-bodied species at lower trophic level become new targets, as is the case of the endemic greenbeak parrotfish, *Scarus trispinosus*. Being the largest of all Brazilian parrotfishes, the greenbeak was the first herbivorous-detritivorous reef fish targeted by spearfishermen, but other large-bodied parrotfishes of the genus *Sparisoma* also became targeted in the last few years. Different species of parrotfishes perform specific functional roles, some reported as critical in maintaining the resistance and resilience of reef ecosystems [63,64]. Along the Brazilian coast, populations of *S. trispinosus* have strongly declined due to spearfishing [17], and the species is considered ecologically extinct in Arraial do Cabo [50]. The species was recently listed by the IUCN [65] as Endangered (EN) yet there is no regulation or national listing to guarantee its protection [25].

Here we have shown the decline of reef fisheries in the Southwestern Atlantic from multiple sources of information: local ecological knowledge from multigear fisheries (line, beach seine and spearfishing), fisheries landing data and even underwater visual census data. The pattern identified for six different target species follows a similar pattern of global deterioration of marine ecosystems reported elsewhere [1,66]. In the Southwestern Atlantic, as elsewhere in the world, fishing has moved from upper to lower trophic levels: groupers used to be fisheries' major target, but today, parrotfishes and even surgeonfishes are targeted. If going back to the early days of the 1960's tell us a very different story on fish abundance and size in the region of Arraial do Cabo, one could only imagine what a 150 or 200-year-old baseline would inform us on those fish species. We emphasize the importance of combining LEK with scientific data in order to improve our understanding of the status of marine habitats and its associated biodiversity, but the occurrence of changes in environmental baselines must be taken into account. Only through a clear picture of the status of natural resources in the past can we help people improve their awareness of the importance of preserving natural resources and ecosystem services.

Supporting Information

File S1 File contains Questionnaire and Figure S1. Questionnaire applied in interviews with fishermen in Arraial do Cabo. Figure S1. Catch trends for *Sparisoma* (Scarinae) species reported by spearfishermen in Arraial do Cabo. Linear regressions of the largest individual caught (kg) and best day's catch (number of individuals) of the parrotfishes *Sparisoma axillare, S. amplum* and *S. frondosum* that spearfishers remembered landing (from top to bottom).

Data S1 Landing data registered in Arraial do Cabo for the bluefish *Pomatomus saltatrix*, and fish abundances (groupers and parrotfishes) recorded in underwater visual census campaigns, also in Arraial do Cabo region.

Author Contributions

Conceived and designed the experiments: CELF CMN GRM PJAS. Performed the experiments: CELF CMN GRM PJAS. Analyzed the data: CELF CMN GRM PJAS MGB OJL SRF. Contributed reagents/materials/analysis tools: CELF CMN GRM PJAS MGB OJL SRF. Wrote the paper: CELF CMN GRM PJAS MGB OJL SRF.

References

1. Jackson JBC, Kirby MX, Berger WH, Bjorndal KA, Botsford LW, et al. (2001) Historical Overfishing and the Recent Collapse of Coastal Ecosystems. Science 293: 629–637.
2. Erlandson JM, Rick TC, Vellanoweth RL (2004) Human impacts on ancient environments: a case study from California's Northern Channel Islands. Voyages of discovery: The archaeology of islands: 51–83.
3. Jackson JB (1997) Reefs since Columbus. Coral reefs 16: S23–S32.
4. Lotze HK, Worm B (2009) Historical baselines for large marine animals. Trends in ecology & evolution 24: 254–262.
5. Myers RA, Worm B (2003) Rapid worldwide depletion of predatory fish communities. Nature 423: 280–283.
6. Worm B, Davis B, Kettemer L, Ward-Paige CA, Chapman D, et al. (2013) Global catches, exploitation rates, and rebuilding options for sharks. Marine Policy 40: 194–204.
7. Barrett JH, Locker AM, Roberts CM (2004) The origins of intensive marine fishing in medieval Europe: the English evidence. Proceedings of the Royal Society of London Series B: Biological Sciences 271: 2417–2421.
8. Barrett JH, Nicholson RA, Cerón-Carrasco R (1999) Archaeo-ichthyological evidence for long-term socioeconomic trends in northern Scotland: 3500 BC to AD 1500. Journal of Archaeological Science 26: 353–388.
9. Erlandson JM, Rick TC (2010) Archaeology Meets Marine Ecology: The Antiquity of Maritime Cultures and Human Impacts on Marine Fisheries and Ecosystems. Annual Review of Marine Science 2: 231–251.
10. Smith I (2005) Retreat and resilience: fur seals and human settlement in New Zealand. The exploitation and cultural importance of sea mammals: 6–18.
11. Morales A, Rosello E, Canas J (1994) Cueva de Nerja (prov. Malaga): a close look at a twelve thousand year ichthyofaunal sequence from southern Spain [Paleolithic, Neolithic, Chalcolithic].
12. Wing S, Wing E (2001) Prehistoric fisheries in the Caribbean. Coral reefs 20: 1–8.
13. Dulvy N, Polunin NV, Mill A, Graham NA (2004) Size structural change in lightly exploited coral reef fish communities: evidence for weak indirect effects. Canadian Journal of Fisheries and Aquatic Sciences 61: 466–475.
14. Jackson JB (2008) Ecological extinction and evolution in the brave new ocean. Proceedings of the National Academy of Sciences 105: 11458–11465.
15. Roberts C, Hawkins JP, Campaign WES (2000) Fully-protected marine reserves: a guide: WWF Endangered seas campaign Washington, DC.
16. Halpern BS, Walbridge S, Selkoe KA, Kappel CV, Micheli F, et al. (2008) A global map of human impact on marine ecosystems. Science 319: 948–952.
17. Floeter S, Halpern BS, Ferreira C (2006) Effects of fishing and protection on Brazilian reef fishes. Biological Conservation 128: 391–402.
18. Frédou T (2004) The fishing activity on coral reefs and adjacent ecosystems: a case study of the Northeast of Brazil. Cybium 28: 274.
19. Rezende SdM, Ferreira BP (2004) Age, growth and mortality of dog snapper Lutjanus jocu (Bloch & Schneider, 1801) in the northeast coast of Brazil. Brazilian Journal of Oceanography 52: 107–121.
20. Ferreira C, Gonçalves J (1999) The unique Abrolhos reef formation (Brazil): need for specific management strategies. Coral reefs 18: 352–352.
21. Freire KM, Pauly D (2010) Fishing down Brazilian marine food webs, with emphasis on the east Brazil large marine ecosystem. Fisheries Research 105: 57–62.
22. Pauly D, Christensen V, Dalsgaard J, Froese R, Torres F (1998) Fishing down marine food webs. Science 279: 860–863.
23. Luiz OJ, Edwards AJ (2011) Extinction of a shark population in the Archipelago of Saint Paul's Rocks (equatorial Atlantic) inferred from the historical record. Biological Conservation 144: 2873–2881.
24. Bender M, Floeter S, Hanazaki N (2013) Do traditional fishers recognise reef fish species declines? Shifting environmental baselines in Eastern Brazil. Fisheries Management and Ecology 20: 58–67.
25. Bender M, Floeter S, Mayer F, Vila-Nova D, Longo G, et al. (2013) Biological attributes and major threats as predictors of the vulnerability of species: a case study with Brazilian reef fishes. Oryx 47: 259–265.
26. Knowlton N, Jackson JB (2008) Shifting baselines, local impacts, and global change on coral reefs. PLoS biology 6: e54.
27. Pauly D (1995) Anecdotes and the shifting baseline syndrome of fisheries. Trends in ecology & evolution 10: 430.
28. Pinnegar JK, Engelhard GH (2008) The 'shifting baseline'phenomenon: a global perspective. Reviews in Fish Biology and Fisheries 18: 1–16.
29. Sáenz-Arroyo A, Roberts C, Torre J, Cariño-Olvera M, Enríquez-Andrade R (2005) Rapidly shifting environmental baselines among fishers of the Gulf of California. Proceedings of the Royal Society B: Biological Sciences 272: 1957–1962.
30. Sáenz-Arroyo A, Roberts CM, Torre J, Cariño-Olvera M (2005) Using fishers' anecdotes, naturalists' observations and grey literature to reassess marine species at risk: the case of the Gulf grouper in the Gulf of California, Mexico. Fish and Fisheries 6: 121–133.
31. Fortibuoni T, Libralato S, Raicevich S, Giovanardi O, Solidoro C (2010) Coding early naturalists' accounts into long-term fish community changes in the Adriatic Sea (1800–2000). PloS one 5: e15502.
32. Sáenz-Arroyo A, Roberts CM, Torre J, Cariño-Olvera M, Hawkins JP (2006) The value of evidence about past abundance: marine fauna of the Gulf of California through the eyes of 16th to 19th century travellers. Fish and Fisheries 7: 128–146.
33. Roberts CM (2003) Our shifting perspectives on the oceans. Oryx 37: 166–177.
34. Drew JA (2005) Use of traditional ecological knowledge in marine conservation. Conservation Biology 19: 1286–1293.
35. Huntington HP (2000) Using traditional ecological knowledge in science: methods and applications. Ecological applications 10: 1270–1274.
36. Berkes F, Colding J, Folke C (2000) Rediscovery of traditional ecological knowledge as adaptive management. Ecological applications 10: 1251–1262.
37. Silvano RA, MacCord PF, Lima RV, Begossi A (2006) When does this fish spawn? Fishermen's local knowledge of migration and reproduction of Brazilian coastal fishes. Environmental Biology of Fishes 76: 371–386.
38. Johannes RE (1981) Working with fishermen to improve coastal tropical fisheries and resource management. Bulletin of Marine Science 31: 673–680.
39. Johannes RE, Freeman MM, Hamilton RJ (2000) Ignore fishers' knowledge and miss the boat. Fish and Fisheries 1: 257–271.
40. Shackeroff JM, Campbell LM (2007) Traditional ecological knowledge in conservation research: problems and prospects for their constructive engagement. Conservation and Society 5: 343.
41. Turvey ST, Barrett LA, Yujiang H, Lei Z, Xinqiao Z, et al. (2010) Rapidly shifting baselines in Yangtze fishing communities and local memory of extinct species. Conservation Biology 24: 778–787.
42. Jones JP, Andriamarovololona MM, Hockley N, Gibbons JM, Milner-Gulland E (2008) Testing the use of interviews as a tool for monitoring trends in the harvesting of wild species. Journal of Applied Ecology 45: 1205–1212.
43. van der Hoeven CA, de Boer WF, Prins HH (2004) Pooling local expert opinions for estimating mammal densities in tropical rainforests. Journal for nature conservation 12: 193–204.
44. Papworth SK, Rist J, Coad L, Milner-Gulland EJ (2009) Evidence for shifting baseline syndrome in conservation. Conservation Letters.
45. Silvano RAM, Begossi A (2010) What can be learned from fishers? An integrated survey of fishers' local ecological knowledge and bluefish (*Pomatomus saltatrix*) biology on the Brazilian coast. Hydrobiologia 637: 3–18.
46. Freire KMF, Carvalho-Filho A (2009) Richness of common names of Brazilian reef fishes. Pan-American Journal of Aquatic Sciences 4: 96–145.
47. Brito R (1999) Modernidade e tradição. Construção da identidade social dos pescadores de Arraial do Cabo, RJ Niterói: EDUFF.
48. Craig MT, Hastings PA (2007) A molecular phylogeny of the groupers of the subfamily Epinephelinae (Serranidae) with a revised classification of the Epinephelini. Ichthyological Research 54: 1–17.
49. Bohnsack JA, Harper DE, Center SF (1988) Length-weight relationships of selected marine reef fishes from the southeastern United States and the Caribbean: National Oceanic and Atmospheric Administration, National Marine Fisheries Service, Southeast Fisheries Center.
50. Floeter SR, Krohling W, Gasparini JL, Ferreira CE, Zalmon IR (2007) Reef fish community structure on coastal islands of the southeastern Brazil: the influence of exposure and benthic cover. Environmental Biology of Fishes 78: 147–160.
51. Underwood AJ (1997) Experiments in ecology: their logical design and interpretation using analysis of variance: Cambridge University Press.
52. Ferreira C, Gonçalves J (2006) Community structure and diet of roving herbivorous reef fishes in the Abrolhos Archipelago, south-western Atlantic. Journal of Fish Biology 69: 1533–1551.
53. Bunce M, Rodwell LD, Gibb R, Mee L (2008) Shifting baselines in fishers' perceptions of island reef fishery degradation. Ocean & Coastal Management 51: 285–302.
54. Humphries P, Winemiller KO (2009) Historical impacts on river fauna, shifting baselines, and challenges for restoration. BioScience 59: 673–684.
55. Rittenhouse CD, Pidgeon AM, Albright TP, Culbert PD, Clayton MK, et al. (2010) Conservation of forest birds: evidence of a shifting baseline in community structure. PloS one 5: e11938.
56. Ainsworth CH, Pitcher TJ, Rotinsulu C (2008) Evidence of fishery depletions and shifting cognitive baselines in Eastern Indonesia. Biological Conservation 141: 848–859.
57. McClenachan L (2009) Documenting loss of large trophy fish from the Florida Keys with historical photographs. Conserv Biol 23: 636–643.
58. O'Donnell K, Pajaro M, Vincent A (2010) How does the accuracy of fisher knowledge affect seahorse conservation status? Animal Conservation 13: 526–533.
59. Paterson B (2010) Integrating fisher knowledge and scientific assessments. Animal Conservation 13: 536–537.

60. Brook RK, McLachlan SM (2008) Trends and prospects for local knowledge in ecological and conservation research and monitoring. Biodiversity and Conservation 17: 3501–3512.
61. Boudreau SA, Worm B (2010) Top-down control of lobster in the Gulf of Maine: insights from local ecological knowledge and research surveys. Marine Ecology Progress Series 403: 181–191.
62. Palumbi SR (2003) Population genetics, demographic connectivity, and the design of marine reserves. Ecological applications 13: 146–158.
63. Hughes TP, Rodrigues MJ, Bellwood DR, Ceccarelli D, Hoegh-Guldberg O, et al. (2007) Phase shifts, herbivory, and the resilience of coral reefs to climate change. Curr Biol 17: 360–365.
64. Mumby PJ, Dahlgren CP, Harborne AR, Kappel CV, Micheli F, et al. (2006) Fishing, trophic cascades, and the process of grazing on coral reefs. Science 311: 98–101.
65. IUCN (2013) IUCN Red List of Threatened Species. Version 2013.2.
66. Worm B, Barbier EB, Beaumont N, Duffy JE, Folke C, et al. (2006) Impacts of biodiversity loss on ocean ecosystem services. Science 314: 787–790.

8

Site Distribution at the Edge of the Palaeolithic World: A Nutritional Niche Approach

Antony G. Brown[1]*, Laura S. Basell[2], Sian Robinson[3], Graham C. Burdge[4]

1 Palaeoenvironmental Laboratories (PLUS), University of Southampton, Southampton, United Kingdom, **2** School of Geography, Archaeology and Palaeoecology, Queen's University Belfast, Belfast, Northern Ireland, **3** MRC Lifecourse Epidemiology Unit, Southampton General Hospital, University of Southampton, Southampton, United Kingdom, **4** Institute of Developmental Sciences Building, Faculty of Medicine, University of Southampton, Southampton, United Kingdom

Abstract

This paper presents data from the English Channel area of Britain and Northern France on the spatial distribution of Lower to early Middle Palaeolithic pre-MIS5 interglacial sites which are used to test the contention that the pattern of the richest sites is a real archaeological distribution and not of taphonomic origin. These sites show a marked concentration in the middle-lower reaches of river valleys with most being upstream of, but close to, estimated interglacial tidal limits. A plant and animal database derived from Middle-Late Pleistocene sites in the region is used to estimate the potentially edible foods and their distribution in the typically undulating landscape of the region. This is then converted into the potential availability of macronutrients (proteins, carbohydrates, fats) and selected micronutrients. The floodplain is shown to be the optimum location in the nutritional landscape (nutriscape). In addition to both absolute and seasonal macronutrient advantages the floodplains could have provided foods rich in key micronutrients, which are linked to better health, the maintenance of fertility and minimization of infant mortality. Such places may have been seen as 'good (or healthy) places' explaining the high number of artefacts accumulated by repeated visitation over long periods of time and possible occupation. The distribution of these sites reflects the richest aquatic and wetland successional habitats along valley floors. Such locations would have provided foods rich in a wide range of nutrients, importantly including those in short supply at these latitudes. When combined with other benefits, the high nutrient diversity made these locations the optimal niche in northwest European mixed temperate woodland environments. It is argued here that the use of these nutritionally advantageous locations as nodal or central points facilitated a healthy variant of the Palaeolithic diet which permitted habitation at the edge of these hominins' range.

Editor: Michael D. Petraglia, University of Oxford, United Kingdom

Funding: The data collection and analysis from SW England was funded by English Heritage (Project PNUM 5696 URL http://www.english-heritage.org.uk/). The funders had no role in study design, data collection and analysis, decision to publish, or preparation of the manuscript.

Competing Interests: The authors have declared that no competing interests exist.

* E-mail: Tony.Brown@soton.ac.uk

Introduction

Two fundamental questions in Palaeolithic archaeology are: 1) do lithic distributions represent activity patterns? And 2) if so what can they tell us about Palaeolithic foraging strategies especially at the edge of the biogeographical range of hominins? Archaeologists accept that site distribution in later periods, both before and after the introduction of farming, can be used to infer subsistence strategies through spatial relationships with environmental factors both for mobile and sedentary societies [1]. Intuitively these relationships should be stronger for hominins prior to the innovation of food storage and durable receptacles. Such relationships should also be most marked at the edge of the geographical range of a species and the archaeological record suggests that hominins only occupied north-western Europe during cool to warm temperate conditions at least up until the last glacial cycle [2,3]. For any research into Palaeolithic site distribution in North West Europe the starting point is that over 90% of Lower and early Middle Palaeolithic finds and sites in Britain and along the north European seaboard come from river gravel deposits [4–6], with the remainder mostly coming from raised beaches, caves or occasionally lakes and sinkholes (particularly dolines in this region). It is not surprising therefore that a strong taphonomic bias has been assumed given that lithic artefacts can be regarded as discoidal clasts and that many artefacts, particularly bifaces do exhibit signs of abrasion due to transport in fluvial environments [6,7]. A second taphonomic factor is the visibility of these sites caused by the widespread practice of extracting sand and gravel (aggregate) by hand during the 19th and early 20th century. This fuelled an antiquarian appetite for lithics from gravel quarries including the payment of workers for 'hand axes' recovered, to the point to which some pits became more valuable for their artefacts than their aggregate [8]. It is very unlikely that rich sites were missed, given the large number of collectors, their particular preference for bifaces, and the large number of quarries operated principally for railway ballast and road construction. Several lines of research in geomorphology and archaeology over the last twenty years suggest that the artefact concentration particularly of bifaces at what has been termed 'super-sites' [9], is not primarily the result of taphonomic processes.

Taphonomic Factors and Site Distribution

For the English Channel River region which is geologically continuous, the first line of evidence comes from the characteristics of the site distribution and artefact density. In this study we have used bifaces (including rough-outs) rather than the entire number of lithics collected as bifaces represent one tool (rather than just knapping activity). Also because they were the most commonly collected artefact type, they should be conservative in statistical terms and the problem of between-site collector bias minimised. The frequency distribution of sites by biface number is highly skewed with most sites producing a small number of artefacts and only a very few sites producing large numbers (Fig. 1). In this study we have used data collected by Wessex Archaeology between 1993 and 1997 as part of The English Rivers Palaeolithic Survey and only that from single quarries (or gravel pits) or adjacent quarries. Although there are problems defining site size there are fewer than 25 Lower to early Middle Palaeolithic (pre-MIS 5) sites in the English Channel region that have yielded over 500 bifaces This pattern holds at a variety of spatial scales as in southwest England one site (Broom) has produced 96% of the bifaces from the region (20,000 km^2). Research focussed on this region involving the location and enumeration of all known artefacts in museum collections has allowed calculations of the relative number recovered from valley floors and slopes (Fig. 2). The result is a valley/slope ratio between 26:1 and 5:1. Moving further east along the southern coast of Britain into the reach of the palaeo-Solent catchment that falls within the county of Hampshire, only 5 out of 281 sites have yielded 79% of all Lower and Middle Palaeolithic bifaces [10]. This is confirmed by the extreme rarity of finds on slopes and plateaus in this region, even in intensively field-walked areas. This spatial bias is illustrated by the cumulative plot of artefacts against area for the South West region (Fig. 2) which is analogous to a variogram with a high 'nugget value' and low range, and conforms to a typical hotspot-type distribution [11]. A very similar spatial pattern has been found for Palaeolithic sites in the Bose Basin, Guangxi, China [12] and in pre-agricultural archaeology in the USA [13]. Since the area covered by this study has been outside the maximum area of glaciation throughout the Pleistocene, glacial erosion cannot be the cause of this strong spatial bias, and although there has been intense and prolonged periglacial activity, solifluction deposits (locally known as 'head') very rarely contain any Palaeolithic artefacts although examples are known including the enigmatic site of Knowle Farm in Wiltshire [13,14]. Another example is a biface from Boxgrove (unit 11) but this was found only 0.4 m away from an unmoved, refitting flake scatter [15], potentially indicating very limited transportation. The examination of many linear kilometres of exposures of head deposits in south western England as part of studies of Pleistocene environmental change has produced virtually no Palaeolithic artefacts [16]. In general solifluction deposits and tributary fluvial gravels deposited during cold marine isotope stages dilute concentrations in mixed and reworked alluvial formations as is clearly evident in the Axe Valley in South West England where background concentrations can be as low as 1 artefact per 200,000 m^{-3} of gravel [17]. Palaeolithic scatters do occur on interfluves in this region especially in association with clay-with-flints which is a Tertiary weathering deposit. However, even at sites with dense scatters of lithics, such as Wood Hill, Kent and Caddington and Round Green in Bedfordshire bifaces are rare [18,19] and none approach the numbers seen at the floodplain super-sites.

A second line of evidence is artefact condition. Whilst many artefacts in river gravel deposits show signs of abrasion such as edge wear and rounding, many do not even those from the same site and stratigraphic level. At Broom using two different systems for recording lithic condition, between 27% and 70% are fresh to lightly abraded (Fig. 2.) [20]. Geomorphological research on clast rounding has further suggested that even those bifaces with 'rolled to very rolled' morphologies may not have travelled far, as the rate of rounding is proportional to clast shape [21] and their tips are frequently not broken. This contention is also supported by budget-based models of river terrace formation which suggest that the majority of gravels in terrace staircases in this region are derived from reworking of previous terrace deposits by lateral channel migration rather than downstream clast-abrasion [22,23]. Even when the lithics are all abraded to heavily-abraded, as reported for Wood Green and Dunbridge (Fig. 1), the longitudinal occurrence of the high concentration is clearly peaked at 21-17 km upstream from the modern river mouth but downstream of the change in bedrock lithology [24]. The occurrence of both fresh and rolled bifaces in the same sedimentary bed, as at Broom and at Chard Junction (both in the Axe Valley) also suggest that the majority of the lithics have not travelled far and most probably are the result of local reworking of artefact spreads on the floodplain from adjacent gravel bars or 'proximal contexts' so although not truly in-situ they are still clustered near to the point of original deposition [25]. This also applies to probable lithic manufacturing sites such as Purfleet [26]. Such dense spreads or 'pavements' of lithics are known from several early Stone Age sites in east Africa such as Olorgesailie [27] and recent excavations at Rubirizi in Uganda [28]. The distinction between dense concentrations of artefacts in-situ or in proximal contexts, and the background lithic artefact content of fluvial gravel, is archaeologically critical as it allows such super-sites to be analysed as part of a behavioural rather than geomorphological distribution and permits meaningful ecological interpretation of site distribution.

Site Distribution and Niche Reconstruction

The pre-MIS 12 hominin occupation of north-west Europe is only known from a small number of sites and may have been an Atlantic phenomenon restricted to mild to warm temperate climates in coastal areas [29]. Post-MIS 12 hominins occupied this region in all interglacials and some interstadials with the apparent exception of Britain in MIS 5e [30]. The distribution of the biface-richest post-MIS 12 but pre-MIS 5 sites in Britain and northern France shows a marked concentration in the lower reaches of river valleys (Fig. 1). The richest sites are defined as those with over 500 bifaces, which frequently occur together with other knapping products including cores and flakes (both prepared and non-prepared) of Acheulean and/or Levallois tradition. The sites can be grouped by catchments all draining into the English Channel or southern North Sea. In the Exe/Axe basin there is only one site at Broom [7,19]; in the palaeo-Solent and East Sussex Plain system there are 7: Milford Hill, Wood Green, Dunbridge and Romsey, Hill Head/Rainbow Bar, Red Barns Warsash-Hook and Boxgrove [31–35]. Although large numbers of lithics have been found in the Bournemouth-Christchurch area they are both widely dispersed and come from a number of terrace units even at the richest site which is Barton on Sea [37]. Similar areas of high background densities occur elsewhere in the region such as Sturry in Kent [14]. In the Thames basin there are 11 sites, Furze Platt and Burnham Beaches in the Maidenhead area, Yiewsley-Drayton-Hayes, Acton and Stoke Newington in the West London area, Croxley Green in the Colne valley, Swanscombe [36,4] and Clacton in the Thames estuary region [4,14,37] and the early Middle Palaeolithic sites at Purfleet, Northfleet, and

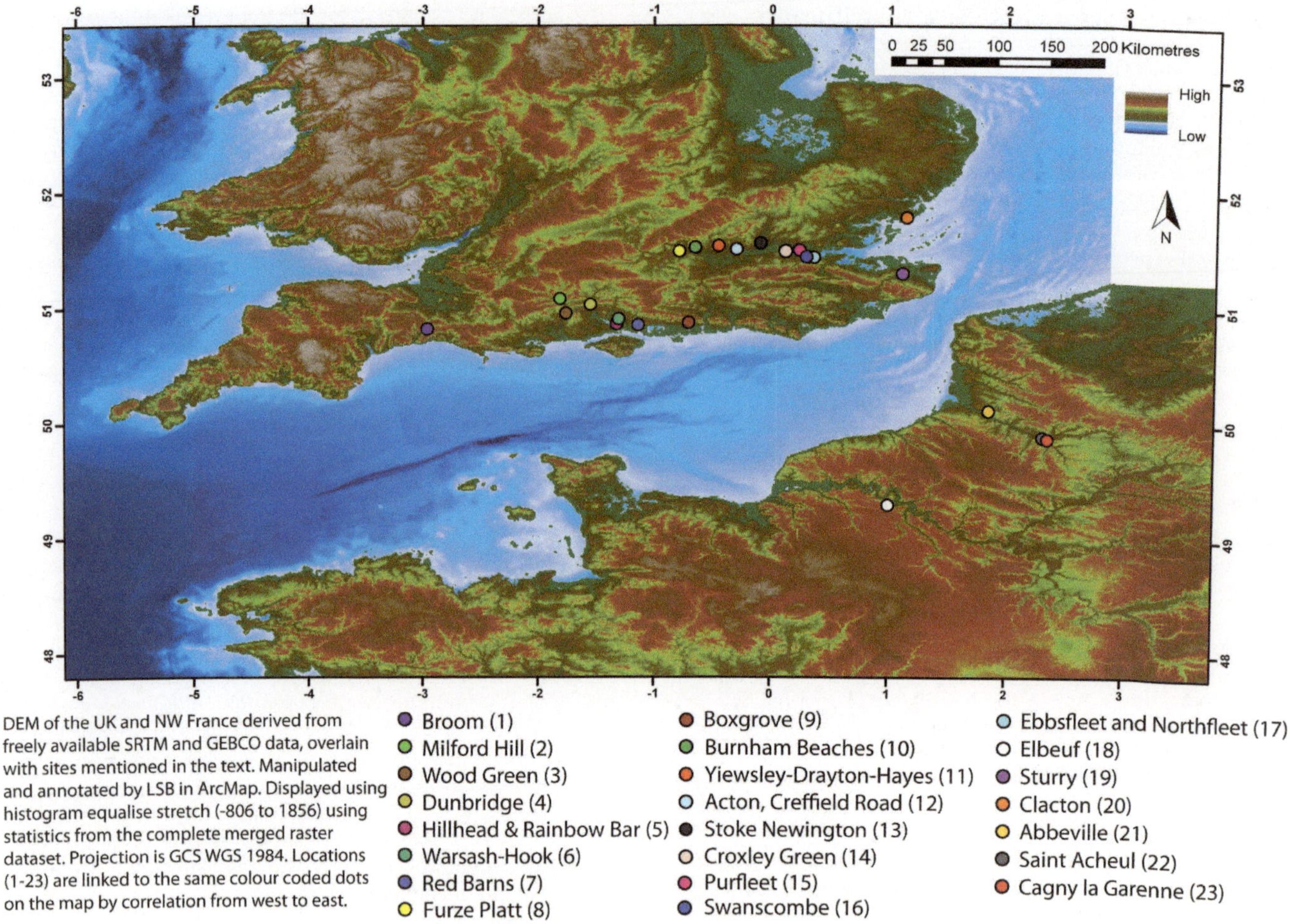

Figure 1. The distribution of Lower to Middle Palaeolithic (pre-MIS 5) sites in Britain and Northern France with over 500 bifaces (including roughouts) and other key sites referred to in the text.

Ebbsfleet also in the Thames estuary region [38,2]. Across the English Channel the major sites are restricted to the Somme valley at the sites of Abbeville, St Acheul and Cagny la Garenne. Due to the low gradient and high tidal range of the Somme, Abbeville is only 3 km upstream of the present natural tidal limit (NTL) and St Acheul and Cagny are just over 40 kms upstream of the NTL. On the Seine, Elbeuf is actually at the present tidal limit [39]. In addition to these high-concentration sites there are a number of recently researched sites which have a small number of artefacts but good environmental and chronostratigraphic data such as Caours in the Somme Valley and La Celle in the Upper Seine-Yonne Valley [39–40]. Caours and La Celle have been included in the palaeoecological database along with Boxgrove as proxies for natural interglacial conditions and due to their comprehensive records.

The distance of the sites upstream of the NTL has been estimated from the corresponding interglacial relative sea level (IRSL) in order to approximate site position at the time of occupation. Estimation of the tidal limits in the last 4 interglacials (MIS 5e, 7, 9 and 11) have been made using the LRO4 stack [41] as a proxy for eustatic sea level adjusted for regional uplift [42] by regarding the associated raised beaches as indicating the interglacial sea level, and using the MIS 1 floodplain gradient (Table S1). This floodplain method was used in preference to the river terrace gradients as these are mostly cold-stage gravels and grade to low glacial sea levels [43,22]. In the absence of any usable Pleistocene data, tidal frames have been assumed as similar to the present day. The results (Table S2) generally places the upstream sites slightly further above the NTL than they are today but the sites in the estuarine zone move closer to the NTL. The additional distances are relatively small being a maximum of 2 km in MIS 9 to a maximum of 23 km in MIS 7. Palaeoecological data suggests that some sites were closer to the NTL than predicted here, such as Swanscombe which has ostracod evidence that it was just above the tidal limit as well as a dolphin vertebra (Ingress Vale Pit) and a marine Gadidae bone [44,45]. Although the estimates of site position in relation to IRSL NTL have only been made for the British sites it is likely that the Somme and Seine sites results would be similar to those from the north side of the English Channel. One uncertainty associated with these calculations is the possible coast to inland variation in uplift rates which could have resulted in gradients different from those used here. A second uncertainty is that occupation phases at some of the sites were probably not associated with peak interglacial sea levels such as Swanscombe and Cagny la Garenne. In these cases the estimation of their position within the valley can only be a minimum distance from tidal waters. Despite these considerations, it is clear from the calculations and the geomorphology of the fluvial systems entering the English Channel that during high or maximum RSLs in each of the last four interglacials all except two of these super-sites

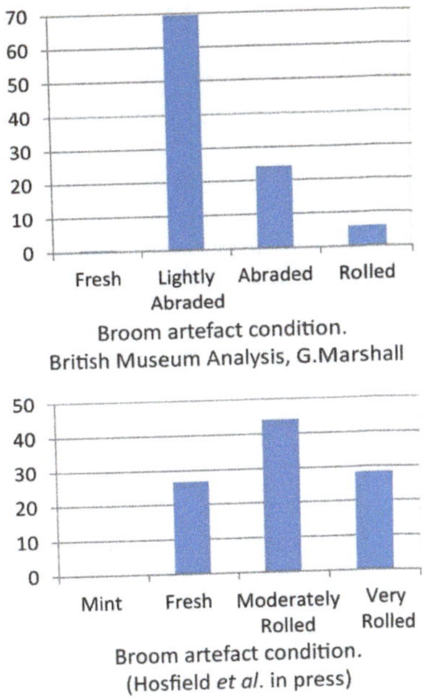

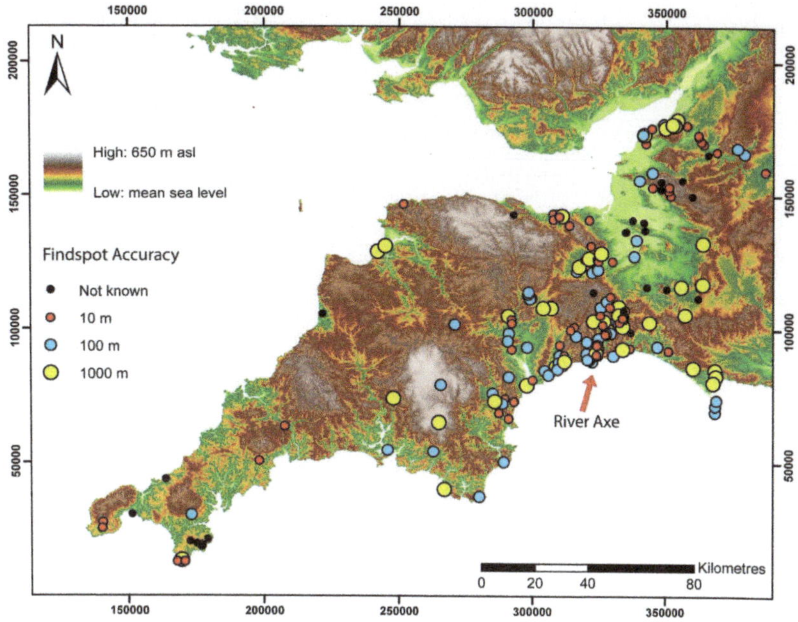

Locations	Bifaces	Cores	% of Total
Gravels/Valley floors	2134	35	85%
Gravels (excluding Broom)	306	34	12%
Solid geology/ slopes/plateau	83	4	3%
(Unknown	122	6	5%)
Total	2523	46	100

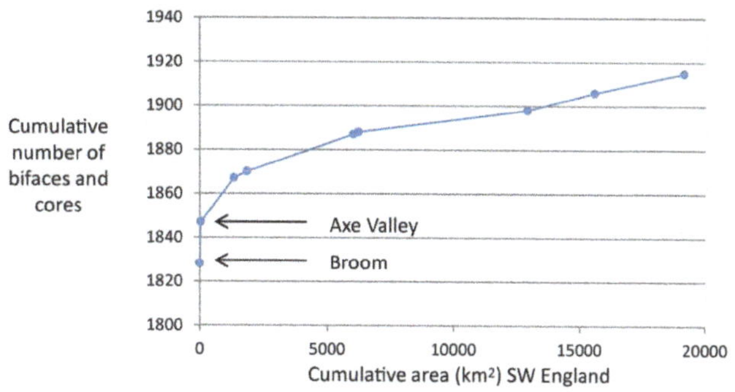

Figure 2. Palaeolithic site and find distribution in SW England from PRoSWEB with the cumulative number of artefacts plotted against area as a measure of the variance of site richness and statistics for find location.

would have been within 40 kms of the tidal limit and most were considerably less. The only other exceptions to this pattern are Boxgrove (unit 4c) which was a coastal plain site close to a freshwater spring [46–47] and Hoxne which was close to an evolving lake shore/riverine setting [48,49]. This 40 kms distance range for the majority of the sites places them within 1 to 2 days walk [50] of access to marine resources. This suggests that although marine resources may have supplemented diets they were not a major source of nutrient intake. It is also noticeable that 12 out of the 19 English sites come from tributary junctions reaches (Broom, Wood Green, Milford Hill, Romsey, Dunbridge, Warsash, Furze Platt, Burnham Beaches, Yiewsley, Croxley Green, Stoke Newington, Purfleet) in which there would have been a greater extent of floodplain and channel habitat. The concentration of these super-sites in this zone raises an important question. Were these lower reaches of valley floors particularly favourable locations or niches for hominin activities and visitation?

In order to test this hypothesis we have attempted to predict the likely distribution of nutrient resources during the last 4 temperate interglacials along a transect from valley floor to the low-altitude plateaus which characterise the topography of this region. This is based upon year-round occupation and the fundamental assumption that adequate nutrient intake is critical to hominin survival throughout the year and in all stages of the life-cycle including pregnancy and infancy. We also make the assumption that that the nutritional requirements of Middle Pleistocene hominins (*Homo heidelbergensis*, *Homo neanderthalensis*) were related to their phenotype and environment in a way which both maximised fitness and is comparable to their nearest living relative (NLR) (i.e. *Homo sapiens*). This is justified by the generally held view that hominin nutrient requirement and digestive physiology appears to be genetically conservative [50–52]. Although the floodplain offers a wide variety of resources (Table 1) adequate nutrient intake is critical for hominin survival and procreation. A third but less critical assumption, particularly pertinent to nutrient requirements, is that energy expenditure patterns in hominins were not unlike those seen today. Support for this assumption comes from studies of total daily energy expenditure in contemporary hunter-gatherers which suggests that metabolic rates were not significantly different when body mass is taken into account [50].

Research by both archaeologists and Pleistocene scientists over the last fifty years has produced a significant record of both fauna and flora from both Palaeolithic archaeological sites and palaeoecological sites without any artefacts. These have included interglacial floral and faunal lists for the British Isles [53,45,54) and from the Netherlands [55] and France [56]. We have used these records (from 37 sites) to compile a list of potentially edible animals and plants that were present in the lowland landscape in southern

Table 1. Potential resources available on the floodplains of medium to large rivers in Britain and N France during interglacial and warmer interstadial periods of the late Pleistocene.

Floodplain Resource	Constitutive components
Raw materials	chert/flint/quartzite/andesite…as cobbles & hard rock exposures in gorges, also natural coppicing & saplings (due to successional growth on floodplains) suitable for spears & sticks
Safety	caves/rock shelters in gorge reaches, tree-throw pits (more common on floodplains due to restricted tree rooting depth)
Shelter	open grass/shrublands (maintained by grazing) in the channel zone & water barriers
nutrients	water: all structural & metabolic processes (flowing (non-stagnant) water) proteins: energy+ tissue development & repair (herbivore flesh, fish, USOs) carbohydrates: energy+ tissue development & repair (USOs, honey) lipids (fats): energy+ tissue development & repair (marrow, animal fats, fish) vitamins: metabolic function (see text) minerals: cell structure, metabolic processes (animal & plant foods, mineral springs & precipitates)

Britain and northern France during temperate periods with a mean annual temperature of 7–12°C, namely MIS 11, 9, 7 and 5e (Fig. 3; Table S3) [57]. Further we have assumed that the only food processing technology available was roasting over an open fire. It is accepted that this cannot be a complete list and there may be some taphonomic bias to floodplain environments. An attempt has been made to counter this by including some species, which on ecological grounds probably were present but for which we have no fossil data as they are not floodplain species (e.g. Pignut; *Conopodium majus*). These are indicated in in the Supplementary Information along with the principal data sources and the food types with selected data in Table 2. The nutrient and energy values of these resources have been collated from a wide variety of sources but mostly nutritional databases (Canadian Ministry of Health, 1999; USDA National Nutritional Database; WHO Nutritional Landscape Information System). The edible status of these resources is taken from modern and historical observations and in effect excludes inedible grasses, woody tissues (except bark) and plants or animals known to be poisonous to humans [58]. Ecological classifications are taken from a wide variety of sources but particularly [59] for plants and [60] for mammals. As a first approximation to estimate nutrient intakes the landscape has been divided into 7 zones from the river to the plateau tops, which typically is a distance of 10 km and 150 to 200 m in relative relief (Fig. 3). The floodplain has further been divided into the river channels, including gravel islands and banks and the remainder of the floodplain including terraces. The rest of the landscape has been subdivided into forested slopes and plateau, and two open or large-gap environments. There are no continuous vegetation records that span the full period from this region but from fragmentary records and correlation with continuous sites to the south an approximate pattern of forest cover has been reconstructed by [61] which correlates reasonably well with human occupation (Fig. 3) [3]. Based on the typical mesocratic vegetation phase of interglacials in this region, forests would have been mixed temperate and deciduous and dominated by oak (*Quercus*), elm (*Ulmus*), lime (*Tilia*) and beech (*Fagus*) further to the south [62]. Although there was both systematic geographical and temporal variation during this period the temperate climate and areas of open grassland would have controlled the diversity of the shrub and herbaceous flora which includes the majority of edible species. It has been argued from a number of sites that the relative openness of interglacial floras was the result primarily of large herbivores and particularly the steppe mammoth (*Mammuthus trogontherii*) [63]. The distribution of large mammals (>30 kg) is assumed to have been even. However, this has been adjusted to reflect the relative ease of ambush predation and scavenging on the floodplain due to open conditions and the congregation of animals at drinking and bathing sites. The open nature of a significant part of the landscape is also suggested by the presence of horses (*Equus*) at nearly all sites. Late Pleistocene horses have a dental crown height ratio of over 5.5 and were therefore adapted to grazing rather than browsing [64]. There is also evidence from movement ecology that some of the species such as deer and horses are preferentially attracted to the forest edge, clearings and wetlands particularly during winter [65,66]. Pending useable data on habitat use and relative kill rates it has been assumed that the availability/kill ratios (probability of encounter and kill rate) for clearings is 2:1 and the floodplain 3:1. Evidence-based estimates are extremely difficult but many of the species are only available to be killed (or gathered) in open habitats; and in terms of closed forest this would be in natural gaps and the floodplain. It is also likely that many large birds and herbivores would be at their most vulnerable when in, or at, water. Differential encounter or kill rates are not used in the summary nutritional values (Fig. 4) as they would primarily affect total intake per unit energy expended whilst the relative differences result from the species patterns. For small mammals ecological data has been used to assign species to principal locations and the herpetofauna and fish have been ascribed solely to the floodplain since these landscapes contain few persistent natural lakes. Although small in biomass, small mammals may have been important for prey-shifting (*sensu* [67]) in times when large mammals were scarce. Birds have also been classified according to habitat and the higher number of species in the floodplain corridor is a reflection of the high number of waterfowl recorded from Pleistocene sites in the region (e.g. Boxgrove [68]). Whilst there may be a bias here the greater ease of hunting waterfowl and egg collection has not been included although this would also be greater in the riparian zones. Edible plants are known from all ecological zones with similar assemblages from open areas and floodplains, although the nutritional value and ease of collection almost certainly varied. These species are then totalled to give a comparison of the zones revealing the greater diversity of edible animals and plants in the floodplain corridor and in clearings than in the intervening mixed temperate woodland. Certain potential nutrient sources have not been included particularly mollusca and insects. This is not because the authors discount them as potential sources, indeed both are potentially high in nutrients, but because of a lack of data on species-specific nutrient contents in this environment.

Nutritional Diversity

There are over 50 nutrients required to sustain human life and diet has long been regarded as a driving force in biophysical, social and cultural evolution within the hominin lineage. Nutrients have

Table 2. Selected food groups from the database of interglacial flora and fauna for the southern England and northern France region with nutritional values.

Group/Type	Genera/species included	kcal per 100 g	Proteinper 100 g	Carb per 100 g	Fat per 100 g	Fibre per 100 g	PUFA per 100 g n3+n6?	Vit A IU per 100 g	Other important nutrients per 100 g
large herbivore (muscle)	**bison** rhino, bear,	223	18.7	0	15.9	0	0.75	0	Cholesterol (83 mg)
deer spp.	(red deer (venison)	111	22	0.2	2.6	0	0.35	0	Iron, Calcium
Marrow	caribou	786	6.7	0	84	0	0.25	240	Fat (86 g)
Marrow	beef/*Aurochs*	900	11	0	95	0	0.25	0.26	Fat (96 g), Sodium (139 mg)
Horse	domesticated horse	133	21	0	4.6	0	0.65	0	Cholesterol (52–63 mg)
small mammals	**rabbit**	114	22	0	2.3	0	0.45	0	Cholesterol (28 mg)
amphibians	**frogs**, toads.	73	16.4	0	0.3	0	0.10	50	
birds/goose/duck	**wild duck**, geese, water fowl	211	17.4	0	15.2	0	2.02	88	Cholesterol (68 mg)
eggs/duck eggs	**chicken***, geese, ducks,	347	17.6	0	29.4	0	1.41	1900	Potassium
fish/eel	**eel**, stickleback, coarse fish	657	65.8	0	41.6	0		124	Potassium, Phosphorus
fish/salmon	**salmon/trout**	141	19.9	0	0	0	1.50	280	Calcium
snails	**raw snails**	90	16.1	2	1.4	0	0.25	100	Iron, Calcium
hazel nuts	**hazel nuts**	655	16.3	6	63.3	7.7	6.45	34	
aquatic plants (seeds)	**yellow waterlily**	361	7.9	80	0.1	19	0.16	0	Niacine (4.2 mg), VE (35.6 mg), Manganese (0.94 mg), Zinc (6.3 mg), Magnesium (86 mg)
aquatic plants (Tubers)	***Typha***, *Phragmites*, *Trapa natans* (fruits), pignut	80	7.7	79.1	4.9	3.3	0.339	2	VC (42%), Folate, Calcium (50 mg)
berries/fruit	blackberries, **cranberry**	52.6	1.3	13.1	0	5.0	0	165	Potassium
leafy veg/	**spinach**, comfrey, scurvy grass.	24	2.9	3.6	0.4	2.2	0.16	9377	VC, VK, Folate, Manganese, Magnesium
aquatics leaves	**watercress**	11	2.3	1.3	0.1	0.5	0.03	3191	Calcium, Folate
salad/lettuce	edible leaf plants	13	1.3	2.6	0	1.3	0	469	VC (2.6 mg), Folate (27 mg)
honey	bees nests	309	0	85.7	0	0	0	0	Manganese (0.9 mg)
EAR(Adult male)	*modern diet*	*1900 kcal*	*56 g day*	*125 g day*	*30 g day*	*38 g day*	*1.6 g day*	*3 g day*	*Calcium (800 mg day), Iron (6 mg day), Vit C (75 mg day), Vit.B12(2 μg day), Niacine (12 mg day), Folate (120 μg day), Phosphorus (580 mg day), Magnesium (330 mg day)*

The animals and plants in bold have been used to estimate the nutrient values in the table.
*indicates a taxa not present in the Early-Middle Pleistocene but used as a nutrient proxy. Carb, carbohydrate; EAR, estimated average requirements; VC, vitamin C; VK, vitamin K; PUFA, polyunsaturated fatty acids;

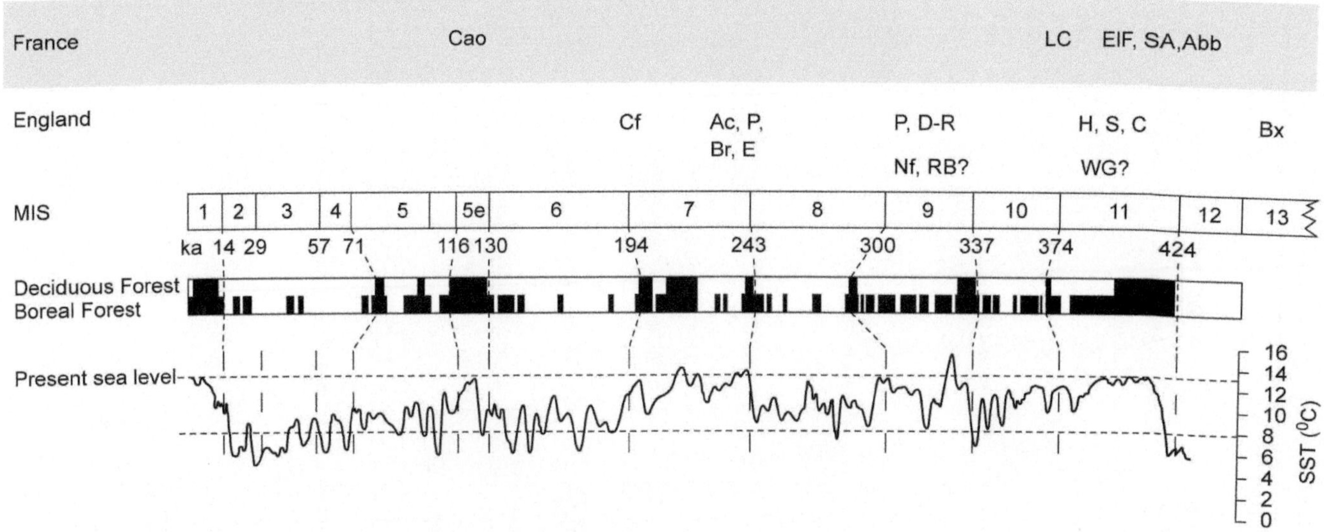

Figure 3. Vegetation reconstruction for southern England with SST from ODP 980 and approximate site chronology. Adapted from Stemerdink et al. (2010). Site code; Ac = Acton, Abb = Abbeville, Br = Broom, Bx = Boxgrove, C = Clacton, Cao = Caours, Cr = Crayford, D-R = Dunbridge-Romsey, E = Ebsfleet, ElF = Elbouf Fm., H = Hill Head, LC = La Celle, Nf = Newfleet, S = Swanscombe, SA = St Acheul, Wg = Woodgreen. Several sites have not been included as dating is too poor to assign them to an individual stage within the period (Furze Platt, Burnham Beaches, Yiewsley, Stoke Newington and Croxley Green).

classically been divided into 6 major groups (proteins, carbohydrates, lipids, vitamins, minerals, water). Whilst animal and plant tissues are the principal sources for all 6 groups the environment can directly supply minerals, including Fe, S, Ca, Na and Se with the principal source locations being springs and the floodplain in the form of Fe, S and Ca bio-precipitates. The relatively higher requirement for mineral intake during pregnancy has been associated with a desire for direct consumption of minerals [69], and mineral springs also attract mammals and plants which bio-accumulate these minerals, and metals in particular (metalophytes). Although most emphasis in reconstructions of Palaeolithic diets has been placed on protein intake, fat is the major energy reserve in hominids and can buffer food scarcity which would have been more common at the edge of the hominin geographical range. A fat-brain trade-off has also been proposed as an alternative to the gut-brain trade-off underlying the expensive-tissue hypothesis of brain enlargement [70,71]. For over fifty years it has been recognised that there are certain nutrients which are essential for both human survival and reproduction. These are primarily micronutrients, vitamins and specific minerals, but also include specific amino acids and the essential polyunsaturated fatty acids (PUFA) which are obtained from plant sources, linoleic acid (LA) and α-linoleic acid (ALNA) [72]. LA and ALNA are essential in the diet because they are the respective precursors of longer chain PUFA including the *n*-6 fatty acid arachidonic acid, a precursor of prostanoids, and the *n*-3 fatty acid - docosahexaenoic acid (DHA) which plays a critical role in the development and function of the central nervous system [73] and which has been shown to modify epigenetic marks directly or indirectly [74]. Because of its critical role in the brain, DHA has been linked to human brain evolution [75]. Humans are poor converters of ALNA to DHA, although women do this more efficiently than men which has implications for the supply of DHA from mother to offspring [76,77], and so preformed DHA derived from the foods at the land/water interface has been associated with the evolution of large brained hominins in Africa [78,79]. Edible plants and animals rich in LA and ALNA, and animals rich in PUFA from plants (LA and ALNA), from meat (arachidonic acid) and fish (DHA) as groups have been totalled and show a marked increase in the floodplain corridor (Fig. 4).

For hominins including Neanderthals which have traditionally been seen as top carnivores [80] there is also a nutritional challenge in the protein ceiling in that protein intake must be balanced by carbohydrates and/or fats. If not so-called 'rabbit starvation' results due to the finite ability of the liver to regulate the rate-limiting enzymes that synthesise urea resulting in very high levels of ammonium ions and acidic amino acids in the blood [81]. This is known to have occurred in modern humans when they have been forced to rely entirely on fat-poor, wild animals [82]. We have used the plant and animal list to calculate the approximate balance of energy (kcals), protein and carbohydrates+fats (CF) as a percentage of the total values for the entire landscape diversity (all zones). As can be seen in Fig. 4 all three are highest in the floodplain zone, however, the floodplain corridor is the only part of the landscape where CF sources are potentially more abundant that protein sources. This is due to the availability of high CF sources including some high fat sources such as fish (particularly eels (*Anguilla*) which are available throughout the year), waterfowl and eggs; and high carbohydrate sources including plants particularly those with underground storage organs (USOs) such as reed mace (*Typha*), common reed (*Phragmites*), water chestnut (*Trapa natans*) and yellow water lily (*Nuphar lutea*). USOs have repeatedly been implicated in hominin evolution and particularly encephalisation and bipedalism in the Africa [83,72,84–85] although this has been challenged [86]. In order to maintain their protein:CF balance it is likely that hominins would have actively sought out high carbohydrate sources such as honey (86 g 100 g^{-1}), aquatic seeds (80 g 100 g^{-1}), berries (13 g 100 g^{-1}), and hazelnuts (6 g 100 g^{-1}). A similar propensity may have been displayed for high fat sources particularly bone marrow (84 g 100 g^{-1}) and brain (70% fat and rich in arachidonic acid and DHA), but also hazelnuts (63 g 100 g^{-1}), eel (42 g 100 g^{-1}), birds eggs (29 g 100 g^{-1}) and certain parts of selected mammals such as the tails of beavers (*Castor fiber*).

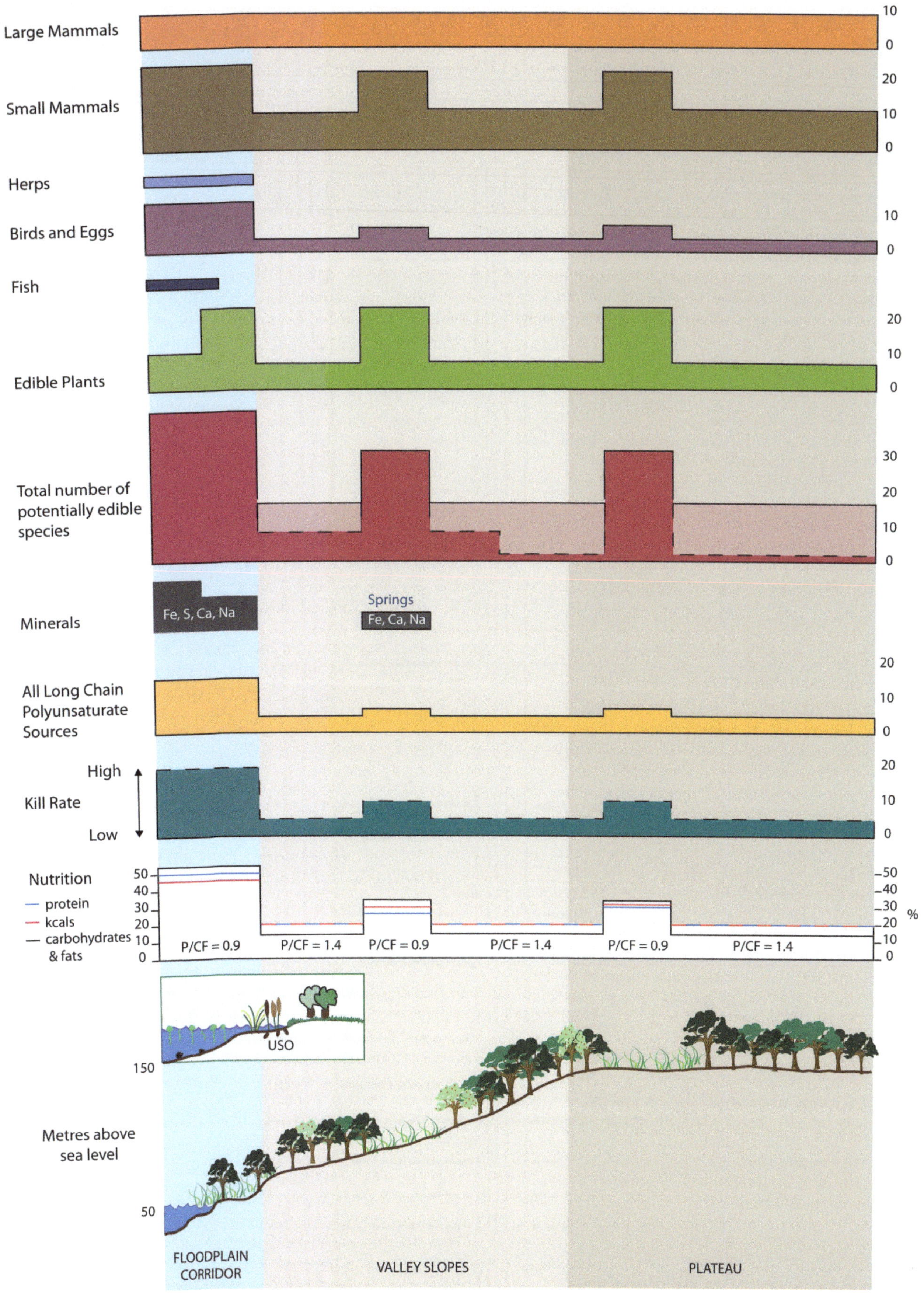
Large Mammals
Small Mammals
Herps
Birds and Eggs
Fish
Edible Plants
Total number of potentially edible species
Minerals
Fe, S, Ca, Na
Springs
Fe, Ca, Na
All Long Chain Polyunsaturate Sources
High
Kill Rate
Low
Nutrition
protein
kcals
carbohydrates & fats
P/CF = 0.9
P/CF = 1.4
%
USO
Metres above sea level
150
50
FLOODPLAIN CORRIDOR
VALLEY SLOPES
PLATEAU

Figure 4. The nutriscape; a schematic representation of dietary diversity in a transect from valley floor to plateau top in the English Channel region. The scale (X axis) for each histogram is the number of species for each of the landscape zones. The variation in the total number of potentially edible species is to allow for more and less open forest cover.

Beavers are particularly interesting as they would have been ubiquitous damming all but the largest channels in this region, and at 20–30 kg they provide as much meat as a small deer whilst their tails can contain as much as 60% fat in the autumn and winter [87]. Presumably the same would apply on a larger scale with the large beaver-like rodent (*Trogontherium cuvieri*) which was present at Boxgrove, Swanscombe and Hoxne [88–90] but extinct after MIS 11 [91]. At Hoxne the remains of the otter (*Lutra*), beaver (*Castor fiber*), and waterfowl directly associated with the Acheulean lithics strongly suggested hominin exploitation, although no butchery marks were found on these species [88]. Bifaces are known to be extremely efficient butchery tools [92] and the emergence of particularly large and heavy bifaces may be related to their importance in marrow extraction in smashing the long-bones of large herbivores as well as for carcass processing. Likewise the high lipid content of brains including macaque and hominins [93] is also a potential driving force for the crushing of skulls and consumption of brain grey and white matter.

In these marginal temperate environments a similar problem arises with essential nutrients such as taurine, vitamins B6, B12, A and C, and folate. As can be seen from Table 2, viscera, especially liver, is particularly important in the supply of vitamin B6, B12, A and C as well as PUFAs, folate and some essential minerals. Fish, including eels are also a rich source of taurine, B6, B12 and vitamin A. Vitamin C is of particular importance as the body has limited capacity for retention (approximately 20 days) after which scurvy will arise if intake falls significantly below the daily requirement [94]. The principal source is herbivore liver, however, since vitamin C in liver is destroyed by cooking higher intakes are achieved if the liver is consumed raw [95]. Other sources are mostly seasonal such as berries and leafy vegetables. There are, however, year-round vitamin C sources within the riparian zone such as many species of the cabbage and carrot families including the wild cabbage (*Brassica oleracea*), watercress (*Nasturtium officinale*), wild carrot (*Daucus carota*), wild celery (*Apium graveolens*) and scurvy-grasses (*Cochlearia* spp.) all of which are found in disturbed floodplain environments and the upper estuarine zone. Also available year-round are sources of folate such as watercress (*Rorippa nasturtium-aquaticum*) and aquatic USOs such as water-chestnut (*Trapa natans*). USOs are particularly valuable in late fall and winter when they have their maximum energy values [58,96]. The same is true for iodine for which there is an easily obtainable year-round source in non-acidic floodplains in the form of watercress although it would have varied with the geology and soils of the catchments [97]. Watercress is likely to have been particularly common around these sites as it benefits from nutrient enrichment in carbonate-rich waters and is tolerant of heavy grazing [59]. Analysis of the vitamin concentrations in these food sources highlights another potential problem in that there is very little vitamin E in the meat or indeed fish. Without adequate intake particularly during lactation there could be a substantial risk of oxidative damage to tissues (Table 3).

Vitamin D could also be a problem at these latitudes as it is only naturally present in a few foods such as fish and egg yolks. However, it must be noted this is difficult to assess as a constraint given the unknown hominin capacity for endogenously produced vitamin D from sunlight. The balance of Fe, I, Se, vitamins D and E and *n*-3 fatty acids are important in both metabolism and inflammation through their role as nuclear transcription factors and this balance is present in land-water ecosystems [85]. In relation to nutrient balance eels are a particularly interesting element of the ecology of the region in that they were ubiquitous, present all year-round, can move over land and would have been naturally attracted to kill and butchery sites on floodplains due to their ability to be able to detect blood in water. Furthermore eels are highly valuable nutritionally having almost equal quantities of protein and fat as well as being high in LCPs, vitamin A and key minerals (P and K). This and the use of other freshwater fish may be a cause of the greater dietary breadth estimated from isotopic values from early modern humans in Europe [99,100] but it does not necessarily follow that this was a unique capability of modern humans. There is also a possible beneficial interaction here between the presence of beaver dams, which both facilitated river crossing (over the dams) and created ponds, but which also attracted both game and fish including eels.

Taken together, the available energy, overall nutritional diversity, and availability of key micro-nutrients the floodplains provide the optimal location in this landscape through easy access to mammals of nearly all types, fish, waterfowl and edible plants including aquatic plants (Fig. 5). Unlike other zones the floodplain provides key potential resources for critical periods both during the year (e.g. for over-wintering [19]) and during critical periods in human life-cycles such as during pregnancy and breastfeeding. Other ecozones can also provide easy access to many of these resources, such as natural forest gaps and clearings, but none can provide such high on-site nutritional diversity. Nutritional diversity can be related to both high fecundity and lower rates of premature birth and infant mortality [100]. Sites in the wider parts of the floodplain such as downstream reaches and at confluence zones would have made beneficial bases or locations of concentrated activity from which other ecological zones could have been exploited and which would have formed hubs or central places in mobile hunter-forager circulatory patterns of seasonal mobility [100–101] over ranges of the order of 1000–2000 km^2 [101]. This has been proposed as the pattern for Middle-Upper Palaeolithic site hierarchy and differentiation in the Rhone valley where the sites on the floodplain or valley floor are the long-term residential camps [102].

Hunter-gatherer Diets, the Palaeodiet and Niche Construction

On the basis of animal bones, artefacts and stable isotope studies Palaeolithic hominins, and particularly Neanderthals, are generally regarded as top carnivores with meat making up the vast majority of their diet [9,100]. As a result it has been argued that the palaeodiet was a low carbohydrate diet (c. 20–40%) but with essential nutrient requirements [103] and that the meat dependency would have increased with increasing latitude to a maximum in circum-Polar groups. However, studies of historical and contemporary hunter-gatherers only partially support this model. One of the problems has been that the vast majority of extant hunter-gatherer groups available for dietary studies are from either the low latitudes (tropics) or from high-latitudes (circum-Polar zone) with very few from the temperate zone and virtually none from the Old World temperate zone. Studies of Arctic hunter-gatherers have unsurprisingly shown levels of meat consumption reaching 90–100% by weight [104,79]. However, it

Table 3. Critical nutrients principal sources and some health implications.

Nutrient	Value/deficiency	Sources
PUFAs	Essential in brain development	Animal liver, brains, eyes, viscera, fish, some plants
Taurine	Important for brain function & eyes	Meat, fish
Vit B6	seborrhoeic dermatitis & if restricted can impair fatty acid synthesis	fish, beef liver and other organ meats, some Underground Storage Organs
Vit B12	Damage to the brain and nervous system, cognitive disfunction, mania, psychosis, pernicious anemia.	liver, kidney, fish, shellfish
Fe, Zn	Anaemia, hypoplasty (eye sockets)/DNA synthesis...	Shellfish, liver, meat, nuts, seaweed
Iodine	thyroid function, goitre, cretinism, learning difficulties	Venison, watercress, leafy vegetables
Folate	Required to avoid birth defects (spinal)	Animal liver, leafy vegetables,
Vit A	Night blindness- erophthalmia, keratomalacia,	Animal liver, leafy vegetables, eggs, fish, berries
Vit C	to avoid scurvy (20 day residence time)	Liver, meat[1], berries, leafy veg

[1]only if eaten raw.

has been argued that there has been under-reporting of the non-meat component in these diets due both to seasonal variations and social factors [105]. Conversely studies of hunter gatherers in the tropics demonstrate that no exclusively vegetarian groups exist, and that the lowest meat intake is 20–30% [83]. Crucially a study of hunter-gathers in the Columbia Plateau [106] revealed a diet with approximately 30% animal food. Using any of these data the projected animal food intake for hunter-gatherers living at latitudes of this study (48°–53°) would be anywhere between 30% and 80% but probably not higher.

Most palaeoecological evidence of diet suffers from major problems of taphonomic bias particularly through the preferential preservation of bone including evidence of butchery. Isotopic analysis of bone collagen of Neanderthals suggests values of −21.8 to −19 and $\delta^{15}N$ values of 8 to 12 [107,97,94]. Evidence from stable isotopic concentrations in hominin bones also has a dietary bias because there is a non-linear relationship between animal food intake and both $\delta^{13}C$ and $\delta^{15}N$ values [107,108] and low-protein foods that may have been critical to survival are invisible in isotopic analyses of bone collagen [97]. The result is that on $\delta^{13}C{:}\delta^{15}N$ plots differentiation between individuals who have 75% as opposed to 95% animal foods is very difficult. This is especially the case if their non-animal intake has been high in $\delta^{15}N$ and $\delta^{13}C$ and intake from other lower-protein foods cannot be measured. This problem of equifinality may be addressed over the next few years by studies of Sr:Ca and Ba:Ca ratios as both Sr and Ba decrease with increasing trophic level and are high in geophytes/USOs [83].

In summary the existing isotope literature may present a biased picture exaggerating meat consumption and under-estimating Palaeolithic consumption of plant food particularly critical, but low-volume, low-protein sources. This is supported by a few studies of independent sources of dietary information such as dental phytoliths. Data from Spy I and II in Belgium have revealed starch grains from USOs and grass seeds (Andropogoneae tribe) and from Shanidar III have produced starch grains of Type 1,

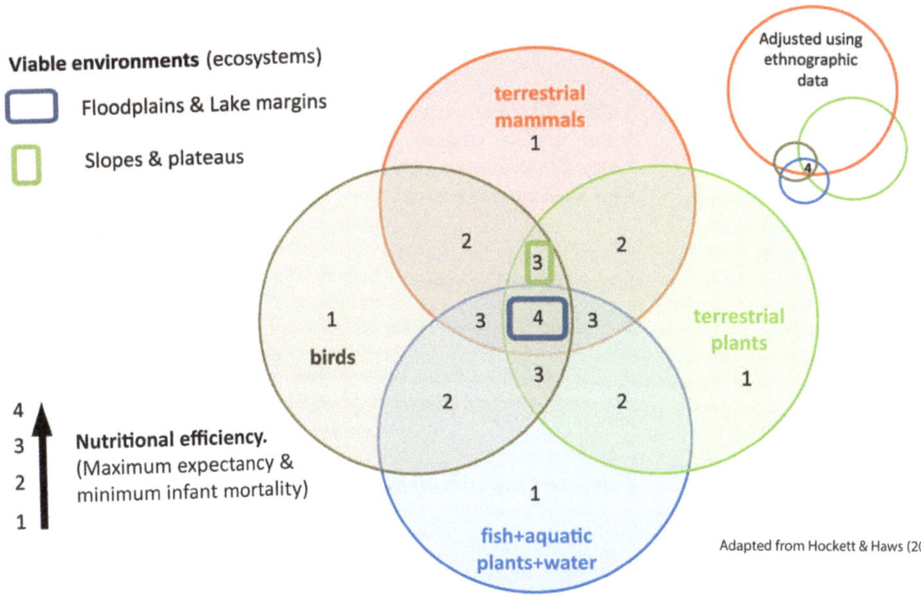

Figure 5. A representation of nutritional landscape ecology for a large temperate floodplain environment. Adapted from [124].

probably from water lily [109]. Molar macro-wear studies [110] have demonstrated high dietary variability in Mediterranean evergreen habitats for both Neanderthals and early modern humans.

So far in this paper the concept of niche has been passive and rooted in twentieth century biology [111] under which environmental factors were seen to act on an organism in a particular place thereby creating its niche. However, modern concepts of the niche are not only multi-dimensional but also include the recognition that organisms, especially hominins, do more than passively survive in that hyperspace but modify their niches and those of other organisms through their metabolism, activities and choices [112]. The recognition of this reciprocal relationship lies at the heart of niche construction theory (NCT) as a key element of evolution [113]. One obvious feedback is the development of clothing allowing occupation further north and persistence in cool-temperate periods such as interstadials, also damping physiological selection in response to extreme temperatures [114]. More subtly the inhabitation and modification of the high-nutrient diversity niche described here may also have had a significant physiological and genetic legacy, analogous to, but very different from, later genetic legacies of agriculture. The nutrient diversity and opportunities to expand nutrient sources may well have propelled both behavioural change such as the increasing intake of marine foods and the development of fishing and of long-distance foraging seen in the later Middle and Upper Palaeolithic in this area [115,116]. It may also have introduced an element of pre-adaptation to increasing diet breadth as epitomised by *Homo sapiens*. It is argued here that the result was that hominins located themselves in nutritionally optimum locations at the edge of their geographical range. Through repeated and/or prolonged occupation of those locales they altered them, probably creating adaptive responses in both the fauna and flora, and they were in turn altered by their constructed niche [113] particularly under migration restriction [117]. This is part of the 'ebb and flow' model of occupation of this region [3] and could also be a selective component of environmental remodelling that favoured adaptive plasticity or variability selection [116–118] resulting in occupation of the area during the last glacial cycle by both Neanderthals and modern humans well outside the climatic envelope considered in this paper.

Conclusions

Based on our analysis of multi-disciplinary data we propose that the distribution of the richest Lower to Middle Palaeolithic archaeological sites in this part of northwest Europe is a real behavioural distribution reflecting hominin activity patterns rather than a taphonomic distribution. During the last four interglacials the location of the richest Palaeolithic sites in terms of biface densities is strongly biased to the lower reaches of river valleys and sites which were above NTL but in proximity to tidal rivers and estuaries. This may also have been the case earlier, as suggested by marine molluscs, barnacles and foramnifera at sites such as Pakefield and Happisburgh III [3] but analysis is prevented by the small number of sites and is complicated by the breaching of the Weald-Artois anticline creating the Straits of Dover and by major changes in fluvial geography [120,28]. There are several reasons why lower reach of river valley floors would have been a favoured habitat in north west Europe including access to water, safety from predators, lowest river crossing points (natural rivers are shallow and frequently anastomose in these reaches facilitating crossing) and food resources. In this paper we have used a database of plant and animals resources known from Pleistocene sites in the region to compile a potential list of nutritional resources. Through ecological classification the nutrient landscape can be estimated and this shows a marked potential locational advantage for floodplain zones as opposed to the forested slopes, plateaus and even clearings. It is argued that this advantage may have included access to plants and animals which provided both essential energy and macronutrients but also critical micronutrients which maintained population health and maximised reproductive success and may have increased cultural complexity [121]. Such Palaeolithic diets with an aquatic component have been implicated in 'healthy aging', an emerging concept in evolutionary nutrition which has as its mantra 'we are what we eat, but we should be what we ate' [122,87]. It is possible these locations were perceived as 'healthy/good places' to which hominins returned on a regular, and prolonged basis, and may have been 'marked' by their assemblage of artefacts [123]. If these were important and revisited locations they provide nodal points in the Palaeolithic geography of north west Europe and support the contention that river valleys provided the nutrient-rich route-ways of exploration and utilization of Palaeolithic landscapes. A possible symbiotic interaction through niche creation on floodplains is postulated between hominins, horses, freshwater fish (particularly eels) and beavers. It is also proposed that this optimal pattern of occupation in the nutritional landscape coupled with low population densities was fundamental to the Palaeolithic diet that was successful in facilitating expansion of hominins outside their evolutionary homelands and the persistent, if episodic, occupation of the less productive higher latitude regions of the northern hemisphere for over half a million years.

Supporting Information

Table S1 Parameters used for the prediction of relative site distance from the natural tidal limit (NTL) during the last 4 interglacials.

Table S2 Distances from estimated NTL during middle-late Pleistocene Interglacials. Calculations have not been undertaken where the sites are on eroded coastal cliffs and for Boxgrove and Hoxne.

Table S3 Summary occurrence and ecological data from edible plants and animals recorded in Middle-Late Pleistocene interglacials from southern England and northern France.

Acknowledgments

Rob Hosfield is thanked for the Broom data and comments on earlier drafts. Valuable assistance and comments on early drafts of this paper were provided by Mike Field, Mark Roberts, Rob Scaife, Vanessa Straker, Jonathon Last, John Stewart and Lyn Ertl. The reviewers are thanked for their very helpful contributions to the paper.

Author Contributions

Conceived and designed the experiments: AGB LSB. Performed the experiments: AGB LSB. Analyzed the data: AGB LSB SR GCB. Contributed reagents/materials/analysis tools: AGB SR GCB. Wrote the paper: AGB LSB.

References

1. Butzer KW (1982) Archaeology as Human Ecology. Cambridge: Cambridge University Press,.
2. White M, Scott B, Ashton N (2006) The Early Middle Palaeolithic in Britain: archaeology, settlement history and human behaviour. Journal of Quaternary Science 21: 525–541.
3. Parfitt SA, Ashton NM, Lewis SG, Abel RL, Coope GR, et al. (2010) Early Pleistocene human occupation at the edge of the boreal zone in northwest Europe. Nature 466: 229–233.
4. Wymer JJ (1999) The Lower Palaeolithic Occupation of Britain. Salisbury: Wessex Archaeology and English Heritage.
5. Tuffreau A, Antoine P (1995) The earliest occupation of Europe: Continental northwestern Europe. In Roebroeks W, van Kolfschoten T (editors) Proceedings of the European Science Foundation Workshop at Tautavel (France). 1993. Leiden: Leiden University Press, 147–163.
6. Brown AG (1997) Alluvial Geoarchaeology. Cambridge: Cambridge University Press.
7. Hosfield RT, Green CP, Toms PS, Scourse J, Scaife R, et al. (2011) The Middle Pleistocene deposits and archaeology at Broom. In Basell LS, Brown AG, Toms PS (editors) 2011. The Quaternary of the Exe and Adjoining Areas. Field Guide. London: Quaternary Research Association, London, 103–127.
8. McNabb J (2012) Dissent with Modification: Human Origins, Palaeolithic Archaeology, and Evolutionary Anthropology in Victorian Britain, 1859–1901, Archaeopress.
9. Ashton NM, Hosfield RT (2010) Mapping the human record in the British early Palaeolithic: Evidence from the Solent river system. Journal of Quaternary Science 25: 737–753.
10. Wilkinson K (2007) The Lower and Middle Palaeolithic of Hampshire. Oxford: Oxford Archaeology.
11. Davis JC (1973) Statistics and Data Analysis in Geology. New York: Wiley International.
12. Xu G, Wang W, Bae CJ, Huang S, Moa Z (2011) Spatial distribution of Paleolithic sites in Bose Basin, Guangxi, China. Quaternary International 281: 10–13.
13. Johnston B (2010) An investigation of the distribution of open archaeological sites in the Upper Kickapoo Valley Archaeological District, Vernon County, Wisconsin. Wisconsin: Report University of Wisconsin.
14. Roe D (1968) A gazetteer of British Lower and Middle Palaeolithic Sites. York: Council for British Archaeology.
15. Roberts MB (1999) Archaeology. In Roberts MB, Parfitt SA (editors) Boxgrove: A Middle Pleistocene hominid site at Eartham Quarry, Boxgrove, West Sussex. London: Archaeological Report 17, English Heritage, 309–426.
16. Brown AG (2013) Gulls, Dates and Axes: Recent Quaternary Research in SW England. Geosciences in South-West England 13: 1–11.
17. Basell LS, Brown AG, Hosfield RT, Toms P (2011) The Geoarchaeology of Palaeolithic Rivers of South West Britain. In Brown AG, Basell LS, Butzer KW (editors) Geoarchaeology, Climate Change and Sustainability. Geological Society of America Special Publication 476: 23–36.
18. Winton V (2004) A Study of Palaeolithic Artefacts from Selected Deposits Mapped as Clay-with-Flints of Southern England. British Archaeological Reports British Series 360.
19. Scott-Jackson JE, Scott-Jackson WB (2008) Alternative location modelling of a Lower/Middle Palaeolithic site on deposits mapped as Clay-with-flints in southern England, UK. Oxford: PADMAC Unit Working Paper. University of Oxford.
20. Hosfield R, Green CP (In press) Quaternary History and Palaeolithic Archaeology in the Axe Valley at Broom, South West England. Oxford: Oxbow.
21. Lewin J, Brewer A (2002) Laboratory simulation of clast abrasion. Earth Surface Processes and Landforms 27: 145–164.
22. Brown AG, Basell LS, Toms PS, Scrivener RC (2009) Towards a budget approach to Pleistocene terraces: preliminary studies using the River Exe in South West England. Proceedings of the Geologists' Association 120: 275–281.
23. Brown AG, Basell LS, Toms PS, Bennett J, Hosfield RT et al. (2010) Late Pleistocene Evolution of the Exe Valley. A Chronstratigraphic Model of Terrace Formation and its Implications for Palaeolithic Archaeology. Quaternary Science Reviews 29: 897–912.
24. Hosfield RT (2001) The lower Palaeolithic of the Solent: 'site' formation and interpretive frameworks. In Wenban-Smith FF, Hosfield RT (editors) Palaeolithic Archaeology of the Solent River. Lithic Studies Society Occasional paper 7: 85–98.
25. Basell LS, Brown AG (2011) The archaeological context of the Exe Valley and adjoining areas. In Brown AG, Basell LS, Toms PT (editors)The Quaternary of the Exe Valley and Adjoining Areas. Field Guide. London: Quaternary Research Association, 2–15.
26. Schreve DC, Bridgland DR, Allen P, Blackford JJ, Gleed-Owen CP et al. (2002) Sedimentology, palaeontology and archaeology of late Middle Pleistocene River Thames deposits at Purfleet Essex, UK. Quaternary Science Reviews 21: 1423–1464.
27. Gowlett JAJ (2005) Seeking the Palaeolithic Individual in East Africa and Europe during the Lower-Middle Pleistocene. In: Gamble CS, Porr M (editors) The hominid individual in context: archaeological investigations of Lower and Middle Palaeolithic landscapes, locales and artefacts. London: Routledge, 50–67.
28. Basell LS, Brown AG (2012) Rubirizi: a New Stone Age Site in Uganda with Giant Bifaces. PAST 71: 3–5.
29. Cohen KM, MacDonald K, Joordens JCA, Roebroeks W, Gibbard P (2012) The earliest occupation of north-west Europe: a coastal perspective. Quaternary International 271: 70–83.
30. Penkman KEH, Preece RC, Bridgland DR, Keen DH, Meijer T, et al. (2011) A chronological framework for the British Quaternary based upon Bithynia opercula. Nature 476: 446–449.
31. Harding P, Bridgland DR, Allen P, Bradley P, Grant MJ, et al. (2012) Chronology of the Lower and Middle Palaeolithic in NW Europe: developer-funded investigations at Dunbridge, Hampshire, southern England. Proceedings of the Geologists Association 123: 584–607.
32. Draper C J (1951) Stone industries from Rainbow Bar, Hants. Archaeological Newsletter 3: 147–149.
33. Wenban-Smith FF, Gamble C S, ApSimon AM (2000) The Lower Palaeolithic site at Red Barns, Portchester: bifacial technology, raw material quality, and the organisation of archaic behaviour. Proceedings of the Prehistoric Society 66: 209–255.
34. Wenban-Smith FF, Bates MR, Bridgland DR, Marshall GD, Schwenninger J-L (2009) The Pleistocene sequence at Priory Bay, Isle of Wight (SZ 635 900). In Briant RM, Bates, MR, Hosfield RT, Wenban-Smith FF (editors) The Quaternary of the Solent basin and West Sussex Raised Beaches Field Guide. London: Quaternary Research Association, 123–137.
35. Briant RM, Wenban-Smith FF, Schwenninger J-L (2009) Solent river gravels at Barton on Sea, Hampshire SZ 230 930. In Briant RM, Bates, MR., Hosfield, RT, Wenban-Smith FF.(editors) The Quaternary of the Solent basin and West Sussex Raised Beaches Field Guide. London: Quaternary Research Association, 161–170.
36. Smith RA (1926) A Guide to the Antiquities of the Stone Age. 3rd Edn., London: British Museum.
37. Bridgland DR, Field MH, Holmes JA, McNabb J, Preece RC, et al. (1999) Middle Pleistocene Thames-Medway deposits at Clacton-on-Sea, England: reconsideration of the biostratigraphical and environmental context of the type Clactonian Paleaolithic industry. Quaternary Science Reviews 18: 109–146.
38. Bridgland DR, Schreve DC (2004) Quaternary lithostratigraphy and mammalian biostratigraphy of the Lower Thames terrace system, South-East England Lithostratigraphie quaternaire et biostratigraphie mammalienne du système de terrasses de la basse vallée de la Tamise (Sud-Est de l'Angleterre). Quaternaire 15 : 29–40.
39. Antoine P, Auguste P, Bahain J-J, Chaussé D, Falguères C, et al. (2010) Chronostratigraphy and palaeoenvironments of Acheulean occupations in Northern France (Somme, Seine and Yonne Valleys). Quaternary International 223–224: 456–461.
40. Limondin-Lozouet N, Nicoud E, Antoine P, Auguste P, Bahain J-J, et al. (2009) Oldest evidence of Acheulean occupation in the Upper Seine valley (France) from an MIS 11 tufa at La Celle, Quaternary International 223–224: 299–311.
41. Lisiecki LE, Raymo ME (2005) A Pliocene-Pleistocene stack of 57 globally distributed benthic D18O records. Paleoceanography 20: PA1003, doi:10.1029/2004PA001071.
42. Westaway R (2011) Quaternary fluvial sequences and landscape evolution in Devon and Somerset. In Basell LS, Brown AG, Toms PS (editors) The Quaternary of the Exe Valley and Adjoining Area. Field Guide. London: Quaternary Research Association, 27–46.
43. Briant RM, Keen DH, Kilfeather AA, Parfitt S, Penkman KEH, et al. (2012) Integrated chronological control on an archaeologically significant Pleistocene river terrace sequence: the Thames-Medway, eastern England. Proceedings of the Geologists Association 123: 87–108.
44. Irving B (1996) The ichthyofauna from the Waechter excavations, Barnfield Pit, Swanscombe. In Conway B, McNabb J, Ashton N (editors) Excavations at Barnfield Pit, Swanscombe, 1968–72:. London: The British Museum, Occasional Paper 94, 145–147.
45. Robinson E (1996) The ostracod fauna from the Waechter excavations. In: Conway B, McNabb J, Ashton, N. (editors), Excavations at Barnfield Pit, Swanscombe, 1968–72. British Museum Occasional Paper 94: 187–191.
46. Roberts MB, Parfitt SA (1999) Boxgrove. A Middle Pleistocene hominid site at Eartham Quarry, Boxgrove, West Sussex. Archaeological Report 17, London: English Heritage.
47. Holmes JA, Atkinson T, Darbyshire FDP, Horne DJ, Joordens J, et al. (2009) Middle Pleistocene climate and hydrological environment at the Boxgrove hominin site (West Sussex) from ostracod records. Quaternary Science Reviews 29: 1515–1527.
48. Singer R, Gladfelter BG, Wymer JJ (1993) The Lower Palaeolithic Site at Hoxne, England. Chicago University Press, Chicago.
49. Ashton N, Lewis SG, Parfitt SA, Penckman KEH, Coope GR (2008) New evidence for complex climate change in MIS 11 from Hoxne, Suffolk, UK. Quaternary Science Reviews 27: 652–668.
50. Pontzer H, Raichlen D, Wood BM, Mabulla AZP, Racette SB, et al. (2012) Hunter-Gatherer Energetics and Human Obesity. PLOS ONE 7, e40503–.

51. Milton K (2000) Hunter-gatherer diets–a different perspective. Am. J. Clin Nutr 71: 665–667.
52. Kuipers RS, Joordens JCA, Muskiet FAJ (2012) A multidisciplinary reconstruction of Palaeolithic nutrition that holds promise for the prevention and treatment of diseases of civilisation. Nutritional Research Reviews 25: 96–129.
53. Godwin H (1975) History of the British Flora. Cambridge: Cambridge University Press.
54. Schreve DC (2001) Differentiation of the British late Middle Pleistocene interglacials: the evidence from mammalian biostratigraphy. Quaternary Science Reviews 20: 1693–1705.
55. Zagwijn WH (1978) A macroflora of Holsteinian age from the northern part of The Netherlands. Rev. Palaeobot. Palynol., 26: 243–248.
56. Antoine P, Limondin-Lozouet N, Auguste P, Locht J-L, Ghaleb B, et al. (2006) La tuf de Caours (Somme, France): mise en evidence d'une sequence Eemienne et d'un site Palaéolithique associé. Quaternaire 17: 281–320.
57. Coope GR (2010) Coleopteran faunas as indicators of interglacial climates in central and southern England. Quaternary Science Reviews 29: 1507–1514.
58. Couplan F (1984) La cuisine sauvage - Comment accommoder 1000 plantes oubliées. Encyclopédie des plantes comestibles de l'Europe - Volume 2. Paris: Debard (réédition Paris : Sang de la Terre, Paris, 2010).
59. Grime JP, Hodgson JG, Hunt R (1990) The Abridged Comparative Plant Ecology. London: Unwin Hymen.
60. Niethammer J, Krapp F (2013) Handbuch der Säugetiere Europas DVD-Rom. Wiesbaden: Akademische Verlagsgesellschaft.
61. Stemerdink C, Maddy D, Bridgland DR, Veldkamp A (2010) The construction of a palaeodischarge time series for use in a study of fluvial system development of the Middle to Late Pleistocene Upper Thames. Journal of Quaternary Science 25: 447–460.
62. Turner C, West RG (1968) The subdivision and zonation of interglacial periods. Eiszeitalter und Gegenwart 19: 93–101.
63. Schreve DC (2001) Mammalian evidence from fluvial sequences for complex environmental change at the oxygen isotope sub-stage level. Quaternary International 79: 65–74.
64. Mihlbachler MC, Rivals F, Solounias N, Semprebon GM (2011) Dietary Change and Evolution of Horses in North America. Science 331: 1178–1181.
65. Torres RT, Carvalho JC, Panzacchi M, Linnell JDC, Fonseca C (2011) Comparative use of forest habitats by roe deer and moose in a human-modified landscape in southeastern Norway during winter. Ecol Res 26: 781–789.
66. Owen-Smith N, Fryxell JM, Merrill EH (2010) Foraging theory upscaled: the behavioural ecology of herbivore movement. Philosophical; Transactions of the Royal Society B, 365: 2267–2278.
67. Fa JE, Stewart JR, Lloveras L, Vargas JM (2013) Rabbits and hominin survival in Iberia. Journal of Human Evolution 64: 233–241.
68. Harrison CJO, Stewart JR (1999) Avifauna. In Roberts MB, Parfitt SA (editors) Boxgrove. A Middle Pleistocene hominid site at Eartham Quarry, Boxgrove, West Sussex. Archaeological Report 17, London: English Heritage, 181–186.
69. Allen L (2005) Multiple micronutrients in pregnancy and lactation: an overview. American Journal of Clinical Nutrition 81: 1206S–1212S.
70. Navarrete A, van Schaik CP, Isler K (2011) Energetics and the evolution of human brain size. Nature 480: 91–93.
71. Potts R (2011) Big brains explained. Nature 480: 43–44.
72. Burdge GC, Calder PC (2006) Dietary alpha-linolenic acid and health-related outcomes: a metabolic perspective. Nutrition Research Reviews 19: 26–52.
73. Cunane SC (2005) Origins and evolution of the western diet: implications of iodine and seafood intakes for the human brain. American Journal of Clinical Nutrition 82: 483–484.
74. Hoile SP, Irvine NA, Kelsall CJ, Sibbons C, Feunteun A et al. (2012) Maternal fat intake in rats alters 20:4n-6 and 22:6n-3 status and the epigenetic regulation of Fads2 in offspring liver. Journal of Nutritional Biochemistry doi:pii: S0955-2863(12)00245-8. 10.1016/j.jnutbio.2012.09.005.
75. Cunane SC, Harbig LS, Crawford MA (1993) The importance of energy and nutrient supply in human brain evolution. Nutrition and Health 9: 219–235.
76. Burdge GC (2004) Alpha-linolenic acid metabolism in men and women: nutritional and biological implications. Current Opinion in Clinical Nutrition and Metabolic Care. 7: 137–44.
77. Godfrey KM, Lillycrop KA, Burdge GC,Gluckman PD, Hanson MA (2007) Epigenetic mechanisms and the mismatch concept of the developmental origins of health and disease. Paediatric Research 61: 5R–10R.
78. Crawford MA, Bloom M, Broadhurst CL, Schmidt WF, Cunnane SC, et al. (1999) Evidence for the Unique Function of Docosahexaenoic Acid During the Evolution of the Modern Hominid Brain. Lipids 34: S39–S47.
79. Cordain L., Watkins BA, Mann NJ (2001) Fatty acid composition and energy density of foods available to African hominids: evolutionary implications for human brain development. World Rev. Nutr. Diet. 90: 144–161.
80. Bocherens H, Drucker DG, Billiou D, Patou-Mathis M, Vandermeersch B (2005) Isotopic evidence for diet and subsistence pattern of the Saint-Césaire I Neanderthal: Review and use of a multi-sourcing mixing model. J.Hum. Evol. 49: 71–87.
81. Cordain L, Brand-Miller J, Eaton SB, Neil Mann N, Holt SHA, et al. 2000. Plant-animal subsistence ratios and macronutrient energy estimations in worldwide hunter-gatherers. American Journal of Clinical Nutrition 71: 682–692.
82. Speth JD, Spielmann KA (1983) Energy source, protein metabolism, and hunter-gatherer subsistence strategies. J. Anthropol Archaeol. 2: 1–31.
83. Sponheimer M, Lee-Thorpe JA (1999) Isotopic evidence for the diet of an early hominid, Australopithecus africanus. Science 283: 368–370.
84. Wrangham R, Cheney D, Seyfarth R, Sarmiento E (2009) Shallow-water habitats as sources of fallback foods for hominins. American Journal of Physical Anthropology 140: 630–642.
85. Kuipers RS, Luxwolda MF, Dijck-Brouwer DAJ, Eaton SB, Crawford MA, et al. (2010) Estimated macronutrient and fatty acid intake from an east African Palaeolithic diet. British Journal of Nutrition 104: 1666–1687.
86. Herries AIR, Hopley PJ, Adams JW, Curnoe D, Maslin MA (2010) Letter to the Editor: Geochronology and palaeoenviroments of southern African hominin-bearing localities – a reply to Wrangham et al, 2009. "Shallow-water habitats as sources of fall back foods for hominins". American Journal of Physical Anthropology 143: 640–646.
87. Coles B (2006) Beavers in Britain's Past. Oxford: Oxbow Books.
88. Parfitt SA (1999) Mammalia. In Roberts MB, Parfitt SA (editors) Boxgrove. A Middle Pleistocene hominid site at Eartham Quarry, Boxgrove, West Sussex. Archaeological Report 17, London: English Heritage, 197–273.
89. Conway B, McNabb J, Ashton N (1996) Excavations at Barnfield Pit, Swanscombe, 1968–72. London: British Museum Occasional Paper 125.
90. Stuart AJ, Wolff RG, Lister AM, Singer R, Egginton JM (1993) Fossil vertebrates. In Singer, R., Gladfelter, B. G. and Wymer, J. J. (editors) The Lower Palaeolithic Site at Hoxne, England. Chicago: University of Chicago Press, 163–206.
91. Mayhew DF (1978) Reinterpretation of the extinct beaver Trogontherium (Mammalia, Rodentia). Philosophical Transactions of the Royal Society of London B281: 407–438.
92. Schick K, Toth N (1993) Making Silent Stones Speak: Human Evolution and the Dawn of Technology. New York: Simon & Schuster.
93. O Brien JS, Sampson EL (1965) Lipid composition of the normal human brain: gray matter, white matter, and myelin. Journal of Lipid Research 6: 537–544.
94. Carpenter KJ (1986) The History of Scurvy and Vitamin C. Cambridge: Cambridge University Press.
95. Li D, Sinclair AJ, Mann NJ, Turner A, Ball MJ (2000) Selected micronutrient intake and status in men with differing meat intakes, vegetarians and vegans. Asia Pac. J. Clin. Nutr. 9: 18–23.
96. Hardy BL (2010) Climatic variability and plant food distribution in Pleistocene Europe: Implications for Neanderthal diet and subsistence. Quaternary Science Reviews 29: 662–679.
97. Mailer MT, Harrison E, Wayne EJ (1964) Absorption of Iodine from ingested watercress. Nature 201: 403–404.
98. Richards MP, Trinkhause E (2009) Isotopic evidence for the diets of European Neanderthals and early modern humans. Proceedings of the national Academy of Science 106: 16034–16039.
99. Pyke GH (1984) Optimal foraging theory: a critical review. Annual Review of Ecological Systematics 15: 523–575.
100. Stiner MC (1994) Honor Among Thieves: A Zoological Study of Neanderthal Ecology. Princeton University Press, Princeton.
101. Gamble CS, Steele J (1999) Hominid ranging patterns and dietary strategies. In Ullrich, H. (Ed.) Hominid Evolution: Lifestyles and survival Strategies. Schwelm, Archea, 396–409.
102. Daujeard C, Fernandes P, Guadelli J-L, Moncel M-H, Santagata C, et al. (2012) Neanderthal subsistence strategies in Southeastern France between the plains of the Rhone Valley and the mid-mountains of the Massif Central (MIS 7 to MIS 3). Quaternary International 252: 32–47.
103. Eaton SB, Konner M (1985) Paleolithic nutrition. A consideration of its nature and current implications. N. Engl. J. Med. 312: 283–289.
104. Hayden B (1981) Subsistence and ecological adaptations of modern hunter/gatherers. In: Hardy R, Teleki, G (editors), Omnivorous Primates: Gathering and Hunting in Human Evolution. New York: Columbia,344–421.
105. Owen LR (2005) Distorting the Past: Gender and Division of Labor in the European Upper Paleolithic. Tübingen: Kerns Verlag,.
106. Hunn ES (1981) On the relative contribution of men and women to subsistence among the hunter-gatherers of the Columbian Plateau: a comparison with ethnographic atlas summaries. J. Ethnobiology 1: 124–134.
107. Richards MP, Hedges RM (2000) Focus: Gough's Cave and Sun Hole Cave human stable isotope values indicate a high animal protein diet in the British Upper Palaeolithic. J. Archaeol. Sci. 27: 1–3.
108. Ambrose AH, Buikstra J, Krueger H (2003) Status and gender differences in diet at Mound 72, Cahokia, revealed by isotopic analysis of bone. J. Anth. Arch. 22: 217–226.
109. Henry AG, Brooks AS, Piperno DR (2011) Microfossils in calculus demonstrate consumption of plants and cooked foods in Neanderthal diets (Shanidar III, Iraq; Spy I and II, Belgium). PNAS 108: 486–491.
110. Fiorenza L, Benazzi S, Tausch J, Kullmer O, Bromage TG, et al. (2011) Molar macrowear reveals Neanderthal eco-geographic dietary variation. PLOS ONE 6(3): e14769. doi:10.1371/journal.pone.0014769.
111. Chase JM, Leibold MA (2003) Ecological Niches. Linking Classical and Contemporary Approaches. University of Chicago Press, Chicago.
112. Laland KN, O'Brien MJ (2010) Niche construction theory and archaeology. J Archaeol Method Theory 17: 303–322.
113. Lewontin RC (1983) The organism as the subject and object of evolution. Scientia 118: 68–82.

114. Erlandson JM (2001) The archaeology of aquatic adaptations: paradigms for a new millennium. 2001. Journal of Archaeological Research 9: 287–350.
115. Holloway RL (1967) The evolution of the human brain: some notes towards a synthesis between neural structure and the evolution of complex behaviour. General Systems 12: 3–19.
116. Via S, Lande R (2012) Genotype-environment interaction and the evolution of phenotypic plasticity. Evolution 39: 505–522.
117. Potts R (1996) Evolution and climate variability. Science 273: 922–923.
118. Potts R (1996) Environmental hypotheses of hominin evolution. Yearbook of Physical Anthropology 41: 93–136.
119. Grove M (2011) Speciation, diversity, and Mode 1 technologies: The impact of variability selection. Journal of Human Evolution 61: 306–319.
120. Rose J (1994) Major river systems of central and southern Britain during the early and middle Pleistocene. Terra Nova 6: 435–443.
121. Powell A, Shennan S, Thomas MG (2009) Late Pleistocene demography and the appearance of modern human behaviour. Science 324: 1298–1301.
122. Wood B, Brooks A (1999) Human evolution. We are what we eat. Nature 400: 219–220.
123. Schick K (1987) Modeling the formation of Early Stone Age artifact concentrations. Journal of Human Evolution 16: 789–807.
124. Hockett B, Haws J (2003) Nutritional Ecology and Diachronic Trends in Paleolithic Diet and Health. Evolutionary Anthropology 12: 211–216.

9

Adaptation and Convergent Evolution within the *Jamesonia-Eriosorus* Complex in High-Elevation Biodiverse Andean Hotspots

Patricia Sánchez-Baracaldo[1]*, Gavin H. Thomas[2]*

1 School of Geographical Sciences, University of Bristol, Bristol, United Kingdom, **2** Department of Animal and Plant Sciences, University of Sheffield, Sheffield, South Yorkshire, United Kingdom

Abstract

The recent uplift of the tropical Andes (since the late Pliocene or early Pleistocene) provided extensive ecological opportunity for evolutionary radiations. We test for phylogenetic and morphological evidence of adaptive radiation and convergent evolution to novel habitats (exposed, high-altitude páramo habitats) in the Andean fern genera *Jamesonia* and *Eriosorus*. We construct time-calibrated phylogenies for the *Jamesonia-Eriosorus* clade. We then use recent phylogenetic comparative methods to test for evolutionary transitions among habitats, associations between habitat and leaf morphology, and ecologically driven variation in the rate of morphological evolution. Páramo species (*Jamesonia*) display morphological adaptations consistent with convergent evolution in response to the demands of a highly exposed environment but these adaptations are associated with microhabitat use rather than the páramo *per se*. Species that are associated with exposed microhabitats (including *Jamesonia* and *Eriorsorus*) are characterized by many but short pinnae per frond whereas species occupying sheltered microhabitats (primarily *Eriosorus*) have few but long pinnae per frond. Pinnae length declines more rapidly with altitude in sheltered species. Rates of speciation are significantly higher among páramo than non-páramo lineages supporting the hypothesis of adaptation and divergence in the unique Páramo biodiversity hotspot.

Editor: Simon Joly, Montreal Botanical Garden, Canada

Funding: This work was supported by a National Science Foundation Doctoral Dissertation Improvement Grant (9801245), the Vice Chancellor for Research Fund Award, and several grants from the Department of Integrative Biology, University of California at Berkeley. PSB has received funding support from the Royal Society Dorothy Hodgkin Fellowship and Colciencias, Colombia. GHT was supported by a NERC Postdoctoral Fellowship NERC NE/G012938/1. The funders had no role in study design, data collection and analysis, decision to publish, or preparation of the manuscript.

Competing Interests: The authors have declared that no competing interests exist.

* Email: p.sanchez-baracaldo@bristol.ac.uk (PSB); gavin.thomas@sheffield.ac.uk (GHT)

Introduction

The Andes are one of the most species-rich biodiversity hotspots in the world [1] with an estimated 45,000 plant species, 44% of which are endemic [2]. As high elevation ecosystems known as páramos emerged during the last major uplifting of the Northern Andes [3–5], during the Pliocene and/or Pleistocene (2–4 Ma) new ecological opportunities became available for species radiations to occur [6–8]. Recent studies suggest that average speciation rates in páramos exceed those of other major plant diversity hotspots [9]. Notable examples of recent Andean radiating lineages include the extremely rapid speciation of the *Lupinus* clade [10], and adaptive radiations of *Valeriana* and *Gentianella* [10–12]. These radiations contribute to the high levels of endemism in the Northern Andean páramo ecosystems, which contribute>60% of the total plant species richness [7,10,13–15]. High levels of endemism are likely associated with glacial cycles during the Pleistocene [3–5]. During this period, the reduction and fragmentation of the páramo into smaller habitat islands during interglacial periods [16] may have contributed to periods of isolation. The fragmented and isolated distribution of páramo habitats is considered akin to island systems and may have enhanced the potential for species diversification [10].

Plant radiations in páramos have several possible biogeographic origins that can broadly be categorised as temperate (both Northern and Southern hemisphere) or tropical montane forest. Northern hemisphere temperate origins are perhaps the dominant source for the rapid origination and recent diversification of numerous endemic plant groups (e.g. *Draba, Gentianella, Lupinus, Valeriana*, and *Viburnum*; [16]). While herbaceous elements mainly diversify in páramo, montane forest tree lineages generally failed to diversify [10,17]. Colonisation from Southern hemisphere temperate regions include *Azorella, Ourisia, Calceolaria* and *Puya* but overall clades with southern origins seem to have contributed less to floral diversity in the páramo than clades from the Northern hemisphere [15,18,19].

Because páramos are characterised by extreme environmental conditions such as strong winds, high levels of insolation, and cool temperatures ranging daily from $-2°C$ to $12°C$ [7,20,21], the success of colonising floras from temperate regions has been attributed to potential exploitation of island-like ecological opportunities by lineages already well adapted to the cooler, more exposed habitats encountered in the high altitude páramo [10]. Examples of morphological adaptations to high altitude include microphyllous, pubescent and sclerophyllous leaves as well as growth-forms such as rosette plants, cushion plants, dwarf shrubs and geophytes [7]. In contrast, lineages with a tropical montane origin are unlikely to be pre-adapted to novel high-altitude habitats. Indeed, few studies have shown plant groups evolving traits *in situ* at high elevations [18]. Although some páramo radiations, such as the species rich genus *Espeletia* in the Asteraceae [6,22] and *Huperzia* [23] likely had origins in the surrounding montane forest [6,23], little is known about patterns of morphological evolution or diversification associated with transitions to high altitude. Here we test for morphological adaptation to, and divergence at, high altitudes in the *Jamesonia-Eriosorus* complex of Andean ferns.

The *Jamesonia-Eriosorus* complex (Fig. 1) has a probable Southern hemisphere Brazilian origin [24,25]. The clade consists of two genera: the paraphyletic *Eriosorus* and polyphyletic *Jamesonia* that have expanded into cloud forest and páramo ecotones throughout the Northern and Central Andes [24,25]. *Jamesonia* are typically found in páramo or exposed parts of the sub-páramo at altitudes ranging from ca. 1500 to 5000 metres. In contrast, *Eriosorus* [26] species are mainly found in cool and moist highlands such as cloud forest and in sheltered and shady microhabitats within the sub-páramo and páramo at altitudes ranging from ca. 600 to 4100 m with most species occurring above 2200 m. Previous studies of the *Jamesonia-Eriosorus* complex focused on phylogeny, specifically, resolving their phylogenetic affinities within the subfamily Taenitidoideae [25], and on the relationship between biogeography and phylogeny of the complex itself [24]. Furthermore, Sánchez-Baracaldo [24] hypothesized that *Jamesonia* species have a distinctive morphology characterized by indeterminate growth (not fully developed leaves) and an increased number of pinnae per frond that was likely favoured in the extreme environmental conditions prevailing in páramo ecosystems. Other potential adaptations include coriaceous pinnae, xeromorphic leaves, and pubescence [27]. However, whether these adaptations are associated with páramo specifically, with exposed rather than sheltered microhabitats, or more generally to upward shifts along an altitudinal gradient is untested.

In this study phylogenetic analyses confirm three Andean clades with high levels of convergent evolution in which *Jamesonia* species evolved independently likely adapting to the extreme environmental characteristics of páramo habitats [24]. Bayesian molecular clock analyses point to a Pleistocene origin for the three main lineages of *Jamesonia-Eriosorus* into both montane and páramo habitats with notably higher speciation rates in the páramo. We test whether putative morphological adaptations to exposed environments represent responses to altitudinal gradients, microhabitat (exposed vs. sheltered) or responses to the unique challenges of páramo ecosystems.

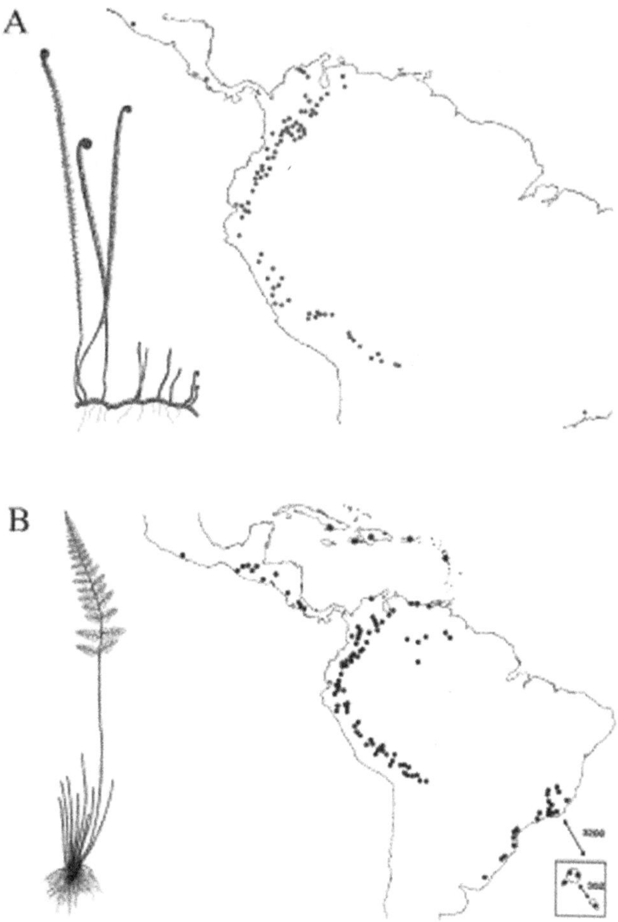

Figure 1. Distribution of known locations of *Jamesonia* (A) and *Eriosorus* (B) [26,27]. The fern diagrams represent the generalized frond architecture of *Jamesonia* and *Eriosorus*. *Jamesonia* exhibit indeterminate growth (retention of fiddle head) with a high number of pinnae per frond; in contrast, the *Eriosorus* have fully developed with approximately 13 to 14 pinnae per frond with the exception of *Eriosorus flexuosus*.

Materials and Methods

Ethics statement

None of the *Jamesonia* and *Eriosorus* species are or were endangered or protected species. All samples were collected during 1998–99. All specimens studied were imported into the United States under a CITES permit (the Convention on International Trade in Endangered Species of Wild Fauna and Flora) granted to the University Herbaria, at UC Berkeley (Director Prof Brent Mishler). Plant specimens were measured before they were mounted in to herbarium specimens. The main collection is housed in the University Herbaria at UC Berkeley.

Venezuela: Plant collections were made in collaboration with Enrique La Marca from Universidad de Los Andes in Merida. The Ministerio de Ambiente granted the permit in Venezuela.

Colombia: All samples were collected in collaboration with the Instituto de Biodiversidad Alexander von Humboldt (Direction at the time Dr Cristián Samper). The Ministerio de Ambiente, Bogota, granted the permit in Colombia.

Ecuador: All collections were made in collaboration with Dr Hugo Navarrete from the Herbario de la Universidad Catolica, Ecuador. The Ministerio de Ambiente, Quito, granted the permit in Ecuador.

Peru: All collections were made in collaboration Asuncion Cano from El Museo de Historia Natural, Lima. The Ministerio de Ambiente, Lima, granted the permit in Peru.

Additional samples from Bolivia were sent by Jasivia Gonzales who at the time was working at Herbario Nacional de Bolivia and samples from Brazil were sent by Jefferson Prado and working at the Instituto Botanico, Sao Paolo, Brazil.

Ecological and morphological data

Current taxonomic treatments [28] list 59 species of *Jamesonia* and *Eriosorus*. However, recent phylogenetic studies [29] show that three species of *Eriosorus* (*E. myriophylla*, *E. schwackeana*, and *E. areniticola*) are more closely to other taenitidoid genera and propose a new genus name, *Tryonia*, for these species along with a fourth species, *E. sellowianus*. *Eriosorus sellowianus* has not yet been phylogenetically sampled but has previously been treated as a subspecies of *E. schwackeana*. We follow the phylogenetic revision resulting in a total of 55 species. For the majority of our analyses we use a subset of species but defining the total richness of the clade is important for our analyses of state dependent diversification (see the section *BiSSE: state dependent diversification* below for details).

We categorised species as present (0) or absent (1) from páramo. We scored species as present in the páramo if they occur in any of three broad páramo habitat zones based on overall altitude, vegetation structure and geographic location: 1) super-páramo (4000–5000 m) is the ecotone between permanent snow and grass páramo below and consists of coarse vegetation growing on rocky scree; 2) grass páramo (3500–4100 m) has continuous vegetation, mainly tussock or bunch grasslands; and 3) sub-páramo (2800/3000–3500 m) is the ecotone between grass páramo and montane forest and is the most floristically diverse zone [21]. Using the above description a summary of *Jamesonia* and *Eriosorus* habitats are found in Table S1. We based categorisation on field observations and monographs for the genera [26,27].

For 25 species we also categorise species according microhabitat based on field observations by PSB from 68 localities. We classified microhabitat as sheltered (e.g., wet or moist shaded forest borders, shaded places at edge of boulders or in caves) or exposed (e.g., among rocks and/or bare soil, open fields among grasses, on very exposed slopes and cliffs); see Table S1 for a more detailed list of microhabitats. Each locality was classified independently. The localities span Costa Rica, Venezuela, Colombia, Ecuador, Peru, Bolivia and Brazil (Fig. 1).

We obtained morphological and altitudinal data for specimens from 55 out of 68 localities, sampling three to four plant specimens per locality. We measured two characters: 1) the number of pinnae along fronds in sexually mature plants (defined as those bearing sporangia), and 2) pinna length (millimetres) of the middle frond. We have chosen these morphological traits since they are variable and likely ecologically significant and we hypothesised that their variation evolved in response to exposed environmental conditions typically found in páramo.

Sequence data and phylogenetic analyses

We re-analysed previously published [24] nucleotide data from samples taken from each of the 68 localities from the nuclear external transcribed spacer (ETS) within the 18S–26S nuclear ribosomal DNA (1152 base pairs), and the plastid *rps4* (576 bp) and intergenic spacer *rps4-trnS* (415 bp). These are rapidly evolving markers that have previously been used to investigate taxonomic relationships at low taxonomic levels [30,31]. The same data set was previously analysed using parsimony [24]; in our reanalysis we use Bayesian relaxed clock methods to simultaneously estimate tree topology and divergence times. The data were aligned by eye with Se-Al v 2.0a11 [32]. We determined the most appropriate partitioning scheme and substitution models among loci using PartitionFinder [33]. We allowed PartitionFinder to search all possible combinations form a total of five possible partitions (ETS, *rps4* divided into codon positions, and *rps4-trnS*). The selected scheme was a TrN+Γ model for the intergenic spacer *rps4-trnS*, a K80+I model for the coding plastid *rps4* region (the best scheme combined the three codon positions), and a TVM+Γ model for ETS.

We used BEAST v. 1.8.0 [34] to reconstruct and date phylogenetic trees using a Bayesian uncorrelated relaxed molecular clock [35]. We calibrated the tree simultaneously with tree estimation using a log-normal uncorrelated relaxed clock [35], a Yule calibration prior [36], and a single fossil constraint. While there are reports for both *Jamesonia* and *Eriosorus* spores in the geological record, there are no morphological characters that can readily tell them apart and so assigning specific spores to specific nodes in the tree for calibration is not possible. We therefore used the first appearance of a spore considered to be *Eriosorus* to calibrate the root of the tree. The spore is dated at 3.5 Ma with an origin from lower elevations [3,37]. We treated this as a minimum age constraint and applied an exponential prior (offset = 3.5, mean = 0.5006). This sets 95% of the prior distribution between 3.5 and 5 Ma, consistent with an origin coincident with the emergence of the páramo. Setting prior distributions on node ages is challenging and due to the lack of robust data with which to estimate a soft maximum constraint, we are cautious in interpretation of the exact timings for major divergences in the *Jamesonia-Eriosorus* clade. We ran four independent analyses, each with 50,000,000 generations. We used Tracer v1.5.0 [38] to assess convergence and mixing of parameters. For all subsequent analyses we use a sample of 1000 trees from the posterior distribution to account for phylogenetic uncertainty.

Predictors of morphological evolution

To visually assess whether *Jamesonia* and *Eriosorus* morphology clusters phylogenetically or by habitat we plotted the number of leaves per frond against pinnae length using the phylomorphospace method [39] implemented in the R package phytools [40]. We then tested the relationship between morphology and habitat and altitude using phylogenetic generalized least squares (pgls) in the R package caper [41]. We fitted the following four sets of explanatory variables to each of pinnae length and leaves per frond: (1) altitude; (2) habitat; (3) altitude + habitat; (4) altitude x habitat. All pgls models were conducted while estimating the strength of phylogenetic signal in the model residuals using Pagel's λ [42,43] and were conducted on full trees. Each analysis was applied to a posterior sample of 1000 trees. We compared the fit of the models using the small-sample Akaike Information Criterion (AICc) and Akaike weights for each tree sample and then took the average AICc and Akaike weights across all 1000 trees to infer the overall best model. We averaged the parameter estimates for the best model across all 1000 trees.

BiSSE: state dependent diversification

We used the Binary State Speciation and Extinction model (BiSSE; [44,45] to test differences in diversification dynamics between páramo and non-páramo lineages. The BiSSE method simultaneously estimates speciation and extinction rates associated with each habitat. A full BiSSE model estimates six parameters: speciation and extinction rates for páramo and non-páramo respectively, and rates of anagenetic character change from non-páramo to páramo and from páramo to non-páramo. We predict that speciation rates should be higher for páramo lineages since this represents a geologically recent habitat. Similarly, we predict that if niche space is limiting in older non-páramo habitats but not in younger páramo habitats, transition rates towards páramo from non-páramo are likely to exceed the reverse.

Our main phylogenetic analyses include specimens from multiple localities from each described species but do not include all described species of *Jamesonia* or *Eriosorus* [26,27]. For the BiSSE analyses we sampled 1000 phylogenies randomly from the posterior distribution. For each phylogeny in turn we randomly sampled one exemplar specimen for each described species. This left 1000 phylogenies with one representative of each species in our data set and allows us to account for phylogenetic uncertainty, particularly in species delimitation. BiSSE requires that all known species are accounted for, either in the phylogeny directly or entered into the analyses as a sampling fraction. We assigned all described species that were absent from our phylogenies to either páramo (22 sampled species from 43 in total) or non-páramo (4 sampled species from 12 in total) habitat based on fieldwork observations, herbarium specimens (University Herbarium at the University of Berkeley, California) and monographs [26,27].

We ran four BiSSE models on each phylogeny. (1) A fully parameterised state dependent model, estimating all six speciation, extinction and character state transition parameters (λ_0, λ_1, μ_0, μ_1, q_{01}, q_{10}); (2) a four parameter model with state independent speciation and extinction but state dependent character transition rates (λ, μ, q_{01}, q_{10}); (3) a four parameter model with state independent extinction and character transition rates but state dependent speciation (λ_0, λ_1, μ, q); and (4) a three parameter state independent model (λ, μ, q). We used maximum likelihood to fit the models. We compared the models using AICc and Akaike weights and averaged these across >1000 tree and data set combinations. We report the distribution of parameter estimates for the best-fit model across the tree sample.

State dependent rates of morphological evolution

To test for differences in the mean and rate of evolution of pinnae length and number of leaves per frond between páramo and non-páramo lineages we used a state dependent trait evolution model [46,47]. Specifically, the model requires that each branch in the tree is assigned to a discrete character state and then estimates the mean and rate of evolution of a continuous variable for each discrete state. The rate-heterogeneous model is then compared to a single rate model using likelihood ratio tests or AIC. To assign a character state to each branch we extracted marginal ancestral state estimates of habitat type from the BiSSE models described above. Goldberg and Igic [48] demonstrate that the BiSSE model frequently performs better than alternative maximum likelihood models (e.g. Mk2; [49,50]), particularly where character transitions are irreversible or extremely rare. This may be the case for *Jamesonia* and *Eriosorus* where we predict transitions to páramo habitats are likely to exceed the reverse. We used the R package diversitree [45] for BiSSE analyses. The ancestral state estimation returns probabilities for each character state at each node. To each node, and to its parent branch, we assign the character state with the highest probability (pendant edges take the character state of the associated species). We then applied four alternative models: (1) a constant rate model with a mean trait value common to the two habitat types (páramo and montane), (2) a two rate model with a common overall mean, (3) a constant rate model with different means for the two habitat types and, (4) a two rate model with different means for the two habitat types. Rates of evolution can be upwardly biased if there is error or noise in the data. To account for this we simultaneously fitted Pagel's λ [42,43] with the rates models. We compared model fit using AICc. Each model was applied to (1) pinnae length and (2) leaves per frond. We repeated all analyses, including ancestral state reconstruction, over the same 1000 sampled phylogenies used in the BiSSE analyses. Analyses of rates of morphological evolution were conducted using the R package motmot [51].

Results

Phylogeny

The phylogeny (Fig. 2) reveals a combination of well-supported clades deep within the tree combined with poor resolution towards the tips. We find very strong support for three previously reported [24] monophyletic groups (clades I, II, III in Fig. 2). However, whilst our analyses included multiple populations of many species, we failed to recover species monophyly in most cases (Fig. 2). Well-supported exceptions include monophyly of *J. bogotensis, J. cinnamomea*, and *J. brasiliensis*, with posterior probabilities of 1, 0.996 and 0.962. The timings of divergence are tentative because they are based on a single fossil spore and are well within the prior distribution, but are consistent with a Pleistocene diversification at a time when páramo ecosystems would have been emerging and expanding across vast areas in the Andes [3–5].

Predictors of morphological evolution

We found that the two axes of leaf morphology (pinnae length and leaves per frond) form two distinct morphological clusters (Fig. 3). The two clusters correspond to habitat: taxa with long pinnae and few leaves (the *Eriosorus* morphotype) are associated with cloud forest and sheltered microhabitats within the páramo whereas taxa with short pinnae and many leaves (the *Jamesonia* morphotype) are associated with exposed (predominantly páramo) habitats.

Models testing the relationship between altitude and morphology confirm these habitat dependent relationships. Two similar models are essentially indistinguishable in explaining variation in leaves per frond. One model includes both habitat and altitude but not their interaction and a simpler model includes only habitat (Table 1). In the more complex model the number of leaves per frond increases with altitude (Fig. 4a). In both models the number of leaves per frond is significantly higher among páramo species. The more complex model explains ~90% of the variation in leaves per frond. In contrast, we find evidence of a significant interaction between habitat and altitude as predictors of pinnae length (Fig. 4b). Pinnae length declines at higher altitudes but the slope of this relationship is steeper among montane species. The shallower slope might be expected if pinnae length is approaching a lower bound in páramo species.

BiSSE: state dependent diversification

The top-ranked diversification model is one with state dependent speciation rates but state independent extinction and character transitions rates (Table 2). That is, we found no difference in transitions into and out of the páramo. Speciation rates were consistently higher for páramo species than for non-páramo species (Fig. 5a). There is a bimodal distribution of speciation rates for non-páramo species across trees. In a small subset of trees, speciation rates for non- páramo are elevated but remain lower than those for páramo species (Fig. 5b).

State dependent rates of morphological evolution

The pattern of evolutionary convergence within habitat types was confirmed by evolutionary rate analyses. The phylogenetically corrected mean of the number of leaves per frond varies markedly with microhabitat (Table 3). However, after accounting for the possible effects of noise in the data, we found no evidence for state dependent variations in rates of evolution of leaves per frond. As with leaves per frond, pinnae length also varies with microhabitat.

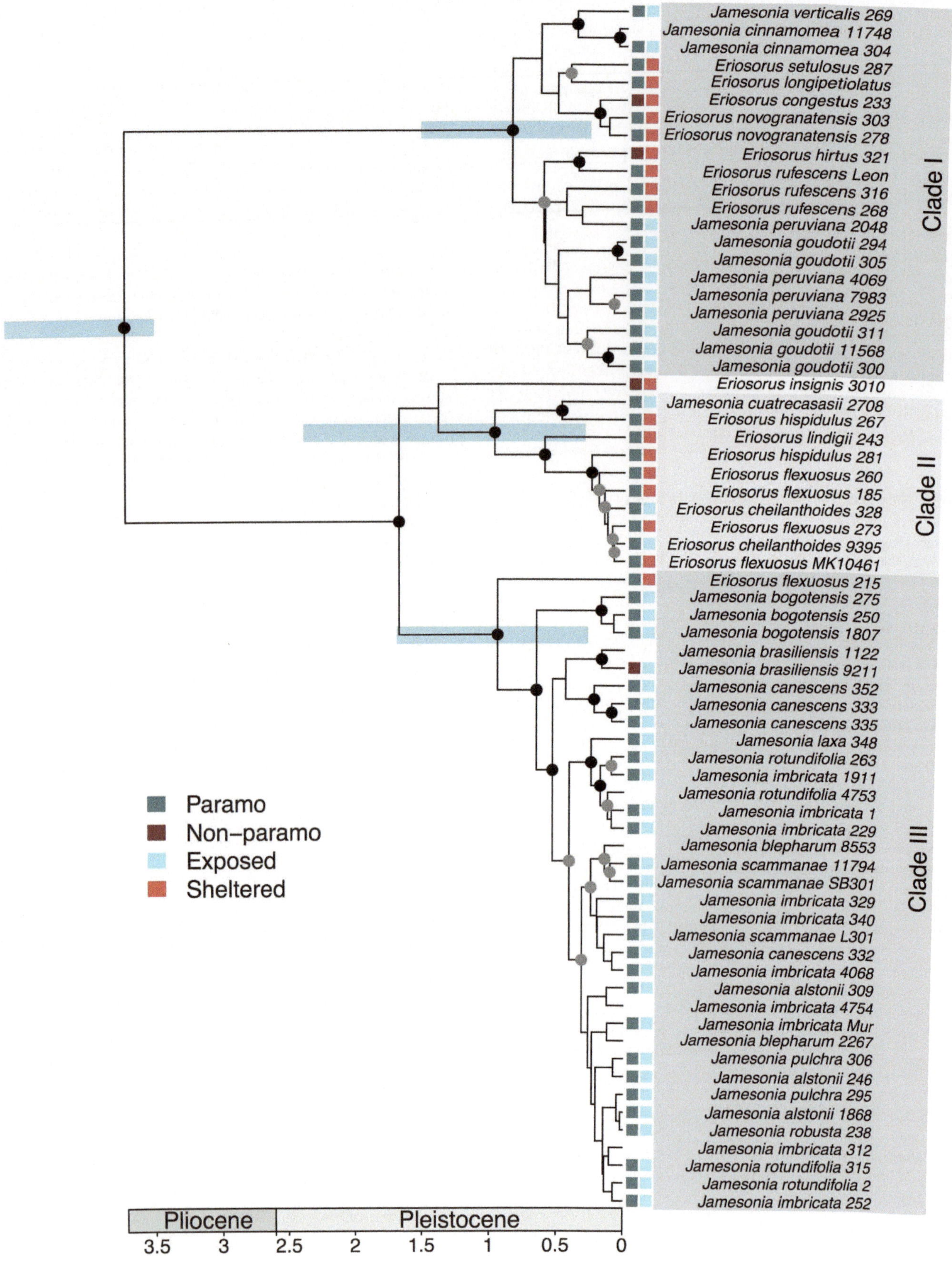
Jamesonia verticalis 269
Jamesonia cinnamomea 11748
Jamesonia cinnamomea 304
Eriosorus setulosus 287
Eriosorus longipetiolatus
Eriosorus congestus 233
Eriosorus novogranatensis 303
Eriosorus novogranatensis 278
Eriosorus hirtus 321
Eriosorus rufescens Leon
Eriosorus rufescens 316
Eriosorus rufescens 268
Jamesonia peruviana 2048
Jamesonia goudotii 294
Jamesonia goudotii 305
Jamesonia peruviana 4069
Jamesonia peruviana 7983
Jamesonia peruviana 2925
Jamesonia goudotii 311
Jamesonia goudotii 11568
Jamesonia goudotii 300
Eriosorus insignis 3010
Jamesonia cuatrecasasii 2708
Eriosorus hispidulus 267
Eriosorus lindigii 243
Eriosorus hispidulus 281
Eriosorus flexuosus 260
Eriosorus flexuosus 185
Eriosorus cheilanthoides 328
Eriosorus flexuosus 273
Eriosorus cheilanthoides 9395
Eriosorus flexuosus MK10461
Eriosorus flexuosus 215
Jamesonia bogotensis 275
Jamesonia bogotensis 250
Jamesonia bogotensis 1807
Jamesonia brasiliensis 1122
Jamesonia brasiliensis 9211
Jamesonia canescens 352
Jamesonia canescens 333
Jamesonia canescens 335
Jamesonia laxa 348
Jamesonia rotundifolia 263
Jamesonia imbricata 1911
Jamesonia rotundifolia 4753
Jamesonia imbricata 1
Jamesonia imbricata 229
Jamesonia blepharum 8553
Jamesonia scammanae 11794
Jamesonia scammanae SB301
Jamesonia imbricata 329
Jamesonia imbricata 340
Jamesonia scammanae L301
Jamesonia canescens 332
Jamesonia imbricata 4068
Jamesonia alstonii 309
Jamesonia imbricata 4754
Jamesonia imbricata Mur
Jamesonia blepharum 2267
Jamesonia pulchra 306
Jamesonia alstonii 246
Jamesonia pulchra 295
Jamesonia alstonii 1868
Jamesonia robusta 238
Jamesonia imbricata 312
Jamesonia rotundifolia 315
Jamesonia rotundifolia 2
Jamesonia imbricata 252
Clade I
Clade II
Clade III
Paramo
Non–paramo
Exposed
Sheltered
Pliocene
Pleistocene
3.5
3
2.5
2
1.5
1
0.5
0

Figure 2. Time-calibrated phylogeny of *Jamesonia – Eriosorus.* Black circles indicate nodes with posterior probabilities>0.95; grey circles indicate posterior probabilities between 0.5 and 0.95. Blue bars show the 95% highest Bayesian probability densities (HPDs) for the age of select well-supported nodes. Coloured squares show presence/absence in páramo (column 1) and microhabitat type (column 2) for samples where morphological measurements were taken.

We also found equivocal evidence for elevated rates of evolution of pinnae length associated with exposed habitats. The Akaike weights marginally favour a state independent over a state dependent model (Table 3) but parameter estimates consistently show that rates of evolution are relatively higher among species occupying exposed habitats. Because the uncertainty in these parameter estimates is very large we suggest that these higher rates should be interpreted with caution.

Discussion

While fern groups are very abundant in cloud forest, there are overall far fewer species in páramos [7,21]. Given the proliferation of other angiosperm clades in the páramo plant hotspot [9], the lack of divergence of ferns is perhaps surprising. The *Jamesonia-Eriosorus* complex provides one counter example, where morphological adaptation to cool and exposed environments has facilitated the transition from sheltered montane habitats. Our analyses suggest that the successful colonization of páramos by the *Jamesonia-Eriosorus* complex is associated with elevated rates of speciation and morphological adaptation.

Morphological evolution and the transition to Páramo

Morphological adaptations among *Jamesonia* and *Eriosorus* colonising the páramo are consistent with evolution as a response to a highly exposed environment. Although we find evidence that the number of leaves per frond increases and pinnae length decreases with altitude, habitat type is the main factor contributing to variation in leaf morphology (Table 1). Among lineages there are two distinct morphological clusters that correspond closely to habitat. Many leaves per frond and short pinnae characterize lineages associated with páramo. The majority of these lineages belong to the genus *Jamesonia* (with the exception of *Eriosorus cheilanthoides*). In contrast, few leaves per frond and long pinnae characterize many species of *Eriosorus* and are associated with montane or sheltered páramo habitats. The evolution of increased numbers of leaves per frond and short pinnae appears to be an example of convergent evolution in which mutliple lineages responsed to the extreme enviromental conditions in similar ways. This renders the *Jamesonia* (exposed) morphotype polyphyletic. We suggest that the well-defined exposed páramo morphology is evidence for the adaptive advantage of increased numbers of leaves per frond and short pinnae provide in extreme environmental conditions characteristic of páramo.

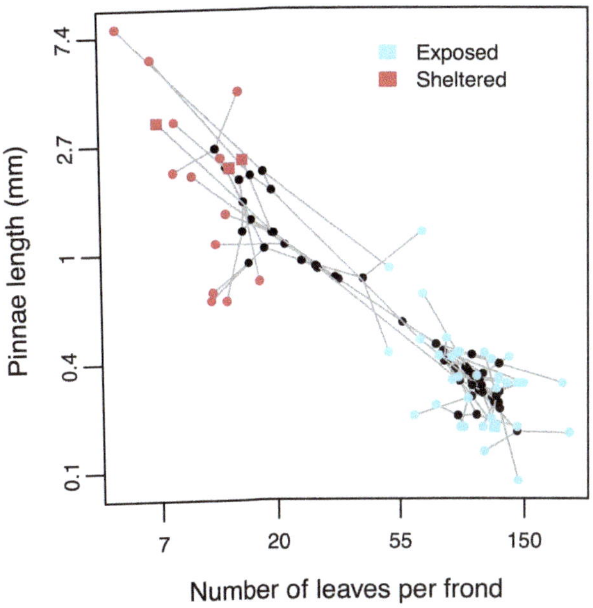

Figure 3. Phylomorphospace plot of log pinnae length and log number of leaves per frond. Black circles are internal nodes and are plotted as estimated ancestral states under a Brownian motion model. Grey lines are branches in the phylogeny.

The effect of freezing temperatures seems to be a major factor limiting species distributions in tropical and subtropical high mountain ecosystems [52]. Plants that grow close to the ground are exposed to more drastic daily leaf temperature changes. There is a temperature gradient from the soil upwards as a result of radiant heating with significantly higher daytime temperatures but notcural freezing closer to the ground [53]. All *Jamesonia* species and *Eriosorus* species associated with exposed microhabitats have evolved erect and taller fronds (Fig. 4) with an apical frond meristem that is therefore further away from the soil compared to *Eriosorus* species in sheltered microhabitats. Under these conditions, a taller frond apical meristem likely helps to protect meristematic tissues from freezing due to extremely low nocturnal temperatures at ground level, a trait shown to be advantageous in angiosperms exhibiting tall aerial stems [54,55]. Moreover as a consequence of increased numbers of pinnae per frond, species in microhabitats bear more sporangia, possibly allowing more spores to be dispersed.

The evolution of the exposed microhabitat morphology may also have been driven by selection for increased photosynthetic capabilities along altitudinal gradients. Previous physiological studies along an altitudinal gradient in the Andes have shown how differences in leaf temperature, influenced both by plant form and microenvironmental conditions, affect photosynthetic capacity [56]. Low temperature is a major abiotic factor limiting photosynthetic carbon acquisition ([52,57]; but see [58] for counter examples), as well as controlling plant growth and survival in high mountain ecosystems [59]. Plants with a herbaceous habit such as *Senecio* exhibited a decrease in optimal leaf temperature for photosynthesis coupled to decreasing air temperature with increasing elevation [56]. Similar to *Senecio, Jamesonia* exhibit an herbaceous habit. A higher number of pinnae per frond could increase photosynthetic capacity, thus counteracting decreased photosynthetic rates resulting from low temperature [7] and reduced CO_2 assimilation rates [60,61].

Although the strongest predictor of leaf morphology is microhabitat, it is also evident that leaf morphology, particularly pinnae length, varies with altitude where shorter pinnae are associated with higher altitudes. Our sample includes both inter and intra specifc variation and the altitudinal correlation may be in part due to phenotypic plasticity within species. Common garden or reciprocal transplant experiments may help to tease apart the mechanisms driving morphological change. Sister species such as *E. cheilanthoides* and *E. flexuosus* with contrasting

Table 1. Models of leaf morphology as a function of habitat and altitude.

(a)	Leave per frond		Pinnae length	
	ΔAICc	Akaike wts	ΔAICc	Akaike wts
Altitude	7.924	<0.001	6.073	<0.001
Habitat	0	**0.463**	2.772	0.030
Alt + Hab	0.254	0.407	0.542	0.427
Alt * Hab	2.538	0.130	**0**	**0.544**
(b)	**Coefficient (est ± SE)**		**95th percentiles**	
Leaves per frond				
Exposed int.	4.157 (0.344)		NA	
Sheltered int.	1.972 (0.116)		NA	
Altitude	1.4e-4 (10e-5)		NA	
Pinnae length				
Exposed int.	−0.480 (0.823)		−0.386 – −0.683	
Sheltered int.	3.225 (0.954)		2.980 – 3.575	
Altitude	−1.77e-4 (2.02e-4)		−1.33e-4 – −2.16e-4	
Alt:Hab	−6.08e-4 (3.02e-4)		−4.41e-4 – −7.17e-4	

Note that all models fitted with interaction terms always include the same variables as main effects. (a) The fits of alternative models are compared by ΔAICc and Akaike weights averaged across 1000 trees. (b) Parameter estimates are the median estimates from pgls models fitted to 1000 trees. *Leaves per frond*: two models including habitat but no interaction term are essentially indistinguishable based on AICc, we present the results from the more complex model (Habitat + Altitude). The median maximum likelihood estimate of Pagel's lambda across trees for this model is 0 and is only non-zero in analyses with 18 from 1000 trees. Consequently, the model is equivalent to one in which phylogeny is not included, and hence there is no variation in parameter estimates attributable to variation in phylogeny. The adjusted R^2 for the model is 0.901. *Pinnae length*: the best-fitting model including an altitude x habitat interaction term explains 56.8% of the among-species variation in pinnae length (95% quantiles for adjusted $R^2 = 0.513–0.735$). The median estimate of Pagel's lambda for the interaction model across 1000 trees is 0.779 (95% quantiles = 0.000–0.957) and greater than 0.5 in analyses on 872 trees.

morphologies and habitat prefences (páramo and sheltered lower elevation sites) would be well suited to assessing whether leaf form can be altered by environmental conditions. Taken together, habitat type seems to be the dominant factor explaining variation in leaf morphology among species while evolution along altitudinal gradients may be an important factor in explaining within species variation in leaf morphology and potentially transitions to páramo habitats. First, accessions of the same species species sampled from different altitudes show a clear reduction in pinnae size with an increase in elevation in clades I and II (e.g. *E. rufescens* accessions

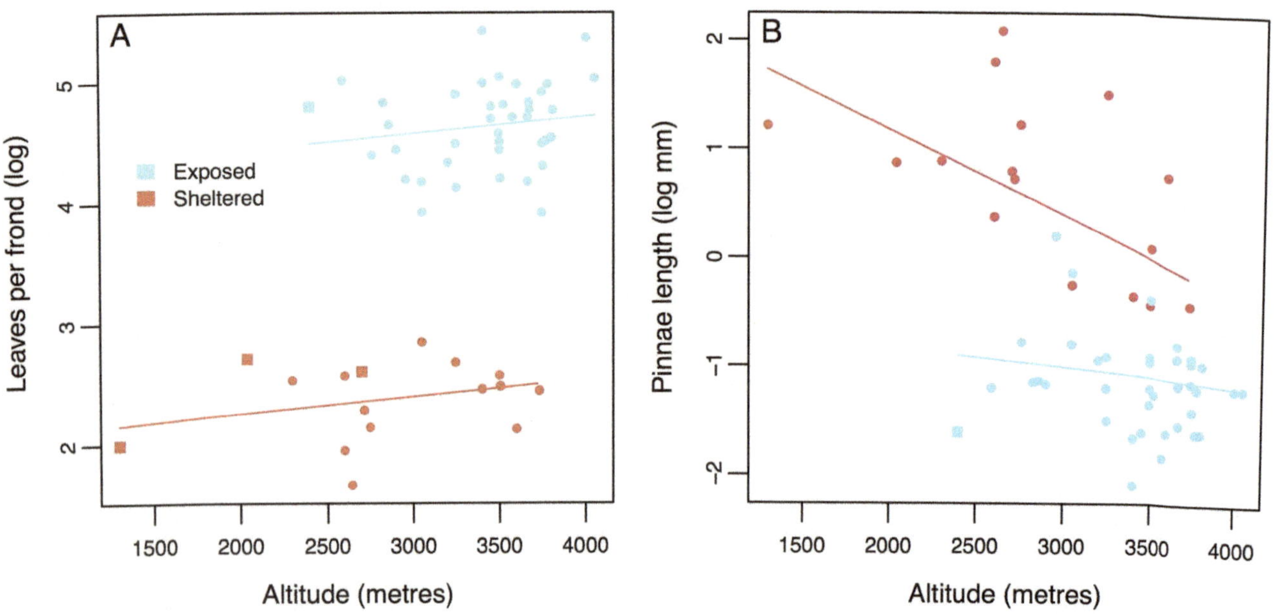

Figure 4. The relationship between morphology and altitude by habitat for (a) leaves per frond and (b) pinnae length. Colour of points indicates observed microhabitats. Samples from species not found in the páramo are shown as squares.

Table 2. BiSSE model comparison.

Model	ΔAICc	Akaike wts
$\lambda_0, \lambda_1, \mu_0, \mu_1, q_{01}, q_{10}$	5.910	0.036
$\lambda, \mu, q_{01}, q_{10}$	3.359	0.196
$\lambda_0, \lambda_1, \mu, q$	0	0.616
λ, μ, q	2.878	0.152

The fits of alternative models are compared by ΔAICc and Akaike weights averaged across 1000 trees.

268 and 316, *E. novogranatensis* accessions 278 and 303, and *E. hispidulus* accessions 267 and 281; [24]). Second, high elevation montane species, such as *E. longipetiolatus* and *E. setulosus*, exhibit an intermediate morphology with highly reduced pinnae (*Jamesonia*-like pinnae). However, they have fully developed fronds rather than indeterminate growth that is typically associated with páramo species [26]. Reduction in pinnae size is associated with high elevation habitats, yet indeterminate growth is a characteristic strictly correlated with extreme and open páramo habitats. Reciprocal transplant experiments amongst habitats would help determine whether phenotypic plasticity has played a role in the evolution of leaf size. Similarly sharp ecological and morphological shifts have been reported in other neotropical plant groups such as *Huperzia* (Lycopodiaceae) where terrestrial páramo forms have evolved from montane epiphytes [23].

Páramo radiation

Our phylogenetic analyses of the *Jamesonia-Eriosorus* clade confirm three well-supported clades inferred in previous studies [24] with multiple transitions into and out of the páramo. The *Jamesonia* ecomorph evolved independently multiple times from within different *Eriosorus* lineages (Fig. 2). The highest diversity of *Jamesonia* is found in the páramo-dominated clade III (Fig. 2), which is mostly restricted to the Northern Andes [24]. Clade I is associated with the Central Andes and contains both páramo and montane lineages whereas clade II is North Andean and dominated by montane lineages.

The páramo radiation of *Jamesonia* and *Eriosorus* shares some features with angiosperms that have radiated in páramo habitats notably including highly reduced leaves (microphylls) and pubescence. Microphyllous leaves are often xeromorphic, possibly protecting tissues from ultraviolet light and/or reducing transpiration [62,63]. Microphyllous leaves are associated with xeric environments and seem to be present in a number of páramo taxa, particularly shrubs such as *Loricaria*, *Baccharis revoluta*, *Diplostephium revolutum*, (Asteraceae), *Aragoa cupressina* (Scrophulariaceae), and *Valeriana microphylla* (Valerianaceae; [7]). Although variable, pubescence is present in all *Jamesonia* species and is consistently present at the frond tip, presumably protecting the meristematic tissues, which could be susceptible to freezing. It has been demonstrated that dense leaf pubescence reduces transpiration and increases leaf temperature [55,64–66]. Pubescence is characteristic of other genera that have presumably radiated in páramo habitats such as *Espeletia*, and it is believed to protect leaves and reduce transpiration [67]. In contrast to *Eriosorus*, all *Jamesonia* species exhibit creeping rhizomes, a feature that facilitates expansion in grasslands as shown in other páramo species such as bunchgrasses [7].

The high speciation rates in the páramo is consistent with predictions of exploitation and adatpation to new ecological

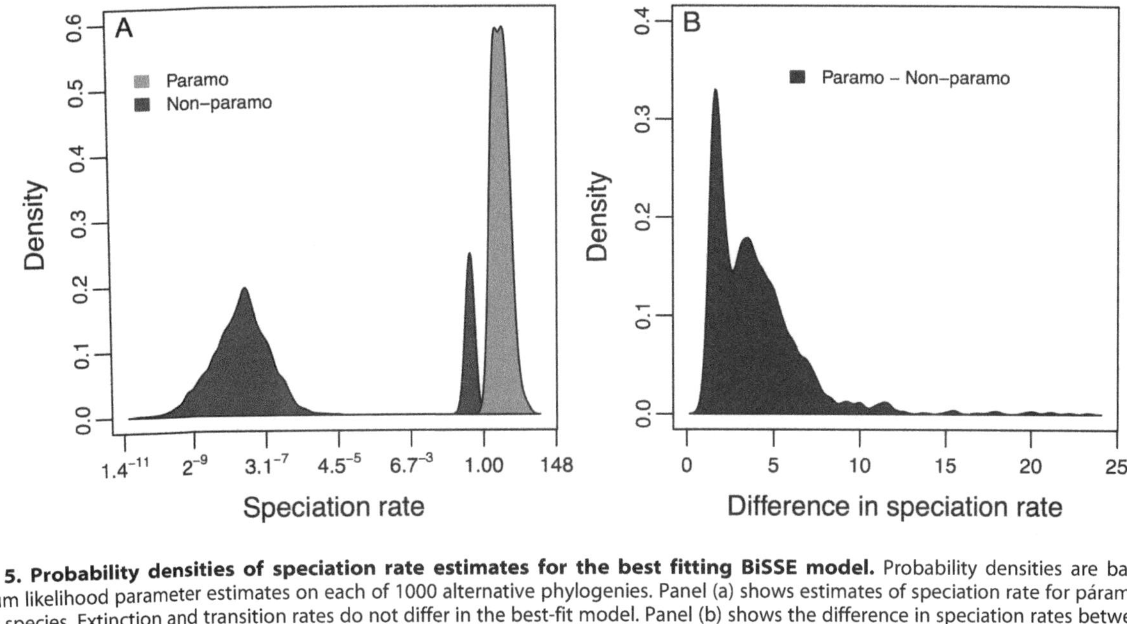

Figure 5. Probability densities of speciation rate estimates for the best fitting BiSSE model. Probability densities are based on the maximum likelihood parameter estimates on each of 1000 alternative phylogenies. Panel (a) shows estimates of speciation rate for páramo and non-páramo species. Extinction and transition rates do not differ in the best-fit model. Panel (b) shows the difference in speciation rates between páramo and non-páramo species plotted as páramo speciation rate – non-páramo speciation rate.

Table 3. State dependent rates of morphological evolution.

Model	ΔAICc	Akaike wts	Mean (sheltered)	Mean (exposed)	Rate (sheltered)	Rate (exposed)
Leaves per frond						
1	33.064	<0.001	3.266		1	-
			(3.102, 3.437)			
2	29.919	<0.001	2.596		1	0.083
			(2.354, 4.389)			(0.012, 3028.817)
3	**0**	**0.753**	**4.254**	**2.506**	**1**	**-**
			(4.028, 4.454)	**(2.365–2.731)**		
4	2.344	0.247	4.204	2.431	1	0.326
			(3.888, 4.430)	(2.252, 2.678)		(<0.001, 1.591)
Pinnae length						
1	14.501	0.005	−0.148	-	1	-
			(−0.315, 0.116)			
2	13.634	0.004	−0.809	-	1	73
			(−1.160, 0.057)			(0.441, 4.262^4)
3	**0**	**0.553**	**−0.963**	**0.501**	**1**	**-**
			(−1.348, −0.638)	**(0.142, 0.795)**		
4	0.409	0.438	−1.014	0.448	1	$>1^5$
			(−1.323, −0.788)	(−0.125, 0.955)		(1.141, $>1^5$)

The fits of alternative models for both leaves per frond and pinnae length are compared by ΔAICc and Akaike weights averaged across 1000 trees. Parameter estimates are reported as the median estimate across the tree distribution with 5th and 95th percentiles. Evolutionary rates are reported as relative rates.

niches. However, the elevated speciaton rates should be treated with caution because there are a number of caveats associated with uncertainty in the phylogeny and incomplete sampling of taxa. We found a lack of phylogenetic resolution within each of the three major clades with repeated polyphyly of traditional (i.e. taxonomic) *Jamesonia* and *Eriosorus* species. While this lack of genetic differentiation amongst the *Jamesonia-Eriosorus* complex potentially challenges the definition of traditional species as described in their monographs [24,26,27], it might alternatively be indicative of an ongoing radiation with continued hybridization and/or incomplete lineage sorting. Species polyphyly is particularly prevalent in clade III, which is dominated by páramo lineages. Species polyphyly could be indicative of ongoing speciation in which case we may have underestimated speciation rates. An alternative explanation for the lack of resolution may be that the molecular markers used here (*rps4* and ETS) evolve too slowly to differentiate species. However, the markers we used have previously been used for phylogenetic studies at lower taxonomic levels and *rps4* shows some of the highest substitution rates within chloroplast genes [68]. Moreover, the nuclear ETS gene has been shown to have higher substitution rates than the more commonly used ITS gene. ETS is therefore expected to be more informative than ITS in resolving phylogenetic relationships at low taxonomic levels in angiosperms [69,70].

In this study we focused on how transitions in ecology influence the evolution of leaf morphology and patterns of diversification. We find evidence of distinct morphologies that are closely associated with transitions to exposed páramo and páramo-like habitats. However, adaptation to these exposed habitats likely involves convergent evolution of a more complex suite of traits. In exposed habitats, an increase in the number of leaves per frond likely arose as a result of retaining the apical frond meristem coupled with the retention of a 'juvenile' morphology (fiddlehead: furled fronds of a young fern) yet bearing sporangia. The repeated evolution of this complex of traits may have been the result of a slowdown in developmental stage at low temperatures. Comparative developmental and physiological studies are needed to better understand how the morphological adaptations associated with the páramo arose and affect survival and physiological performance in extreme and exposed habitats. Explicit tests of limits to clade diversity will require further sampling and resolution of species limits both in páramo and montane regions. Nonetheless, we find support for rapid speciation in the páramo that is consistent with diversification in a novel and often island-like habitat landscape.

Supporting Information

Table S1 Habitats of *Jamesonia* and *Eriosorus* species. Habitats include: 1) super-páramo (4000–5000 m); 2) grass páramo (3500–4100 m); 3) sub-páramo (2800/3000–3500 m), and 4) montane forest (1150–2800/3000 m). Abbreviations of geographical distribution are: BO, Bolivia; BR, Brazil; CO, Colombia; CR, Costa Rica; EC, Ecuador; ME, Mexico; PA, Panama; PE, Peru; UR, Uruguay; and VE, Venezuela.

Acknowledgments

We thank K. Evans and two anonymous reviewers for thoughtful and constructive comments on earlier versions of this manuscript.

Author Contributions

Conceived and designed the experiments: PSB GHT. Performed the experiments: PSB GHT. Analyzed the data: PSB GHT. Contributed reagents/materials/analysis tools: PSB GHT. Wrote the paper: PSB GHT. Collected morphological and ecological data: PSB.

References

1. Pennington RT, Lavin M, Särkinen T, Lewis GP, Klitgaard BB, et al. (2010) Contrasting plant diversification histories within the Andean biodiversity hotspot. Proc Natl Acad Sci USA 107: 13783–13787. doi:10.1073/pnas.1001317107.
2. Myers N, Mittermeier RA, Mittermeier CG, da Fonseca GA, Kent J (2000) Biodiversity hotspots for conservation priorities. Nature 403: 853–858. doi:10.1038/35002501.
3. Van der Hammen T, Gonzalez E (1960) Upper Pleistocene and Holocene Climate and Vegetation of the Sabana de Bogota (Colombia, South America). Leidse Geol Mededel. 25: 261–315.
4. Van der Hammen T (1985) The Plio-Pleistocene climatic record of the tropical Andes. J Geol Soc 142: 483–489. doi:10.1144/gsjgs.142.3.0483.
5. Hooghiemstra H (1989) Quaternary and upper-pliocene glaciations and forest development in the tropical Andes: Evidence from a long high-resolution pollen record from the sedimentary basin of Bogotá, Colombia. Palaeo3 72: 11–26. doi:10.1016/0031-0182(89)90129-6.
6. Rauscher JT (2002) Molecular phylogenetics of the *Espeletia* complex (Asteraceae): evidence from nrDNA ITS sequences on the closest relatives of an Andean adaptive radiation. Am J Bot 89: 1074–1084. doi:10.3732/ajb.89.7.1074.
7. Luteyn JL (1999) New York Botanical Garden Pr Dept. Botanical Garden Press, New York.
8. Yoder JB, Clancey E, Roches Des S, Eastman JM, Gentry L, et al. (2010) Ecological opportunity and the origin of adaptive radiations. J Evol Biol 23: 1581–1596. doi:10.1111/j.1420-9101.2010.02029.x.
9. Madriñán S, Cortés AJ, Richardson JE (2013) Páramo is the world's fastest evolving and coolest biodiversity hotspot. Front Genet 4: 192. doi:10.3389/fgene.2013.00192.
10. Hughes C, Eastwood R (2006) Island radiation on a continental scale: Exceptional rates of plant diversification after uplift of the Andes. Proc Natl Acad Sci USA, 103: 10334–10339.
11. Hagen Bernhard von, Joachim K (2001) The phylogeny of (Gentianaceae) and its colonization of the southern hemisphere as revealed by nuclear and chloroplast DNA sequence variation. Org Divers Evol 1: 61–79. doi:10.1078/1439-6092-00005.
12. Bell CD, Donoghue MJ (2005) Phylogeny and biogeography of Valerianaceae (Dipsacales) with special reference to the South American valerians. Org Divers Evol 5: 147–159. doi:10.1016/j.ode.2004.10.014.
13. Karl R, Koch MA (2013) A world-wide perspective on crucifer speciation and evolution: phylogenetics, biogeography and trait evolution in tribe Arabideae. Ann Bot 112: 983–1001. doi:10.1093/aob/mct165.
14. Nürk NM, Scheriau C, Madriñán S (2013) Explosive radiation in high Andean *Hypericum*-rates of diversification among New World lineages. Front Genet 4: 175. doi:10.3389/fgene.2013.00175.
15. Jabaily RS, Sytsma KJ (2012) Historical biogeography and life-history evolution of Andean Puya (Bromeliaceae). Bot J Linn Soc 171: 201–224. doi:10.1111/j.1095-8339.2012.01307.x.
16. Sklenář P, Dušková E, Balslev H (2011) Tropical and temperate: evolutionary history of páramo flora. Bot Rev 77: 71–108.
17. Burnham RJ, Graham A (1999) The history of neotropical vegetation: new developments and status. Ann Missouri Bot Gard 546–589.
18. Donoghue MJ (2008) Colloquium Paper: A phylogenetic perspective on the distribution of plant diversity. Proc Natl Acad Sci USA 105: 11549–11555. doi:10.1073/pnas.0801962105.
19. Cosacov A, Sérsic AN, Sosa V (2010) Multiple periglacial refugia in the Patagonian steppe and post-glacial colonization of the Andes: the phylogeography of *Calceolaria polyrhiza*. Biogeog 8: 1463–1477.
20. Sarmiento G (1986) Ecological features of climate in high tropical mountains. High Altitude Tropical Biogeography. Oxford Univ. Press, New York.
21. Luteyn JL, Baslev H (1992) Páramos: why study them?. Pp. 1–14 in H. Balslev, and J. L. Luteyn, eds. Páramo an Andean ecosystem under human influence. Academic Press, London.
22. Monasterio M, Sarmiento L (1991) Adaptive radiation of *Espeletia* in the cold Andean tropics. Trends Ecol Evol 6: 387–391. doi:10.1016/0169-5347(91)90159-U.
23. Wikström N, Kenrick P, Chase M (1999) Epiphytism and terrestrialization in tropical *Huperzia* (Lycopodiaceae). Pl Syst Evol 218: 221–243.
24. Sánchez-Baracaldo P (2004) Phylogenetics and biogeography of the neotropical fern genera *Jamesonia* and *Eriosorus* (Pteridaceae). Am J Bot 91: 274–284. doi:10.3732/ajb.91.2.274.
25. Sánchez-Baracaldo P (2004) Phylogenetic relationships of the subfamily Taenitidoideae, Pteridaceae. Am J Bot 91, 274–284.
26. Tryon AF (1970) A monograph of the fern genus *Eriosorus*. Contrib Gray Herb. 200, 54–174.
27. Tryon AF (1962) A monograph of the fern genus *Jamesonia*. Contrib Gray Herb. 1991, 109–203.
28. Tropicos.org (2014) Missouri Botanical Garden. Available: http://www.tropicos.org. Accessed 2014 Jul 21.
29. Cochran AT, Prado J, Schuettpelz E (2014) Tryonia, a new taenitidoid fern genus segregated from *Jamesonia* and *Eriosorus* (Pteridaceae). PhytoKeys 35: 23–43, doi: 10.3897/phytokeys.35.6886
30. Logacheva MD, Valiejo-Roman CM, Degtjareva GV, Stratton JM, Downie SR, et al. (2010) A comparison of nrDNA ITS and ETS loci for phylogenetic inference in the Umbelliferae: an example from tribe Tordylieae. Mol Phylogenet Evol 57: 471–476. doi:10.1016/j.ympev.2010.06.001.
31. de Lange PJ, Smissen RD, Wagstaff SJ, Keeling DJ, Murray BG, et al. (2010) A molecular phylogeny and infrageneric classification for *Kunzea* (Myrtaceae) inferred from rDNA ITS and ETS sequences. Aust Systematic Bot 23: 309–319. doi:10.1071/SB10019.
32. Rambaut A (1996) Rambaut: Se-Al v2. 0a11: sequence alignment editor. University of Oxford.
33. Lanfear R, Calcott B, Ho S, Guindon S (2012) PartitionFinder: combined selection of partitioning schemes and substitution models for phylogenetic analyses. Mol Biol Evol 6:1695–701. doi: 10.1093/molbev/mss020.
34. Drummond AJ, Rambaut A (2007) BEAST: Bayesian evolutionary analysis by sampling trees. BMC Evol Biol 7: 214. doi:10.1186/1471-2148-7-214.
35. Drummond AJ, Ho SYW, Phillips MJ, Rambaut A (2006) Relaxed Phylogenetics and Dating with Confidence. Plos Biol 4: e88. doi:10.1371/journal.pbio.0040088.sd001.
36. Heled J, Drummond AJ (2012) Calibrated tree priors for relaxed phylogenetics and divergence time estimation. Syst Biol 61: 138–149. doi:10.1093/sysbio/syr087.
37. Tryon RM, Tryon AF (1982) Pteridaceae. Ferns and Allied Plants. Springer New York, 213–354.
38. Rambaut A, Drummond AJ (2009) Rambaut: Tracer v1. 4. Edinburgh: University of Edinburgh.
39. Sidlauskas B (2008) Continuous and arrested morphological diversification in sister clades of characiform fishes: a phylomorphospace approach. Evolution 62: 3135–3156. doi:10.1111/j.1558-5646.2008.00519.x.
40. Revell LJ (2012) phytools: an R package for phylogenetic comparative biology (and other things). Method Ecol Evol 3: 217–223.
41. Orme CDL, Freckleton RP, Thomas GH, Petzoldt T, Fritz S, et al. (2012) caper: Comparative Analyses of Phylogenetics and Evolution in R. R package version 0.5. Available: http://CRAN.R-project.org/package=caper.
42. Pagel M (1997) Inferring evolutionary processes from phylogenies. Zool Scripta 26: 331–348. doi:10.1111/j.1463-6409.1997.tb00423.x.
43. Pagel M (1999) Inferring the historical patterns of biological evolution. Nature 401: 877–884. doi:10.1038/44766.
44. Maddison WP, Midford PE, Otto SP (2007) Estimating a binary character's effect on speciation and extinction. Syst Biol 56: 701–710. doi:10.1080/10635150701607033.
45. FitzJohn RG (2012) Diversitree: comparative phylogenetic analyses of diversification in R. Method Ecol Evol 3: 1084–1092.
46. Thomas GH, Freckleton RP, Székely T (2006) Comparative analyses of the influence of developmental mode on phenotypic diversification rates in shorebirds. Proc Biol Sci 273: 1619–1624. doi:10.1098/rspb.2006.3488.
47. Thomas GH, Meiri S, Phillimore AB (2009) Body size diversification in *Anolis*: novel environment and island effects. Evolution 63: 2017–2030. doi:10.1111/j.1558-5646.2009.00694.x.
48. Goldberg EE, Igić B (2008) On phylogenetic tests of irreversible evolution. Evolution 62: 2727–2741. doi:10.1111/j.1558-5646.2008.00505.x.
49. Pagel M (1994) Detecting correlated evolution on phylogenies: a general method for the comparative analysis of discrete characters. Proc R Soc Lond B 255: 37–45.
50. Lewis PO (2001) A likelihood approach to estimating phylogeny from discrete morphological character data. Syst Biol 50: 913–925.
51. Thomas GH, Freckleton RP (2011) MOTMOT: models of trait macroevolution on trees. Method Ecol Evol 3: 145–151. doi:10.1111/j.2041-210X.2011.00132.x.
52. Cabrera HM (1996) Low temperature and altitudinal limits in plant ecosystems: Species responses to cold in tropical and subtropical mountains. Revista Chilena De Historia Natural 69: 309–320.
53. Azócar A, Monasterio M (1980) Estudio de la variabilidad meso y microclimática en el Páramo de Mucubají. Estudios ecológicos de los páramos. Universidad de los Andes, Mérida.
54. Smith AP (1980) The paradox of plant height in an Andean giant rosette species. J Ecol 68, 63–73.
55. Meinzer F, Goldstein G, Rada F (1994) Páramo microclimate and leaf thermal balance of Andean giant rosette plants. p. 45–59 *in* Rundel, P.W., Smith, A.P., and Meinzer, F.C., eds. Tropical alpine environments: plant form and function. Cambridge Univ. Press, Cambridge
56. Cabrera HM, Rada F, Cavieres L (1998) Effects of temperature on photosynthesis of two morphologically contrasting plant species along an altitudinal gradient in the tropical high Andes. Oecologia 114: 145–152.
57. Johnson DM, Germino MJ, Smith WK (2004) Abiotic factors limiting photosynthesis in *Abies lasiocarpa* and *Picea engelmannii* seedlings below and above the alpine timberline. Tree Physiol 24: 377–386.
58. Körner C (2003) Alpine Plant Life: Functional Plant Ecology of High Mountain Ecosystems. Springer.
59. Sakai A, Larcher W (1987) Frost survival of plants. Responses and adaptation to freezing stress. Springer-Verlag.

60. Keeley JE, Osmond CB, Raven JA (1984) Stylites, a vascular land plant without stomata absorbs CO2 via its roots. Nature 310: 694–695. doi:10.1038/310694a0.
61. Smith AP, Young TP (1987) Annual Review of Ecology and Systematics, Vol. 18 (1987), pp. 137–158. Annu Rev Ecol Syst. 18, 137–158.
62. Hedberg O (1964) Features of Afroalpine plant ecology. Acta Phytogeography Suecica 49:1–144.
63. Beck E (1994) Tropical Alpine Environments: Plant Form and Function. Tropical alpine environments Plant form and function. Cambridge University Press, Cambridge, 77–110.
64. Baruch Z (1979) Elevation differentiation in *Espeletia schultzii* (Compositae), a giant rosette plant of the Venezuelan páramos. Ecology, 85–98.
65. Meinzer F, Goldstein G (1985) Some consequences of leaf pubescence in the Andean giant rosette plant *Espeletia timotensis*. Ecology 66: 512–520.
66. Miller GA (1986) Pubescence, floral temperature and fecundity in species of Puya (Bromeliaceae) in the Ecuadorian Andes. Oecologia 70: 155–160. doi:10.1007/BF00377126.
67. Rada F, Azócar A, González J, Briceño B (1998) Leaf gas exchange in Espeletia schultzii Wedd, a giant caulescent rosette species, along an altitudinal gradient in the Venezuelan Andes. Acta Oecologica. 19: 73–79.
68. Wolf PG, Der JP, Duffy AM, Davidson JB, Grusz AL, et al. (2011) The evolution of chloroplast genes and genomes in ferns. Plant Mol Biol 76: 251–261. doi:10.1007/s11103-010-9706-4.
69. Baldwin BG, Markos S (1998) Phylogenetic Utility of the External Transcribed Spacer (ETS) of 18S–26S rDNA: Congruence of ETS and ITS Trees of *Calycadenia* (Compositae). Mol Phylogenet Evol 10: 449–63.
70. Zacharias EH, Baldwin BG (2010) A molecular phylogeny of North American Atripliceae (Chenopodiaceae), with implications for floral and photosynthetic pathway evolution. Syst Bot 35: 839–857.

10

Region-Specific Sensitivity of Anemophilous Pollen Deposition to Temperature and Precipitation

Timme H. Donders[1]*, Kimberley Hagemans[1,2], Stefan C. Dekker[2], Letty A. de Weger[3], Pim de Klerk[4], Friederike Wagner-Cremer[1]

1 Palaeoecology, Department of Physical Geography, Faculty of Geosciences, Utrecht University, Laboratory of Palaeobotany and Palynology, Utrecht, The Netherlands, **2** Department of Environmental Sciences, Copernicus Institute, Faculty of Geosciences, Utrecht University, Utrecht, The Netherlands, **3** Department of Pulmonology, Leiden University Medical Centre, Leiden, The Netherlands, **4** Botany section, Staatliches Museum für Naturkunde Karlsruhe, Karlsruhe, Germany

Abstract

Understanding relations between climate and pollen production is important for several societal and ecological challenges, importantly pollen forecasting for pollinosis treatment, forensic studies, global change biology, and high-resolution palaeoecological studies of past vegetation and climate fluctuations. For these purposes, we investigate the role of climate variables on annual-scale variations in pollen influx, test the regional consistency of observed patterns, and evaluate the potential to reconstruct high-frequency signals from sediment archives. A 43-year pollen-trap record from the Netherlands is used to investigate relations between annual pollen influx, climate variables (monthly and seasonal temperature and precipitation values), and the North Atlantic Oscillation climate index. Spearman rank correlation analysis shows that specifically in *Alnus*, *Betula*, *Corylus*, *Fraxinus*, *Quercus* and *Plantago* both temperature in the year prior to (T_{-1}), as well as in the growing season (T), are highly significant factors (T_{April} r_s between 0.30 [$P<0.05$[and 0.58 [$P<0.0001$]; T_{Juli-1} rs between 0.32 [$P<0.05$[and 0.56 [$P<0.0001$]) in the annual pollen influx of wind-pollinated plants. Total annual pollen prediction models based on multiple climate variables yield R^2 between 0.38 and 0.62 ($P<0.0001$). The effect of precipitation is minimal. A second trapping station in the SE Netherlands, shows consistent trends and annual variability, suggesting the climate factors are regionally relevant. Summer temperature is thought to influence the formation of reproductive structures, while temperature during the flowering season influences pollen release. This study provides a first predictive model for seasonal pollen forecasting, and also aides forensic studies. Furthermore, variations in pollen accumulation rates from a sub-fossil peat deposit are comparable with the pollen trap data. This suggests that high frequency variability pollen records from natural archives reflect annual past climate variability, and can be used in palaeoecological and -climatological studies to bridge between population- and species-scale responses to climate forcing.

Editor: Robert Guralnick, University of Colorado, United States of America

Funding: This work is supported by Utrecht University and Leiden University Medical Centre. The funders had no role in study design, data collection and analysis, decision to publish, or preparation of the manuscript.

Competing Interests: The authors have declared that no competing interests exist.

* Email: t.h.donders@uu.nl

Introduction

Pollen production by wind pollinated (anemophilous) plants is characterized by inter-annual variation [1]. This variation is not random but related to biological processes (e.g. mast cycles, [2] and changes in vegetation dynamics such as tree line fluctuations [3]. Part of the variation is caused by climatic conditions, particularly the character of the seasonal cycle [3,4] but also lower-frequency climate variability [5]. Identification of the strength, sign and type of effect of climatic variables on the annual pollen deposition is significant in multiple fields of research, ranging from palaeoecology and palaeoclimate [6], global change biology [7], plant ecology [8], to allergology [9], and forensics [10].

Firstly, if annual pollen production is influenced by climate variables, then high resolution records of fossil pollen have the potential for quantitative reconstruction of past climate conditions on annual to decadal time scales [6,11–13]. Traditionally, palynological studies focus on population-scale successional changes to study variations in past vegetation cover and associated climate change [14]. A higher resolution is needed for understanding past variability of climate systems that vary on annual to decadal timescales such as the North Atlantic Oscillation (NAO) for Europe and North America [15], and climatic oscillations in other parts of the world. Beside vegetation succession, annual-scale studies of pollen production and deposition rates allow us to assess the influence of climatic factors on such short timescales, and identify how such signals are preserved in natural archives that are the source for vegetation and derived climate reconstructions.

Secondly, in allergology accurate and timely prediction of seasonal pollen production is important for sufferers of pollinosis (hay fever) [9]. Tree pollen allergens affect health in up to 15% of the population [2].While much effort is directed at observed and future changes on the timing of pollen release e.g. [16,17], or masting effects [2], it is vital to also establish the exact relation between total annual pollen production and climate variables to improve predictions of seasonal pollen concentrations [18]. Global climate changes will likely affect the intensity of the pollen season e.g. [7,19] whereby region-specific studies of climate parameters relevant for pollen production and deposition will lead to better long-term pollinosis scenarios. This involves quantification of both the annual-scale variability and decadal-scale climatic trends and pollen production to improve the long-term pollen predictions and region-specific seasonal forecast of pollinosis.

Thirdly, forensic studies benefit from well-documented relations between pollen deposition and climate parameters to reduce uncertainty between concentration and composition of pollen deposition and pollen assemblages collected from crime scenes [10]. As for pollen-based climate reconstructions, forensic palynological studies need to take into account both annual climatic conditions, as well as differential preservation (taphonomy) of pollen in soils compared to the atmospheric composition as factors that influence sample comparison.

In this study we aim to improve our understanding of the region-specific response of annual pollen deposition to climate variables for common anemophilous pollen types in NW Europe. We assess the regional variability between pollen trap records, the strength and type of climate-pollen correlations, the robustness of the correlations for the purpose of seasonal forecasts and climate reconstructions, and the preservation of the annual variability in natural peat archives in comparison to the trap data.

While timing of pollen release is commonly known to be climate dependent [17,20–22], the annual pollen production is also affected in several ways [3–5,23,24]. In most trees, reproductive structures are formed during the summer in the year prior to flowering [1,4]. Warmer conditions at that time can positively influence pollen production and thereby influence the intensity of the pollen season in the following year [4–6,25]. During the flowering season, weather conditions influence pollen release in tree species that typically flower in late winter and early spring. For example frost may damage the flower buds before pollen release and can therefore reduce the total annual pollen production [1]. These relations differ regionally, as pollen production can be limited by summer temperatures and temperature sum (boreal Finland and Denmark [4–6], winter temperature (Jura Mountains [26]), or humidity (Australia [24], Central Europe and Caucasus [25]). Depending on the species, correlations can have opposite signs [25], and reveal autocyclic biological variations in time [2,23].

To achieve our aims and improve our understanding on climate-pollen relations between different regions and in different types of archives, multiple records of atmospheric and sediment-derived pollen deposition are needed [14]. Also, information on local land use changes and vegetation composition [26], as well as reliable precipitation and temperature data are required. A rare combination of such time series is available in The Netherlands, which presents a unique opportunity to investigate the impact and consistency of climate on annual pollen deposition with exact time control. The correlations and patterns between the different records allow better explanation and independent testing of the annual variability, and thereby help to strengthen long-term allergy forecasts, aid development and understanding of annual-resolved palaeoclimate records, and increase reliability of forensic studies.

Material and Methods

At Leiden University Medical Centre (LUMC, 52°23'N; 4°29E) in W Netherlands (Fig. 1, Table S1 in File S1) atmospheric pollen concentrations have been recorded daily since AD 1969 [17,27], resulting in one of the longest and most continuous pollen trapping stations worldwide. Since AD 1975 a second locality in The Netherlands, the Elkerliek Hospital in Helmond (SE Netherlands, 51°29'N; 5°38E, Fig. 1, Table S2 in File S1), produced daily pollen counts, although counts were not year round until AD 2008. Thus far studies on the pollen-climate relations in the Netherlands have only focused on the influence of climate on the long-term trends and timing of pollen release [17,27]. The records present a unique opportunity to investigate the impact of seasonal climate variables on annual pollen production as well as assess regional and plant-specific variability in the Netherlands. For example, the degree of consistency between records determines how regionally applicable a pollen forecast is and whether the climatic interpretation of high-resolution palaeovegetation records are robust, and ecosystem dependent. We test effects of seasonal variations in temperature and precipitation on the total annual pollen deposition of anemophilous plants in pollen traps from the Netherlands, and assess regional variations by comparison of both trapping stations and a natural sediment archive.

Comparison with a climate-pollen correlation study from southern Denmark [4] provides a supra-regional view on the most important climate drivers on pollen deposition in NW Europe. Extending the Danish study, we hypothesize that temperature of the growing season in the previous year regionally is the most important factor for annual pollen deposition (used here as the combined result of pollen production and transport processes), and that this factor will be consistent between the different pollen trap records in NW Europe. We expect the natural archive to contain elevated levels of local site-specific vegetation relative to the trapping stations, but the trends and annual-scale patterns to be comparable. Following recommendations by Joosten and De Klerk [28] we indicate pollen taxa in small capitals (e.g., BETULA) to distinguish them from taxonomic plant species.

We use simple linear correlation and multiple linear regression to determine, and statistically model, the influence of monthly and seasonal temperature and precipitation on total annual pollen production. Winter temperature and precipitation in the Netherlands (Fig. 2) are influenced by the NAO, defined by the difference of normalized sea level pressure between Lisbon, Portugal and Reykjavik, Iceland [15]. Therefore, the effect of the winter NAO (December through March) on pollen production is also analysed, as well as testing for cyclic patterns and autocorrelations in the data. Further, a well-dated near-annually resolved pollen record (±1953–1992) from a peat deposit in the vicinity of Helmond [29] provides insight on the reflection of annual-scale variations in pollen production and species-specific offsets in natural archives compared to the trap records. If similar climate-forced variations in annual pollen production can be identified for both the pollen traps and the natural sediment archives, the forcing is regionally relevant and preserves well in natural archives implying that similar high-frequency climate variations can be reconstructed for pre-instrumental periods.

Source data description

The LUMC trap is located in an urban area with non-natural vegetation in nearby parks, gardens and along streets. In these parks, *Alnus* and *Populus* are the most dominant trees [30]. The Elkerliek area is mostly surrounded by parks and gardens. Pollen was collected with a Hirst-type [31] volumetric, continuous pollen trap [27]. Initially a daily operated type (Castella) was used, while in 1976 (LUMC) and 1980 (Elkerliek) a weekly (Burkard) trap was installed on the roofs ~200 m from the original location [30]. Daily pollen concentrations are expressed as n m^{-3} day^{-1} in the air, and are corrected for changes in counting and trapping method. Daily pollen counts were summed to determine total annual pollen influx (n m^{-3} $year^{-1}$). The pollen concentration in

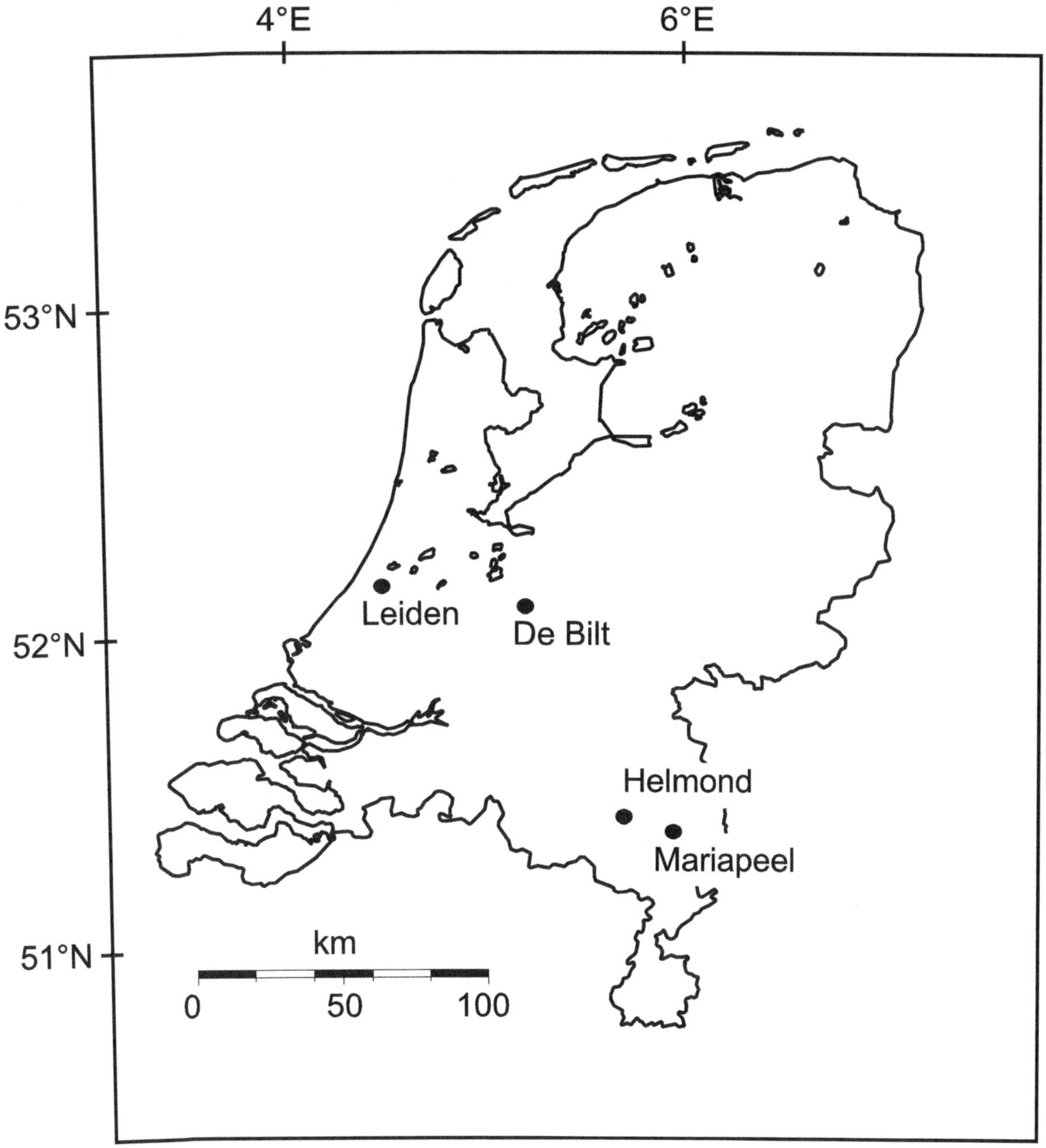

Figure 1. Map of the Netherlands showing the location of the LUMC (Leiden) and Elkerliek (Helmond) pollen traps, De Bilt meteorological station, and the peat deposit from Mariapeel Natural reserve.

the air was not measured whole year round at LUMC until 1977 (Table 1), and in Elkerliek until 2008. At LUMC, this caused missing or incomplete data for Corylus counts between 1969 and 1974, and in 1976, and for Alnus counts in 1969, 1973, and 1976 since these species flowered before pollen counts started in those years. For calculation of correlations the missing values on annual pollen influx were replaced by the series mean. Plots of pollen influx for individual taxa against time (Fig. S1) were evaluated for data consistency and missing values, and if needed, corrected based on the original count sheets.

A summary pollen diagram from the peat deposit from Mariapeel Nature Reserve nearby Helmond (51° 25' 03.04" N, 5.55' 31.94" E, Fig. 1, Table S3 in File S1) was reported earlier [29], and includes age information and site description. The profile represents secondary peat growth AD ±1953 to 1992 in a small pit after the site was mined in the early 20th century and shows near-annual variability of local and regional vegetation. Age assessment was based on linear interpolation between three local well-documented land-use changes (AD 1961, 1973 and 1980), resulting in a time-scale with an estimated accuracy of ±1 year [29]. Pollen accumulation rates were calculated based on an added spike with a known amount of *Lycopodium clavatum* spores and a fixed sample volume. Initially, counts of 20 pollen types were selected based on anemophily and presence in all three data

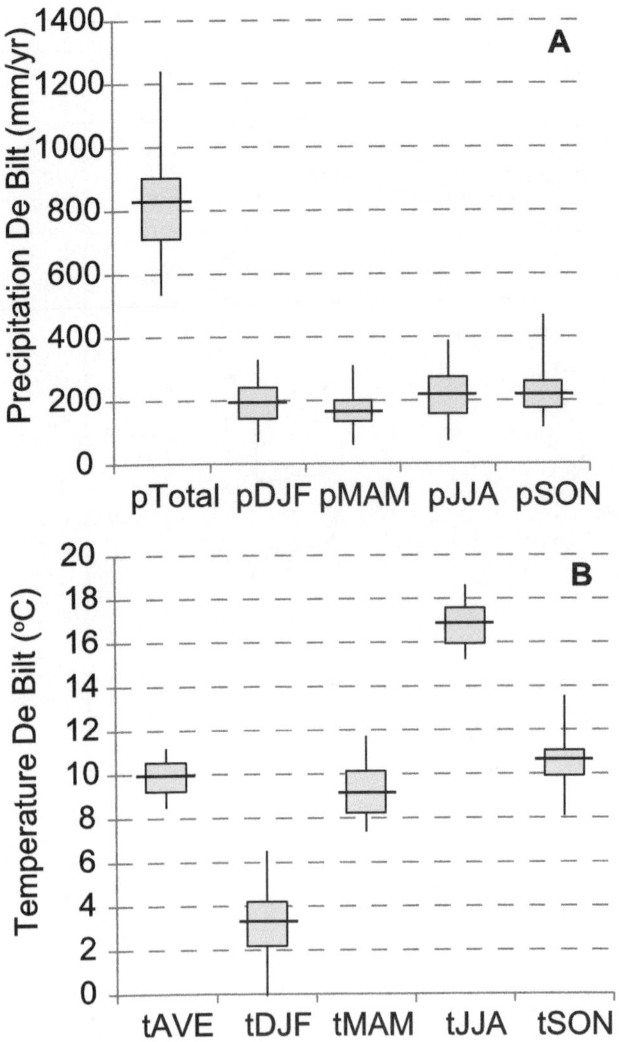

Figure 2. Annual and seasonal variability of the meteorological data from De Bilt. Values are expressed as the upper and lower quartiles, median, and minimum and maximum values of (a) cumulative precipitation and (b) mean temperature for AD 1969–2012.

Table 1. Availability of LUMC (day of the year, January 1st = day 1).

LUMC pollen data	
Year AD	**Measurement period (days)**
1969	99–259
1970	75–259
1971	75–259
1972	69–245
1973	71–167
1974	60–167
1975	61–260
1976	63–238
>1977	Year round

sources, eight of which are associated with hay fever. LUMC data were used to test correlations with climatic parameters as it is the most complete and consistent, the Elkerliek data are used to assess the regional consistency, and Mariapeel data for comparison between trapping stations and a natural sediment archive. Temperature and precipitation data were obtained from the Royal Netherlands Meteorological Institute (KNMI) at the automatic meteorological station in De Bilt, a central location in the Netherlands (52°06'N; 5°11'E) representative of mean conditions and at intermediate distance between both trapping sites (Figs. 1, 2). The NAO-index data were obtained from the Climate Analysis Section of the National Center for Atmospheric Research, Boulder, USA [32].

Statistical analysis

Pollen counts were square-root transformed as Q–Q plots and non-parametric Kolmogorov-Smirnov testing showed a non-normal distribution for most pollen types. A principal component analysis (PCA) of the LUMC square-root transformed pollen counts was performed to assess the internal relations of the pollen assemblages using CANOCO software version 4.5 [33]. A two-tailed Spearman rank correlation was performed to assess which pollen types (n = 20), as well as the PCA 1st axis values representative of the pollen assemblage, correlate significantly with climate. All variables, monthly and seasonal temperature (T_{month} and T_{season}), precipitation (P_{month} and P_{season}), and the NAO index, were tested for both the flowering year and the prior year (subscript $_{variable-1}$). Due to the large number of comparisons, care should be taken for correlations that are only significant at the 95% level as they are likely randomly occurring [4]. Common pollen types with significant correlation to multiple climate variables (see results, Tables 2, 3) were analysed further with a linear regression analysis with time as a predictor, and a multiple regression analysis with climatological variables as predictors.

An initial model was created for each selected pollen type (n = 6), including monthly temperature values of the year before flowering (i.e. lag −1), and monthly temperature during the current year until one month after the end of the flowering season (following ref. [34]) (Table 3). Precipitation variables were only included when the Spearman rank correlation showed a significant correlation to pollen influx. To prevent statistical over fitting, a stepwise backward multiple regression analysis was performed based on the initial models to reduce the number of redundant parameters. While stability of weather patterns can cause several consecutive months to correlate with pollen production, only a single or few months are relevant for pollen production [25]. Prediction skill of the resulting model was tested by random division of the dataset in a calibration and an independent test set. Predicted values were plotted against observed values, whereby R^2 and associated P-values <0.05 were considered to be statistically significant models. Cyclicity patterns and autocorrelations were evaluated through multitaper spectral analysis method incorporated in PAST software version 2.17 [35], which produces an F-value statistic for significance testing.

Results

General trends

The total pollen influx of all pollen types (n = 20) at LUMC ranges from 12.5×10^3 to 45.7×10^3 m^{-3} $year^{-1}$ with an average pollen influx of 23.5×10^3 m^{-3} $year^{-1}$(Fig. 3a). Long-term influx of Alnus, Carpinus, Corylus, Quercus, Juglans and Fraxinus

Table 2. Selected pollen types of tree and herb species and their flowering season.

Family	Pollen type	Flowering season
Betulaceae	*Alnus**	January, February, March
	*Betula**	April, May, June
	*Corylus**	January, February, March
Fagaceae	*Quercus*	May, June
Oleaceae	*Fraxinus*	April, May
Plantaginaceae	*Plantago**	May, June, July, August, September

Pollen types associated with pollinosis are indicated by *.

pollen increases from about AD 1990, while TILIA is more dominant before AD 1990 (see Fig. 3b and Fig. S1). The herb pollen types POACEAE, ARTEMISIA, CHENOPODIACEAE, PLANTAGO and URTICACEAE show a phase of maximal influx between approximately AD 1990 and 2000. For 20 pollen types tested, only 6 show no significant correlation to any climate parameter (Tables S4–8 in File S2). Correlations are clearly stronger and more frequent for the temperature variables. In contrast to precipitation, within-season variability is much smaller than between-season variability for temperature (Fig. 2), but the within-season effectively controls the length of the seasons. Six common pollen types with multiple significant correlation coefficients and consistent records at both pollen trap stations, ALNUS, BETULA, CORYLUS, FRAXINUS, QUERCUS and PLANTAGO (Table 2), were selected for further regression analysis (Tables 4–6, Fig. 4). Year-to-year deposition of the six pollen types at LUMC and Elkerliek vary consistently with highly significant correlations

Table 3. Initial model parameters for multiple regression analysis using backward selection.

Pollen	Predictors
ALNUS	January - December Temperature$_{-1}$
	January - April Temperature
BETULA	January - December Temperature$_{-1}$
	January - June Temperature
	April Precipitation$_{-1}$
	June Precipitation$_{-1}$
	April Precipitation
CORYLUS	January - December Temperature$_{-1}$
	January - April Temperature
	Precipitation March$_{-1}$
FRAXINUS	January - December Temperature$_{-1}$
	January - June Temperature
QUERCUS	January - December Temperature$_{-1}$
	January - June Temperature
	Precipitation April
PLANTAGO	January - December Temperature$_{-1}$
	January - September Temperature
PCA 1st axis	January - December Temperature$_{-1}$
	January - September Temperature

Temperature$_{-1}$ stands for temperature in the year before flowering.

of the square root-transformed data (Fig. 4). A linear regression of the total influx shows significant long-term increase with time for all six selected pollen types except CORYLUS for the LUMC site, while only FRAXINUS increases significantly at Elkerliek (Table 4).

Correlation to climate

Positive significant relations were found between the NAO index and the annual pollen influx of QUERCUS and PLANTAGO, as well as the first axis of the PCA analysis, which represents mean composition of the pollen assemblage. Positive significant influence of mean growing season temperature, of the year prior to the flowering season ($T_{summer-1}$ and /or $T_{spring-1}$) is evident in all six pollen types (Table 5), where ALNUS, FRAXINUS and PLANTAGO influx show particularly strong relations with $T_{spring-1}$, while $T_{summer-1}$ is important for BETULA and, to a lesser degree, CORYLUS and QUERCUS. During the year of flowering, T_{spring} is dominant except in BETULA. Based on the monthly correlations, April and March are the most important months in T_{spring} and $T_{spring-1}$, while T_{jul-1} is the dominant summer month. The only herbaceous type, PLANTAGO is also significantly influenced by T_{may} and T_{may-1}. Correlations show much less influence of precipitation on total pollen influx values (Table 6). Only BETULA influx is positively influenced by $P_{spring-1}$ (mainly P_{apr-1}), while in both BETULA and CORYLUS P_{spring} (for BETULA: P_{apr} and CORYLUS: P_{mar}) has a negative influence, although not highly significant and possibly a result of random correlations.

Predictive regression model

The effect of a combination of meteorological variables on pollen influx, assessed through multiple regression analysis with backward selection, is shown in Table 7 for the LUMC data. Calibration and prediction skill of the resulting optimal model of pollen influx are shown in Fig. 5. The optimal regression models show a good performance based on their reported R^2. In BETULA and FRAXINUS some redundant parameters were discarded after the data splitting step, creating a model with slightly lower performance but less parameters, thereby reducing the complexity and statistical overfitting. Growing season temperature of the previous year again shows to be the most important factor, especially T_{mar-1}. In addition, winter temperatures prior to flowering show to be significant as well (T_{nov-1}, T_{dec-1}, T_{jan}, T_{feb}) and mostly have a negative loading (Table 7). Depending on the genus, T_{mar} has a negative (BETULA) or positive (ALNUS, PLANTAGO) loading in the model. Plotted against time (Fig. 5), the trends and phase relations between the observed and predicted influx values correspond well, especially for BETULA and PLANTAGO, although the variability in the predicted values is usually slightly lower than the observations. The multitaper spectral analysis showed signif-

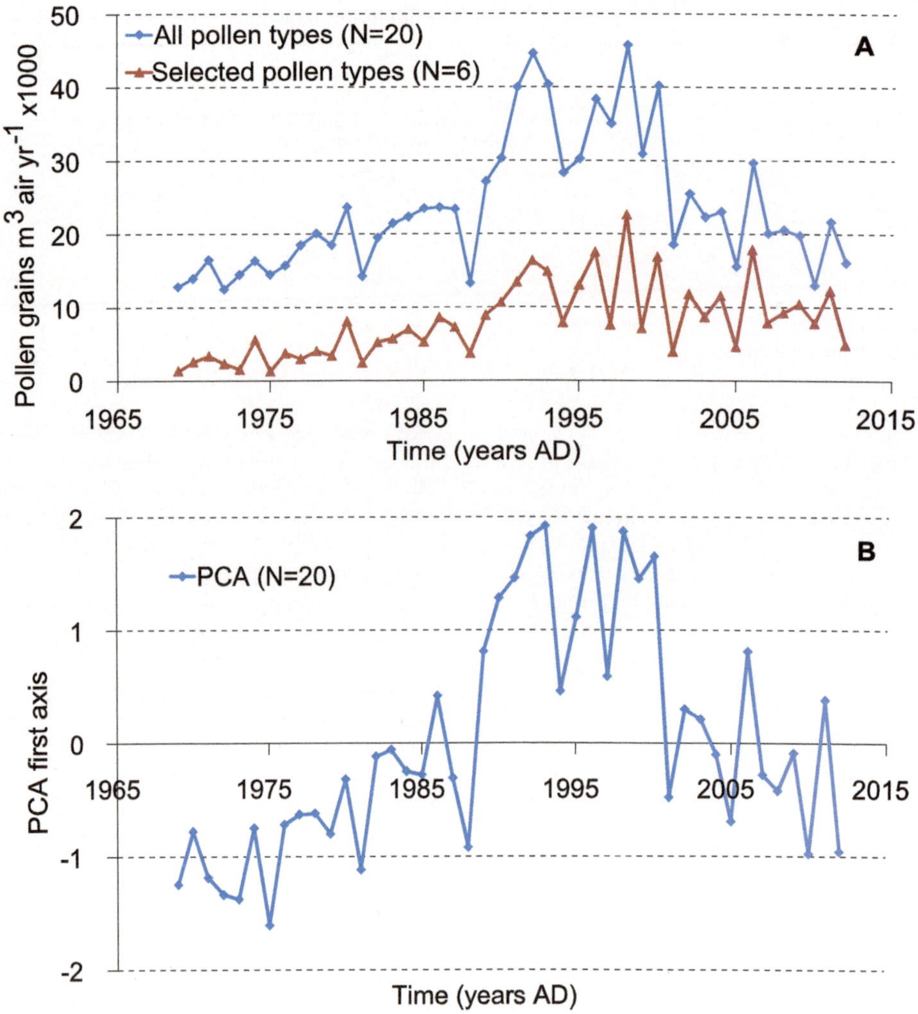

Figure 3. Total annual pollen influx values 1969–2012. (a) Total LUMC pollen influx for all pollen types (n = 20) and for selected pollen types (n = 6), and (b) sample values of the 1st PCA axis based on square-root transformed pollen count data.

icant cyclic patterns for Betula (Fig. 6b) at a frequency of 0.19 (5.3 year) and 0.45 (2.2 year). Also Corylus (3.3 and 2 year, Fig. 6c) and Plantago (3.2 and 2.6 year, Fig. 6f) show significant variability at short periodicities.

Discussion

General observations

The primary aim of this study was to investigate the possible effects of temperature and precipitation variables on the annual pollen production of wind-pollinated plants in the Netherlands. Earlier evaluations showed no significant long-term increase for Betula and Quercus [30], however, ten years of extra data has revised this conclusion (Table 4). The long-term trend observed in Fig. 3b can in part be attributed to construction work at the LUMC site from 1985 until the mid-1990s, which provided favourable conditions for early successional herb species on cleared land. Increased planting in recent years of non-native *Alnus spaethi*, which is well-adapted to urban environments and has an early flowering season, probably adds to the long-term increase as well. Explanation of the increased pollen influx observed in most taxa (see Fig. S1) in terms of long-term temperature change is tempting, considering the sensitivity to annual-scale temperature variations shown in Table 5, but other factors such as long-term increase of CO_2 [7,36], potentially increased production due to long-term atmospheric nutrient deposition [37,38], as well as non-documented changes in composition and management of the urban vegetation cannot be discarded.

The results of this study showed that the annual pollen influx of Alnus, Betula, Corylus, Fraxinus and Quercus, was positively influenced by summer temperature in the year before flowering, which coincides with the production of flowering buds for most tree species [1,4] and confirms our hypothesis. The reported relation between annual influx of Alnus, Betula, Fraxinus, Quercus and Plantago and temperature during the flowering season is likely more related to the final stages of pollen ripening and deposition speed. In that light, it is surprising that precipitation has little effect on most taxa (Table 6), in contrast to the conclusions for central Europe [25]. Only for Betula the effect of additional precipitation is beneficial for the following year, while spring rain during pollen release reduces atmospheric pollen content [4], although it is unclear why this is only for Betula and Corylus in our data. The Netherlands rarely has significant water

Table 4. Linear regression of total annual of pollen influx with time (year) as predictor.

Pollen	LUMC, Leiden				Elkerliek, Helmond			
	M	*S.E.*	r^2	*β*	*M*	*S.E.*	r^2	*β*
ALNUS	2926	354	.278	.527***	2241	234	.037	.191
BETULA	2643	287	.144	.379*	7880	963	.010	.100
CORYLUS	147	16	.072	.268	233	22	.010	.102
FRAXINUS	841	110	.387	.622***	653	96	.456	.675**
QUERCUS	1469	183	.225	.474***	3558	361	.091	.302
PLANTAGO	180	11	.201	.448**	106	10	.095	−.308

Significance levels are indicated: * $P<0.05$; ** $P<0.01$; *** $P<0.001$. S.E = standard error, M = mean, β = slope.

shortage (Fig. 2), and water stress is therefore an unlikely limiting factor for pollen production for most taxa. Comparison of the correlation coefficients in Tables 5 and 6 with results from Southern Denmark [4] show that particularly pollen deposition of BETULA, FRAXINUS and QUERCUS are driven by largely the same climatic variables, while CORYLUS agrees in sensitivity to T_{apr} and T_{apr-1}, but not $T_{jan/feb}$ (Denmark) and T_{jul-1} (LUMC). Contrary to the LUMC data, ALNUS in Denmark shows surprisingly little correlation with climatic variables. Clearly, these results highlight the need for region-specific correlation models for pollen-climate relationships, but do identify climate factors of regional relevance. Specifically for the medical treatment of pollinosis, the linear models (Table 7 and Fig. 5) are a first step in the development of a seasonal predictive model for pollen forecasting. A subsequent step should involve combining the total season pollen load with knowledge on the timing of pollen release, which is climate dependent [17,39]. For forensic palynology, the most important aspect of the study is the fact that the pollentrap record shows variability that is consistent across large distances (Fig.4), which corroborates a study from the Jura Mountains in which pollen composition from several high-resolution peat records were compared [26]. Annual variations in pollen deposition and composition are largely climate driven and, hence, it is meaningful to compare samples from a crime scene to a pollen trap record to estimate ages or relative timing (season).

ALNUS. In *Alnus*, catkins are initiated in early July and meteorological conditions before this period can be expected to influence their formation [40]. In this study, the positive influence of T_{mar-1} suggests that warmer March temperatures stimulate catkin formation in *Alnus*. Catkin formation is followed by a dormancy period during autumn and winter, which protects them from frost damage, until temperature increases at the start of pollen season [40,41]. For the United Kingdom, it has been suggested that bud dormancy starts in August and lasts until February [40]. However, the results of this study indicated that T_{aug-1} affects ALNUS production, suggesting that catkin formation continues in August and bud dormancy starts in the period after August in the Netherlands. A study on the Iberian Peninsula suggested that the chilling period for *Alnus* starts in November and continues until January [41].

In the Netherlands, anthesis (opening of flower buds) and pollination takes place from January until March [34]. The relation of T_{mar} with ALNUS influx found here suggests that warmer temperatures during the flowering season stimulate pollen release or lengthens the flowering season. In the Netherlands, it has been demonstrated that T_{jan} affects the start of the ALNUS pollen season [39], although the Danish results showed no clear relation with temperature [4]. Despite a reported biannual cyclicity for ALNUS [23,41], our study (Fig. 6a) and results from Denmark and Spain showed no significant cyclic patterns [4,41], although some autocorrelation is present with a 2 year time lag (Fig. S2), pointing to a limited role of autocyclic processes in *Alnus* flowering intensity.

BETULA. In the Netherlands *Betula* species occur: *Betula pendula* and *B. pubescens*. *B. pubescens* is closely related to *B. papyrifera* [18] and the phenology of the latter has been thoroughly investigated [42–44]. The phenology of *B. papyrifera* is also characteristic to *B. pendula* and *B. pubescens* [18] and can thus be used as a general description of flower formation of *Betula* trees in the Netherlands. The initiation of male catkins starts just before leaves develop in the year before flowering. They become visible in June and July and develop until August after which dormancy begins [18,43,44]. In the Netherlands, leaf unfolding generally begins in April [45]. It can thus be expected that the

Table 5. Correlation coefficients between annual total pollen influx per pollen type, temperature and NAO index as shown by a Spearman Rank Correlation.

			Spearman rank correlation (r_s), 2-tailed						
Climate variable			**Alnus**	**Betula**	**Fraxinus**	**Corylus**	**Quercus**	**Plantago**	**PCA 1st axis**
Monthly temperature	Previous year	T_{jan-1}	.240	.134	.129	−.024	.128	.248	.229
		T_{feb-1}	.195	−.006	**.324***	.158	.046	**.318***	.167
		T_{mar-1}	**.530*****	.197	**.512*****	.309	**.360***	**.487*****	**.459****
		T_{apr-1}	**.360***	−.014	**.513*****	.048	.173	**.360***	.145
		T_{may-1}	.170	**.308***	.193	.224	.144	**.444****	.297
		T_{jun-1}	.124	**.317***	.037	−.037	.110	.019	.090
		T_{jul-1}	**.318***	**.564*****	**.371***	**.562*****	**.440****	.192	**.442****
		T_{aug-1}	**.363***	.237	.290	**.348***	.135	.048	.260
		T_{sept-1}	.054	.072	.064	.079	.124	.175	.062
		T_{oct-1}	.016	.093	−.010	.152	−.046	−.002	.031
		T_{nov-1}	.138	.200	.081	−.032	.005	−.036	.035
		T_{dec-1}	−.044	.109	.030	−.062	.070	**.318***	.098
	Flowering year	T_{jan}	−.040	.078	.130	−.090	.201	.209	.084
		T_{feb}	.117	**.320***	.167	.143	**.494*****	.278	.245
		T_{mar}	**.329***	−.022	**.385****	.277	.147	**.500*****	**.330***
		T_{apr}	**.507*****	**.463****	**.393****	**.325***	**.576*****	**.303***	**.430****
		T_{may}		.093	.224		**.510*****	**.440****	**.314***
		T_{jun}		.171	.241		.272	.244	.206
		T_{jul}						**.318***	.143
		T_{aug}						.179	.152
		T_{sept}						.131	.150
Seasonal temperature	Previous year	$T_{winter-1}$	.225	.136	.227	.083	.169	.293	.260
		$T_{spring-1}$	**.506*****	.254	**.597*****	.295	.290	**.592*****	**.422****
		$T_{summer-1}$	**.365***	**.534****	**.306***	**.399***	**.362***	.113	**.372***
		$T_{autumn-1}$	.146	.175	.135	.150	.031	.096	.080
	Flowering year	T_{winter}	−.004	.219	.153	−.027	.328*	**.355***	.176
		T_{spring}	**.534*****	.200	**.452****	**.350***	**.572*****	**.573*****	**.476*****
		T_{summer}		.114	**.314***		.139	**.349***	.223
NAO	Previous year	NAO_{-1}	.178	.085	.054	−.117	−.041	.199	.168
		NAO_{DJFM-1}	.272	.242	.202	−.029	.184	**.448****	**.376***
	Flowering year	NAO	.197	.251	.296	.279	**.373***	**.466****	**.401****
		NAO_{DJFM}	.100	.160	.114	.019	.209	.248	.290

Significance levels are indicated: * P<0.05; ** P<0.01; ***P<0.001.

Table 6. Correlation coefficients between annual total pollen influx per pollen type and precipitation as shown by a Spearman Rank Correlation.

			Spearman rank correlation (r_s), 2-tailed						
Climate variable			ALNUS	BETULA	FRAXINUS	CORYLUS	QUERCUS	PLANTAGO	PCA 1st axis
Monthly Precipitation	Previous year	P_{jan-1}	.161	−.051	.030	−.153	−.124	−.043	.017
		P_{feb-1}	.163	.095	.203	.190	.044	.144	.074
		P_{mar-1}	−.037	.238	−.074	−.042	.241	.110	.147
		P_{apr-1}	−.246	**.407****	−.042	.288	.186	.044	.110
		P_{may-1}	−.089	.138	.029	−.172	.114	−.048	−.083
		P_{jun-1}	−.054	**−.326***	.130	−.041	−.119	.173	−.098
		P_{jul-1}	−.038	−.283	.095	−.205	−.074	.030	−.208
		P_{aug-1}	.002	.064	−.021	.226	.181	.079	.016
		P_{sept-1}	.211	.090	.215	−.037	.218	.065	.185
		P_{oct-1}	.006	−.045	−.075	−.241	.056	.024	.070
		P_{nov-1}	−.198	−.090	−.102	.111	.048	−.048	−.064
		P_{dec-1}	.019	.196	.253	−.138	−.012	.176	.114
	Flowering year	P_{jan}	−.241	.066	.006	−.216	.058	−.053	−.068
		P_{feb}	−.117	.186	.057	.053	.144	.029	.021
		P_{mar}	−.085	−.052	−.139	**−.340***	.195	.183	−.005
		P_{apr}	−.080	**−.308***	−.148	−.143	**−.333***	.175	−.072
		P_{may}		−.154	−.138		−.129	−.030	−.160
		P_{jun}		.037	−.120		.123	−.008	.104
		P_{jul}						−.136	−.114
		P_{aug}						.030	−.029
		P_{sept}						.229	.168
Seasonal Precipitation	Previous year	$P_{winter-1}$	.189	−.014	.083	−.076	−.033	.049	.061
		$P_{spring-1}$	−.204	**.375***	−.054	−.004	**.312***	.060	.086
		$P_{summer-1}$	−.025	−.208	.094	.034	.083	.208	−.091
		$P_{autumn-1}$	.066	−.014	.032	−.116	.195	.037	.130
	Flowering year	P_{winter}	−.188	.194	.143	−.166	.068	.069	.012
		P_{spring}	−.142	**−.303***	−.259	**−.334***	−.126	.162	−.139
		P_{summer}		.068	.052		.271	.003	.001

Significance levels are indicated: * $P<0.05$; ** $P<0.01$; *** $P<0.001$.

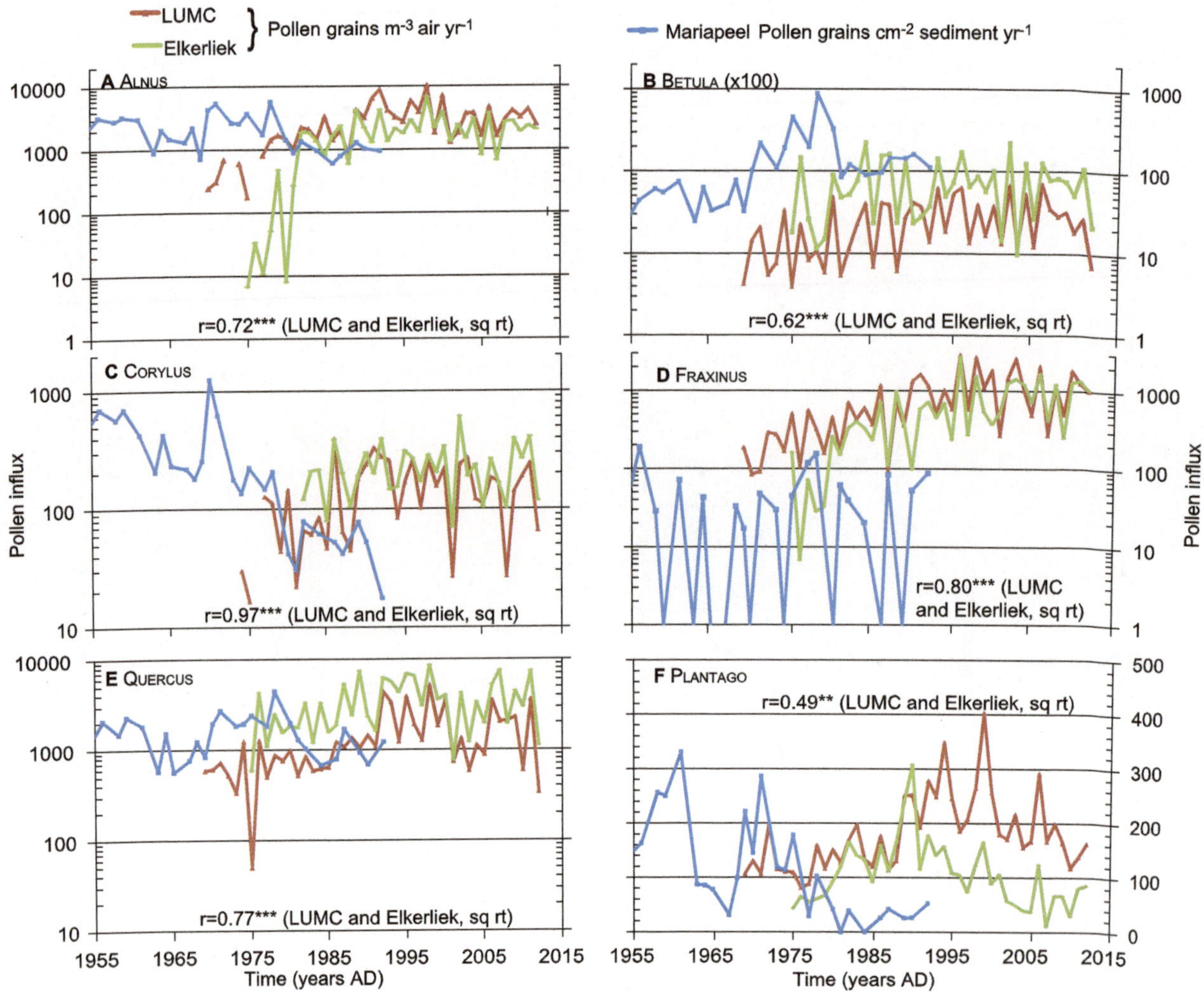

Figure 4. Annual pollen influx values 1969–2012 for LUMC, Elkerliek (n m^{-3} air yr^{-1}) and Mariapeel (n cm^{-2} sediment yr^{-1}). Values are shown for (a) ALNUS, (b) BETULA, (c) CORYLUS, (d) FRAXINUS, (e) QUERCUS, and (f) PLANTAGO. Note that a–e are on a logarithmic y-axis.

formation of male catkins begins in April and that the complete catkin development likely continues until August.

The here reported relation of BETULA influx with T_{mar-1}, T_{may-1} and T_{jul-1} coincides with the main period of catkin formation. Studies in Denmark show correlations to T_{may-1}, T_{jun-1}, T_{jul-1} [4,48], and in northern Finland to T_{jun-1} [6], suggesting an earlier start of catkin formation more to the south. The sensitivity of BETULA to precipitation ($P_{apr/jun-1}$) is identical to that in Denmark [4], and similar to central Europe [25], and might be related to its preference for relatively wet habitats, particularly in *B. pubescens*. The 5-year cyclicity found in BETULA only, possibly relates to precipitation as well (Fig. 5b). Here, we hypothesize that *Betula* invests in reproduction strategies (i.e. pollen production) during dry conditions, while in wetter conditions these trees invest in vegetative growth.

Including the temperature values of the October - March winter dormancy period in *Betula* [18] improves the predictive model (Table 7). It can be hypothesized that warmer temperatures in November may disturb this winter dormancy but as the sign of the correlation differs between November and December this needs independent confirmation. The negative relation with T_{mar} is remarkable as *Betula* flowering season is between April and June [34]. A possible explanation might be that although warmer temperatures in early spring can advance the *Betula* pollen season [17], the increased chance of frost damage results in lower annual pollen influx.

Beside the strong 5-year cyclicity, a 2.4 year cycle in BETULA influx confirms earlier finds of bi-annual variability [18] and a three-year cycle [21]. As high pollen production likely results in a high energy-intensive fruit production this might inhibit the development of reproductive structures for the flowering season in the next year [30]. Ranta et al. [2] indeed conclude that "masting of birch species is regulated by weather factors together with the system of resource allocation among years". High inflorescence numbers might result in smaller and fewer leaves, lowering overall photosynthetic capacity of the tree including development of new flower buds. As a consequence, few inflorescences in the following year will relocate more energy for the development of leaves, and in turn stimulates the development of flower buds [18]. Pollen production is indeed correlated to catkin formation, and year-to year changes have previously been shown to be similar across large distances (up to 500 km) [49].

As temperature and precipitation influence photosynthetic capacity, climatic conditions are likely to exert control over this autocyclic pattern, which might explain variations between 2- and 3-year cycle lengths.

Table 7. Statistical parameter values of the regression model with backward selection method and the optimal regression model after data splitting.

Pollen type	Regression model after backward selection			Optimal regression model after data splitting analysis		
	R^2	Variable	β	R^2	Variable	β
Alnus	.538***	$T_{March-1}$	.549***	.538***	$T_{March-1}$	.549***
		$T_{August-1}$	.303**		$T_{August-1}$	.303**
		$T_{January}$	−.265*		$T_{January}$	−.265*
		T_{March}	.230		T_{March}	.230
Betula	.677***	$T_{January-1}$	−.258*	.565***	$T_{March-1}$	.286*
		$T_{March-1}$	.432***		T_{May-1}	.333**
		T_{May-1}	.411***		T_{July-1}	.486***
		T_{July-1}	.410**		$T_{August-1}$	.195
		$T_{August-1}$	.263*		$T_{November-1}$	−.248
		$T_{October-1}$	.209		$T_{December-1}$	.261*
		$T_{November-1}$	−.327*		T_{March}	−.358**
		$T_{December-1}$	.298*		P_{June-1}	−.357**
		$T_{February}$	.225			
		T_{March}	−.513***			
		P_{June-1}	−.363**			
Corylus	.616***	$T_{January-1}$	−.377**	.616***	$T_{January-1}$	−.377**
		$T_{March-1}$	.330**		$T_{March-1}$	.330**
		T_{May-1}	.276*		T_{May-1}	.276*
		T_{July-1}	.437***		T_{July-1}	.437***
		$T_{August-1}$	.219		$T_{August-1}$	.219
		$T_{October-1}$	.261*		$T_{October-1}$	.261*
		$T_{November-1}$	−.364**		$T_{November-1}$	−.364**
		P_{march}	−.224*		P_{march}	−.244*
Fraxinus	.629***	$T_{January-1}$	−.298*	.386***	$T_{February-1}$	.325*
		$T_{February-1}$	.438**		$T_{April-1}$	.376**
		$T_{March-1}$	.392**		T_{July-1}	.298*
		$T_{April-1}$	.414***			
		T_{July-1}	.475***			
		$T_{November-1}$	−.232			
		$T_{January}$	−.340**			
Quercus	.703***	$T_{March-1}$	.296**	.523***	T_{July-1}	.493***
		T_{July-1}	.498***		T_{May}	.544***
		$T_{November-1}$	−.219*			
		$T_{January}$	−.242*			
		T_{April}	.242*			
		T_{May}	.514***			
Plantago	.579***	$T_{March-1}$	.254*	.463***	$T_{March-1}$	.277*
		T_{May-1}	.211		T_{May-1}	.196
		$T_{November-1}$	−.212		T_{March}	.298*
		T_{March}	.230		T_{July}	.276*
		T_{May}	.264*			
		T_{July}	.289*			

Significance levels are indicated: * $P<0.05$; ** $P<0.01$; *** $P<0.001$.

Corylus. As in Betula, Corylus pollen influx broadly depends on growing season temperature of the year before and shows an earlier start compared to Denmark [4]. *Corylus* flowers early in the year, from January until March [34], and unlike observations in Denmark [4], Corylus influx has no (negative) relation with respect to winter temperatures, which confirms their

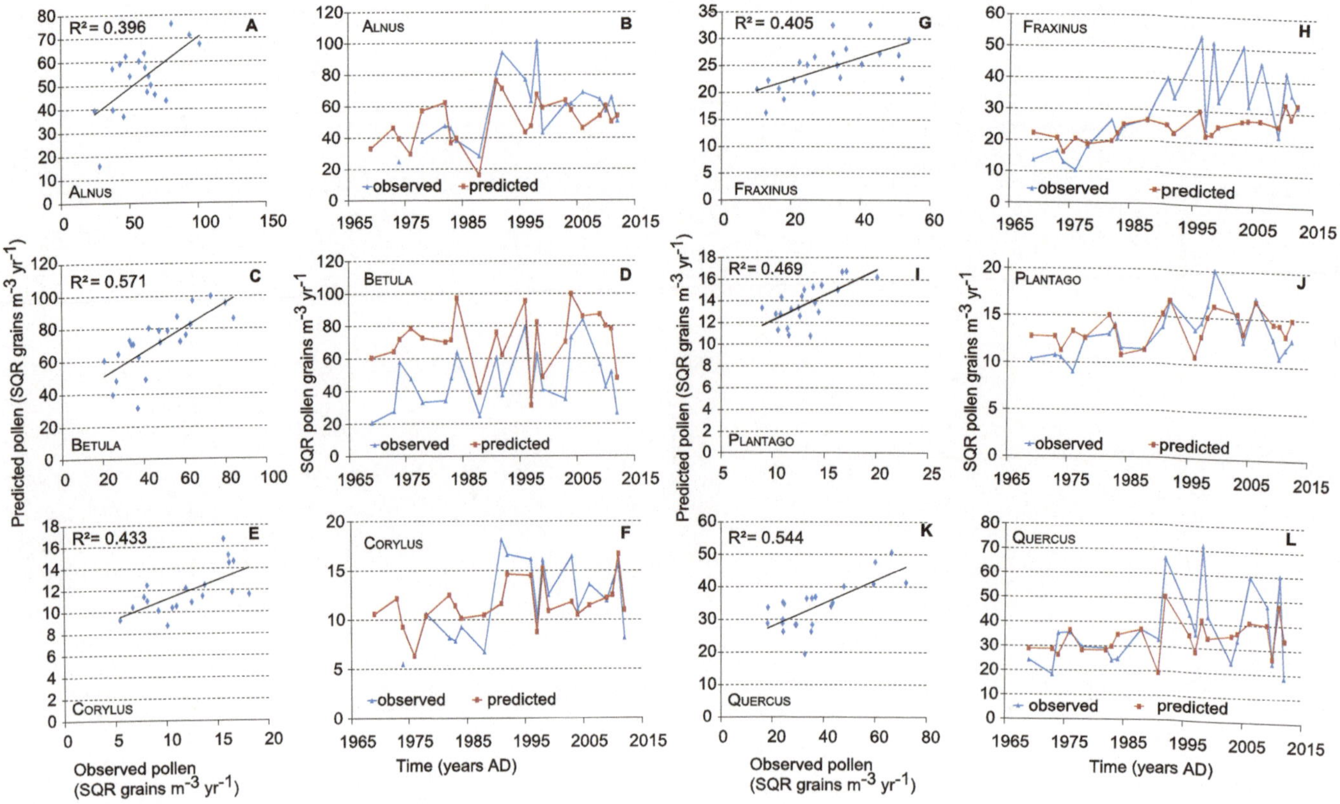

Figure 5. Observed versus climate predicted LUMC pollen influx based on the split data sets. Values are shown for (a,b) ALNUS, (c,d) BETULA, (e,f) CORYLUS, (g,h) FRAXINUS, (i,j) QUERCUS, and (k,l) PLANTAGO.

cold adaptation and relative insensitivity to temperature in that period in this region. The negative effect of P_{mar} on annual pollen influx is probably the consequence of rain washing the pollen out of the atmosphere during the pollination period. The 2.3 year cyclicity in the signal is of the same character as in BETULA (Fig. 6c), and has also been observed in the United Kingdom [22]. Although no specific studies confirm this, we infer a similar mechanism as that described for *Betula* as both are Betulaceae.

FRAXINUS. The factors T_{feb-1}, T_{apr-1} and T_{jul-1} best predict the annual FRAXINUS influx suggesting a particularly long period of flower bud formation and sensitivity. Highly similar results from Denmark confirm this [4], while a study from Galicia (northwest Spain) involves also precipitation as an important factor for *Fraxinus* flowering [8]. The predictive model does not explain the majority of the variability (39%), suggesting edaphic factors and other internal biological processes play a significant role.

QUERCUS. In the temperate climate zone, *Quercus* forms flower buds during the summer of the previous year and enters a dormancy period during autumn and winter [46].The sensitivity to T_{jul-1} at LUMC and T_{aug-1} in Denmark [4] suggest a short critical period for bud formation. In cork-oak (*Q. suber*), reports of steep temperature drop during initial stages of microsporogenesis resulted in catkin mortality and much lower pollen quantities [16].

PLANTAGO. In perennial herbaceous plants, warm springs and summers in previous years are important for plant growth and flower development, as shown by the influence of T_{mar-1} and T_{may-1}in our analysis, whereas summer droughts may kill plants and thereby affect the pollen production [23]. Warmth during early spring and summer stimulates plant development and flower differentiation [23] and continued pollen generation typical of herbs [47], which can explain the positive impact of T_{mar} and T_{Jul} on the annual influx of PLANTAGO. Earlier studies of *P. lanceolata* from Poland did not reveal climate/pollen relationships, likely limited by the length of the record (10 years) [25]. The results of this study also show a positive correlation between the NAO index and the total annual PLANTAGO influx. The multi-taper analysis of the annual PLANTAGO influx reveals a particularly strong 3.2- year cycle (Fig. 6f), which, in combination with a positive correlation with the NAO index, points to a multi-annual climatic influence.

Palaeoecological interpretation

Our aim to compare trends in pollen trap data from LUMC and Elkerliek with those in the palaeoecological record are complicated by human management of the Mariapeel area and, despite the accurate depth-age model, slight age uncertainties that preclude a year-to-year correlation between the trapping and sub-fossil data. The two pollen trap sites agree surprisingly well in terms of phase relation and absolute numbers of pollen (Fig. 3). The Elkerliek site has generally higher amounts of BETULA and QUERCUS, which is in agreement with the greater amount of tree cover, sandy soils and vicinity to natural reserves of that site. LUMC has a greater amount of PLANTAGO, in agreement with a more urban setting. The offset in ALNUS of the early part of the record in Elkerliek is related to incomplete counts in the first years. The consistency and significant correlations between both trap data (Fig. 4) shows that the stations produce data that are representative of a broad region. This suggests that the correlation with climate quantified at LUMC is a regionally relevant ecological factor that is responsible for the majority of the observed annual-scale pollen influx variability. This observation

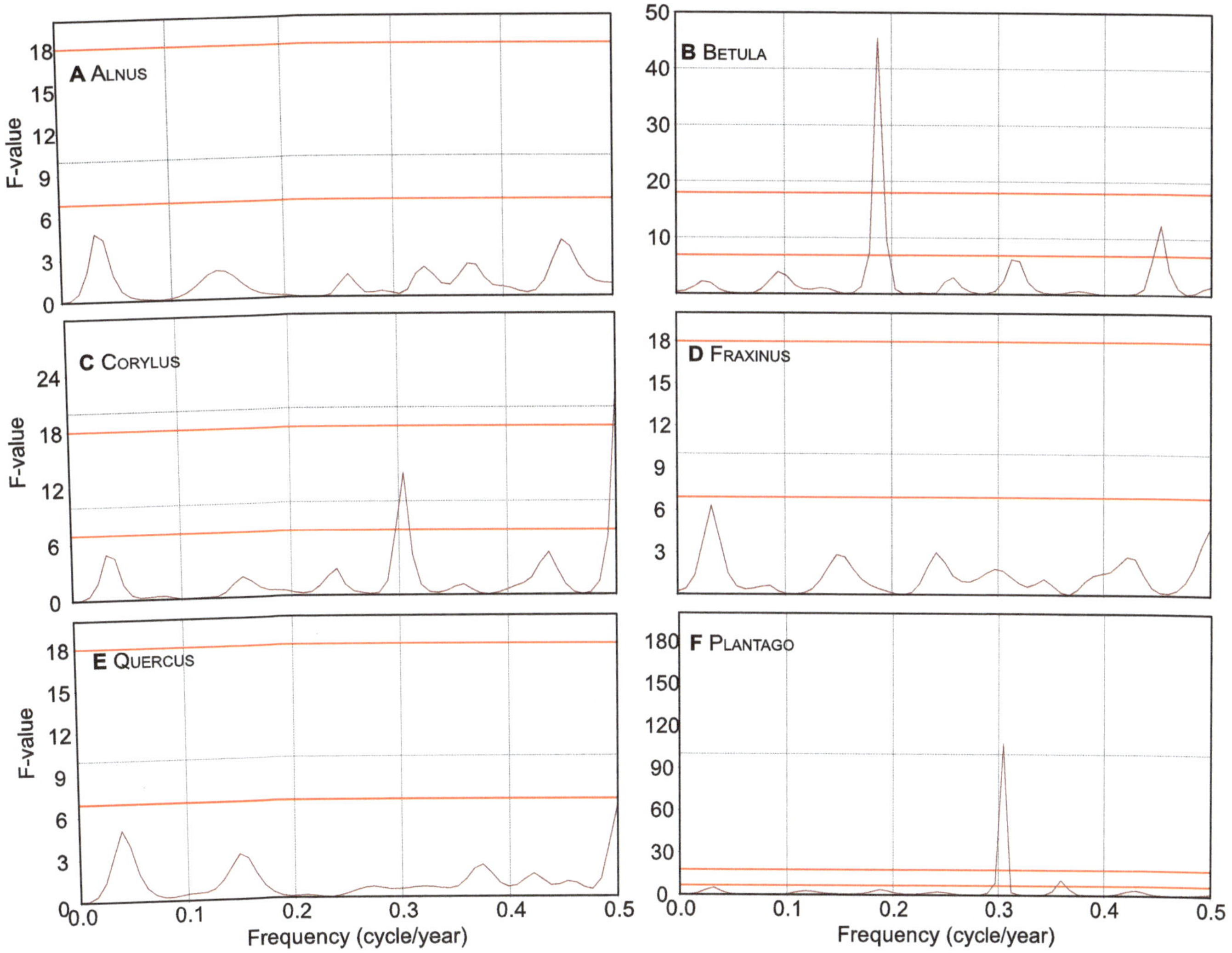

Figure 6. Spectral analysis output with 95% and 99% significance levels (upper and lower red line, respectively) of the 1969-2012 annual pollen influx time series in the LUMC trap. (a) Alnus, (b) Betula, (c) Corylus, (d) Fraxinus, (e) Quercus, and (f) Plantago.

also aides comparison of pollen assemblages with the trapping stations in forensic studies (e.g. for the purpose of dating a crime scene).

The Mariapeel site presently is a semi-open *Betula* forest with secondary *Sphagnum* growth in pits formerly mined for peat. The pollen accumulation rates from the sub-fossil peat deposit show a large decrease in all taxa around AD 1980 when the area was declared nature reserve and much of the local standing *Betula* tree vegetation has been cut and water levels were raised to stimulate peat regrowth. Visual inspection of the pre-1980 data (Fig. 3) from Mariapeel show variations of the same order (frequency and amplitude) as the variability present in the pollen trap records. Pollen from the more regionally occurring trees, such as *Quercus* and *Corylus*, show comparable trends and short-term changes as the trap data, but the local changes and short overlap period between unimpacted local vegetation and the traps preclude detailed correlation. The comparison does demonstrate that natural (peat) archives record and preserve high-order variations, with similar amplitude and frequencies as in the pollen trap data. Given the preserved high frequency variations and the identified climatic drivers (Tables 5 and 6) of annual pollen deposition, natural archives can provide significant insights in past climatic variations at near annual scales (see e.g. [13]). Hence, our analysis provides a regional interpretation framework for climatic interpretations of high-frequency changes in pollen records in NW Europe.

Conclusion

This study showed that climate is an important factor in annual pollen production in the Netherlands and that annual pollen influx shows highly similar variability across a broad geographical area, which is driven by largely the same variables. Summer temperatures in the year before flowering, as well as temperature during the flowering season, are the primary climate variables that determine the annual pollen influx of wind-pollinated plants, while the effect of precipitation is minimal, except for Betula. Summer temperature influences the formation of reproductive structures, while temperature during the flowering season is thought to influence pollen release. The importance of long observational records is evident as shorter series often contain too much scatter to determine pollen-climate relations. Our results provide a first developmental step toward a region-specific predictive model for seasonal pollen forecasting for hay fever patients and forensic studies, and on short timescales (years to decades) predicts the likely impact of changing temperatures on annual pollen production due to global change.

The similar-scale high frequency variations observed in the peat record compared to the pollen traps suggest that, although influenced by local edaphic factors, natural archives can provide a proxy for quantitative reconstruction at high (e.g. annual) temporal resolution. However, the observed relation between climate and pollen production found for the pollen data from Leiden cannot be directly tested on the samples of from the Mariapeel, due to local landscape management changes and small differences in age.

Supporting Information

Figure S1 Total annual pollen accumulation rates of all recorded taxa in the pollen traps from LUMC, Leiden, and climate variables from De Bilt, The Netherlands from 1969 to 2012.

Figure S2 Autocorrelation of LUMC pollen accumulation rates for 0 to 20 years lags. Curved lines represent 95% significance level.

File S1 This file contains Table S1–Table S3. Table S1. Pollen count data for LUMC pollen traps. Table S2. Pollen count data for Elkerliek pollen traps. Table S3. The peat core from Mariapeel Natural Reserve, The Netherlands. For locations see main text and Fig. 1.

File S2 Spearman rank correlation coefficients between annual pollen influx and climate variables of the pollen types not shown in Tables 5 and 6 of the main text (Tables S4–S8). Table S4. North Atlantic Oscillation Index (annual and winter). Table S5. Temperature in the flowering year. Table S6. Temperature in the year before flowering. Table S7. Precipitation in the flowering year. Table S8. Precipitation in the year before flowering.

Acknowledgments

The pollen data from Helmond were kindly made available by dr. A. O. de Graaf of the Elkerliek Hospital in Helmond. We thank two anonymous reviewers for the constructive comments that improved the manuscript, and thank Arnold van Vliet (Wageningen University) for his advice on the topic.

Author Contributions

Conceived and designed the experiments: THD FWC SD. Performed the experiments: KH THD FWC SD. Analyzed the data: THD KH SD. Contributed reagents/materials/analysis tools: LdW PdK. Wrote the paper: THD KH FWC SD LdW PdK.

References

1. Rogers CA (1993) Application of aeropalynological principles in palaeoecology. Rev Palaeobot Palynol 79: 133–140. http://dx.doi.org/10.1016/0034-6667(93)90043-T.
2. Ranta H, Oksanen A, Hokkanen T, Bondestam K, Heino S (2005) Masting by *Betula*-species; applying the resource budget model to north European data sets. Int J Biometeorol 49: 146–151. 10.1007/s00484-004-0228-0. Available: http://dx.doi.org/10.1007/s00484-004-0228-0.
3. Hicks S (2001) The use of annual arboreal pollen deposition values for delimiting tree-lines in the landscape and exploring models of pollen dispersal. Review of Palaeobotany and Palynology 117: 1–29.
4. Nielsen A, Möller P, Giesecke T, Stavngaard B, Fontana S, et al. (2010) The effect of climate conditions on inter-annual flowering variability monitored by pollen traps below the canopy in Draved forest, Denmark. Vegetation History and Archaeobotany 19: 309–323. 10.1007/s00334-010-0253-3. Available: http://dx.doi.org/10.1007/s00334-010-0253-3.
5. Barnekow L, Loader NJ, Hicks S, Froyd CA, Goslar T (2007) Strong correlation between summer temperature and pollen accumulation rates for *Pinus sylvestris, Picea abies* and *Betula* spp. in a high-resolution record from northern Sweden. Journal of Quaternary Science 22: 653–658. 10.1002/jqs.1096.
6. Autio J, Hicks S (2004) Annual variations in pollen deposition and meteorological conditions on the fell Aakenustunturi in northern Finland: Potential for using fossil pollen as a climate proxy. Grana 43: 31–47.
7. Ziska LH, Caulfield FA (2000) Rising CO_2 and pollen production of common ragweed (*Ambrosia artemisiifolia* L.), a known allergy-inducing species: Implications for public health. Functional Plant Biol 27: 893-898. Available: http://www.publish.csiro.au/paper/PP00032.
8. Jato V, Rodríguez-Rajo J, Dacosta N, Aira M (2004) Heat and chill requirements of *Fraxinus* flowering in Galicia (NW Spain). Grana 43: 217–223.
9. Shea KM, Truckner RT, Weber RW, Peden DB (2008) Climate change and allergic disease. J Allergy Clin Immunol 122: 443–453. Available: http://linkinghub.elsevier.com/retrieve/pii/S0091674908011810?showall = true
10. Walsh KAJ, Horrocks M (2008) Palynology: Its position in the field of forensic science. J Forensic Sci 53: 1053–1060. 10.1111/j.1556-4029.2008.00802.x.
11. Kuoppamaa M, Huusko A, Hicks S (2009) *Pinus* and *Betula* pollen accumulation rates from the northern boreal forest as a record of interannual variation in July temperature. Journal of Quaternary Science 24: 513–521. 10.1002/jqs.1276.
12. Donders TH, Punyasena SW, de Boer HJ, Wagner-Cremer F (2013) ENSO signature in botanical proxy time series extends terrestrial el niño record into the (sub)tropics. Geophys Res Lett 40: - 2013GL058038. 10.1002/2013GL058038.
13. Finsinger W, Schoning K, Hicks S, Lücke A, Goslar T, et al. (2013) Climate change during the past 1000 years: A high-temporal-resolution multiproxy record from a mire in northern Finland. Journal of Quaternary Science 28: 152–164. 10.1002/jqs.2598.
14. Brewer S, Guiot J, Barboni D (2007) Pollen methods and studies, use of pollen as climate proxies. In: Elias SA, editor.Encyclopedia of Quaternary Science.Ox-Oxford: Elsevier. pp. 2497–2508. Available: http://dx.doi.org/10.1016/B0-44-452747-8/00177-0.
15. Hurrell JW, van Loon H (1997) Decadal variations in climate associated with the North Atlantic Oscillation. Climatic change 36: 301–326.
16. García-Mozo H, Hidalgo PJ, Galán C, Maria TG, Domínguez E (2001) Catkin frost damage in mediterranean cork-oak (*Quercus suber* L.). Isr J Plant Sci 49: 42–47. 10.1560/JR25-EB31-9JTG-WTD6. Available: http://www.tandfonline.com/doi/abs/10.1560/JR25-EB31-9JTG-WTD6.
17. Van Vliet AJH, Overeem A, De Groot RS, Jacobs AFG, Spieksma FTM (2002) The influence of temperature and climate change on the timing of pollen release in the Netherlands. Int J Climatol 22: 1757–1767. 10.1002/joc.820.
18. Dahl-Strandhede ÅS (1996) Predicting the intensity of the birch pollen season. Aerobiologia 12: 97–106. 10.1007/BF02248133. Available: http://dx.doi.org/10.1007/BF02248133
19. Beggs PJ (2004) Impacts of climate change on aeroallergens: Past and future. Clinical & Experimental Allergy 34: 1507–1513. 10.1111/j.1365-2222.2004.02061.x.
20. Spieksma FT, Emberlin JC, Hjelmroos M, Jäger S, Leuschner RM (1995) Atmospheric birch (*Betula*) pollen in europe: Trends and fluctuations in annual quantities and the starting dates of the seasons. Grana 34: 51–57. 10.1080/00173139509429033. Available: http://dx.doi.org/10.1080/00173139509429033.
21. Detandt M, Nolard N (2000) The fluctuations of the allergenic pollen content of the air in Brussels (1982 to 1997). Aerobiologia 16: 55–61. 10.1023/A:1007619724282. Available: http://dx.doi.org/10.1023/A%3A1007619724282.
22. Emberlin J, Smith M, Close R, Adams-Groom B (2007) Changes in the pollen seasons of the early flowering trees *Alnus* spp. and *Corylus* spp. in Worcester, United Kingdom, 1996–2005. Int J Biometeorol 51: 181–191. 10.1007/s00484-006-0059-2. Available: http://dx.doi.org/10.1007/s00484-006-0059-2.
23. Andersen ST (1980) Influence of climatic variation on pollen season severity in wind-pollinated trees and herbs. Grana 19: 47–52. 10.1080/00173138009424986. Available: http://www.tandfonline.com/doi/abs/10.1080/00173138009424986.
24. Green D, Singh G, Polach H, Moss D, Banks J, et al. (1988) A fine-resolution palaeoecology and palaeoclimatology from south-eastern Australia. Journal of Ecology 76: 790–806.
25. Knaap WO, Leeuwen JN, Svitavská-Svobodová H, Pidek I, Kvavadze E, et al. (2010) Annual pollen traps reveal the complexity of climatic control on pollen productivity in Europe and the Caucasus. Vegetation History and Archaeobotany 19: 285–307. 10.1007/s00334-010-0250-6. Available: http://dx.doi.org/10.1007/s00334-010-0250-6.

26. Sjögren P, van Leeuwen JFN, van der Knaap WO, van der Borg K (2006) The effect of climate variability on pollen productivity, AD 1975–2000, recorded in a *Sphagnum* peat hummock. The Holocene 16: 277–286. 10.1191/0959683606hl924rp.
27. Spieksma FM, Nikkels AH (1998) Airborne grass pollen in Leiden, the Netherlands: Annual variations and trends in quantities and season starts over 26 years. Aerobiologia 14: 347–358. 10.1007/BF02694304. Available: http://dx.doi.org/10.1007/BF02694304.
28. Joosten H, de Klerk P (2002) What's in a name?: Some thoughts on pollen classification, identification, and nomenclature in quaternary palynology. Rev Palaeobot Palynol 122: 29–45. Available: http://dx.doi.org/10.1016/S0034-6667(02)00090-8.
29. Wagner F, Below R, Klerk PD, Dilcher DL, Joosten H, et al. (1996) A natural experiment on plant acclimation: Lifetime stomatal frequency response of an individual tree to annual atmospheric CO_2 increase. Proceedings of the National Academy of Sciences 93: 11705–11708.
30. Spieksma FT, Corden JM, Detandt M, Millington WM, Nikkels H, et al. (2003) Quantitative trends in annual totals of five common airborne pollen types (*Betula, Quercus, Poaceae, Urtica*, and *Artemisia*), at five pollen-monitoring stations in western Europe. Aerobiologia 19: 171–184. 10.1023/B:AERO.0000006528.37447.15. Available: http://dx.doi.org/10.1023/B%3AAERO.0000006528.37447.15.
31. Hirst JM (1952) An automatic volumetric spore trap. Annals Applied Biology 39: 257–265.
32. Anonymous (2012) North Atlantic Oscillation index data provided by the Climate Analysis Section, NCAR, Boulder, USA. Available: http://climatedataguide.ucar.edu/guidance/hurrell-north-atlantic-oscillation-nao-index-station-based.
33. Ter Braak CJF, Smilauer P (2002) CANOCO reference manual and CanoDraw for windows user's guide: Software for canonical community ordination (version 4.5). Ithaca, New York: Microcomputer Power.
34. van der Meijden R (2005) Heukels' flora van Nederland. Groningen, the Netherlands.: Wolters-Noordhoff.
35. Hammer O, Harper D, Ryan P (2001) PAST: Paleontological statistics software package for education and data analysis. Paleontologia Electronica 1: 1–9.
36. Ziello C, Sparks TH, Estrella N, Belmonte J, Bergmann KC, et al. (2012) Changes to airborne pollen counts across Europe. PLoS ONE 7: e34076. Available: http://dx.doi.org/10.1371%2Fjournal.pone.0034076.
37. Lau T, Stephenson AG (1993) Effects of soil nitrogen on pollen production, pollen grain size, and pollen performance in *Cucurbita pepo* (Cucurbitaceae). Am J Bot 80: 763–768. Available: http://www.jstor.org/stable/2445596.
38. Bobbink R, Hicks K, Galloway J, Spranger T, Alkemade R, et al. (2010) Global assessment of nitrogen deposition effects on terrestrial plant diversity: A synthesis. Ecol Appl 20: 30–59. 10.1890/08-1140.1. Available: http://dx.doi.org/10.1890/08-1140.1
39. Spieksma FT, Frenguelli G, Nikkels AH, Mincigrucci G, Smithuis LOMJ, et al. (1989) Comparative study of airborne pollen concentrations in central Italy and the Netherlands (1982–1985). Grana 28: 25–36. 10.1080/00173138909431009. Available: http://dx.doi.org/10.1080/00173138909431009.
40. McVean DN (1955) Ecology of *Alnus glutinosa* (L.) gaertn.: I. fruit formation. J Ecol 43: 46–60. Available: http://www.jstor.org/stable/2257118.
41. Rodríguez-Rajo F, Dopazo A, Jato V (2004) Environmental factors affecting the start of the pollen season and concentrations of airborne *Alnus* pollen in two localities of Galicia (NW Spain). Annals of Agricultural and Environmental Medicine 11: 35–44.
42. Caesar JC, Macdonald AD (1984) Shoot development in *Betula papyrifera*. V. effect of male inflorescence formation and flowering on long shoot development. Can J Bot 62: 1708–1713. 10.1139/b84-231. http://dx.doi.org/10.1139/b84-231.
43. Macdonald AD, Mothersill DH, Caesar JC (1984) Shoot development in *Betula papyrifera*. III. long-shoot organogenesis. Can J Bot 62: 437–445. 10.1139/b84-066. Available: http://dx.doi.org/10.1139/b84-066.
44. Macdonald AD, Mothersill DH (1983) Shoot development in *Betula papyrifera*. I. short-shoot organogenesis. Can J Bot 61: 3049-3065. 10.1139/b83-342. Available: http://www.nrcresearchpress.com/doi/abs/10.1139/b83-342.
45. Anonymous (2014) Netherlands phenological observation network (transl.). Environmental Systems Analysis Group, Wageningen University, The Netherlands. Available: www.natuurkalender.nl
46. Kasprzyk I (2009) Forecasting the start of *Quercus* pollen season using several methods - the evaluation of their efficiency. Int J Biometeorol 53: 345–353. 10.1007/s00484-009-0221-8. Available: http://dx.doi.org/10.1007/s00484-009-0221-8.
47. Tyler G (2001) Relationships between climate and flowering of eight herbs in a Swedish deciduous forest. Annals of Botany 87: 623–630. 10.1006/anbo.2001.1383.
48. Rasmussen A (2002). The effect of climate change on the birch pollen season in Denmark. Aerobiologia 18: 253–265.
49. Ranta H, Hokkanen T, Linkosalo T, Laukkanen L, Bondestam K, et al. (2008). Male flowering of birch: Spatial synchronization, year-to-year variation and relation of catkin numbers and airborne pollen counts. Forest Ecology and Management 255: 643–650.

11

Quantifying Assemblage Turnover and Species Contributions at Ecologic Boundaries

Lee-Ann C. Hayek[1], Brent Wilson[2*]

1 Smithsonian Institution Mathematics and Statistics NMNH MRC-121, Washington D.C., United States of America, 2 Petroleum Geoscience Programme, Department of Chemical Engineering, The University of the West Indies, Saint Augustine, Trinidad, Trinidad and Tobago

Abstract

Not all boundaries, whether stratigraphical or geographical, are marked by species-level changes in community composition. For example, paleodata for some sites do not show readily discernible glacial-interglacial contrasts. Rather, the proportional abundances of species can vary subtly between glacials and interglacials. This paper presents a simple quantitative measure of assemblage turnover (assemblage turnover index, ATI) that uses changes in species' proportional abundances to identify intervals of community change. A second, functionally-related index (conditioned-on-boundary index, CoBI) identifies species contributions to the total assemblage turnover. With these measures we examine benthonic foraminiferal assemblages to assess glacial/interglacial contrasts at abyssal depths. Our results indicate that these measures, ATI and CoBI, have potential as sequence stratigraphic tools in abyssal depth deposits. Many peaks in the set of values of ATI coincide with terminations at the end of glaciations and delineate peak-bounded ATI intervals (PATIs) separated by boundaries that approximate to glacial terminations and to transgressions at neritic depths. These measures, however, can be used to evaluate the assemblage turnover and composition at any defined ecological or paleoecological boundary. The section used is from Ocean Drilling Program (OPD) Hole 994C, drilled on the Blake Ridge, offshore SE USA.

Editor: Vanesa Magar, Plymouth University, United Kingdom

Funding: These authors have no support or funding to report.

Competing Interests: The authors have declared that no competing interests exist.

* E-mail: brent.wilson@sta.uwi.edu

Introduction

Biostratigraphers historically have sought means to subdivide the sedimentary stratigraphic record as finely as possible. There is, however, evidently a limit to the zonal resolution that can be attained using a single fossil group [1]. For example, the majority of the Pleistocene, the base of which is placed at 2.588 Ma [2], is currently ascribed to the *Globorotalia truncatulinoides truncatulinoides* (dOrbigny) planktonic foraminiferal Zone [3] or to Zones PT1a and PTIb based on planktonic foraminifera [4] (author names are given at the first mention of any species).

It has long been appreciated, however, that glacial and interglacial fauna and flora within the Pleistocene differ [5], both on land [6] and in the oceans [7]. In Chile, for example, plant leaf architecture changes from a mixture of species belonging to the cool temperate North Patagonian Forest and more thermophilous rain forest vegetation between glacials and interglacials [8]. Glacial-interglacial contrasts in the insect community have been recorded in Greenland [9]. Such changes have been recorded among some foraminifera, but not at all sites. Schott [10] found that *Globorotalia menardii* (d'Orbigny) in the Indian Ocean was more abundant in interglacial than in glacial sediments. Phleger et al. [11], in a study of North Atlantic foraminifera, noted the presence in piston cores of "layers of faunas normal for their latitude alternating with faunas typical of a higher latitude," while Bandy [12] was able to distinguish glacials from interglacials off southern California using the ratio between populations of sinistrally and dextrally coiled *Neogloboquadrina pachyderma* (Ehrenberg). Streeter [13] found that Atlantic benthonic foraminifera at depths >2500 m varied greatly over the last 150 ka and suggested that this arose because of depression and elevation of faunas through a depth range of several hundred meters between glacials and interglacials [14]. Streeter and Lavery [15] recorded that uppermost Pleistocene faunas in cores from the western North Atlantic slope and rise north of 35°N were dominated by *Uvigerina*, but that *Hoeglundina* dominated in the Holocene. This faunal transition was apparently diachronous, occurring at ~12 ka at 3,000 m, but at ~8 ka at 4,000 m. Thomas et al. [16] examined benthonic foraminifera in two lower bathyal (~1700 m) and abyssal (~3500 m) piston cores spanning the last 45 ka in the northeastern Atlantic Ocean. They found that *Epistominella exigua* (Brady) and *Alabaminella weddellensis* (Earland), which bloom opportunistically where a spring plankton bloom produces a pulse of phytodetritus, were rare during the last glacial maximum but abundant in the Holocene. In contrast, *Cassidulina*, *Pullenia*, bolivinids, buliminids and uvigerinids were common during glacial MIS (marine isotope stages) 2, 3 and 4, although the interglacial MIS 3 was not as warm as other Late Quaternary interglacials [17]. Thomas et al. [16] suggested that this reflects an enhanced organic carbon flux during glacials, rather than sluggish glacial bottom circulation leading to poorly oxygenated bottom water. This may be related to the plankton multiplier effect proposed by Woods and Barkmann [18], in which a diminished greenhouse effect during glacials reduces radiative forcing of the ocean, increasing the depth of winter convection. This in turn increases the annual resupply of nutrients to the euphotic zone, which leads to increased annual primary production. Gaby and Sen Gupta [19] found glacial and postglacial assemblages of the abyssal Venezuela Basin to differ, the Holocene fauna containing abundant *Cibicides wuellerstorfi* (Schwager), *Melonis pompilioides* (Fichtel and Moll), *Nuttallides umbonifera* (Cushman), and *Pullenia* sp., while the fauna in the last glacial was dominated by *Massilina* sp., *Globocassidulina subglobosa* (Brady) and *Nummoloculina irregularis* (d'Orbigny).

Glacial-interglacial contrasts in the benthonic foraminiferal fauna are not everywhere marked, however. Streeter and Lavery [15] wrote that on the western North Atlantic continental rise below 4,000 m, "the glacial to modern faunal shift is subtle, but it clearly occurs later than on the upper rise." Sen Gupta et al. [20] examined benthonic foraminifera over the past 127 ka in three bathyal cores (depths near 2000 m) from the western Grenada Basin, eastern Caribbean Sea. They found only subtle changes, rather than drastic turnovers, at glacial-interglacial boundaries based on the abundance of *Globorotalia menardii*. They stated that neither species richness S nor the information function $H = -\sum p_i \cdot \ln(p_i)$, where p_i is the proportional abundance of the *i*th species) showed any distinct stratigraphic trend (although H is not expected to show such a trend [21]). However, they suggested *Nuttallides umbonifera, Bulimina buchiana* dOrbigny and *Chilostomella oolina* Schwager to be rarer in the last glacial than in the two bounding glacials. Wilson [22,23] examined the benthonic foraminifera in two bathyal piston cores near the northern Leeward Islands, eastern Caribbean Sea. He did not find any marked faunal changes at the Pleistocene-Holocene boundary, but showed that the organic flux in one core decreased gradually through the entire core. Wilson [24] found only weak evidence of Milankovich cycles in the Upper Quaternary of ODP Hole 1006A (Santaren Channel, offshore western Bahamas), where *Globocassidulina subglobosa* and *Cibicidoides* aff. *C. io* (Cushman) were smaller assemblage components during most glacial MISs. However, the percentages of these species varied between odd-numbered MISs and they were insignificantly correlated with one another, *G. subglobosa* being rare in MIS 9 while *C.* aff. *C. io* was common.

The inability to detect glacial-interglacial contrasts at all sites appears to arise because not all sites show marked changes in community composition at the species level at glacial-interglacial boundaries. Rather, the proportional abundances of species vary between glacials and interglacials to differing degrees. This paper presents a simple quantitative measure, the assemblage turnover index (ATI), which uses changes in species' proportional abundances to identify intervals of marked community change. Whittaker [25] distinguished two categories of diversity: inventory diversity, which calculates the diversity of associations within samples (point diversity) or habitats (α diversity); and differentiation diversity, which examines the change in diversity between samples (pattern diversity) or habitats (β diversity). The assemblage turnover index presented here is a form of differentiation diversity. A conditioned-on-boundary index (CoBI), developed as a function of the ATI, identifies species that contribute most to maxima and minima in sets of values of the ATI. The section used in this demonstration is from Ocean Drilling Program (OPD) Hole 994C, drilled on the Blake Ridge, offshore SE USA (Figure 1). Although Bhaumik and Gupta [26,27] and Mohan et al. [28] have examined Neogene benthonic foraminifera from this and nearby ODP Holes, glacial-interglacial contrasts have not been recorded at the Blake Ridge before this study.

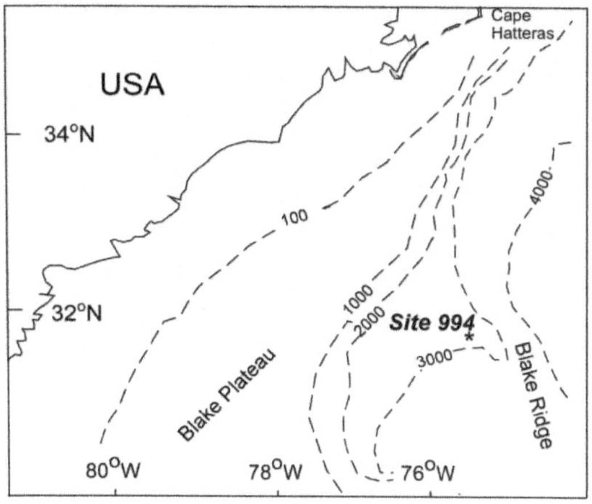

Figure 1. The location of ODP Hole 994C, Blake Ridge.

Site Description

ODP Site 994 (31°47.139′N, 75°32.753′W; water depth 2799 m) is situated on the sediment drift deposit that forms Blake Ridge [29,30], a drift deposit consisting of current-lain sediment east of the Blake Plateau. The Blake Ridge was deposited by the Western Boundary Undercurrent, this being a thermohaline-induced contour current (a current that flows parallel to bathymetric contours), which flows southward along the U.S. continental margin [31]. The Western Boundary Undercurrent transported clays eroded from eastern North America north of 40°N, at least as far south as Puerto Rico [32]. ODP Hole 994C cored a 700 m-thick succession of clays with calcareous nannofossils within which there are no obvious depositional hiatuses [30]. In this paper we examine the topmost ~14 m of sediment in ODP Hole 994C. Mass sediment transport complexes are absent. The studied section, which is a distinct lithostratigraphic unit termed Unit 1 [30], consists of light gray to gray and greenish gray nannofossil-rich clays in beds up to 1.20 m thick. The presence of the trace fossil *Zoophycos* indicates that some bioturbation has taken place, but mostly at bedding planes. Biostratigraphic correlation within the Quaternary of ODP Hole 994C is limited. Okada [33] found the first occurrence of *Emiliania huxleyi* (Lohmann) at 8.05–9.05 meters below the seafloor (mbsf), between OPD Hole 994C Core 2H-3, 65 cm and 2H-4, 15 cm, for which he suggested an age of 0.26 mya. This indicates a depositional rate of 3.5 cm/ka in the uppermost part of the Hole. The first appearance of *E. huxleyi* has subsequently been placed between 0.262–0.264 Myr [34].

Bhaumik and Gupta [27] studied the benthonic foraminifera in nearby ODP Hole 997A, and found *Brizalina paula* (Cushman and Cahill), *Cibicidoides kullenbergi* (Parker), *Uvigerina hispidocostata* Cushman and Todd and *Uvigerina peregrina* Cushman to be abundant in that part of the section within the gas hydrate zone. Bhaumik and Gupta [26] examined benthic foraminiferal assemblages (>125 mm size fraction), the information function H and total organic carbon in Hole 997A during the late Neogene (last 5.4 Ma). They concluded that *B. paula, Cassidulina carinata* Silvestri, *Chilostomella oolina, Fursekoina fusiformis* (Cushman), *Globobulimina pacifica* Cushman, *Nonionella auris* (d'Orbigny) and *Trifarina bradyi* are potential methane seep-related foraminifera, while *U. hispidocostata, U. peregrina, U. proboscidea* (Schwager) and *Melonis barleeanus* (Williamson) indicate a high organic carbon flux independent of deep-sea oxygenation.

Materials and Methods

Sixty seven samples of 20 cm^3 were taken at 20 cm intervals from ODP Hole 994C, Cores 1 and 2, between 0.08–13.25 mbsf. They were provided by the Ocean Drilling Program (ODP) that is sponsored by the U.S. National Science Foundation (NSF) and participating countries. The cored site being in international waters, no specific permissions were required for these locations/ activities. The material comprising fossils, sampling did not involve endangered or protected species. Each sample was ~2 cm thick and represents ~600 years. Samples were soaked in water until disaggregated, washed over a 63 μm mesh to remove silt and clay, and dried over a gentle heat. An attempt was made to pick N = 250–300 specimens of benthonic foraminifera from the >63 μm fraction from each sample. However, only 42 samples yielded >250 specimens (mean, 251 specimens per sample, minimum 104). The methods on assemblages used here have been reported by Wilson [35]. The foraminifera were sorted into species and identified using Cushman [36,37,38,39,40,41,42], Cushman and Henbest [43], Phleger and Parker [44], Phleger et al. [11], Parker [45], Pflum and Frericks [46] and Mohan et al. [28]. The number of specimens (n_i) was recorded for each species or species group (i.e., rare species in the same genus that were left in open nomenclature and grouped together).

Elphidium excavatum (Terquem), *Epistominella takayanagii* Iwasa, *Quinqueloculina poeyana* d'Orbigny and *Quinqueloculina* ex gr. *lamarckiana* d'Orbigny, which are typical of neritic water, were regarded as allochthonous and excluded from this analysis. *Elphidium* is typically regarded as a shallow-water genus [47] that has to be removed from data sets of studies of bathyal foraminifera [48]. Sen Gupta and others [49] recorded abundant *E. excavatum* on the Louisiana continental shelf. In the southern North Sea, it dominates the foraminiferal fauna at depths between 25–30 m [50]. *Epistominella takayanagii* has been recorded from Chaleur Bay eastern Canada, mostly in waters <100 m deep [51], and may have been transported southwards to ODP Site 994C. The proportional abundance of both *E. excavatum* and *E. takayanagii* peaked in MIS 10, glacial cycle E, as did the percentage of overall allochthonous, shallow-water species. A single specimen of *Stilostomella lepidula* (Schwager) recovered from 5.45 mbsf was presumed to be reworked, this species having gone extinct during middle Pleistocene times [52], and was excluded from further analysis. This left a presumed predominantly in situ abyssal assemblage, within which there may have been some slight downslope transport of *Angulogerina occidentalis* (Cushman), *Bulimina aculeata* d'Orbigny, *B. alazanensis* Cushman, *Cibicides* sp., *Fursenkoina fusiformis* (Williamson), *Globocassidulina obtusa* (Williamson) and *Nuttallides rugosa* (Phleger and Parker) [35]. This presumed in situ assemblage forms the subject of the remainder of this paper. To examine turnover of an entire assemblage quantitatively across a delineated boundary we developed the ATI index. For a set of samples from a given site, the Assemblage Turnover Index for each pair of adjacent samples is defined as

$$ATI = |p_{i2} - p_{i1}| \qquad (1)$$

in which p_{i1} and p_{i2} are the proportional abundances of the *i*th species, i = 1,..., s, in the lower and upper samples (see Appendix S1 for a glossary of terms). This assemblage turnover index between samples will be denoted as ATI_s. Note that although for each sample $\sum p_i = 1$, the measure ATI_s can be >1. Thus, ATI_s gives the proportion or percent of turnover or change specifically across a defined or particular boundary. We calculated the mean $\bar{x}$, and standard deviation, σ, of values of ATI_s over all samples within the core. To develop our control chart we determined all points with $ATI_s > (\bar{x} + \sigma)$, which were then deemed to be positions of major turnover. Oba et al. [53] presented a δ18O curve for ODP Hole 994C (see Figure 2). Their samples were taken at irregular intervals (sample spacing 7–49 cm; mean 22.6 cm, sd 10.9 cm). The values of δ18O for the samples used here were interpolated from Oba et al. 's [53] curve and correlation between ATI_s and interpolated δ18O was calculated. Because Oba et al. 's [53] uppermost sample was taken at 0.14 mbsf, it was not possible to estimate the δ18O value for the uppermost sample picked for this study. Point (sample) values of species richness S and the information function H were calculated. Dominance was determined using max(p_i), the proportional abundance of the most abundant species in each sample [54]. We chose to calculate correlations between ATI_s, S, H and max(p_i) using the upper (younger) sample in each sample pair.

Peaks in ATI_s, those values larger than our designated control value of $(\bar{x} + \sigma)$, were used to divide the succession into peak-bounded ATI_s (PATI-) intervals. These intervals were numbered, commencing from PATI-1 for the most recent. PATI-1 and the oldest PATI are incomplete, their upper and lower boundaries respectively not being bounded by ATI_s peaks.

To assess which species contributed most to the ATI at the PATI boundaries, a conditioned –on-boundary index CoBI was derived. CoBI provides the proportion that each species within an assemblage contributed to the change or turnover specifically across the PATI boundary. For each species at any PATI boundary

$$CoBI = |p_{i2} - p_{i1}|/ATI \qquad (2)$$

where p_{ij}, j = 1,2 are the *i*th species proportions on either side of the selected boundary of interest and at which the ATI is calculated.

There are two forms of CoBI:

1. Partial conditioned-on-boundary index, $CoBI_p$, in which the assemblage turnover index ATI_s was calculated between the entire set of samples within the PATI below the ATI peak and the first sample immediately above the peak. In this case, the ATI is designated as ATI_p. The value of ATI (= ATI_p) was substituted into equation (2), as were p_{i1}, the proportional abundance of the *i*th species in the entire PATI below the peak in ATI_p, and p_{i2}, the proportional abundance of that *i*th species in the first sample above the peak in ATI_s. The proportional contribution of each species to ATI_p was assessed from the vector of $CoBI_p$ values at each ATI_s peak.
2. Thorough conditioned-on-boundary index $CoBI_t$, in which the ATI is denoted as ATI_t, was calculated between the values in two complete PATIs separated by the peak in ATI_s (see Figure 2). The value of ATI_t was substituted into equation (2), as were p_{i1} and p_{i2}, the proportional abundance of the *i*th species in the two PATIs separated by the peak in ATI_t. The proportional contribution of each species to the value of ATI_t was assessed from the vector of partial CoBI for each ATI_s peak.

Thus, to detect change between the total set of samples from the assemblages within two distinct PATIs we evaluate the ATI = ATI_t at the boundary between these two. The partial indices are

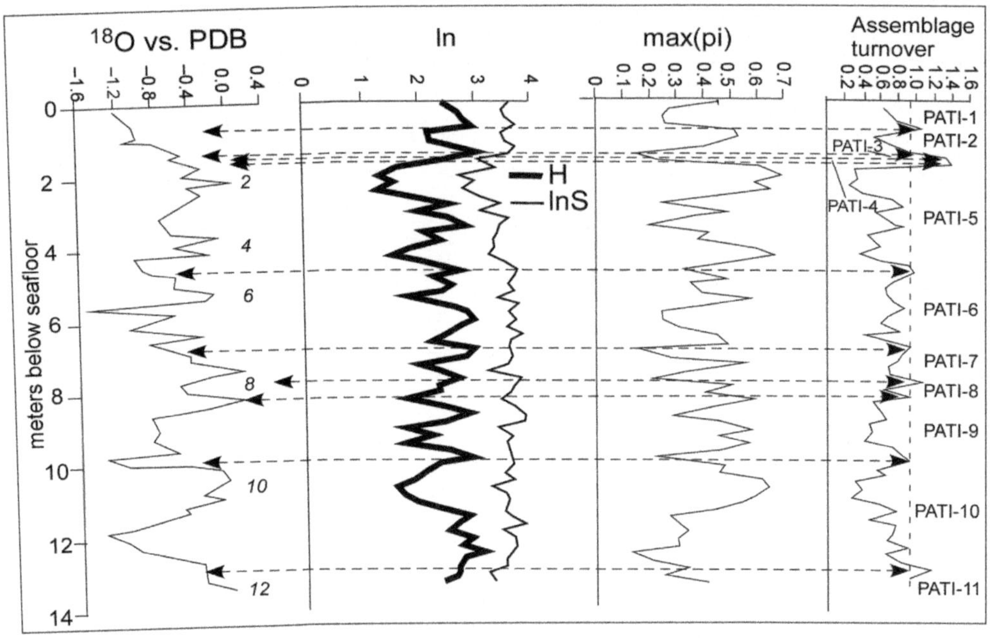

Figure 2. δ18O from Oba et al. **[53], the information function H, species richness S (graphed as lnS to facilitate comparison with H), max(p_i) and between sample assemblage turnover index (ATI_s) [from this paper] in the Upper Quaternary, ODP Hole 994C, Blake Ridge.** Numbers in italics indicate marine isotope stages; PATI are peak-bounded ATI_s intervals.

used to detect assemblage change exactly at the boundary between a single PATI and the next contiguous sample.

Results

Assemblage Turnover Index ($ATI = ATI_s$) Between Samples

Our presumed in situ, abyssal fauna comprised 16,184 specimens in 157 species (see File S1), and was dominated by *Brizalina lowmani* (Phleger and Parker) (42.1% of total recovery; range 6.9–70.0% per sample) with subdominant *Globocassidulina obtusa* (Williamson) (8.9% of total recovery; range 0–21.9% per sample). Thirteen other species formed >1% of total recovery: *Bulimina aculeata* d'Orbigny, *Cassidulina laevigata* d'Orbigny, *Cibicides wuellerstorfi*, *Cibicidoides robertsonianus* (Brady), *Epistominella exigua*, *Gyroidina lamarckiana* (d'Orbigny), *Hoeglundina elegans* (d'Orbigny), *Melonis baarleeanus* (Williamson), *M. pompilioides* (Fichtel and Moll), *Oridorsalis umbonatus* (Reuss), *Pullenia bulloides* (d'Orbigny), *Pyrgo lucernula* (Schwager) and *Uvigerina hispidocostata* Cushman and Todd. The distributions of selected species are shown in Figure 3.

The assemblage turnover index between adjacent samples ranged from $ATI_s = 0.263–1.421(\bar{x} = 0.710, \sigma = 0.233)$ (Figure 2), indicating total assemblage change from 26% to 142%. The value of ATI_s exceeded $\bar{x}+ \sigma = 0.943$ across nine pairs of samples. We chose to include the borderline value of ATI_s at 9.85 mbsf, where it nevertheless formed a pronounced peak. We computed correlations of ATI with the indices H and max (p_i) and with δ18O. Although the formulae for these measures utilize the relative abundances, there is no linear functional relationship among them that necessitates a significant correlation. The ATI_s was positively correlated with the information function H for the younger sample in the pair ($r = 0.62$, $p < 0.0001$). ATI_s was negatively correlated with max(p_i) ($r = -0.65$, $p < 0.0001$), which indicates a change in dominance across peaks in ATI_s, and negatively correlated with δ18O ($r = -0.32$, $p < 0.01$). H and δ18O were in turn significantly negatively correlated ($r = -0.53$, $p < 0.0001$).

Partial Conditioned-on-Boundary Index ($CoBI_p$)

The accepted peaks in the ATI_s delimited eleven peak-bounded ATI intervals (PATI-). These intervals contained between one (PATI-3 and PATI-4) and fifteen (PATI-5 and PATI-10) samples. The ATI_p used for $CoBI_p$ ranged between 0.83–1.33 (PATI-6/5 and PATI-4/3 boundaries respectively; Table 1), which indicates assemblage turnover between 83% and 133% at the boundaries. Only 31 species showed $CoBI_p \geq 0.02$ (i.e., accounted for >2% of ATI_p) at any one PATI boundary (Table 1) and these species differed over the PATI boundaries.

Thirteen species had a $CoBI_p$ of 0.02–0.05 at any one boundary, while 18 (~11.5% of all species recorded) had a $CoBI_p \geq 0.05$ across any one PATI boundary. Seven species (*Cassidulina laevigata*, *Epistominella exigua*, *Eponides regularis* Phleger and Parker, *Globocassidulina obtusa*, *Hoeglundina elegans*, *Pyrgo murrhina* (Schwager), *Uvigerina celtica* Schönfeld) had a $CoBI_p \geq 0.05$ across one PATI boundary only. Two species (*Melonis baarleeanus*, *Uvigerina hispidocostata*) had a $CoBI_p \geq 0.05$ across four PATI boundaries. *Brizalina lowmani* fluctuated most markedly, having the $CoBI_p \geq 0.05$ across eight PATI boundaries. The maximum $CoBI_p$ was 0.47, indicating 47% change in *B. lowmani* abundance across the PATI-7/6 boundary. Four other species (*Bulimina aculeata*, *Gyroidina lamarckiana*, *Melonis baarleeanus*, *M. pompilioides*) had a maximum $CoBI_p \geq 0.20$ across any one PATI boundary. Using $CoBI_p$, *Brizalina lowmani* decreased in proportional abundance across the boundaries between PATI-8/7 and 2/1, *M. pompilioides* between PATI-7/6 and 2/1, and *U. hispidocostata* between PATI-7/6 and 3/2.

Thorough Conditioned-on-Boundary Index ($CoBI_t$)

The ATI_t used for $CoBI_t$ ranged between 0.36–1.33 (PATI-10/9 and PATI-4/3 boundaries respectively; Table 2), or an observed

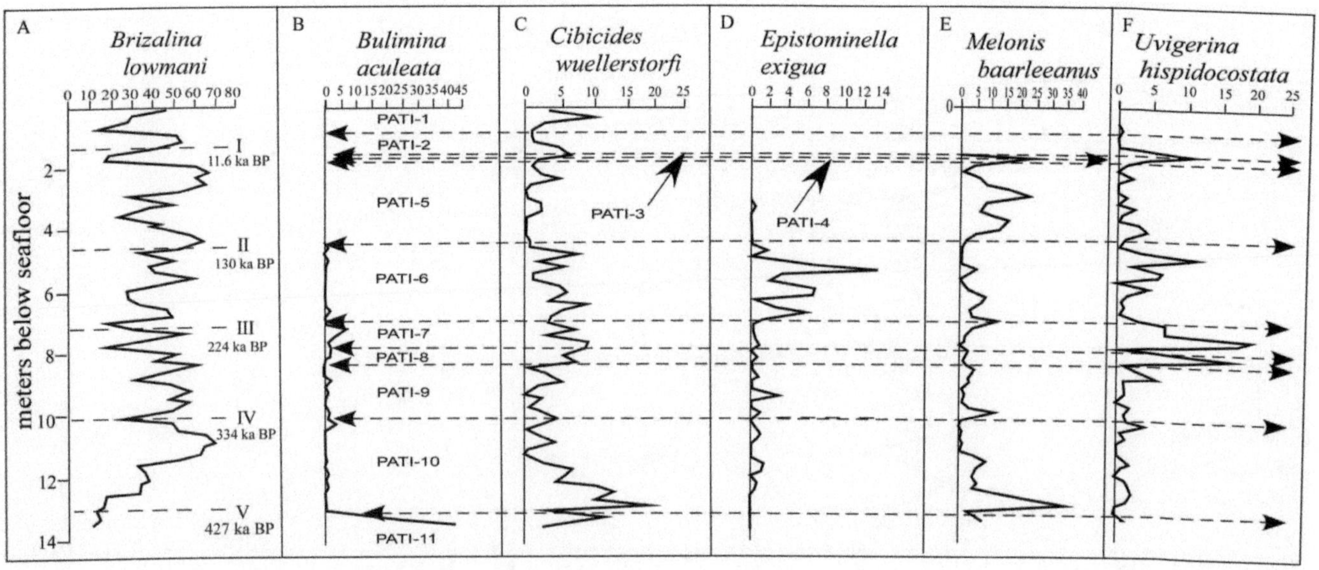

Figure 3. Percentage abundances of selected species in the Upper Quaternary of ODP Hole 994C, species presented in alphabetical order. A. *Bulimina aculeata*. B. *Brizalina lowmani*. C. *Cibicides wuellerstorfi*. D. *Epistominella exigua*. E. *Melonis baarleeanus*. F. *Uvigerina hispidocostata*. Dashed lines in A indicate positions of Terminations I – V, with ages in ka BP. Dashed lines in B – F indicate boundaries between PATIs 1 to 11. Terminations are indicated by Roman numerals in order of increasing age.

total assemblage change ranging from 36% to 133% between the PATI pairs from PATI-1/2 through PATI-10/11. Thirty three species resulted in a value of $CoBI_t \geq 0.02$ (i.e., accounted for >2% of ATI_t) at any one PATI boundary. Twenty eight species contributed to ≥ 0.02 of both ATI_p and ATI_t. Seventeen species had a $CoBI_t$ of 0.02–0.05 at any one boundary. Thus, only sixteen species (~10% of all species recorded) had a $CoBI_t \geq 0.05$ across any one PATI boundary. The highest $CoBI_t$ was 0.31 for *Brizalina lowmani*, indicating pronounced change in dominance across the PATI-3/2 boundary, while *Bulimina aculeata* presented a comparable $CoBI_t$ of 0.29 across the PATI-11/10 boundary. All other species contributed a maximum $CoBI_t$ <0.20, although the maxima for *Globocassidulina obtusa* and *Melonis baarleeanus* were 0.19 and 0.17 respectively. The $CoBI_t$ for *B. lowmani* was >0.02 across eight PATI boundaries, while those for *G. obtusa* were >0.02 across six boundaries and for *M. baarleeanus* and *Cibicidoides robertsonianus* across five.

Discussion

The assemblage turnover index is a form of differentiation diversity sensu Whittaker [25]. It is here presented at two scales, of which the end members are (a) ATI_s, which corresponds to Whittaker's (1972) pattern or between-sample diversity, and (b) ATI_t, which is here taken as corresponding to his between-habitat or β-diversity. ATI_s was strongly positively correlated with the information function H and negatively correlated with max(p_i) for the younger of the samples in the sample pairs used in its calculation. This indicates that assemblage turnover – the sum of the changes in proportional abundances of species – increases with increasing diversity and with decreasing dominance (i.e. increasing equitability).

The correlation between ATI_s, and interpolated values of δ18O was significant and negative (Figure 2). Oxygen isotopes have been used to erect a paleotemperature record of marine isotope stages (MISs) that is reliable back to MIS 16, 650 ka [17], within which odd numbered MISs are interglacials [55] and against which faunal changes can be compared. Much of the time between MIS 1–12 (the interval examined in this study) consists of 100 ka MIS couplets [56,57]. Broecker and Van Donk [58] grouped the MISs into glacial cycles (segments of two to four MISs) that were separated by terminations (i.e., pronounced boundaries between isotopic maxima and minima). Because warming during deglaciation occurs more rapidly than does cooling during the development of glaciation [59], each termination separates a preceding glacial from a succeeding interglacial. MIS 3 being a subdued glaciation, there is no termination between MIS 4 and 3, but Termination T-I occurs between MIS 2 and 1, and T-II between MIS 6 and 5. The interval examined during this study encompasses terminations T-V to T-I, which separated Glacial Cycles F to A (Figure 3). Cheng et al. [60] positioned terminations at the mid-point between the peaks and troughs in the graph of δ18O. However, because the δ18O curve for ODP Hole 994C presented by Oba et al. [53] was based on irregularly spaced samples their technique cannot be used here. It does appear, however, that the boundaries between PATI-11/10, 10/9, 7/6 and 6/5 and 2/1 approximate to terminations T-V through T-I, respectively. Thus, the slower onsets of glacials are marked by low levels of turnover, ATI_s, while the more rapid transitions to interglacials are marked by peaks in ATI_s. Because the peaks in ATI_s occurred within terminations, the correlation between ATI_s and δ18O is low.

Not all peaks in ATI_s detected across our samples coincide with terminations. The boundary between PATI-9/8 and 8/7 occurred within MIS 8 and indicates an increase in the flux of organic carbon through that glacial MIS. This may reflect increasing efficiency of the plankton multiplier of Woods and Barkmann (1993).

The close grouping of the boundaries between PATI-5 through PATI-1, all of which occurred during the transition from MIS 2 to MIS 1, show this to have been an interval of protracted environmental change at ODP Site 994. Sea level rose by ~120 m during termination T-1 [61], but did so in several decimeter steps [62]. It is possible that the closely spaced

Table 1. Partial conditioned-on-boundary index ($CoBI_p$) for peak-bounded ATI_s (PATI-) intervals in the Upper Quaternary in ODP Hole 994C, Blake Ridge.

Boundary	PATI-11/10	PATI-10/9	PATI-9/8	PATI-8/7	PATI-7/6	PATI-6/5	PATI-5/4	PATI-4/3	PATI-3/2	PATI-2/1
ATI	**1.09**	**0.89**	**0.88**	**0.93**	**0.91**	**0.83**	**0.96**	**1.33**	**0.95**	**0.97**
Alabaminella weddellensis	-	-	-	-	-	-	-	-	-	0.02
Brizalina lowmani	-	**0.20**	**0.35**	*0.33*	**0.47**	**0.27**	**0.08**	-	**0.26**	***0.21***
Bulimina aculeata	*0.31*	-	-	-	-	-	-	-	-	-
Cassidulina laevigata	-	0.02	-	-	0.04	-	**0.09**	0.03	*0.04*	-
Cibicides wuellerstorfi	*0.03*	**0.05**	**0.09**	0.03	-	**0.09**	-	0.04	0.02	-
Cibicidoides bradyi	-	-	-	-	-	-	0.04	0.03	-	0.03
Cibicidoides robertsonianus	-	-	-	-	-	0.03	-	0.02	0.03	-
Epistominella exigua	-	-	-	-	**0.06**	-	-	-	-	-
Eponides regularis	-	-	-	-	-	-	-	-	-	***0.10***
Gavelinopsis praegeri	-	-	-	-	-	-	-	-	-	0.03
Globobulimina pacifica	-	-	-	-	-	-	-	0.03	0.03	-
Globocassidulina murrhina	-	-	-	-	-	-	-	0.04	-	*0.02*
Globocassidulina obtusa	0.03	0.03	0.02	-	**0.09**	-	-	-	-	-
Globocassidulina subglobosa	-	0.04	-	-	-	-	**0.07**	**0.05**	-	-
Gyroidina lamarckiana	-	**0.13**	-	-	-	**0.05**	-	0.03	-	**0.22**
Gyroidinoides soldanii	-	0.03	-	-	-	-	-	-	-	-
Hoeglundina elegans	-	-	-	-	-	**0.15**	-	-	-	0.03
Martinottiella communis	-	-	-	-	-	-	**0.17**	*0.12*	-	-
Melonis baarleeanus	**0.27**	**0.11**	-	*0.02*	0.02	-	**0.24**	*0.17*	-	-
Melonis pompilioides	0.04	-	-	**0.21**	*0.06*	-	-	-	**0.08**	*0.03*
Nonionella bradyi	-	-	-	-	-	0.02	-	-	-	-
Oridorsalis umbonatus	-	**0.08**	**0.05**	-	-	-	-	-	-	-
Pullenia bulloides	-	0.03	0.04	-	-	-	0.02	-	-	-
Pullenia spp.	-	-	-	-	-	-	-	0.03	*0.03*	-
Pyrgo lucernula	-	-	**0.05**	*0.03*	-	-	0.03	-	*0.03*	-
Pyrgo murrhina	-	0.03	-	-	-	**0.06**	-	-	-	-
Robertinoides bradyi	-	-	-	-	-	0.02	-	-	-	-
Sigmoilopsis schlumbergeri	*0.02*	-	-	-	-	-	-	*0.08*	**0.09**	-
Trifarina angulosa	-	-	-	-	-	0.02	-	-	-	0.03
Uvigerina celtica	-	-	0.03	*0.04*	-	-	**0.07**	0.04	-	-
Uvigerina hispidocostata	-	-	**0.18**	**0.09**	*0.04*	0.02	0.04	**0.05**	*0.11*	-

Dashes indicate $CoBI_p$ <0.02, bold indicates $CoBI_p$ >0.05. Italics indicate that a species' proportional abundance decreased over a PATI boundary.

boundaries from PATI-5 through PATI-1 reflect these steps. *Brizalina lowmani* did not decrease in proportional abundance across all boundaries between PATI-5 through PATI-1, but increased across the PATI-3/2 boundary.

Some data suggest that the changes in the fauna across the peaks in values of ATI_s reflect changes in either (a) dissolved oxygen levels, (b) the organic carbon flux and (c) bottom current strength, although the first two of these factors are frequently correlated [63]. For a paleoenvironmental summary, see Table 3.

den Dulk et al. [64] studied benthonic foraminifera under an upwelling system in the northern Arabian Sea. They recognised two groups of foraminifera:

1. Species that prefer high dissolved oxygen and low organic carbon levels: (*Bulimina striata, Gavelinopsis lobatula, Chilostomella oolina, Sphaeroidina bulloides, Cibicides ungerianus, Hyalina balthica, Hoeglundina elegans, Melonis barleeanus, Quinqueloculina* spp., *Globocassidulina subglobosa and Cassidulina carinata*;
2. Species preferring low dissolved oxygen and high organic carbon: *Bulimina exilis, Rotaliatinopsis semiinvoluta, Brizalina alata, B. pygmaea, Globobulimina spp., and Bulimina* sp.1.

The species of *Brizalina* and *Bulimina*, which dominate in ODP Hole 994C, are limited to group 2. de Rijk et al. [65] found that *Bulimina aculeata*, which dominated in PATI-11 (Figure 3B), lives primarily where the flux of organic carbon exceeds 3 g m^{-2} yr^{-1}. *Melonis baarleeanus* (Figure 3E), although placed in group 1 by den Dulk et al. [64], was shown by Qvale and Van Weering [66] to prefer a fine-grained substrate with a relatively high organic carbon content [67]. Mackensen et al. [68] suggested that in the South Atlantic Ocean it prefers seasonally varying productivity. Taldenkova et al. [69] found this species to be more abundant in

Table 2. Thorough conditioned-on-boundary index ($CoBI_t$) for peak-bounded ATI_s (PATI-) intervals in the Upper Quaternary in ODP Hole 994C, Blake Ridge.

PATI boundary	PATI-11/10	PATI-10/9	PATI-9/8	PATI-8/7	PATI-7/6	PATI-6/5	PATI-5/4	PATI-4/3	PATI-3/2	PATI-2/1
ATI_t	**1.19**	**0.36**	**0.55**	**0.54**	**0.56**	**0.56**	**1.23**	**1.33**	**1.03**	**0.67**
Alabaminella weddellensis	-	0.04	-	-	-	0.02	-	-	-	-
Brizalina lowmani	**0.30**	**0.06**	0.03	***0.28***	**0.18**	**0.17**	***0.29***	-	**0.31**	***0.29***
Bulimina aculeata	***0.29***	-	-	0.04	***0.05***	-	-	-	-	-
Bulimina alazanensis	-	*0.03*	-	-	-	-	-	-	-	-
Bulimina striata mexicana	-	*0.03*	-	-	-	-	-	-	-	*0.02*
Cassidulina laevigata	0.02	***0.05***	-	-	-	0.04	0.04	*0.03*	*0.04*	-
Cibicides wuellerstorfi	-	***0.08***	**0.09**	-	*0.04*	***0.05***	-	0.04	*0.04*	0.03
Cibicidoides bradyi	-	-	-	-	-	-	0.03	*0.03*	-	-
Cibicidoides robertsonianus	-	-	-	-	-	0.02	*0.02*	0.02	*0.02*	0.02
Epistominella exigua	-	-	-	-	**0.08**	***0.09***	-	-	-	-
Eponides regularis	-	-	-	-	-	-	-	-	-	**0.05**
Gavelinopsis praegeri	-	-	-	-	-	-	-	-	-	0.02
Globobulimina pacifica	-	-	-	-	-	-	-	0.03	*0.03*	-
Globocassidulina murrhina	-	-	-	-	0.02	*0.02*	-	0.04	*0.03*	**0.06**
Globocassidulina obtusa	**0.07**	**0.17**	*0.19*	**0.05**	-	**0.05**	***0.08***	0.02	-	-
Globocassidulina subglobosa	-	-	-	-	-	-	**0.05**	***0.05***	-	-
Gyroidina lamarckiana	-	-	*0.03*	-	**0.05**	*0.04*	-	0.03	-	**0.11**
Gyroidinoides soldanii	-	-	*0.02*	0.02	-	-	-	-	-	-
Hoeglundina elegans	-	-	-	**0.05**	*0.04*	-	-	-	0.02	*0.03*
Lagena spp.	-	-	-	-	-	-	-	-	0.02	*0.03*
Martinottiella communis	-	-	-	-	-	-	**0.13**	***0.12***	-	-
Melonis baarleeanus	-	**0.08**	0.02	0.03	-	**0.09**	**0.12**	***0.17***	-	-
Melonis pompilioides	*0.03*	*0.04*	0.03	**0.06**	***0.07***	*0.02*	-	-	0.02	-
Nonionella bradyi	-	-	-	-	-	-	-	-	-	0.03
Nuttallides rugosa	-	-	*0.03*	-	-	-	-	-	-	0.02
Oridorsalis umbonatus	*0.03*	0.03	-	-	-	*0.03*	*0.02*	-	-	-
Pullenia bulloides	*0.04*	-	0.03	-	*0.03*	-	-	-	-	-
Pullenia quinqueloba	-	-	-	0.03	-	-	-	-	0.02	0.02
Pullenia spp.	-	-	-	-	-	-	-	0.03	*0.03*	*0.02*
Pyrgo lucernula	-	-	0.04	*0.04*	-	0.03	-	-	*0.03*	-
Sigmoilopsis schlumbergeri	-	-	0.02	-	-	-	-	**0.08**	***0.07***	***0.05***
Uvigerina celtica	-	-	**0.06**	***0.05***	0.02	-	**0.05**	*0.04*	*0.02*	-
Uvigerina hispidocostata	-	0.02	**0.13**	0.03	***0.13***	*0.04*	0.02	**0.05**	***0.10***	-

Dashes indicate $CoBI_t$ <0.02, bold indicates $CoBI_t$ >0.05. Italics indicate that a species' proportional abundance decreased over a PATI boundary.

the upper bathyal Holocene of the Arctic Ocean than the latest Pleistocene, and ascribed it to a distal-river group of relatively deep-water species that thrive on slightly altered organic matter and is therefore restricted to areas with periodic delivery of organic matter. Murray [70] noted that M. *baarleeanus* has been recorded live in all oceans except the Indian Ocean. In ODP Hole 994C this species accounted for >0.02 of the $CoBI_t$ across six of the ten PATI boundaries, and was abundant in the early part of PATI-10 and in PATI-5 and PATI-4. It was rare to absent in PATI-3 through PATI-1. This suggests that seasonality varied through the Late Quaternary at Blake Ridge. Unlike in the Arctic Ocean [69], at Blake Ridge seasonality was much reduced in the Holocene, after termination T-1.

Globocassidulina subglobosa, which is found throughout the Atlantic, Pacific and Southern Oceans (Murray, 2013), has been suggested to be an oxic indicator [71] that prefers an elevated mean organic carbon flux of 15 g m^{-2} yr^{-1} [72]. This species was abundant in PATI-4 (which equates to a brief episode in the glacial MIS 2). Smart and Gooday [73] examined trends in benthonic foraminiferal abundances along an organic enrichment gradient on the continental slope off North Carolina, eastern Atlantic Ocean. They found *Bulimina aculeata* and *Globocassidulina subglobosa* to be equally abundant at all sites, suggesting that these cannot be used as proxies for the organic flux. It is unclear, however, if *Globocassidulina obtusa* and *G. murrhina* have the same tolerances.

Table 3. Environmental interpretation of peak-bounded ATI_s (PATI-) intervals in the Upper Quaternary in ODP Hole 994C, Blake Ridge.

ATI interval	Isotope stage	Notable species presence/absence	Paleoenvironmental interpretation
PATI-1	MIS 1	Few *Melonis baarleeanus*, abundant *Cibicides wuellerstorfi*	decreased seasonality, low organic carbon flux, enhanced current action
PATI-2	? MIS 2/1	Few *Melonis baarleeanus*	decreased seasonality
PATI-3	? MIS 2/1	Few *Melonis baarleeanus*, abundant *Cibicides wuellerstorfi*	decreased seasonality, low organic carbon flux, enhanced current action
PATI-4	? MIS 2/1	Abundant *Melonis baarleeanus*	enhanced seasonality
PATI-5	MIS 4/3/2	Abundant *Melonis baarleeanus*	enhanced seasonality
PATI-6	MIS 6/5	Abundant *Cibicides wuellerstorfi*, *Epistominella exigua*	low organic carbon flux, enhanced current action and seasonality
PATI-7	MIS 8	Abundant *Uvigerina hispidocostata*	enhanced carbon flux
PATI-8	MIS 8	Abundant *Uvigerina hispidocostata*	enhanced carbon flux
PATI-9	MIS 9	Abundant *Cibicides wuellerstorfi*	low organic carbon flux, enhanced current action
PATI-10	MIS 11/10	Abundant *Melonis baarleeanus*	enhanced seasonality
PATI-11	MIS 12	Dominant *Bulimina aculeata*	organic flux >3 g C m^{-2} yr^{-1}

Kaiho [74] suggested that many of the species recovered from ODP Hole 994C are indicative of suboxic bottom waters, although he also suggested that *Cibicides wuellerstorfi* (figure 3C) and *Cibicidoides robertsonianus* are indicative of oxic water [75]. The abundance of *C. wuellerstorfi* was relatively high during PATI-6. Altenbach et al. [72] recorded the annual organic carbon flux levels best tolerated by some live benthonic foraminifera. Most species recorded by Altenbach et al. [72] that were also recovered from ODP Hole 994C lived under a flux rate of ~2–6 g m^{-2} yr^{-1} (*Bulimina striata mexicana*, *C. robertsonianus*, *Pullenia bulloides*, *P. quinqueloba*). Schönfeld [71] recorded these four species as living infaunally within the sediment at a variety of depths down to 4.5 cm. However, Altenbach et al. [72] recorded live *C. wuellerstorfi* primarily at low organic carbon flux rates of 1.5–3 g m^{-2} yr^{-1} and *Oridorsalis umbonatus* at 2–3.5 g m^{-2} yr^{-1}. *Cibicides wuellerstorfi* is an epiphytal species living on raised substrate particles that prefers active bottom currents [76,77]. $CoBI_t$ showed in ODP Hole 994C that *C. wuellerstorfi* was recovered primarily from PATI-9, 6, 3 and 1, during which the organic carbon flux may have been low, the strength of the Western Boundary Undercurrent enhanced, or both. Altenbach et al. [72] recovered *Hoeglundina elegans* mainly from areas with a flux rate of 4.5–15 g m^{-2} yr^{-1}, although Schönfeld [71] suggested it to be an oxic indicator. The costate species *Uvigerina mediterranea* and *U. peregina*, which have morphologies comparable to *U. hispidocostata*, primarily under a carbon flux of 3–9 g m^{-2} yr^{-2}. Schönfeld [71] recorded *U. peregrina* as living mostly at depths of 0–1 cm below the sediment water interface, but *U. mediterranea* as occurring down to 6 cm below the interface. *Uvigerina hispidocostata* in ODP Hole 994C was recovered mainly from PATI-8 and PATI-7, which coincide with MIS 8, for which it could indicate an interval of enhanced organic carbon flux but might also highlight an interlude in which uvigerinids penetrated deeper into the sediment. Schoenfeld and Altenbach [78] found that *Uvigerina* spp. in the north-eastern Atlantic Ocean were more abundant during glacial MIS 2 than interglacial MIS 1, and ascribed this to a widespread change from glacial to modern productivity characteristics across termination T-I. MIS 8 is similarly a glacial stage. Seiglie [79] noted *B. lowmani* to be indicative of high organic carbon levels and Sen Gupta and Strickert [80] found it to be dominant on the continental slope off Florida at depths >100 m, below the Gulf Stream. In ODP Hole 994C *B. lowmani*, *Cassidulina laevigata*, *Cibicides wuellerstorfi*, *Melonis baarleeanus*, *M. pompilioides* and *Uvigerina hispidocostata* had the highest number of PATI boundaries across which $CoBI_p$ >0.02, while *B. lowmani*, *Cibicidoides robertsonianus*, *Globocassidulina obtusa*, *M. baarleeanus*, *M. pompilioides*, *U. celtica* and *U. hispidocostata* had the highest number PATI boundaries across which $CoBI_t$ >0.02. These indicate that the organic carbon flux, dissolved oxygen levels and bottom current strength varied between PATIs, rather than between glacials and interglacials.

Gooday et al. [81] found *Epistominella exigua* (Figure 3D to be abundant in well-oxygenated, abyssal water below the oxygen minimum zone of the Arabian Sea. Smart et al. [82] showed that *E. exigua* colonises aggregates of phytodetritus and they speculated that this opportunistic, epifaunal species may represent a proxy for seasonal phytodetritus pulses originating from surface primary productivity in open ocean eutrophic areas. They suggested that inputs added over a geologically prolonged period of time would be reflected in peaks of *E. exigua*. This species was at its most abundant in PATI-6, having a ATI_p >0.05 across the PATI-7/6 boundary and a ATI_t >0.05 across both the PATI-7/6 and PATI-6/5 boundaries. This implies a brief interlude of enhanced seasonality in MIS 7 and 6, and may be related to a change in surface circulation and the position of the Gulf Stream at that time.

Sequence stratigraphy is the correlation of sedimentary rock successions using key events produced by worldwide changes in sea level [1]. These events are used to divide the succession into packages (systems tracts) that are bounded by characteristic surfaces [83,84]. Benthonic foraminifera have long been used in sequence stratigraphy at neritic paleodepths [85]. However, the use of benthonic foraminiferal assemblage characteristics for sequence stratigraphic purposes at abyssal depths has thus far been problematic. For example, at neritic depths the planktonic/benthonic foraminiferal ratio has been used to determine changes in sea level [86,87]. However, this index cannot be readily applied at depths of more than ~500 m, at which planktonic foraminifera typically form >99% of the assemblage. At neritic depths, maximum flooding surfaces are reflected by peaks in uvigerinid abundance that have been ascribed to sluggish circulation at times of maximum transgression [88]. This is contrary, however, to the enhanced abundance of bathyal and abyssal *Uvigerina* during glacial lowstands. Nagy et al. [89] suggested that at neritic depths the information function H is low on interglacial maximum flooding surfaces. In ODP Hole 994C, however, H is negatively correlated with δ18O and the index is lower during glacial, even numbered MIS than it is during interglacial MIS.

Therefore, we propose that ATI_s peaks show strong potential as a sequence stratigraphic tool for abyssal deposits, some peaks at the PATI boundaries coinciding with terminations that are marked by transgressive systems tracts. However, the apparent coincidence between peaks in ATI_s and terminations must be applied with caution, since not all peaks coincide with terminations; two peaks occurred within glacial MIS 8. However, this can be avoided by judicious use of the control limits.

Conclusions

Assemblages are not constant entities, but change over time as the proportional abundance of each species within a community changes. As one species acquires a higher proportional abundance, one or more others must decrease in abundance. Peaks in ATI_s, the ATI between successive samples, delimit peak-bounded intervals (PATIs-) within which the community is relatively stable. The current inability to detect glacial-interglacial contrasts in general appears to arise because not all sites show marked changes in community composition at the species level at glacial-interglacial boundaries. Both ATI and CoBI can be applied to successions for which there is no immediately obvious differentiation of glacial and interglacial assemblages. Peaks in ATI_s in the Upper Quaternary of ODP Hole 994C, Blake Ridge, define eleven PATIs. Eight of the PATI boundaries approximate to terminations, although, as shown by termination T-I, a termination can be marked by more than one PATI boundary if, like termination T-I, it consists of a series of events marked by decimeter changes in sea level. While it appears that for our data set all terminations were marked by at least one PATI boundary, not all PATI boundaries coincided with terminations; two PATI boundaries were recorded within MIS 8. Nevertheless, this suggests that PATIs and peaks in assemblage turnover as measured by our index, ATI_s, have potential as a sequence stratigraphic tool. Our quantitative approach allows some sequence stratigraphic concepts to be extended into the abyssal environment.

Both $CoBI_p$ and $CoBI_t$ suggest that species that changed markedly across PATI boundaries were responding to changes in paleo-oxygenation, the organic matter flux, or bottom current strength. A transitory peak in *Epistominella exigua* within PATI-6 implies a brief interlude of enhanced seasonality in MIS 7 and 6, and may be related to a change in surface circulation and the position of the Gulf Stream at that time.

The assemblage turnover (ATI) and conditioned-on-boundary (CoBI) indices have here been applied to the ecostratigraphy of abyssal benthonic foraminifera. However, these measures can also be used to detect and characterise boundaries for any taxon and applied in both paleoecological and ecological studies.

Acknowledgments

This research used samples provided by the Ocean Drilling Program (ODP) that is sponsored by the U.S. National Science Foundation (NSF) and participating countries under management of Joint Oceanographic Institutions (JOI), Inc. Thanks are due to two anonymous reviewers, whose suggestions improved this paper markedly.

Author Contributions

Performed the experiments: BW L-ACH. Analyzed the data: BW L-ACH. Contributed reagents/materials/analysis tools: BW L-ACH. Wrote the paper: BW L-ACH. Collected data used in this analysis: BW. Developed the statistical measures: L-ACH.

References

1. Torrens HS (2002) Some personal thoughts on stratigraphic precision in the twentieth century. In: Oldroyd DR, editor. The earth Inside and Out: Some Major Contributions to Geology in the Twentieth Century: Geological Society of London Special Publication. pp. 251–272.
2. Gibbard PL, Head MJ, Walker MJC (2009) Formal ratification of the Quaternary System/Period and the Pleistocene Series/Epoch with a base at 2.58 Ma. Journal of Quaternary Science 25: 96–102.
3. Bolli HM, Saunders JB (1985) Oligocene to Holocene low latitude planktic foraminifera. In: Bolli HM, Saunders JB, Perch-Nielsen K, editors. Plankton Stratigraphy. Cambridge, England: Cambridge University Press. pp. 155–262.
4. Wade BS, Pearson PN, Berggren WA, Pälike H (2011) Review and revision of Cenozoic tropical planktonic foraminiferal biostratigraphy and calibration to the geomagnetic polarity and astronomical time scale. Earth Science Reviews 104: 111–142.
5. Geikie J (1874) The Great Ice Age and its Relation to the Antiquity of Man. London: W. Isbister and Co. 575 p.
6. Lewis FJ (1906) The history of the Scottish peat mosses and their relation to the glacial period. Scottish Geographical Magazine 22: 241–252.
7. Baden-Powell DFW (1956) The correlation of the Pliocene and Pleistocene marine beds of Britain and the Mediterranean. Proceedings of the Geologists' Association 66: 271–292.
8. Astorga G, Pino M (2011) Fossil leaves from the last interglacial in Central-Southern Chile: Inferences regarding the vegetation and paleoclimate. Geological Acta 9: 45–54.
9. Böcher J (2012) Interglacial insects and their possible survival in Greenland during the last glacial stage. Boreas 41: 644–659.
10. Schott W (1935) Deep-sea sediments of the Indian Ocean. In: Trask PD, editor. Recent Marine Sediments. Tulsa, Oklahoma: Society of Economic Paleontologists and Mineralogists. pp. 396–408.
11. Phleger FB, Parker FL, Peirson JF (1953) North Atlantic Foraminifera Reports of the Swedish Deep-Sea Expedition 7: 1–122.
12. Bandy OL (1960) The geologic significance of coiling ratios in the foraminifer *Globigerina pachyderma* (Ehrenberg). Journal of Paleontology 34: 671–681.
13. Streeter SS (1973) Bottom water and benthonic foraminifera in the North Atlantic – Glacial-interglacial contrasts. Quaternary Research 3: 141–141.
14. Schnitker D (1974) West Atlantic abyssal circulation during the past 120,000 years. Nature 248: 385–387.
15. Streeter SS, Lavery SA (1982) Holocene and latest glacial benthic foraminifera from the slope and rise off eastern North America. Geological Society of America Bulletin 93: 190–199.
16. Thomas E, Booth L, Maslin M, Shackleton NJ (1995) Northeastern Atlantic benthic foraminifera during the last 45,000 years: Changes in productivity seen from the bottom up. Paleoceanography 10: 545–562.
17. Berger WH (2012) Miklankovitch theory – hits and misses. Scripps Institution of Oceanography Technical report. 1–36.
18. Woods J, Barkmann W (1993) The plankton multiplier – positive feedback in the greenhouse. Journal of Plankton Research 15: 1053–1074.
19. Gaby ML, Sen Gupta BK (1985) Late Quaternary benthic foraminifera of the Venezuela Basin. Marine Geology 68: 125–144.
20. Sen Gupta BK, Temples TJ, Dallmeyer MDG (1982) Late Quaternary benthic foraminifera of the Grenada Basin: Stratigraphy and Paleoceanography. Marine Micropaleontology 7: 297–309.
21. Hayek LC, Buzas MA (2013) On the proper and efficient use of diversity measures for individual field samples. Journal of Foraminiferal Research 43: 305–313.
22. Wilson B (2008) Late Quaternary benthonic foraminifera in a bathyal core from the Leeward Islands, Lesser Antilles, NE Caribbean Sea. Journal of Micropalaeontology 27: 177–188.

23. Wilson B (2011) Alpha and beta diversities of Late Quaternary bathyal benthonic foraminiferal communities in the NE Caribbean Sea. Journal of Foraminiferal Research 41: 40–47.
24. Wilson B (2013) Ecostratigraphic regime shift during Late quaternary Marine Isotope Stages 8–9 in Santaren Channel, western tropical Atlantic Ocean: benthonic foraminiferal evidence from ODP Hole 1006A. Journal of Foraminiferal Research 43: 143–153.
25. Whittaker RH (1972) Evolution and measurement of species diversity. Taxon 21: 213–251.
26. Bhaumik AK, Gupta AK (2007) Evidence of methane release from Blake Ridge ODP Hole 997A during the Plio-Pleistocene: Benthic foraminifer fauna and total organic carbon. Current Science 92: 192–199.
27. Bhaumik AK, Gupta AK (2005) Deep-sea benthic foraminifera from gas hydrate-rich zone, Blake Ridge, Northwest Atlantic (ODP Hole 997A). Current Science 88: 1969–1973.
28. Mohan K, Gupta AK, Bhaumik AK (2011) Distribution of deep-sea benthic foraminifera in the Neogene of Blake Ridge, NW Atlantic Ocean. Journal of Micropalaeontology 30: 22–74.
29. Dillon WP, Hutchinson DR, Drury RM (1996) Seismic reflection profiles on the Blake Ridge near Sites 994, 995, and 997. In: Paull CK, Matsumoto R, Wallace PJ, editors. Proceedings of the Ocean Drilling Program, Initial Reports. pp. 47–56.
30. Shipboard Scientific Party (1996) Site 994. In: Paull CK, Matsumoto R, Wallace P, editors. Proceedings of the Ocean Drilling Program, Initial Reports. pp. 99–174.
31. Heezen BC, Hollister CD, Ruddiman WF (1966) Shaping of the continental rise by deep geostrophic contour currents. Science 152: 502–508.
32. Tucholke BE (2002) The Greater Antilles Outer Ridge: development of a distal sedimentary drift by deposition of fine-grained contourites. In: Stow DAV, Pudsey CJ, Howe JA, Faugeres J-C, Viana AR, editors. Deep-Water Contourite Systems: Modern Drifts and Ancient Series, Seismic and Sedimentary Characteristics: Geologicical Society of London, Memoir. pp. 39–55.
33. Okada H (2000) Neogene and Quaternary calcareous nannofossils from the Blake Ridge, Sites 994, 995 and 997. In: Paull CK, Matsumoto R, Wallace PJ, Dillon WP, editors. Proceedings of the Ocean Drilling Program, Scientific Results. pp. 331–341.
34. Sun H, Li T, Sun R, Yu X, Chang F, et al. (2011) Calcareous nannofossil bioevents and microtektite stratigraphy in the Western Philippine Sea during the Quaternary Chinese Science Bulletin 56: 2732–2738.
35. Wilson B (in press) Abundance Biozones and Community Structures in the Upper Quaternary of ODP Hole 994C (Blake Ridge, western Atlantic Ocean) compared with Marine Isotope Stages, Glacial Cycles and Terminations. Journal of Foraminiferal Research.
36. Cushman JA (1920) The Foraminifera of the Atlantic Ocean. Part 2: Lituolidae. United States National Museum Bulletin 104: 1–89.
37. Cushman JA (1922) The Foraminifera of the Atlantic Ocean. Part 3: Textulariidae. United States National Museum Bulletin 104: 1–149.
38. Cushman JA (1923) The Foraminifera of the Atlantic Ocean, Part 4: Lagenidae. United States National Museum Bulletin 104: 1–228.
39. Cushman JA (1924) The Foraminifera of the Atlantic Ocean, Part 5: Chilostomellidae and Globigerinidae. United States National Museum Bulletin 104: 1–55.
40. Cushman JA (1929) The Foraminfiera of the Atlantic Ocean, Part 6: Miliolidae. United States National Museum Bulletin 104: 1–129.
41. Cushman JA (1930) The Foraminfiera of the Atlantic Ocean, Part 7: Nonionidae, Camerinidae, Peneroplidae and Alveolinellidae. United States National Museum Bulletin 104: 1–179.
42. Cushman JA (1931) The Foraminifera of the Atlantic Ocean, Part 8: Rotaliidae, Amphisteginidae, Calcarinidae, Cymbalopoerttidae, Globorotaliidae, Anomalinidae, Planorbulinidae, Rupertiidae, and Homotremidae. United States National Museum Bulletin 104: 1–179.
43. Cushman JA, Henbest LG (1940) Geology and Biology of North Atlantic Deep-Sea Cores, Part 2: Foraminifera. United States Geolgical Survey Professional Paper 196-A: 35–56.
44. Phleger FB, Parker FL (1951) Foraminifera Species. Ecology of Foraminifera, Northwest Gulf of Mexico Geological Society of America Memoir. 1–64.
45. Parker FL (1954) Distribution of foraminifera in the northeastern Gulf of Mexico. Bulletin of the Museum of Comparative Zoology 111: 454–588.
46. Pflum CE, Frerichs WE (1976) Gulf of Mexico deep-water foraminifers. Cushman Foundation for Foraminiferal Research, Special Publication 14: 1–125.
47. Murray JW (2006) Ecology and Applications of Benthic Foraminifera. CambridgeUK: Cambridge University Press. 438 p.
48. Wright R (1978) Neogene paleobathymetry of the Mediterranean based on benthic foraminifers from DSDP Leg 42A. Initial Reports Deep Sea Drilling Project 48: 837–846.
49. Sen Gupta BK, Turner RE, Rabalais NN (1996) Seasonal oxygen depletion in continental-shelf waters of Louisiana: Historical record of benthic foraminifers. Geology 24: 227–230.
50. Murray JW (1992) Distribution and population dynamics of benthic foraminifera from the southern North Sea. Journal of Foraminiferal Research 22: 114–128.
51. Schafer CT, Cole FE (1978) Distribution of foraminifera in Chaleur Bay, Gulf of St. Lawrence. Geological Survey of Canada Paper 77–30: 1–55.
52. Gupta AK (1993) Biostratigraphic vs. paleoceanographic importance of Stilostomella lepidula (Schwager) in the Indian Ocean. Micropaleontology 39: 47–51.
53. Oba T, Shikama A, Okada H (2000) Data report: Oxygen Isotopic record of the last 0.8 m.y. at the Blake Ridge, Site 994C. In: Paull CK, Matsumoto R, Wallace PJ, Dillon WP, editors. Proceedings of the Ocean Drilling Program, Scientific Results.
54. Berger WH, Parker FL (1970) Diversity of planktonic foraminifera in deep-sea sediments. Science 168: 1345–1347.
55. Emiliani C (1955) Pleistocene temperatures. Journal of Geology 63: 538–578.
56. Hays JD, Imbrie J, Shackleton NJ (1976) Variations in the Earth's orbit: Pacemaker of the ice ages. Science 194: 1121–1132.
57. Huybers PJ, Wunsch C (2005) Obliquity pacing of the late Pleistocene glacial terminations. Nature 434: 491–494.
58. Broecker WS, van Donk J (1970) Insolation changes, ice volumes and the O^{18} record in deep-sea cores. Reviews of Geophysics 8: 169–198.
59. Waelbroeck CT, Labeyrie L, Michel E, Duplessy JC, McManus JF, et al. (2002) Sea-level and deep water temperature changes derived from benthic foraminifera isotopic records. Quaternary Science Reviews 21: 295–305.
60. Cheng H, Edwards RL, Broecker WS, Denton GH, Kong X, et al. (2009) Ice Age Terminations. Science 326: 248–252.
61. Poag CW, Valentine PC (1976) Biostratigraphy and ecostratigraphy of the Pleistocene basin Texas-Louisiana continental shelf. Gulf Coast Association of Geological Societies Transactions 26: 185–256.
62. Blanchon P, Shaw J (1995) Reef drowning during the last deglaciation: Evidence for catastrophic sea-level rise and ice-sheet collapse. Geology 23: 4–8.
63. Smart CW (2002) Environmental applications of deep-sea benthic foraminifera. In: Haslett SK, editor. Quaternary Environmental Micropalaeontology. London, UK: Arnold. pp. 14–58.
64. den Dulk M, Reichart CJ, Memon GM, Roelofs EMP, Zachariasse WJ, et al. (1998) Benthic foraminiferal response to variations in surface water productivity and oxygenation in the northern Arabian Sea. Marine Micropaleontology 35: 43–66.
65. de Rijk S, Jorissen FJ, Rohling EJ, Troelstra SR (2000) Organic flux control on bathymetric zonation of Mediterranean benthic foraminifera. Marine Micropaleontology 40: 151–166.
66. Qvale G, Van Weering TCE (1985) Relationship of surface sediments and benthic foraminiferal distribution patterns in the Norwegian Channel (northern North Sea). Marine Micropaleontology 9: 469–488.
67. Bornmalm L (1997) Taxonomy and paleoecology of late Neogene benthic foraminifera from the Caribbean Sea and eastern equatorial Pacific Ocean. Fossils and Strata 41: 1–96.
68. Mackensen A, Schmid ME, Harloff J, Giese M (1995) Deep-sea foraminifera in the South Atlantic Ocean: Ecology and assemblage generation. Micropaleontology 41: 342–358.
69. Taldenkova E, Bauch HA, Stepanova A, Ovsepyan Y, Pogodina I, et al. (2012) Benthic and planktic community changes at the North Siberian margin in response to Atlantic water mass variability since last deglacial times. Marine Micropaleontology 96–97: 13–28.
70. Murray JW (2013) Living benthic foraminifera: biogeographical distributions and the significance of rare morphospecies. Journal of Micropalaeontology 32: 1–26.
71. Schonfeld J (2001) Benthic foraminifera and pore-water oxygen profiles: A re-assessment of species boundary conditions at the Western Iberian Margin. Journal of Foraminiferal Research 31: 86–107.
72. Altenbach AV, Pflaumann U, Schiebel R, Thies A, Timm S, et al. (1999) Scaling percentages and distributional patterns of benthic foraminifera with flux rates of organic carbon. Journal of Foraminiferal Research 29: 173–185.
73. Smart CW, Gooday AJ (2006) Benthic foraminiferal trends in an relation to an organic enrichment gradient on the continental slope (850 m water depth) Off North Carolina (USA). Journal of Foraminiferal Research 36: 34–43.
74. Kaiho K (1994) Benthic foraminiferal dissolved-oxygen index and dissolved-oxygen levels in the modern ocean. Geology 22: 719–722.
75. Schoenfeld J (2001) Benthic foraminifera and pore-water oxygen profiles: a re-assessment of species boundary conditions at the western Iberian margin. Journal of Foraminiferal Research 31: 86–107.
76. Linke P, Lutze GF (1993) Microhabitat preferences of benthic foraminifera – a static concept or a dynamic adaptation to optimize food acquisition. Marine Micropaleontology 20: 215–234.
77. Rai AK, Singh VB (2012) Response of eastern Indian Ocean (ODP Site 762B) benthic foraminiferal assemblages to the closure of the Indonesian seaway. Oceanologia 54: 449–472.
78. Schönfeld J, Altenbach AV (2005) Late Glacial to Recent distribution pattern of deep-water *Uvigerina* species in the north-eastern Atlantic. Marine Micropaleontology 57: 1–24.
79. Seiglie GA (1968) Foraminiferal Assemblages as Indicators of High Organic Carbon Content in Sediments and of Polluted Waters. AAPG Bulletin 52: 2231–2241.
80. Sen Gupta BK, Strickert DP (1982) Living benthic foraminifera of the Florida-Hatteras slope: distribution trends and anomalies Geological Society of America Bulletin 93: 218–224.
81. Gooday AJ, Bernhard JM, Levin L, Suhr SB (2000) Foraminifera in the Arabian Sea oxygen minimum zone and other oxygen-deficient settings: taxonomic

composition, diversity, and relation to metazoan faunas Deep-Sea Research II 47: 25–54.
82. Smart CW, King SC, Gooday AJ, Murray JW, Thomas E (1994) A benthic foraminiferal proxy of pulsed organic matter paleofluxes. Marine Micropaleontology 23: 89–99.
83. Catuneanu O (2006) Principles of Sequence Stratigraphy Amsterdam, The Netherlands: Elsevier Science. 386 p.
84. Catuneanu O, Galloway WE, Kendall CGSC, Miall AD, Posamentier HW, et al. (2011) Sequence stratigraphy: methodology and nomenclature. Newsletters on Stratigaphy 44/3: 173–245.
85. Armentrout JM (1996) High resolution sequence biostratigrpahy: examples from the Gulf of Mexico Plio-Pleistocene. In: Hwell JA, Aitken JF, editors. High Resolution Sequence Stratigraphy: Innovations and Applications:: Geological Society of London: Special Publications 104. pp. 65–86.
86. Lüning S, Marzouk AM, Kuss J (1998) The Paleocene of central East Sinai, Egypt: 'sequence stratigraphy' in monotonous hemipelagites. Journal of Foraminferal Research 28: 19–39.
87. Wilson B (2003) Foraminifera and Paleodepths in a Section of the Early to Middle Miocene Brasso Formation, Central Trinidad. Caribbean Journal of Science 39: 209–214.
88. Pekar SF, Kominz MA (2001) Two-Dimensional Paleoslope Modeling: A New Method for Estimating Water Depths of Benthic Foraminiferal Biofacies and Paleoshelf Margins. Journal of Sedimentary Research 71: 608–620.
89. Nagy J, Finstad EK, Dypvik H, Bremer MGA (2001) Response of foraminiferal facies to transgressive-regressive cycles in the Callovian of northeast Scotland. Journal of Foraminiferal Research 31: 324–349.

12

When Did *Carcharocles megalodon* Become Extinct? A New Analysis of the Fossil Record

Catalina Pimiento[1]*, Christopher F. Clements[2]

1 Florida Museum of Natural History, Gainesville, Florida, United States of America; Department of Biology, University of Florida, Gainesville, Florida, United States of America; Smithsonian Tropical Research Institute, Balboa, Panama, **2** Institute of Evolutionary Biology and Environmental Studies, The University of Zurich, Zurich, Switzerland

Abstract

Carcharocles megalodon ("Megalodon") is the largest shark that ever lived. Based on its distribution, dental morphology, and associated fauna, it has been suggested that this species was a cosmopolitan apex predator that fed on marine mammals from the middle Miocene to the Pliocene (15.9–2.6 Ma). Prevailing theory suggests that the extinction of apex predators affects ecosystem dynamics. Accordingly, knowing the time of extinction of *C. megalodon* is a fundamental step towards understanding the effects of such an event in ancient communities. However, the time of extinction of this important species has never been quantitatively assessed. Here, we synthesize the most recent records of *C. megalodon* from the literature and scientific collections and infer the date of its extinction by making a novel use of the Optimal Linear Estimation (OLE) model. Our results suggest that *C. megalodon* went extinct around 2.6 Ma. Furthermore, when contrasting our results with known ecological and macroevolutionary trends in marine mammals, it became evident that the modern composition and function of modern gigantic filter-feeding whales was established after the extinction of *C. megalodon*. Consequently, the study of the time of extinction of *C. megalodon* provides the basis to improve our understanding of the responses of marine species to the removal of apex predators, presenting a deep-time perspective for the conservation of modern ecosystems.

Editor: Brian Lee Beatty, New York Institute of Technology College of Osteopathic Medicine, United States of America

Funding: This project was funded by the National Science Foundation NSF EAR 0418042, PIRE 0966884 (http://www.nsf.gov/funding). The funders had no role in study design, data collection and analysis, decision to publish, or preparation of the manuscript.

Competing Interests: The authors have declared that no competing interests exist.

* Email: pimientoc@ufl.edu

Introduction

Carcharocles megalodon ("Megalodon") was the largest shark that ever lived [1]. Based on its dentition, distribution and associated fauna, it has been suggested that this species could reach up to 18 m of total length, was a cosmopolitan apex predator and fed on cetaceans [1–7]. Its gigantic size and abundant fossil record has made this shark a charismatic example of extinct marine megafauna. However, despite its popularity and widespread fossil record, remarkably little is known about its extinction.

It has been widely stated in the literature that the extinction of apex predators can trigger cascading effects through entire food webs and impact ecosystem composition and function [8–9]. Concurrently, it has been demonstrated that the elimination of large sharks produces broad marine ecosystem degradation [10]. In modern marine systems, apex predators, especially large sharks, are significantly declining throughout the global oceans [11–12]. The study of the extinction of apex predatory sharks is therefore of significant interest.

A fundamental step towards understanding the effects of an extinction event is to know when it occurred. However, identifying the exact time of extinction of a species is notoriously difficult because the fossil record is inherently incomplete. Hence, the absence of records of a species does not necessarily mean that it is extinct, but could instead reflect preservation bias, spatially heterogeneous populations, sampling effort, or as yet uncovered fossil remains [13–15]. Such issues imply that the last recorded occurrence of a species, or "Last Appearance Date" (LAD) as a proxy for time of extinction provides an inherently biased estimate [16].

In order to identify the time of extinction of a species, multiple methods based on the temporal distribution of the most recent sighting events (or analogously the fossil record), have been proposed (see [14] for a review). Many of these remain poorly tested; however, the Optimal Linear Estimation (OLE) model [13] has been shown to provide accurate estimates of when a species can be considered to have become extinct [15].

Given that *C. megalodon* has an abundant fossil record, cosmopolitan distribution and high trophic level, its extinction is an ideal case study to better understand the ecological and macroevolutionary responses of marine species to top-down control release. It is generally reported that fossils of this species range from the middle Miocene (15.9–11.6 Ma) to the Pliocene (5.3–2.6 Ma), with some unconfirmed reports (i.e. considered to be unreliable, see Supplementary Information) from the Pleistocene (2.6–0.01 Ma) [1–7]. However, the time of extinction of *C. megalodon* has never been quantitatively assessed before.

Here, we synthesize the most recent records of *C. megalodon* from the literature and scientific collections, and probabilistically infer the time by which it became extinct by using the OLE model. Because OLE has only been previously used to tackle the problem of inferring the time of extinction of modern species [13], in this study we extend its use to the analogous issue of inferring extinction events in deep time (for a more detailed explanation of the reasons why we used this method, see the Materials and Methods section). Based on our estimated time of extinction and known ecological and macroevolutionary trends in cetaceans, we further outline the possible effects of this event, providing the basis for better understanding of the responses of marine species to the removal of apex predators.

Results

Based on the known age range of *Carcharocles megalodon*, we considered any post-Miocene occurrence as part of the most recent records of the species, and treated them as historic sightings to be used in the OLE calculation. Accordingly, we identified a total of 53 most recent records, which are made available in the Paleobiology Database (PaleoBioDB, http://paleobiodb.org). However, we only included in our analysis a subset of data consisting of the 42 records deemed to be reliable (Table S1).

The age of the fossils is not absolute, but instead falls within a range, with an upper and lower time estimate. To account for this unbiased uncertainty, we re-sampled the fossil data 10,000 times, bootstrapping the timing of each record from a uniform distribution between its upper and lower age. Hence, the results presented will to some extent be sensitive to the uniform distribution used in this analysis. Nevertheless, given that the actual age of the fossils is equally likely to have occurred anywhere within their age span, the use of this distribution is justified.

Because OLE infers the time by which a species can be considered extinct from the temporal distribution of the most recent sighting events [13], we regarded the time of the extinction event to most likely have occurred between the oldest inferred date of extinction (the first point in time at which the species can be considered as extinct), and the modal value (the most frequently inferred time of extinction). We use the modal value, as opposed to the mean, as it gives a more accurate reflection of the skewed distribution of inferred dates of extinction.

Results from applying OLE to our set of most recent records show that the modal inferred time of extinction is 2.6 Ma. This suggests that *C. megalodon* was unlikely to have survived beyond this time, with extinction time thus likely to have occurred between 3.5 Ma (oldest inferred extinction date), and this modal value (Figure 1). Approximately 50% of the simulations inferred the extinction event to have occurred before 2.6 Ma (modal value), with the remaining 50% of simulations being roughly uniformly distributed between 2.6 Ma and 0.1 Ma (Figure 1). In a very small proportion of simulations (1.5%), the inferred date of extinction fell after 0.1 Ma. In six simulations (0.06%) the inferred date of extinction fell after the present day (and thus the species could not be considered as extinct). However, because in the vast majority of the 10,000 simulations (>99.9%) the extinction time was inferred to have occurred before the present day, we reject the null hypothesis (that the species is extant) and the popular claims of present day survival of *C. megalodon*.

Discussion

Our analysis suggests that the extinction of *Carcharocles megalodon* most likely occurred around Pliocene–Pleistocene boundary (~2.6 Ma, modal value). Interestingly, subsequent to this time and throughout the Pleistocene, baleen whales (Cetacea, Mysticeti) reached modern gigantic sizes [17–19]. This mysticete composition contrasts with assemblages that were contemporaneous with *C. megalodon*, and that included mostly small-bodied species [17]. Because body size correlates with ecological functions, it has been further proposed that the faunal turnover observed in cetaceans during the Pliocene-Pleistocene boundary resulted in additional niches occupied by baleen whales [17,20–21].

Fossils of mysticetes are frequently found along with *C. megalodon* teeth. This has led to the hypothesis that they interacted in ancient marine communities [e.g. 5]. Whether or not *C. megalodon* preyed upon mysticetes needs further investigation. However, based on the inferred time of extinction of *C. megalodon* and the known ecological and macroevolutionary trends of cetaceans, we propose that the modern composition and function of gigantic filter-feeding whales established after the extinction of *C. megalodon*.

The controlling factors within ecosystems are not limited to top-down processes. Bottom-up effects are also important drivers in marine ecosystem dynamics [e.g. 22] and should be taken into consideration when studying the ecological and macroevolutionary trends of marine organisms. Of relevance to this work, diatom diversity and temperature changes (indicated by oxygen stable isotopes) have been associated with the changes in diversity of marine mammals throughout the Cenozoic [23]. However, it is not clear if these bottom-up processes drove the evolution of modern gigantic sizes in filter-feeding whales. Future work contrasting top-down and bottom-up processes with mysticetes body size trends are needed to discern the drivers of the evolution of gigantism in cetaceans.

Even though ecosystems are not driven entirely by only one type of control, the study of the extinction of apex predators has great potential to advance the understanding of the responses of marine species to top-down control release. Despite the limitations and uncertainties of the fossil record, the study of the time of extinction of *C. megalodon* provides a baseline to understand the establishment of the modern structure and function of gigantic filter-feeding whales. Furthermore, the methods used here could be applied to quantitatively assess the time of extinction of other fossil organisms, ultimately helping elucidate the causes and effects of extinction events.

Materials and Methods

Data collection

Based on the known fossil record of *Carcharocles megalodon*, we considered any post-Miocene occurrence as part of the most recent records of the species. Accordingly, we gathered all known post-Miocene records from the literature and scientific collections. First, we searched for *C. megalodon* occurrences in the PaleoBioDB using the following parameter: Species name = *Carcharocles megalodon*. The PaleoBioDB includes all known synonyms (e.g. *Carcharodon megalodon*) in the search. There, each data record represents a fossil collection. A collection is any set of fossils whose exact geographic and stratigraphic position and date of collection cannot be distinguished, regardless of when they were described, and by whom. With relevance to this study, a data record or collection in the PaleoBioDB is treated as an occurrence

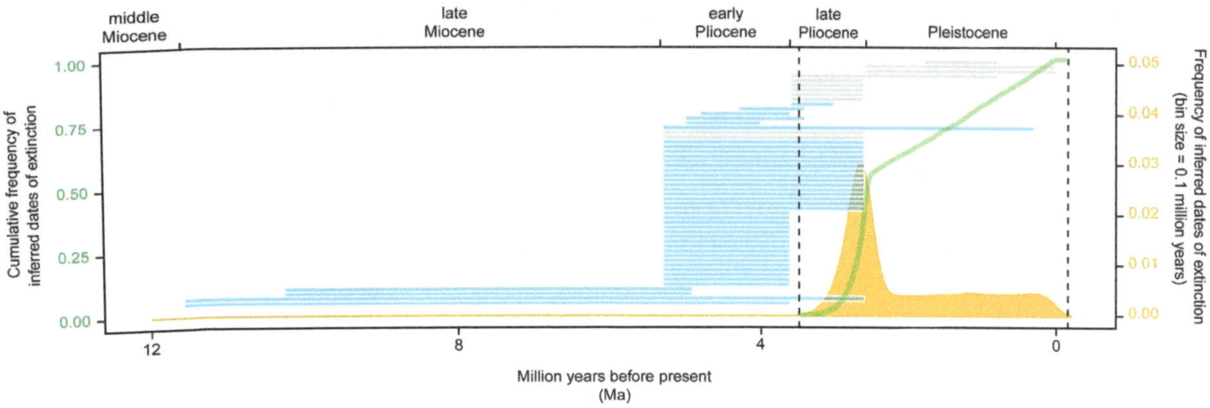

Figure 1. Temporal distribution of the inferred dates of extinction of *Carcharocles megalodon* using the Optimal Linear Estimation (OLE) model bootstrapped 10,000 times. The orange area shows the distribution of inferred dates of extinction through time, whereas the green line shows the cumulative frequency of inferred dates of extinction. The modal peak represents the point in time by which the species was most likely to have gone extinct (2.6 Ma). Approximately 50% of simulations fell before the modal peak of inferred dates of extinction (2.6 Ma), whereas the remaining 50% are roughly evenly distributed between the mode and the present day. The two vertical dashed lines indicate the most recent and oldest inferred dates of extinction (160,000 years in the future and 3.5 Ma respectively). The horizontal bars represent the time range of each fossil occurrence. The blue bars are the occurrences used in OLE. The grey bar represents the occurrences that failed the age evaluation process and were not used in the analysis.

(i.e. data record = collection = occurrence) and each of them are linked to one or more supporting references. Second, we searched for additional *C. megalodon* reports in ISI Web of Science (http://webofknowledge.com) GeoRef (http://geoscienceworld.org), Shark-References (http://shark-references.com) and Google Scholar (http://scholar.google.com) using the search terms: Megalodon OR *Carcharodon* AND *megalodon* OR *Carcharocles* AND *megalodon*.

Dataset construction

With the data collected, we created a comprehensive and updated dataset of the most recent records of *C. megalodon*. These include 15 pre-existing PaleoBioDB records and 38 new records, for a total of 53 post-Miocene records. They accessible in the PaleoBioDB, and can be found when doing a "Full Search" of "Fossil collection records" using the species name *Carcharocles megalodon* under "Taxon Name", and restricting the search to Pliocene to Holocene under "Time interval". These records are part of the ongoing PaleoBioDB Data Archive # 20 (http://goo.gl/Pplh0G) and are also presented in the Supporting Information.

Data filtering

To determine which of these records to include in our analyses, we performed a standardized evaluation as follows: for each reference reporting a post-Miocene occurrence of *C. megalodon*, we assessed if the age of the record was clearly stated in the text. In addition, we studied a number of supplementary references that further documented or refined the age of the localities from which some of the specimens were recovered. Finally, we visited the most relevant museum collections housing the specimens referenced in the literature. There we analyzed their morphology to assess for signs of re-deposition, examined their labels, and explored the collection databases to verify the age assignments. Whenever a reference did not meet the requirements of our evaluation process, we did not include that record in our analysis (Table S2). Additional information, including all records and their evaluations can be found in the Supporting Information section (Text S1). Furthermore, more detailed information and supporting references can be accessed when clicking on the PaleoBioDB# of each record.

Analysis

As a result of our age evaluation process (Text S1), we selected the records deemed to be reliable (Table S1) and treated them as historic sightings to apply the OLE model. We employed this method, rather than other methods that have previously been used to infer extinction from the fossil record [24–27] because of two main reasons: (1) OLE has been thoroughly tested and shown to produce accurate estimates of the date of extinction of a species, and to outperform many similar methods [14–15,28]. (2) The problem of inferring the time of extinction of a species from the fossil record is mathematically analogous to inferring the extinction of modern species from sighting record [e.g. 14, 29–31].

OLE infers the time by which a species can be considered as extinct from the temporal distribution of the most recent sighting events [13]. This method assumes that, regardless of the distribution of the complete sighting record, the most recent k sighting events have the form of a Weibull extreme value distribution, and infers the shape parameter, and thus the timing of extinction, from the joint distribution of these k sightings [13]. In effect, the method uses the spacing between recorded sighting events to infer the time at which, probabilistically, no more sightings will occur (the species can be considered extinct).

From Roberts and Solow [13], the Optimal Linear Estimation takes k sighting events and estimates $\hat{T}_E$, the time at which the extinction can be inferred, using the form:

$$\hat{T}_E = \sum_{k}^{i=1} w_i t_{n-i+1}$$

Where the weight vector w, length k, is given by:

$$w = \left(e'\Lambda^{-1}e\right)^{-1}\Lambda^{-1}e$$

e being a vector of k 1′s and Λ is a symmetric k by k matrix with typical element:

$$\Lambda_{ij} = \frac{\Gamma(2\hat{v}+i)\Gamma(2\hat{v}+j)}{\Gamma(\hat{v}+i)\Gamma(j)} j \leq i$$

where Γ is the gamma function and $\hat{v}$ is an estimate of the shape parameter of the Weibull extreme value distribution given by

$$\hat{v} = \frac{1}{k-1}\sum_{i=1}^{k-2} \log\frac{t_n - t_{n-k+1}}{t_n - t_{i+1}}$$

t_n being the n times a species is observed over the period of time t. So under the assumption a species is extinct, the upper bound of an approximate $1-\alpha$ confidence interval for $\hat{T}_E$ is

$$T_E^u = \frac{t_n - c(\alpha)t_{n-k+1}}{1-c(\alpha)}$$

Where

$$c(\alpha) = \left(\frac{k}{-\log(\alpha)}\right)^{-\hat{v}}$$

Note that in Solow [32] there is an error in the final equation, but presented above is the correct version.

There has been some debate as to what should be considered the k most recent sighting events [31–32], however recent work has shown that increasing the number of sighting events used in the calculation provides increasing accuracy of inferred dates of extinction [31]. Because of this, we use the most recent sighting events as presented in the Supporting information.

As in Clements et al. [15], we disregarded the confidence intervals produced by the technique as they have been shown to be very wide (Figure S1) and hence, concentrated on the distribution of the inferred extinction dates. Therefore, we regarded the modal value of the 10,000 estimates as the time by which *C. megalodon* is most likely to have gone extinct, with the exact date to have occurred between the mode and the oldest estimate. All simulations were made using the R statistical software [33] and the code is made available in the Supporting Information section (Files S1–S2).

Supporting Information

Figure S1 Distribution of inferred extinction date (red line), as well as upper (blue) and lower (green) 95% confidence intervals through time. The modal peaks of the upper and lower 95% confidence intervals fall close to the modal peak of the inferred date of extinction; however, the tail of the upper 95% confidence interval extends far beyond the present day, with the latest estimate falling 2.6 million years in future.

File S1 Code to calculate the distribution of inferred dates of extinction of *Carcharocles megalodon* using the Optimal Linear Estimation (OLE) model.

File S2 Data set of the Post-Miocene occurrences of *Carcharocles megalodon* used in the analysis.

Table S1 Post-Miocene records of *Carcharocles megalodon* included in the Optimal Linear Estimation (OLE) model (click on the PaleoBioDB# for more details).

Table S2 Post-Miocene records of *Carcharocles megalodon* excluded from Optimal Linear Estimation (OLE) model (click on the PaleoBioDB# for more details).

Text S1 Age evaluation of *Carcharocles megalodon* post-Miocene records (click on the PaleoBioDB# for more details).

Acknowledgments

CP thanks B. MacFadden for his constant encouragement and feedback. We would like to thank C. Jaramillo, J. Velez-Juarbe and J. Griffin for commenting on earlier versions of this manuscript; M. Uhen for support with the PaleoBioDB; A. Hendy for support with the NPC Database; R. Boessenecker and G. Morgan for insight regarding the data from California and Florida respectively; and M. Balk for assisting in visiting collections. We thank the Florida Museum of Natural History, the British Museum of Natural History, the National Museum of Natural Histroy, the Natural History Museum of Los Angeles County, the Museo Nacional de Historia Natural de Chile, the Muséum national d'Histoire naturelle (special thanks to Gäel Clément), the University of California Museum of Paleontology, and the San Diego Natural History Museum for allowing us accessing their collections and databases. Finally, we are grateful for the reviews made by N. D. Pyenson and D. Roberts, which substantially improved the original version of the manuscript. This is University of Florida Contribution to Paleobiology number 672 and Paleobiology Database publication number 210.

Author Contributions

Analyzed the data: CC. Wrote the paper: CP. Collected and processed the data: CP. Preparation of the final manuscript: CP CC.

References

1. Gottfried MD, Compagno LJV, Bowman SC (1996) Size and skeletal anatomy of the giant megatooth shark *Carcharodon megalodon*. In: Klimley AP, Ainley DG, editors. Great white sharks: the biology of *Carcharodon carcharias*. San Diego: Academic Press. pp. 55–89.
2. Pimiento C, Ehret DJ, MacFadden BJ, Hubbell G (2010) Ancient nursery area for the extinct giant shark Megalodon from the Miocene of Panama. Plos One 5: e10552. doi: 10.1371/journal.pone.0010552
3. Pimiento C, González-Barba G, Ehret DJ, Hendy AJW, MacFadden BJ, et al. (2013) Sharks and Rays (Chondrichthyes, Elasmobranchii) from the Late Miocene Gatun Formation of Panama. Journal of Paleontology 87: 755–774. doi: http://dx.doi.org/10.1666/12-117
4. Applegate SP, Espinosa-Arrubarrena L (1996) The fossil history of *Carcharodon* and its possible ancestor, *Cretolamna*: a study in tooth identification. In: Klimley AP, Ainley DG, editors. Great white sharks: the biology of *Carcharodon carcharias*. San Diego: Academic Press. pp. 19–36.
5. Purdy R (1996) Paleoecology of fossil white sharks. In: Klimley AP, Ainley DG, editors. Great white sharks: the biology of *Carcharodon carcharias*. San Diego: Academic Press. pp. 67–78.

6. Purdy RW, Schneider VP, Applegate SP, McLellan JH, Meyer RL, et al. (2001) The Neogene sharks, rays, and bony fishes from Lee Creek Mine, Aurora, North Carolina. Smithsonian Contributions to Paleobiology 90: 71–202.
7. Cappetta H (2012) Handbook of paleoichthyology. Volume 3E. Chondrichthyes. Mesozoic and Cenozoic Elasmobranchii: teeth. Stuttgart; New York: Gustav Fisher Verlag. 512 p.
8. Terborgh J, Holt RD, Estes JA (2010) Trophic cascades: predators, prey, and the changing dynamics of nature. Washington, D.C.: Island Press. 465 p.
9. Estes JA, Terborgh J, Brashares JS, Power ME, Berger J, et al. (2011) Trophic downgrading of planet Earth. Science 333: 301–306. doi: 10.1126/science.1205106
10. Myers RA, Baum JK, Shepherd TD, Powers SP, Peterson CH (2007) Cascading Effects of the Loss of Apex Predatory Sharks from a Coastal Ocean. Science 315: 1846–1850. doi: 10.1126/science.1138657
11. Myers RA, Worm B (2003) Rapid worldwide depletion of predatory fish communities. Nature 423: 280–283. doi: 10.1038/nature01610
12. Dulvy NK, Fowler SL, Musick JA, Cavanagh RD, Kyne PM, et al. (2014) Extinction risk and conservation of the world's sharks and rays. Elife 3: e00590. doi: 10.7554/eLife.00590
13. Roberts DL, Solow AR (2003) Flightless birds – When did the dodo become extinct? Nature 426: 245–245. doi: 10.1038/426245a
14. Rivadeneira MM, Hunt G, Roy K (2009) The use of sighting records to infer species extinctions: an evaluation of different methods. Ecology 90: 1291–1300. doi: 10.1890/08-0316.1
15. Clements CF, Worsfold NT, Warren PH, Collen B, Clark N, et al. (2013) Experimentally testing the accuracy of an extinction estimator: Solow's optimal linear estimation model. J. Anim. Ecol. 82: 345–354. doi: 10.1111/1365-2656.12005
16. Steadman DW, Martin PS, MacPhee RDE, Jull AJT, McDonald HG, et al. (2005) Asynchronous extinction of late Quaternary sloths on continents and islands. PNAS 102: 11763–11768. doi: 10.1073/pnas.0502777102
17. Boessenecker RW (2013) Pleistocene survival of an archaic dwarf baleen whale (Mysticeti: Cetotheriidae). Naturwissenschaften 100: 365–371. doi: 10.1007/s00114-013-1037-2
18. Pyenson ND, Sponberg SN (2011) Reconstructing body size in extinct crown cetacea (Neoceti) using allometry, phylogenetic methods and tests from the fossil record. J. Mamm. Evol. 18: 269–288. doi: 10.1007/s10914-011-9170-1
19. Lambert O, Bianucci G, Post K, de Muizon C, Salas-Gismondi R, et al. (2010) The giant bite of a new raptorial sperm whale from the Miocene epoch of Peru. Nature 466. doi: 10.1038/nature09067
20. El Adli JJ, Demere TA, Boessenecker RW (2014) *Herpetocetus morrowi* (Cetacea: Mysticeti), a new species of diminutive baleen whale from the Upper Pliocene (Piacenzian) of California, USA, with observations on the evolution and relationships of the Cetotheriidae. Zool. J. Linn. Soc. 170: 400–466. doi: 10.1111/zoj.12108
21. Bouetel V (2005) Phylogenetic implications of skull structure and feeding behavior in balaenopterids (Cetacea, Mysticeti). J. Mamm 86: 139–146. doi: 10.1644/1545-1542
22. Micheli F (1999) Eutrophication, fisheries, and consumer-resource dynamics in marine pelagic ecosystems. Science 285: 1396–1398. doi: 10.1126/science.285.5432.1396
23. Marx FG, Uhen MD (2010) Climate, critters, and cetaceans: Cenozoic drivers of the evolution of modern whales. Science 327: 993–996. doi: 10.1126/science.1185581
24. Marshall CR (1997) Confidence intervals on stratigraphic ranges with nonrandom distributions of fossil horizons. Paleobiology 23: 165–173.
25. Turney CSM, Flannery TF, Roberts RG, Reid C, Fifield LK, et al. (2008) Late-surviving megafauna in Tasmania, Australia, implicate human involvement in their extinction. PNAS 105: 12150–12153. doi: 10.1073/pnas.0801360105
26. Collen B, Purvis A, Mace GM (2010) When is a species really extinct? Testing extinction inference from a sighting record to inform conservation assessment. Divers. Distrib. 16: 755–764. doi: 10.1111/j.1472-4642.2010.00689.x
27. Solow AR, Roberts DL, Robbirt KM (2006) On the Pleistocene extinctions of Alaskan mammoths and horses. PNAS 103: 7351–7353. doi: 10.1073/pnas.0509480103
28. Solow AR, Roberts DL, Robbirt KM (2006) On the Pleistocene extinctions of Alaskan mammoths and horses. PNAS 103: 7351–7353. doi: 10.1073/pnas.0509480103
29. Robson DS, Whitlock JH (1964) Estimation of a truncation point. Biometrika, 31(1/2): 33–39. doi: 10.1093/biomet/51.1-2.33
30. Strauss D, Sadler PM (1989). Classical confidence intervals and Bayesian probability estimates for ends of local taxon ranges. Mathematical Geology, 21: 411–427. doi: 10.1007/BF00897326
31. Clements CF, Collen B, Blackburn TM, Petchey OL (2014) Effects of recent environmental change on accuracy of inferences of extinction status. Conservation Biology 28: 971–981. doi: 10.1111/cobi.12329
32. Solow AR (2005) Inferring extinction from a sighting record. Mathematical Biosciences, 195(1): 47–55. doi: 10.1016/j.mbs.2005.02.001
33. R Development Core Team (2012) R: A language and environment for statistical computing. Vienna, Austria: R foundation for statistical computing.

13

Species Richness and Evidence of Random Patterns in Assemblages of South American Titanosauria during the Late Cretaceous (Campanian–Maastrichtian)

Washington Luiz Silva Vieira[1]*, Kleber Silva Vieira[1], Rômulo Pantoja Nóbrega[1], Paulo Fernandes Guedes Pereira Montenegro[1], Gentil Alves Pereira Filho[2], Gindomar Gomes Santana[3], Rômulo Romeu Nóbrega Alves[3,4], Waltécio Oliveira Almeida[5], Alexandre Vasconcellos[6]

1 Laboratório de Ecofisiologia Animal, Departamento de Sistemática e Ecologia, Universidade Federal da Paraíba, João Pessoa, PB, Brazil, 2 Museu de Zoologia da Universidade de São Paulo, Ipiranga, São Paulo, SP, Brazil, 3 Programa de Pós-Graduação em Ecologia e Conservação (PPGEC)/Departamento de Biologia, Universidade Estadual da Paraíba, Campina Grande, Paraíba, Brazil, 4 Departamento de Biologia, Universidade Estadual da Paraíba, Campina Grande, Paraíba, Brazil, 5 Departamento de Química Biológica, Centro de Ciências Biológicas e da Saúde, Universidade Regional do Cariri - URCA, Campus do Pimenta, Crato, CE, Brazil, 6 Departamento de Sistemática e Ecologia, Universidade Federal da Paraíba, João Pessoa, PB, Brazil

Abstract

The Titanosauria were much diversified during the Late Cretaceous, but paleobiological information concerning these sauropods continues to be scarce and no studies have been conducted utilizing modern methods of community analysis to infer possible structural patterns of extinct assemblages. The present study sought to estimate species richness and to investigate the existence of structures in assemblages of the South American Titanosauria during the Late Cretaceous. Estimates of species richness were made utilizing a nonparametric estimator and null models of species co-occurrences and overlapping body sizes were applied to determine the occurrence of structuring in this assemblages. The high estimate of species richness (n = 57) may have been influenced by ecological processes associated with extinction events of sauropod groups and with the structures of the habitats that provided abundant support to the maintenance of large numbers of species. The pseudocommunity analysis did not differ from that expected by chance, indicating the lack of structure in these assemblages. It is possible that these processes originated from phylogenetic inertia, associated with the occurrence of stabilized selection. Additionally, stochastic extinction events and historical factors may also have influenced the formation of the titanosaurian assemblages, in detriment to ecological factors during the Late Cretaceous. However, diagenetic and biostratinomic processes, influenced by the nature of the sedimentary paleoenvironment, could have rendered a random arrangement that would make assemblage structure undetectable.

Editor: Leon Claessens, College of the Holy Cross, United States of America

Funding: The authors have no support or funding to report.

Competing Interests: The authors have declared that no competing interests exist.

* Email: wlsvieira@yahoo.com.br

Introduction

Sauropods constituted a group of Saurischian dinosaurs that were highly diversified, attaining large dimensions and wide geographic distributions [1]. These herbivorous giants appeared in the fossil record during the Late Triassic and persisted up until the Late Cretaceous [2–3]. These dinosaurs probably had a common ancestor related to the basal sauropodomorphs, or "prosauropods" – a globally widespread paraphyletic group of dinosaurs from the Late Triassic and Early Jurassic [2–5].

Titanosauria constituted the most diverse sauropod lineage, represented by more than 30 known genera, widely distributed on nearly all continental landmasses during the Late Cretaceous [6–7]. However, it is in South America where titanosaurs remains are more common and most diverse in terms of species richness in relation to the other continents [7]. Its diversification and radiation were probably influenced by the global extinction of diplodocoid sauropods in the Late Coniacian during the fragmentation of Gondwana [8–12]. Included in this radiation were, for example, saltasaurids, nemegtosaurids, and related forms, such as the genera *Isisaurus* and *Diamantinasaurus* [8–10].

Studies of Late Cretaceous South American Titanosauria have dealt predominantly with taxonomic and biochronological aspects, paleogeographic distributions, strategies of locomotion and behavior, reproductive and developmental biology, appendicular morphology, cranial morphology and phylogenetic systematics [4,7,9,13–20]. Research on the paleoecology of Titanosauria has been scarce and no studies have been carried out utilizing modern methods of ecological analysis (with estimators of species richness and null models, based on a pseudocommunity analysis) to infer the occurrence of structural patterns in assemblages of South American sauropods or other groups of extinct vertebrates. In general, ecological considerations such as species richness, morphological patterns, strategies of resource utilization, distributions over time and space, and historical and biogeographical factors that could structure species assemblages in a given area have not been examined [21]. The size and the overlapp of

Table 1. Matrix of occurrence of Titanosauria species recorded in the stratigraphic formations of the Late Cretaceous.

Species/Formation	**A**	**B**	**C**	**D**	**E**	**F**	**G**	**H**	**I**	**J**	**K**	**L**	**M**	**N**
Adamantisaurus mezzalirai	1	0	0	0	0	0	0	0	0	0	0	0	0	0
Aeolosaurus colhuehuapiensis	0	0	0	0	0	1	0	0	0	0	0	0	0	0
Aeolosaurus rionegrinus	0	0	1	0	1	0	0	0	0	0	0	0	0	0
Antarctosaurus wichmannianus	1	1	0	0	0	0	0	1	1	0	0	0	0	0
Argyrosaurus superbus	0	0	0	0	0	1	0	0	0	0	0	0	0	0
Atacamatitan chilensis	0	0	0	0	0	0	0	0	0	0	0	0	0	1
Barrosasaurus casamiquelai	0	1	0	0	0	0	0	0	0	0	0	0	0	0
Baurutitan britoi	0	0	0	0	0	0	1	0	0	0	0	0	0	0
Bonitasaura salgadoi	0	0	0	0	0	0	0	0	0	0	1	0	0	0
Bonatitan reigi	0	0	0	1	0	0	0	0	0	0	0	0	0	0
Gondwanatitan faustoi	1	0	0	0	0	0	0	0	0	1	0	0	0	0
Laplatasaurus araukanicus	0	1	0	1	0	0	0	1	0	0	0	0	0	0
Maxakalisaurus topai	1	0	0	0	0	0	0	0	0	0	0	0	0	0
Narambuenatitan palomoi	0	1	0	0	0	0	0	0	0	0	0	0	0	0
Neuquensaurus australis	0	1	0	1	0	0	0	1	0	0	1	0	0	0
Panamericansaurus schroederi	0	0	0	1	0	0	0	0	0	0	0	0	0	0
Pellegrinisaurus powelli	0	1	0	0	0	0	0	0	0	0	0	0	0	0
Pitekunsaurus macayai	0	1	0	0	0	0	0	0	0	0	0	0	0	0
Puertasaurus reuili	0	0	0	0	0	0	0	0	0	0	0	1	0	0
Rocasaurus muniozi	0	0	0	1	0	0	0	0	0	0	0	0	0	0
Saltasaurus loricatus	0	0	0	0	0	0	0	0	0	0	0	0	1	0
Trigonosaurus pricei	0	0	0	0	0	0	1	0	0	0	0	0	0	0
Uberabatitan ribeiroi	0	0	0	0	0	0	1	0	0	0	0	0	0	0
Total species richness	4	7	1	5	1	2	3	3	1	1	2	1	1	1

A: Adamantina Formation, B: Anacleto Formation, C: Angostura Colorada Formation, D: Allen Formation, E: Los Alamitos Formation, F: Bajo Barreal Formation, G: Marília Formation, H: Palacio Formation, I: Plottier Formation, J: Cambambe Formation, K: Bajo de la Carpa Formation, L: Pari Aike Formation; M: Lecho Formation, N: Tolar Formation.

ecological niches among sympatric and syntopic species can play important roles in this structuring [22–24].

The structure of a vertebrate assemblage can be defined as nonrandom patterns in resource utilization among individuals that coexist in time and space [25–26], with species composition and richness having important roles in this structuring [27]. The co-occurrence of vertebrate species in time and space will therefore be determined and structured by their interspecific relationships [28].

Within this context, studies have shown that co-occurrence patterns are commonly attributed to competitive interactions or to environmental filters, and these patterns can be generated by historical factors, habitat associations, and/or species dispersal limits [29–31]. The development of null models using pseudo-communities generated by randomizations of the original information by Monte Carlo simulations has provided an important tool for studying biological assemblage structures [32–36].

Considering the hypothesis that species co-occurrence in time and space will be determined and structured by their interspecific relationships, the present work sought to estimate the species richness of the South American Titanosauria during the Campanian and Maastrichtian ages (83.5–65 Mya) and to investigate the occurrence of structuring in this assemblage with respect to species co-occurrences and overlapping body sizes.

Materials and Methods

The list of sauropods presented in this work (Table S1) was prepared based on records of species that coexisted geographically and stratigraphically in the South American fossiliferous formations of the Late Cretaceous (Lower Campanian to the Upper Maastrichtian) as described in the literature ([1–9,37–41] and see references in Table S1) and in the databanks of the website Paleobiology Database [42]. We chose to utilize only those records identified to the species level, excluding citations of supraspecific taxonomic groups – thus constructing an assemblage of sympatric and presumably competitive species for analysis. The body size estimates of the different sauropod species (Table S1) were obtained from the literature [1–3,5–8,42–46].

The nonparametric estimator Chao1 was utilized to estimate sauropod species richness based on abundance data: $S_{Chao1} = Sobs + F^2_1/2F_2$, where *Sobs* is the number of species recorded in the assemblage sampled, F_1 is the number of species represented by only one individual ("singletons"), and F_2 is the number of species represented by two individuals ("doubletons") [47–49]. The values obtained from the richness estimator, based on 1,000 randomizations without replacement, were plotted indicating the estimated species richness (with 95% confidence intervals) that probably exists in the stratigraphic formations of the Lower Campanian to the Upper Maastrichtian (83.5–65 Mya). The analyses were performed using Estimates 7.5 software [49].

The EcoSim Module of co-occurrence was utilized to test the occurrence of nonrandom patterns of co-occurrence of the Titanosauria species recorded in the stratigraphic formations corresponding to the Lower Campanian to the Upper Maastrichtian in South America [27,50]. The data for this analysis consisted of a presence (1) absence (0) matrix in which each species corresponded to a line and each stratigraphic formation a column (Table 1). The presence/absence data of the matrix were randomized to produce patterns that would be expected in the absence of competitive interactions between the species. The following EcoSim options were utilized: *C-score* index [50–52] as a quantitative co-occurrence index, fixed row and column totals and column constraints, and algorithms of the "Sequential Swap" matrix randomization, with 10,000 randomizations. *C-scores* measure the mean numbers of units in a single block (checkerboard units – CU) for all pairs of species [27,36,51–52].

In a structured assemblage, the mean numbers of units in a single block should be significantly higher than the score expected by chance, according to a null model [51–52]. The number of units in a single block for any pair of species is calculated by: $CU = (r_i–S)(r_j–S)$, where r_i and r_j correspond to the totals in a row, and S is the number of sites occupied by both species. The utilization of fixed row and column totals and column restrictions generate null matrices with the same number of occurrences of sites per species (row totals) and the same number of species per stratigraphic formation (column totals) as observed in the original data. The algorithm of sequential change repeatedly rearranges the original matrix, changing the sub-matrices that preserve the row and column totals, and is not very inclined toward type I or type II errors [27,36]. The stratigraphic formations utilized for this analysis of co-occurrence are found in Table 1 and Table S1.

The EcoSim Module of Size Overlap was utilized to determine the presence of nonrandom patterns of body size overlapping among species [27]. In this analysis, the estimated total size utilized ($\log_{10}$ transformed) for each species was derived from data available in literature [1–3,5–9,42–46]. The original matrix was then reformulated to produce random patterns that would be expected in the absence of competitive interactions. The following options were utilized in EcoSim: Variance in segment length as a size-overlap metric, logarithmic transformation, no rounding, and all species in the matrix included in the source pool, with the colonization weights set to 1. Because of the occurrence of variations in segment length, minimum segment length values were utilized.

Segment length was calculated by the ordination of size estimates of the different species. These values represent the differences in body size between two consecutive species. Utilizing the variance in segment length as the size overlap metric, the overall tendency for the observations can be measured. A structured assemblage would have an observed variance significantly smaller than that seen in random assemblages (pseudo-communities). When the minimum segment length values (in meters) are selected, the smallest segment of the assemblage can be calculated by measuring the difference between the closest pair of species. This measure determines whether a minimum space between species is necessary for their coexistence in an assemblage. Thus, in a structured assemblage, the minimum segment length should be significantly greater than that expected by chance [27,36,53].

Results

A total of 23 species were recorded in fourteen fossiliferous strata ranging from the Lower Campanian to the Upper Maastrichtian in different localities in South America (Table 1 and Table S1). In relation to the total number of species, eight (34.78%) were restricted to the Campanian, five (21.73%) to the Maastrichtian, and 10 (43.47%) were distributed in both the Campanian and Maastrichtian (Table S1). The Chao1 richness estimator indicated a richness of 57 species during the Late Cretaceous, with a 95% confidence interval of 36 to 115 species (Fig. 1).

Evaluations of the species encountered in each stratigraphic formation indicated that the Anacleto and Allen formations were the richest, with seven and six species distributed among the Campanian and Maastrichtian, respectively (Table 1 and Table S1). The analyses of species co-occurrence in fourteen stratigraphic formations indicated that all of the species formed checkerboard

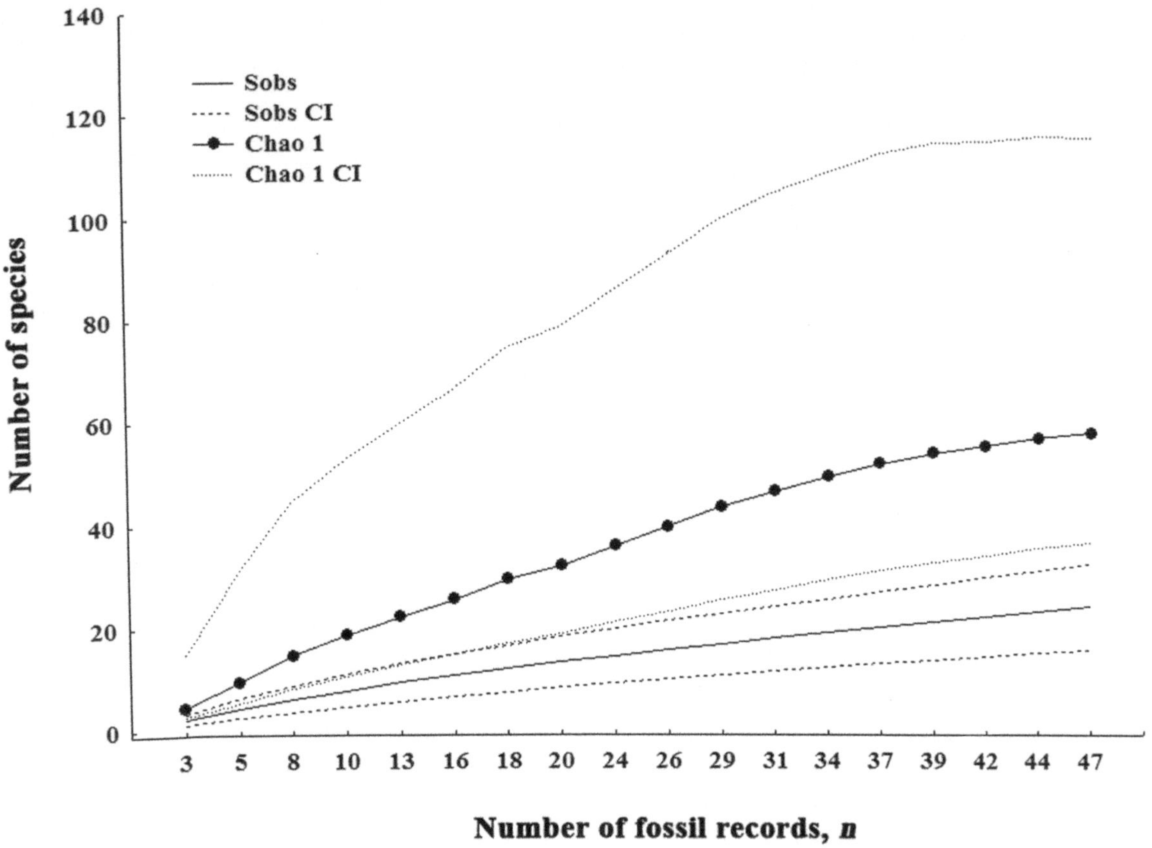

Figure 1. Estimates of species richness represented by rarefaction curves calculate with data of fossil records. The curves were generated from 1000 randomizations with replacement and the sampling units corresponding to the total of fossil record (*n*) in the stratigraphic formations of Late Cretaceous in South America (Campanian–Maastrichtian). Sobs: richness observed.

units in the presence-absence matrix. The observed *C-score* index was 1.96, which did not differ significantly from the mean expected by chance (1.91, Fig. 2). The result is consistent with the hypothesis that coexistence of South American Titanosauria species during the Late Cretaceous was not structured by deterministic processes.

The species themselves showed great variability with respect to their estimated sizes. *Rocasaurus muniozi* had the lowest estimated size (8 m long), whereas *Puertasaurus reuili* was estimated to be 30 m long [14,37] (Table S1). Size-overlap analyses, based on minimum segment lengths, indicated that no mean overlap significantly larger than expected was observed in the titanosaurian assemblages (Fig. 3). The size-overlap analysis based on the variance of segment length showed that no mean overlap significantly smaller than random occurred in the titanosaurian assemblages (Fig. 3). Therefore, both analyses indicated a lack of structuring in the assemblages.

Discussion

The fossil record points to the South American continent as having had a diverse assemblage of Titanosauria, and estimators of species richness indicate an even greater species richness during the Late Cretaceous (Early Campanian-Late Maastrichtian). This high species richness was possibly influenced by the availability and occupation of ecological niches left by the diplodocoids sauropods after their extinction in the Late Coniacian, resulting in a rich diversity of forms and sizes within the clade Titanosauria [10,12]. Additionally, other ecological factors such as the association of titanosaurs with inland environments and a diet adapted to ingesting angiosperms may have contributed to clade Titanosauria diversification during this period [54].

The species richness could also have been related to the structural complexity of the habitats occupied by titanosaurian assemblages during the Late Cretaceous. Dinosaurs that coexisted during the Late Jurassic exhibit close associations with the characteristics of environments in which they lived, indicating the occurrence of structural patterns in those assemblages [55]. In modern ecosystems, structurally more complex habitats provided greater support for the maintenance of larger varieties of coexisting species as compared to more spatially homogeneous environments [55–56], and the richness and distributions of species that coexisted in time and space, when associated with environmental characteristics, can give rise to nonrandom patterns in ecological interactions and structured assemblages [25–26].

The results obtained by co-occurrence analyses of species richness demonstrated that the observed numbers of checkerboard units did not differ from random. This pattern is consistent with the hypothesis that the local coexistence of Titanosauria species during the Campanian and Maastrichtian in South America was not structured by ecological factors existing during the Late Cretaceous, such as resource limitations in the environment, interspecific competition, or predator-prey relationships.

The coexistence of species in an assemblage can be limited by ecological interactions known to be negative, such as interspecific competition for spatial and trophic niches, the occurrence of

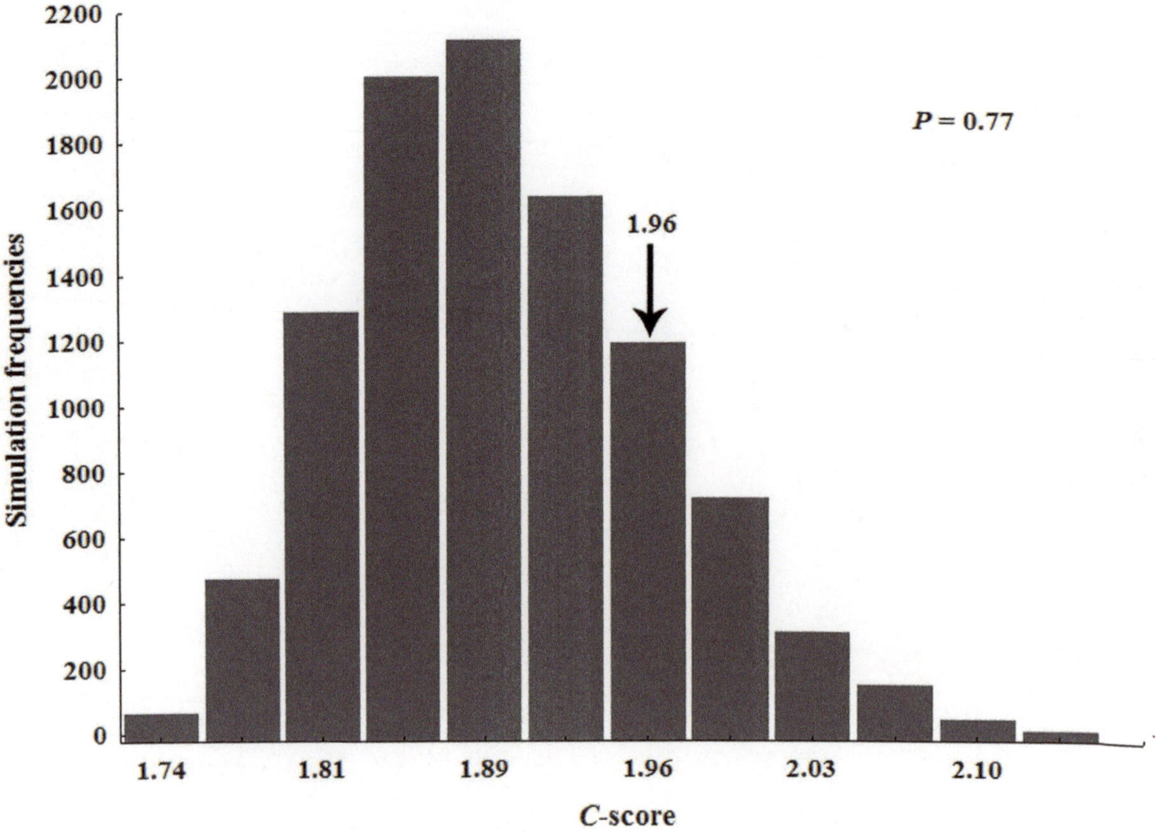

Figure 2. Frequency distribution of checkerboard *C*-scores obtained from 10,000 simulations produced by randomizations of titanosaurian assemblages. Arrow indicates the observed mean and *P* is the probability of the observed mean to be statistically greater than that expected.

aggressiveness between individuals of the assemblage, species that develop specific preferences for certain types of habitats on a wide geographic scale, or predator-prey relationships [54–55,57–61].

Random patterns in coexistence of Titanosauria species during the Late Cretaceous could have originated not through competitive interactions between species but through the influence of species showing specificity for a particular habitat type on a reduced spatial scale, distributions restricted to a particular period, endemism, or low abundances of some species [55,57–58]. Additionally, the lack of structure in species assemblages can be driven by the stochastic nature of extinction events, and would be demonstrated by decreasing numbers of checkerboard units in the co-occurrence analyses [36].

Another important issue regarding the lack of structure in South American titanosaurian assemblages is the fact that species records in any particular formation can be influenced by factors such as taphonomic processes, the types of sedimentary paleoenvironment, and sampling efforts in fossil collecting. This latter aspect will be influenced by the numbers of paleontological explorations in each stratigraphic formation and by the environmental conditions at the sites where the fossiliferous strata are found (for example, sites located in forest areas that make fossil discovery more difficult in contrast to sites in arid environments with scarce vegetation cover that facilitates exploration).

Paleoecological studies should emphasize the importance of taphonomic processes for the different types of sedimentary paleoenvironments, since these factors can influence the fossil records of one or more species in stratigraphic formations. Diagenetic and biostratinomic processes, influenced by the nature of the sedimentary paleoenvironment at the site where the animal died (which can hinder fossilization) and the transport carcasses to different assemblages, will determine the number of specimens preserved in place and, consequently, estimated species richness [62–65] – but may also generate random patterns of species distributions. Thus, fossiliferous formations with low species richness or a set of under-sampled taxa could provide insufficient paleoecological data, making any structuring of assemblages of extinct species undetectable.

The analysis of size-overlapping in this study indicated a lack of structure in the Titanosauria assemblage, suggesting that the sizes of these dinosaurs were not a determinant factor for species coexistence in time and space. The random patterns attributed to body size overlapping among vertebrates in general may be due to local extinctions, non-limited food availability, or reduced population sizes [36]. Other factors such as genetic drift, invasion, or colonization by other species can also cause these patterns in terrestrial vertebrate assemblages [66].

Ecological differences between sauropod lineages could also have been associated with certain morphological attributes, such as body size and differences in dentition, shape of the necks and cranial morphology [7,14,67–68]. Morphological variations between species that coexisted in a given area can direct the utilization of certain types of resources, establishing guilds of morphologically similar species and determining the levels of overlap between them [61].

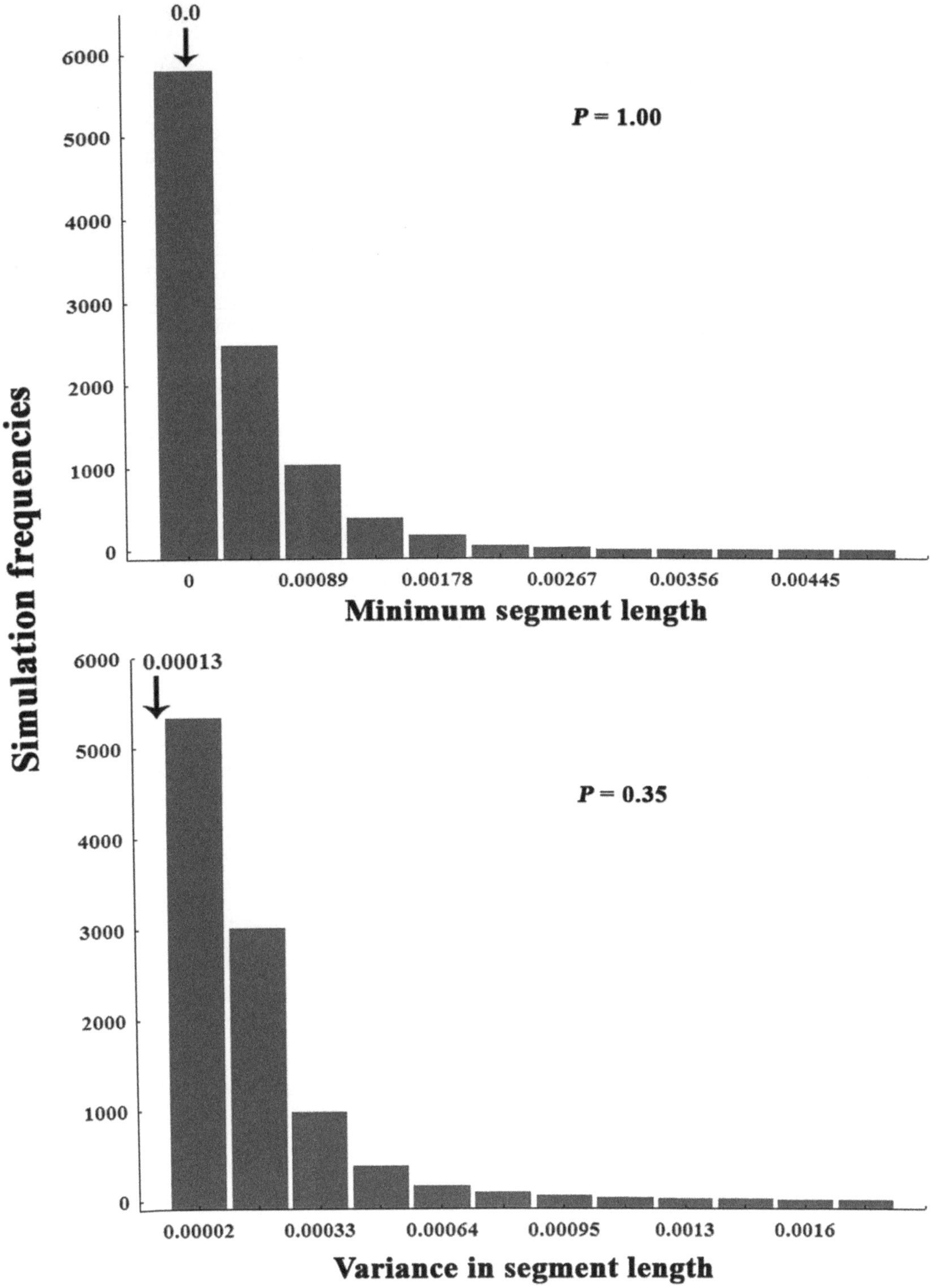

Figure 3. Observed and expected size overlap of Titanosauria species during the Late Cretaceous in South America. The dimensions used were minimum segment length and variance in segment length, with values log transformed. The arrows indicate observed means and *P* indicates the probability that the observed value is greater than the expected value (10,000 simulations).

Competitive interactions for food resources and habitat utilization in bird assemblages can become reduced through morphometric variations related to the size and shape of the beak, the length of metatarsus, or body size [69]. The morphological differences between phylogenetically closely related species in lizard assemblages in Central America contributed to their segregation with respect to microhabitat utilization [70]. Thus, morphological variations will significantly contribute to niche segregation between species, making assemblage coexistence possible and determining how available resources in the environment will be utilized by each species [71].

Another important aspect that should be taken into consideration concerns possible historical effects on South American titanosaurian assemblages that occurred during the Late Cretaceous. The preference for, and utilization of, particular resources by Titanosauria lineages could have been strongly influenced by the evolutionary histories of the different clades. Phylogenetic effects include important processes that will determine the ecologies of large numbers of species (as opposed to putative interactions between members of the assemblage in terms of the utilization of available resources), reflecting the evolutionary histories of different lineages that diverged over millions of years [72–73]. Body size is strictly related to phylogenetic structures in different clades, suggesting that stabilizing selection processes may have been involved in the evolution of this character – which would be expected, as body size correlates with various ecological attributes and the life histories of the species [74].

Conclusions

It is possible to conclude that the species richness of Titanosauria during the Late Cretaceous in South America was influenced by various ecological processes associated with the extinction events of various sauropods groups during this period and habitat structures that provided support for the maintenance of high species diversity in the assemblage. The observed patterns of co-occurrence and size overlapping suggest the existence of random processes and a lack of structuring in this assemblage. It is likely that these processes originated from phylogenetic inertia, associated with the occurrence of stabilizing selection, and that extinction events and historical factors had important roles in the formation of titanosaurian assemblages during the Late Cretaceous, more than did strictly ecological factors. Nonetheless, diagenetic and biostratinomic processes (influenced by the nature of the sedimentary paleoenvironment) can cause random species distribution patterns, making structuring of those undetectable.

Supporting Information

Table S1 Species of Titanosauria recorded in the stratigraphic formations of the Late Cretaceous in South America. The number of recorded fossils (n) and whole information were obtained from the matrix of data available in the Paleobiology Database [1] and in the literature.

Acknowledgments

We thank the editor, Paul Upchurch (Department of Earth Sciences, University College London) and anonymous reviewers for their detailed comments on the manuscript. A. Leyva and Roy Funch helped with the English translation and the editing of the manuscript.

Author Contributions

Conceived and designed the experiments: WLSV. Performed the experiments: WLSV KSV. Analyzed the data: WLSV KSV AV. Contributed reagents/materials/analysis tools: WLSV KSV RPN AV PFGPM GAPF GGS RRNA WOA. Wrote the paper: WLSV KSV.

References

1. Fastovsky DE, Weishampel DB (2012) Dinosaurs: A concise natural history. 2nd ed. Cambridge: Cambridge University Press. 408 p.
2. Upchurch P, Barrett PM, Dodson P (2004) Sauropoda. In: Weishampel DB, Dodson P, Osmólska H, editors. The Dinosauria, 2nd ed. Berkeley and Los Angeles: University of California Press. 259–322.
3. Mannion PD, Upchurch P, Carrano MT, Barrett PM (2011) Testing the effect of the rock record on diversity: a multidisciplinary approach to elucidating the generic richness of sauropodomorph dinosaurs through time. Biol Rev 86: 157–181.
4. Wilson JA (2005) Overview of sauropod phylogeny and evolution. In: Rogers K.A, Wilson JA, editors. The Sauropods: Evolution and Paleobiology. 15–49.
5. Sertich JJW, Loewen MA (2010) A New Basal Sauropodomorph Dinosaur from the Lower Jurassic Navajo Sandstone of Southern Utah. PLoS ONE 5(3): doi:10.1371/journal.pone.0009789.
6. Powell JE (2003) Revision of South American titanosaurid dinosaurs: paleobiological, paleogeographical and phylogenetic aspects. Records Queen Vict Mus 111: 1–173.
7. García R, Salgado L, Fernández M, Cerda I, Carabajal A, et al. (2014) Paleobiology of titanosaurs: reproduction, development, histology, pneumaticity, locomotion and neuroanatomy from the South American fossil record. Ameghiniana, Preprint doi:10.5710/AMGH.16.07.2014.829.
8. Hocknull SA, White MA, Tischler TR, Cook AG, Calleja ND, et al. (2009) New Mid-Cretaceous (Latest Albian) Dinosaurs from Winton, Queensland, Australia. PLoS ONE 4(7): doi:10.1371/journal.pone.0006190.
9. Poropat SF, Upchurch P, Mannion PD, Hocknull SA, Kear BP, et al. (2014). Revision of the sauropod dinosaur *Diamantinasaurus matildae* Hocknull, et al. 2009 from the mid Cretaceous of Australia: Implications for Gondwanan titanosauriform dispersal. Gondwana Research (2014), http://dx.doi.org/10.1016/j.gr.2014.03.014.
10. Barrett PM, Upchurch P (2005) Sauropodomorph diversity through time. In: CurryRogers KA, Wilson JA, editors. The Sauropods: Evolution and Paleobiology. Berkeley: University of California Press. 125–156.
11. Upchurch P, Barrett PM (2005) Phylogenetic and taxic perspectives on sauropod diversity. In: CurryRogers KA, Wilson JA, editors. The Sauropods: Evolution and Paleobiology. Berkeley: University of California Press. 104–124.
12. Whitlock JA (2011) A phylogenetic analysis of Diplodocoidea (Saurischia: Sauropoda). Zool J Linn Soc 161: 872–915.
13. CurryRogers KA (2005). Titanosauria: A Phylogenetic Overview. In: CurryRogers KA, Wilson JA, editors. The Sauropods: Evolution and Paleobiology. University of California Press. 50–103.
14. Gonzalez Riga BJ (2010) Paleobiology of South American titanosaurs. In: Calvo J, Porfiri J, Gonzalez Riga B, Dos Santos D, editors. Paleontología y dinosaurios desde America Latina, EDIUNC, Universidad Nacional de Cuyo. 125–141.
15. Salgado L, Coria RA, Calvo J (1997) Evolution of titanosaurid sauropods I: Phylogenetic analysis based on the postcranial evidence. Ameghiniana 34: 3–32.
16. Kellner AWA, Campos DA (2000) Brief review of dinosaur studies and perspectives in Brazil. An Ac Bras Cienc 72(4): 509–538.
17. Santucci RM, Bertini RJ (2001) Distribuição paleogeográfica e biocronológica dos titanossauros (Saurischia, Sauropoda) do Grupo Bauru, Cretáceo Superior do Sudeste Brasileiro. Rev Bras Geociênc 31: 307–314.
18. Zaher H, Pol D, Carvalho AB, Nascimento PM, Riccomini C, et al. (2011) A complete skull of an Early Cretaceous sauropod and the evolution of advanced titanosaurians. PLoS One 6(2): 1–10.
19. Wilson JA, Upchurch P (2003) A revision of Titanosaurus Lydekker (Dinosauria-Sauropoda), the first dinosaur genus with a 'Gondwanan' distribution. J Sys Palaeontol 1(3): 125–160.
20. Apesteguía S (2007) The sauropod diversity of the La Amarga Formation (Barremian), Neuquén (Argentina). Gondwana Res 12: 533–546.
21. Novas F (2009) The age of dinosaurs in South American (Life of the past). Bloomington: Indiana University Press. 480 p.
22. MacArthur RH (1972) Geographical ecology. New Jersey: Princeton University Press. 269 p.
23. Case TJ, Gilpin ME (1974) Interference Competition and Niche Theory. Proc Natl Acad Sci 71(8): 3073–3077.
24. Vandermeer JH (1972) Niche Theory. Annu Rev Ecol Syst 3: 107–132.
25. Luiselli L (2006) Resource partitioning and interspecific competition in snakes: the search for general geographical and guild patterns. Oikos 114: 193–211.
26. Winemiller KO, Pianka ER (1990) Organization in natural assemblages of desert lizards and tropical fishes. Ecol Monogr 60: 27–55.
27. Gotelli NJ, Entsminger GL (2001) EcoSim: Null models software for ecology. (Acquired Intelligence Inc. and Kesey–Bear, Burlington, Vermont).
28. Toft CA (1985) Resource partitioning in amphibians and reptiles. Copeia 1: 1–21.
29. Peres-Neto PR, Olden JD, Jackson DA (2001) Enviromentally constrained null models: site suitability as occupancy criterion. Oikos 93: 110–120.
30. Urich W (2004) Species co-occurrences and neutral models: reassessing J.M. Diamond's assembly rules. Oikos 107: 603–609.
31. Bloch CP, Higgins CL, Willing MR (2007) Effects of large-scale disturbance on metacommunity structure of terrestrial gastropods: temporal trends in nestedness. Oikos 116: 395–406.
32. Gotelli N (2001) Research frontiers in null model analysis. Global Ecol and Biogeogr 10: 337–343.

33. Gotelli NJ, Colwell RK (2001) Quantifying biodiversity: procedures and pitfalls in the measurement and comparison of species richness Ecol Lett 4: 379–391.
34. Luiselli L (2008) Resource partitioning in freshwater turtle communities: A null model meta-analisis of available data. Acta Oecol 34: 80–88.
35. Morreira LF, Maltichk L (2012) Assessing patterns of nestedness and co-occurrence in coastal pond anuran assemblages. Amphibia-Reptilia 33: 261–271.
36. Gainsbury AM, Colli GR (2003) Lizard assemblages from natural cerrado enclaves in southwestern Amazonia: the role of stochastic extinctions and isolation. Biotropica 35(4): 503–519.
37. Salgado L, Azpilicueta C (2000) Un nuevo saltasaurino (Sauropoda, Titanosauridae) de la provincia de Río Negro (Formación Allen, Cretácico Superior), Patagonia, Argentina. Ameghiniana 37(3): 259–264.
38. Kellner AWA, Campos DA, Trotta MNF (2005) Description of a titanosaurid caudal series from the Bauru Group, Late Cretaceous of Brazil. Arq Mus Nac 63(3): 529–564.
39. Campos DA, Kellner AWA, Bertini RJ, Santucci RM (2005) On a titanosaurid (Dinosauria, Sauropoda) vertebral column from the Bauru Group, Late Cretaceous of Brazil. Arq Mus Nac 63(3): 565–593.
40. Candeiro CRA, Santos AR, Rich TH, Marinho TS, Oliveira EC (2006) Vertebrate fossils from the Adamantina Formation (Late Cretaceous), Prata paleontological district, Minas Gerais state, Brazil. Géobios 39: 319–327.
41. Martinelli A, Forasiepi AM (2004) Late Cretaceous vertebrates from Bajo de Santa Rosa (Allen Formation), Río Negro province, Argentina, with the description of a new sauropod dinosaur (Titanosauridae). Rev Mus Argen Cienc Nat 6(2): 257–305.
42. Paleobiology Database (2000) Available: http://www.paleodb.org/cgi-bin/bridge.pl. Accessed 25 January 2013.
43. Peczkis J (1995) Implications for body-mass estimates for dinosaurs. J Vert Paleontol 14: 520–533.
44. Mazzetta GV, Christiansen P, Fariña RA (2004) Giants and bizarres: Body size of some southern South American Cretaceous dinosaurs. Hist Biol: 1–13.
45. Wedel MJ (2005) Postcranial pneumaticity in sauropods and its implications for mass. In: Rogers KA, Wilson JA, Editors. The Sauropods: Evolution and Paleobiology. 201–228.
46. Sander PM, Christian A, Clauss M, Fechner R, Gee CT, et al. (2011) Biology of the sauropod dinosaurs: the evolution of gigantism. Biol Rev 86: 117–155.
47. Colwell RK, Coddington J (1994) Estimating terrestrial biodiversity through estrapolation. Phil Trans R Soc B 345: 101–118.
48. Walther BA, Moore JL (2005) The concepts of bias, precision and accuracy, and their use testing the performance of species richness estimators, with literature review of estimator performance. Ecogaphry 28: 815–829.
49. Colwell RK (2005) EstimateS: Statistical estimation of species richness and shared species from samples, version 7.5. Available: http://purl.oclc.org/estimates. Accessed 20 June 2013.
50. Behrensmeyer AK, Kidwell SM, Gastaldo RA (2000) Taphonomy and paleobiology. Paleobiology, 26: 103–147.
51. Stone L, Roberts A (1990) The checkerboard score and species distributions. Oecologia 85: 74–79.
52. Gotelli NJ (2000) Null model analysis of species co-occurrence patterns. Ecology 81(9): 2606–2621.
53. Simberloff DS, Boecklen W (1981) Santa Rosalia reconsidered: Size ratio and competition. Evolution 35: 1206–1228.
54. Mannion PD, Upchurch P (2010) A quantitative analysis of environmental associations in sauropod dinosaurs. Paleobiology 36(2): 253–282.
55. Noto CR, Grossman A (2010) Broad-Scale Patterns of Late Jurassic Dinosaur Paleoecology. PLoS ONE 5(9): 1–11.
56. Lieberman SS (1986) Ecology of the leaf litter herpetofauna of a neotropical rain forest: La Selva, Costa Rica. Acta Zool Mex 15: 1–72.
57. Connor EF, Simberloff D (1979) The assembly of species communities: Chance or competition. Ecology 60: 1132–1140.
58. Gotelli NJ, Buckley NJ, Wiens JA (1997) Co-occurrence of Australian land birds: Diamond's assembly rules revisited. Oikos 80(2): 311–324.
59. Farlow JO, Pianka ER (2002) Body size overlap, habitat partitioning and living space requirements of terrestrial vertebrate predators: implications for the paleoecology of large theropod dinosaurs. Hist Biol 16(1): 21–40.
60. Horner-Devine MC, Silver JM, Mathew AL, Bohannan BJM, Colwell RK, et al. (2007) A comparison of taxon co-occurrence patterns for macro- and microorganisms. Ecology 88(6): 1345–1353.
61. Tiho S, Josens G (2007) Co-occurrence of earthworms in urban surroundings: A null model analysis of community structure. Euro J Soil Biol 43: 84–90.
62. Behrensmeyer AK, Kidwell SM, Gastaldo RA (2000) Taphonomy and paleobiology. Paleobiology, 26: 103–147.
63. Fiorillo AR, Eberth DA (2004) Dinosaur taphonomy. In: Weishampel DB, Dodson P, Osmólska H, editors. The Dinosauria, 2nd ed. Berkeley and Los Angeles: University of California Press. 607–613.
64. Edinger EN (2003) Fossilization Processes: Bioerosion. In: Briggs DEG, Crowther PR, editors. Palaeobiology II: Blackwell Publishing company, Berlin, Germany. 273–277.
65. Behrensmeyer AK (2003) Taphonomy: Terrestrial Vertebrates. In: Briggs DEG, Crowther PR, editors. Palaeobiology II: Blackwell Publishing company, Berlin, Germany. 318–321.
66. Strong DR, Szyska LA, Simberloff DS (1979) Tests of community-wide character displacement against null hypotheses. Evolution 33: 897–913.
67. Sereno PC, Wilson JA, Witmer LM, Whitlock JA, Maga A, et al. (2007) Structural Extremes in a Cretaceous Dinosaur. PLoS ONE (11): 1–9.
68. Whitlock JA (2011) Inferences of Diplodocoid (Sauropoda: Dinosauria) Feeding Behavior from Snout Shape and Microwear Analyses. PLoS ONE 6(4): 1–20.
69. Piratelli AJ, Melo FP, Caliri RF (2001) Dados morfométricos de aves de sub-bosque da região leste de Mato Grosso do Sul. Rev Bras Zool 18(2): 305–317.
70. Losos JB (1992) The evolution of convergent structure in Caribbean *Anolis* communities. Syst Biol 41: 403–420.
71. Miller JR, Cale P (2000) Behavioral mechanisms and habitat use by birds in a fragmented agricultural landscape. Ecol Appl 10: 1732–1748.
72. Vitt J, Pianka ER (2005) Deep history impacts present-day ecology and biodiversity. PNAS 102(22): 7877–7881.
73. Colston TJ, Costa GC, Vitt LJ (2010). Snake diets and the deep history hypothesis. Biol J Linn Soc 101: 476–486.
74. Diniz-Filho JAF, Vieira CM (1998) Padrões e processos na evolução do tamanho do corpo em carnívoros (Mamalia) da América do Sul. Rev Bras Biol 58(4): 649–657.

14

Modelling Growth-Competition Relationships in Trembling Aspen and White Spruce Mixed Boreal Forests of Western Canada

Jian-Guo Huang[1*¤], Kenneth J. Stadt[2], Andria Dawson[1], Philip G. Comeau[1]

1 Department of Renewable Resources, University of Alberta, Edmonton, Alberta, Canada, **2** Forest Management Branch, Sustainable Resource Development, Edmonton, Alberta, Canada

Abstract

We examined the effect of competition on stem growth of *Picea glauca* and *Populus tremuloides* in boreal mixedwood stands during the stem exclusion stage. We combined traditional approaches of collecting competition data with dendrochronology to provide retrospective measurements of stem diameter growth. Several competition indices including stand basal area (BA), the sum of stem diameter at breast height (SDBH), and density (N) for the broadleaf and coniferous species, as well as similar indices considering only trees with diameters greater than each subject (BAGR, SDBHGR, and NGR), were evaluated. We used a nonlinear mixed model to characterize the basal area increment over the past 5, 10, 15, 20, 25, 30, and 35 years as a function of growth of nearby dominant trees, the size of the subject trees, deciduous and coniferous competition indices, and ecoregions. SDBHGR and BAGR were better predictors for spruce, and SDBHGR and NGR were better for aspen, respectively, than other indices. Results showed strongest correlations with long-term stem growth, as the best models integrated growth for 10–25 years for aspen and ≥25 for spruce. Our model demonstrated a remarkable capability (adjusted $R^2>0.67$) to represent this complex variation in growth as a function of site, size and competition.

Editor: Harald Auge, Helmholtz Centre for Environmental Research - UFZ, Germany

Funding: This project was funded by the Natural Science and Engineering Research Council of Canada, the Mixedwood Management Association, and the Forest Resource Improvement Association of Alberta. The funders had no role in study design, data collection and analysis, decision to publish, or preparation of the manuscript.

Competing Interests: The authors have declared that no competing interests exist.

* E-mail: jianguo.huang@ualberta.ca

¤ Current address: Département de Sciences Fondamentales, Université du Québec à Chicoutimi, Chicoutimi, Québec, Canada

Introduction

Both inter- and intraspecific competition play important roles in affecting forest growth, composition, structure, and succession. Following a stand-replacing disturbance, boreal mixedwood sites are typically dominated by shade-intolerant species such as trembling aspen (*Populus tremuloides* Michx.), white birch (*Betula papyrifera* Marsh.) and lodgepole or jack pine (*Pinus contorta* Dougl. ex. Loud. and *Pinus banksiana* Lamb.). Shade-tolerant white spruce [*Picea glauca* (Moench.) Voss] may co-establish with these pioneer species or establish later; these spruce grow steadily in the understory, reaching the canopy after 50–90 years [1,2]. Late-successional balsam fir [*Abies balsamea* (L.) Mill.] establishes and emerges last, if fire does not occur [3,4]. In Canada, mixedwood stands composed of trembling aspen and white spruce are a major component of the boreal forest, especially in the prairie provinces of western Canada [5]. In these mixedwood forests, competition is particularly critical during the stem exclusion or self-thinning stage (typically 20−40 years of stand age; [6]) and during the subsequent understory reinitiation stage. In these stages, critical transitions occur, such as crown separation of the pioneer species following self-thinning episodes, or spruce crown closure in stands with higher conifer density [4,7]. Intense competition among trees for resources such as light, soil moisture and nutrients occurs as they expand in size, leading to logarithmic decreases in stem density [4,8]. Spruce shows a strong response in radial and height growth to increases in light [9,10]. In the boreal mixedwood forest of western Canada, poor conifer height growth under dense hardwood canopies was attributed to light below 20% of full sunlight, well below levels needed for optimum growth [9,11]. Hence it is particularly critical to quantify the effects of competition on growth at this stage, to help us better understand forest growth and dynamics, and to improve forest management decisions. At a stand level, the emergent property of these tree dynamics may account for greater volume yields in mixed vs. pure stands [12,13].

Previous studies of the links between competition and growth typically rely on permanent sample plot (PSP) data which consists of periodic re-measurements at 2–10 year intervals (e.g., breast height diameter, height, and conditions of trees) [14,15,16,17]. PSP programs vary by jurisdiction and are often not well-structured in terms of age, productivity and composition. In the western boreal, data are particularly lacking for mixed species stands aged 30−70 years. The sparse nature of existing mid-rotation PSP data makes it difficult to confidently address questions regarding growth dynamics during the stem exclusion and understory re-initiation stages.

A dendrochronological approach may be more effective than PSPs to address this lack of data. Tree-ring analysis provides growth data without re-measurement of plots, and is therefore more efficient than PSP data [15]. Dendrochronology relies on inter-annual ring-width growth patterns to model tree growth dynamics as well as endogenous (e.g. competition) and exogenous (e.g., disturbance, climate) factors [18]. For example, Biondi [19] demonstrated how dendrochronological data can be used to obtain insight into stand growth patterns over longer periods than is possible using PSPs. Metsaranta and Lieffers [15] compared dendrochronological stand reconstruction techniques with boreal pine PSPs and found these techniques can provide similar information on stand development patterns, though dendrochronology provided annual resolution that is rarely obtained from PSP data. Numerous other studies [10,20] have also used short sequences of the most recent growth rings to characterize growth rates in response to current competition.

To quantify the relationship between growth and competition during the stem exclusion and understory reinitiation period, we surveyed 51 stands in the boreal mixedwood region of Alberta using an efficient sampling technique which combined traditional approaches of collecting competition data with dendrochronology. The specific objectives of the study were to 1) model the relationships between stem growth, tree size and competition in the boreal mixedwood forest over different time intervals and ecoregions; 2) develop a sub-model to estimate the growth potential of dominant trees, and 3) identify competition indices suitable for long-term growth prediction of sub-dominants. We hypothesize that the growth response to competition can be more effectively predicted when conditioned by the growth of nearby dominant trees, that the pattern of growth rate vs. tree size can be simply modeled, and if competition effects are carefully characterized, provide a powerful model for predicting tree growth.

Materials and Methods

Ethics Statement

No specific permissions were required to conduct field research in our study areas; these are public forests, where scientific research is encouraged. We confirm that our field studies did not involve endangered or protected species.

Study Area

Our study area included the mixedwood ecoregions in Alberta, Canada, which occupy 75% of the forested area of the province (Fig. 1). The upland forests in these regions are dominated by even- or uneven-aged mixtures of trembling aspen, white spruce, balsam poplar (*Populus balsamifera* L.), paper birch, and balsam fir [21,22]. Lodgepole pine also occurs within these mixtures on the eastern slopes of the Rocky Mountains. Our sampling covered most of these regions, from Rocky Mountain House in the south to High Level in the north, and east from Fort McMurray to Fairview and Hinton in the west (Fig. 1). Sampling locations were distributed throughout the Central Mixedwood, Dry Mixedwood, Lower Foothills, and Lower Boreal Highlands ecoregions [23,24], which contain the majority of the productive mixedwoods. Elevations of the sampled sites ranged from 266 m to 1308 m.

The region is dominated by a typically dry continental climate, with cold winters and warm summers. Mean annual temperature decreases with increasing latitude and elevation (i.e. northward and westward); precipitation increases with elevation. The 1971–2000 climate normals indicate mean annual temperatures, growing degree days (GDD>5°C), and mean annual total precipitation were 0.7°C, 1376.4°C, and 455 mm with 25% in the form of snow, respectively, in the easternmost portion of the study area [Fort McMurray, 56°39′N, 111°13′W, 369 m above sea level (a.s.l.)], and 2.1°C, 1377.5°C, and 471.6 mm with 30% in the form of snow, respectively, for the westernmost study region (Fairview, 56°04′N, 118°23′W, 670 m a.s.l.). At the southern extent of the study area (Rocky Mountain House, 52°26′N, 114°55′W, 988 m a.s.l.), these climate variables were 2.3°C, 1163.6°C, and 535.4 mm with 30% in the form of snow, respectively, and −1.3°C, 1225.7°C, and 394.1 mm (34% as snow) in the northern area (High Level, 58°37′N, 117°10′W, 338 m a.s.l.) [25]. Soils in these regions were mainly orthic gray luvisols and brunisols with primarily silty or clay loam texture, due to glacial till and glaciolacustrine parent material [23,24].

Data Collection

Field sampling was conducted in 2007, 2008, 2010 and 2011, and was designed to obtain data for this project as well as a retrospective white spruce mortality study (Dawson et al. in prep.). We randomly sampled accessible trembling aspen and white spruce dominated mixedwood stands, which ranged from 25 to 100 years according to the Phase 3 inventory database [26]. We excluded stands where the aspen cohort was clearly uneven-aged or stands with indications of ground fires or insect outbreaks. In each stand, a belt transect was employed to assess growth-competition relationships for trembling aspen and/or white spruce vs competitors. Transects ranged from 5 m to 785 m long and 5 m to 20 m wide. The transect area ranged from 25 to 1600 m^2 (mean = 415 m^2), and depended primarily on spruce density (low numbers of spruce required larger areas to obtain an adequate sample). Within each transect, all trees over 1 m tall were counted by species and classified into three size classes [1−3 m height (H), 3 m H −10 cm stem diameter at breast height (DBH; defined as 1.3 m above-ground), and over 10 cm DBH]. Eight to 15 live spruce and ten aspen were then randomly chosen in each transect as subject trees. Height, DBH, and a competition assessment were taken before each subject tree was felled and a disk of each subject tree was taken at stump height (0.3 m). For competitor assessment, the DBH of all trees and shrubs taller than 1.3 m and within a 1.78 m radius plot (10 m^2) centered at the subject tree were measured. These local competition plots were then aggregated to obtain a transect-level assessment of competition for each species (usually>100 m^2 of each transect's area was sampled for competitors) during calculation of competition indices. Larger plots would be desirable for local neighbourhood competition assessments at each tree, but forest growth models are typically applied at the stand (transect) level [27]; this sampling was designed to assess competition at a similar scale.

We also sampled approximately ten of the largest aspen and five of the largest white spruce trees near the transect in the same ecosite to provide a measure of potential growth on the site. At most sites, these largest aspen and spruce were canopy dominants. However, at some sites, there were no dominant spruce present. In these cases, large spruce trees were still sampled, and their competition was assessed. The DBH of each dominant or large spruce tree was measured and two 5.1 mm increment cores from each of these trees were collected at 1.3 m height for annual ring-width measurement in the lab. Ecosite classification was performed for each transect, following Beckingham and Archibald [24] and Beckingham et al. [23]. In total, 51 transects were sampled for spruce growth-competition assessment; 42 of these were also sampled for aspen growth-competition assessment, though all were mixed stands (Fig. 1). In total, stem growth and competition was measured for 1670 trees (spruce: 620 subject trees and 227 dominant or large spruce trees; aspen: 420 subject trees and 403 dominant trees) throughout the mixedwood forests of Alberta (Table 1).

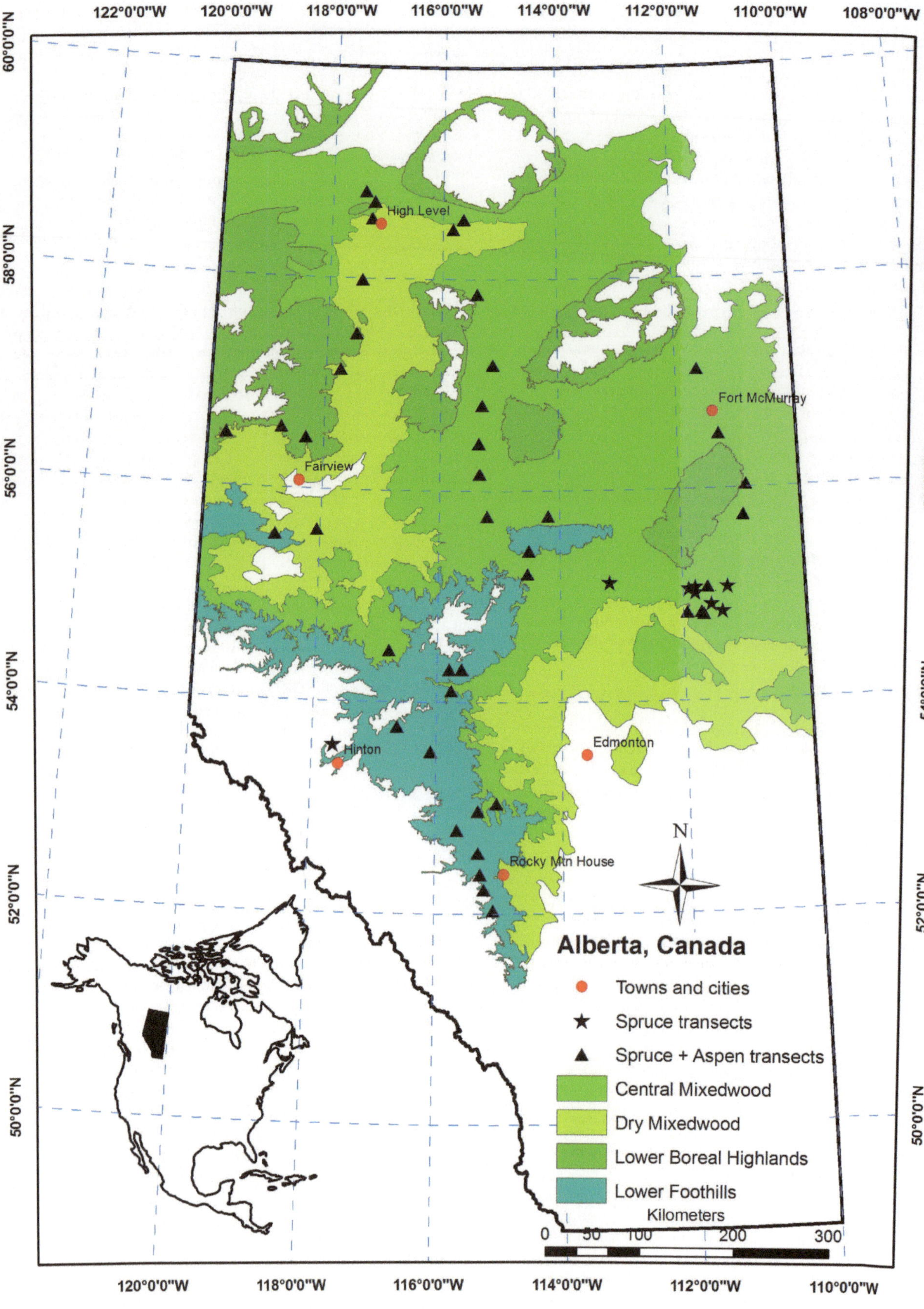

Figure 1. Map showing the locations of transects in relation to the natural ecoregions of Alberta, Canada.

Table 1. Characteristics of the transects sampled in Alberta mixedwood forest, western Canada.

Natural subregion	Transects	Spruce																	
		Deciduous competition																	
		BA (m^2/ha)			SDBH (m/ha)			N (stems/ha)			BAGR (m^2/ha)			SDBHGR (m/ha)			NGR (stems/ha)		
		Max	Mean	Min	Max	Mean	Min	Max	Mean	Min	Max	Mean	Min	Max	Mean	Min	Max	Mean	Min
CM+DM	34	49	20	2	394	187	12	7600	1955	67	49	15	0	361	128	0	2600	1064	0
LF+LBH	17	28	15	4	332	159	43	19692	2738	400	28	12	3	206	110	31	8385	1282	250
		Spruce competition																	
		BA (m^2/ha)			SDBH (m/ha)			N (stems/ha)			BAGR (m^2/ha)			SDBHGR (m/ha)			NGR (stems/ha)		
		Max	Mean	Min	Max	Mean	Min	Max	Mean	Min	Max	Mean	Min	Max	Mean	Min	Max	Mean	Min
CM+DM	34	81	26	5	548	249	98	11500	3342	917	24	7	0	228	64	0	4600	683	0
LF+LBH	17	49	23	4	471	271	89	8400	4100	1133	20	7	0	135	66	2	2308	664	67
Natural subregion	**Transects**	**Aspen**																	
		Deciduous competition																	
		BA (m^2/ha)			SDBH (m/ha)			N (stems/ha)			BAGR (m^2/ha)			SDBHGR (m/ha)			NGR (stems/ha)		
		Max	Mean	Min	Max	Mean	Min	Max	Mean	Min	Max	Mean	Min	Max	Mean	Min	Max	Mean	Min
CM+DM	26	49	24	9	423	245	82	7600	2726	643	45	13	0	362	101	0	3000	717	0
LF+LBH	16	28	17	4	308	173	43	4867	2080	400	28	11	0	264	80	0	2133	529	0
		Spruce competition																	
		BA (m^2/ha)			SDBH (m/ha)			N (stems/ha)			BAGR (m^2/ha)			SDBHGR (m/ha)			NGR (stems/ha)		
		Max	Mean	Min	Max	Mean	Min	Max	Mean	Min	Max	Mean	Min	Max	Mean	Min	Max	Mean	Min
CM+DM	26	48	25	9	610	305	108	13000	4495	1438	36	12	0	369	92	0	3750	650	0
LF+LBH	16	49	23	5	392	262	97	8500	3868	1200	40	11	0	276	77	0	2125	478	0

Abbreviations: CM+DM: Central Mixed wood and Dry Mixed wood; LF + LBH: Lower Foothills and Lower Boreal Highlands; BA: sum of basal area; SDBH: sum of diameter at breast height; N: density; BAGR: sum of basal area for trees thicker than each subject; SDBHGR: sum of diameter at breast height for trees thicker than each subject; NGR: density for trees thicker than each subject.

In the laboratory, all tree-ring samples (discs and cores) were dried and carefully polished with successively finer grits of sandpaper. The dry stump height diameter with bark (*DSH*) was also measured on stump height disks for use in the calculation of the inside-bark DSH (see Compiling stem growth data below). The DSH was also used in developing models linking DSH with DBH to allow others to link our model to their studies, since most forest models use DBH as response or driving variables whereas our model used stump height ring-width data. Two radii separated by an angle of 90−180° (avoiding knots or severe reaction wood) were chosen from each disc. Visual cross-dating for each radii and core collected was conducted under a binocular microscope, and focused on the pointer years (wide and narrow rings) observed in each transect and among transects. The dated cores and radii were all carefully measured using a Velmex measuring system interfaced with the 'Times Series Analysis Program' (TSAP; Frank Rinntech, Heidelberg, Germany) to a precision of 0.001 mm. Visual cross-dating was verified using COFECHA [28]. The average correlation between individual series and the master chronology within each transect was well above 0.55 ($P<0.01$), indicating strong similarities in inter-annual ring growth pattern among trees within transects. Master chronologies between nearby transects were also well correlated ($r>0.48$, $P<0.01$), which suggests that our crossdating was reliable. Stand age for each transect was determined using the aspen stump height ring count. This may be slightly younger than the maximum stand age which can be dated from the root collar [29]. We avoided obvious multi-aged stands, nonetheless it is possible that some of our stands are older than the age of the dominant aspen [21].

Compiling Stem Growth Data

To compute the past growth of individual trees, we averaged the two ring-width measurements corresponding to the same year for a given tree to represent annual ring growth of that tree. We then took the cumulative sum of its annual ring growth over several time intervals t extending back to the year of the outermost complete ring ($t = 5$, 10, 15, 20, 25, 30 and 35 years from the year of the outermost complete ring) to obtain the total radial increment in time interval t (RW_t), for analysis. We then calculated the basal area increment (*BAI*, cm^2) of each tree over these time intervals. To avoid approximating the pith location in the dominant tree cores, we anchored our measurements at the cambium rather than the pith. To do this we factored and substituted the cambium offset into the normal BAI equation (1) (which needs measurements from the pith), to determine BAI.

$$BAI_t = \pi\left(R_{y1}^2 - R_{y2}^2\right) = \pi(R_{y1} + R_{y2})(R_{y1} - R_{y2}) \quad (1)$$

$$BAI_t = \pi\left(R_{y1} + R_{y1} - \Delta R_t\right)\Delta R_t = \pi(DSH_{subi} - \Delta R_t)\Delta R_t \quad (2)$$

In Equations (1) and (2), we used that $R_{y1} = DSH_{subi}/2$, and $R_{y2} = DSH_{subi}/2 - \Delta R_t$, where $\Delta R_t = R_{y1} - R_{y2}$, y_1 is the calendar year of the outermost complete ring and $y_2 = y_1 - t$. Lab measurements of *DSH* were converted to inside-bark DSH (DSH_{subi}) using the published provincial equations [30] to facilitate this BAI calculation without needing the pith.

Growth vs. Size Trend for Large and Dominant Trees

Selecting large or dominant trees to identify growth potential is an effective way to condition a growth model for site, population genetics and climate effects. Additional growth reduction in suppressed trees can then be attributed to competition. However, small and very large trees generally do not have the same growth potential as intermediate trees [31]. During their early development, trees build leaf and root area, increasing their ability to acquire resources, fix carbon and increase in size. As size increases, however, maintenance respiration costs for supporting root and stem architecture also increase, and more energy is required to raise water to a taller crown, reducing photosynthetic efficiency [32]. Canham et al. [31] described this trend in radial growth using a log-normal function: the growth rate increased initially with increasing size (DBH), reached a maximum, then declined as DBH exceeded 30 to 40 cm. Others have likewise found it important to model the effect of size on growth of the subject tree, either directly [16,33] or as part of the competition index [34]. An independent measure, such as leaf or crown surface area is desirable for assessing size-related growth potential [20], but is difficult to obtain routinely [22]. We tested Canham et al.'s [31] log-normal function, as well as quadratic, logarithmic and several other functions to model this maximum growth – size relation using the ring width sequences from the large or dominant trees (Fig. 2).

Based on these analyses, the best model to capture the growth rate vs. initial size trend for spruce and aspen was a simple parabolic function, with no linear term is given by equation (3),

$$G_{\max} = G_L - b_1(DSH_{\max} - DSH_{sub})^2 \quad (3)$$

where G_{max} is the maximum potential basal area growth rate over the time interval (5, 10, etc. years), adjusted for a subject tree which is smaller than the dominant tree, G_L is the average basal area growth rate of the large (usually dominant) trees, DSH_{max} is the average diameter at stump height (0.3 m) of the large trees, DSH_{sub} is the diameter of the subject tree, and b_1 is a parameter.

Competition Indices

Distance-independent competition indices are easily obtained in the field with less cost than distance-dependent versions, but have been shown to be nearly as effective in quantifying the growth-competition relationship [22,31,35,36]. There may be some scale-dependence of distance-independent indices due to the interaction of plot size and the underlying spatial pattern [37]; however, for the near-random dispersion patterns found in these forests [38], scale-dependence will be minor. We calculated six simple distance-independent competition indices for each subject tree which have been found to be effective in quantifying the growth-competition relationship [22,34,35]: density [N (stems/ha)], the sum of stem diameter at breast height [SDBH (m/ha)], and stand basal area [BA (m^2/ha)]. These indices impose a series of exponent weights, *k*, on competitor size as shown in Equation (4) ($k = 0$ for N, $k = 1$ for SDBH, $k = 2$ for BA). Competition indices for each subject tree were calculated from plots aggregated within each transect, and were calculated using Equation (4)

$$C_i = \frac{m}{A}\sum_{j=1}^{n_i}[dbh/100]^k \quad (4)$$

where *m* is an adjustment to scale BA correctly (i.e. $m = \pi/4$ when $k = 2$, $m = 1$ otherwise), *A* is the sum of the plot area within each transect in ha, n_i is the number of trees of competitor species group *i* in the plots, *j* is the competitor tree, dbh is in cm and *k* is the size-weighting exponent. We also evaluated the same competition indices considering only trees thicker (GR) than each subject, denoted as NGR, SDBHGR, and BAGR. Indices using thicker trees assume competition is primarily one-sided, i.e. for light. Competition indices were computed for both deciduous and coniferous groups. We initially calculated indices for pine separately from the more shade-tolerant and dense-crowned spruce and occasional fir, but there were too few to obtain significant effects; we therefore redefined a conifer group which included spruce, fir, and pine.

There is a considerable body of literature on competition indices [20,22,39], most of it using empirical evidence (e.g. best RMSE or R^2) to support the choice of a competition metric. In choosing this series of distance-independent indices, we considered the ecophysiological tenet that leaf area is generally proportional to sapwood area, which appears to hold at both a tree and stand scale [40]. We surmised that since the most conductive sapwood is laid down in the most recent growth rings [41], a good starting point would be to assume that the thickness of the sapwood stays roughly constant over time. The basal area of young trees will therefore be nearly entirely sapwood, but the ratio of sapwood basal area to total basal area will decline as the tree ages. If sapwood was an annulus of constant (and relatively small) depth, then sapwood area would scale with stem radius or DBH rather than with basal area. At a stand level then, leaf area should be proportional to the sum of DBH on a per hectare basis (SDBH). We plotted SDBH and BA against leaf area index for a chronosequence of pure aspen stands (the only boreal species for which we had such data; obtained from Lieffers et al. [6]), and found a pattern (Fig. 3) which demonstrates correspondence between SDBH and leaf area. Stand basal area, on the other hand, behaves quite differently, continuing to increase well past the 20-year age at which leaf area peaks. For aspen, we suggest SDBH should therefore be a superior long-term index of competition than basal area; we hypothesize this holds true for other tree species as well. We tested a series of indices, from density (N) through sum of DBH (SDBH) to basal area (BA) (Equation 4) for both aspen and spruce. If one index is more accurate at characterizing competition, it will be reflected in better fit statistics in the overall model (Equation 6).

We aggregated the plot competition data within transects to calculate the competition indices. Because each subject tree was

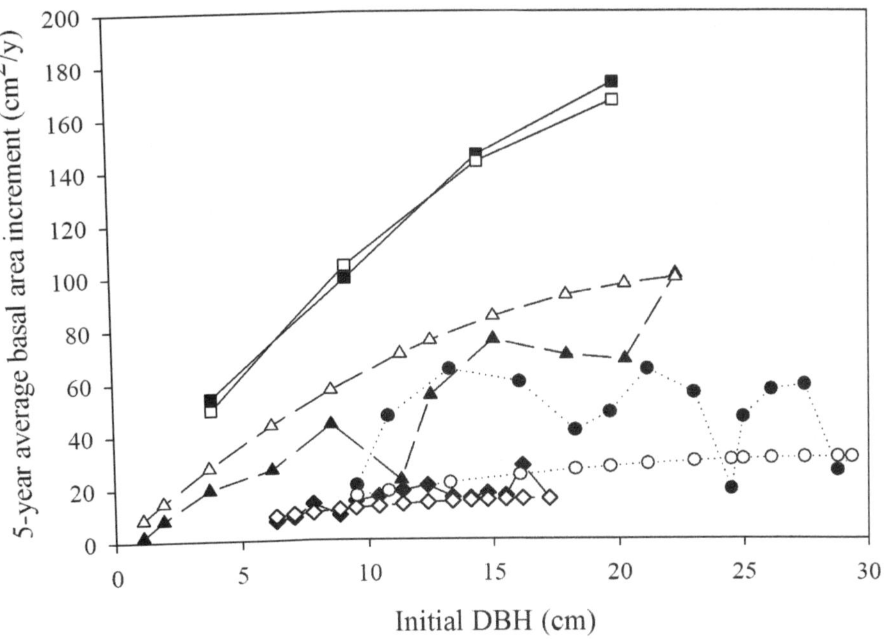

Figure 2. Effect of tree size (DSH) on the maximum basal area growth potential (G_L, cm^2/y). Data are from a random of sample large, usually dominant trees, and were calculated from the full ring chronology of these trees, from pith to bark. Each different symbol is a different tree, and the lines linking the symbols show their growth rates over time. Smooth lines (open symbols) show the modeled maximum growth potential (G_{max}, Equation 3) for each tree.

randomly selected in each transect, we used the small 10 m^2 assessment plots as samples of transect level structure. To accomplish this, we combined all competitors from all subject trees (including all subject trees except the current one), then calculated competition indices at the transect level. For the indices for thicker trees, all competitors from each transect which were thicker than the subject tree were considered. Plot area (Equation 4) was the sum of all small plot areas within the transect. Occasionally, the areas of the plots overlapped, resulting in a lack of independence among plots. However, this is roughly equivalent to sampling with replacement; the stand-level competition index values were still unbiased.

For some transects without dominant spruce nearby, we had to adjust the competition levels to recognize that our maximum spruce growth rate estimates were not competition-free. In most cases (all aspen, many spruce), we could assume the dominants are growing at their maximum potential for the site. This provides the "zero competition index" growth benchmark for the site; for the subordinate trees, this value was then reduced by the competition due to deciduous and conifer neighbours. Where there were no dominant spruce nearby, we recognized this by assessing the competition levels of these largest spruce (C_{x0}), and used these values as an offset to properly locate these in our growth vs. competition model (Equation 5, Fig. 4).

$$C'_x = C_x - C_{x0} \quad (5)$$

Here C'_x is the competition index for species group x (deciduous or coniferous) adjusted for the competition level of the largest spruce, and C_x is the competition index for subject tree. For true dominants without thicker competitors, $C_{x0} = 0$. This adjustment moves the growth vs competition response to the left, while the use of the growth of the largest spruce as the maximum, removes site effects and puts all trees within all sites on the same growth – competition response (Fig. 4). This worked for the competition indices for thicker trees (NGR, SDBHGR, BAGR) since these indices result in a range of competition levels for the trees in each stand, depending on their size relative to their competitors. For the other indices (N, SDBH, BA), all subordinate trees (technically this would include the large sub-dominant spruce) had the same deciduous and the same coniferous stand level competition index. Offsetting these would result in zero competition for all trees. We examined two alternatives: first, that competitors could be ignored for these large trees (consider these true dominants), and second that these few sites (six) were better left out of the data. We found little difference in parameter values or adjusted R^2 among these alternatives, so report results for the N, SDBH and BA indices from a model which considered these large trees as dominants and assumed they had no competition (i.e. $C_{x0} = 0$).

Natural Sub-regions

In our preliminary analyses, we found no significant differences in our growth vs competition parameters (Equation 6) between Central Mixedwood and Dry Mixedwood, and between Lower Foothills and Lower Boreal Highlands forests. We therefore grouped the Central Mixedwood and Dry Mixedwood sites together as a group (CM+DM), and Lower Foothills and Lower Boreal Highlands sites together as another group (LF+LBH). These regional groupings correspond to climate patterns [23,24]. A dummy variable was coded as 0 and 1, respectively for the CM+DM and LF+LBH, for final modeling analysis.

Overall Model

The full model used in our analysis predicts the basal area growth of a subject tree over recent years as a function of 'competition-free' basal area growth from nearby dominant trees, size (*DSH*) of the nearby dominant trees and of the subject tree,

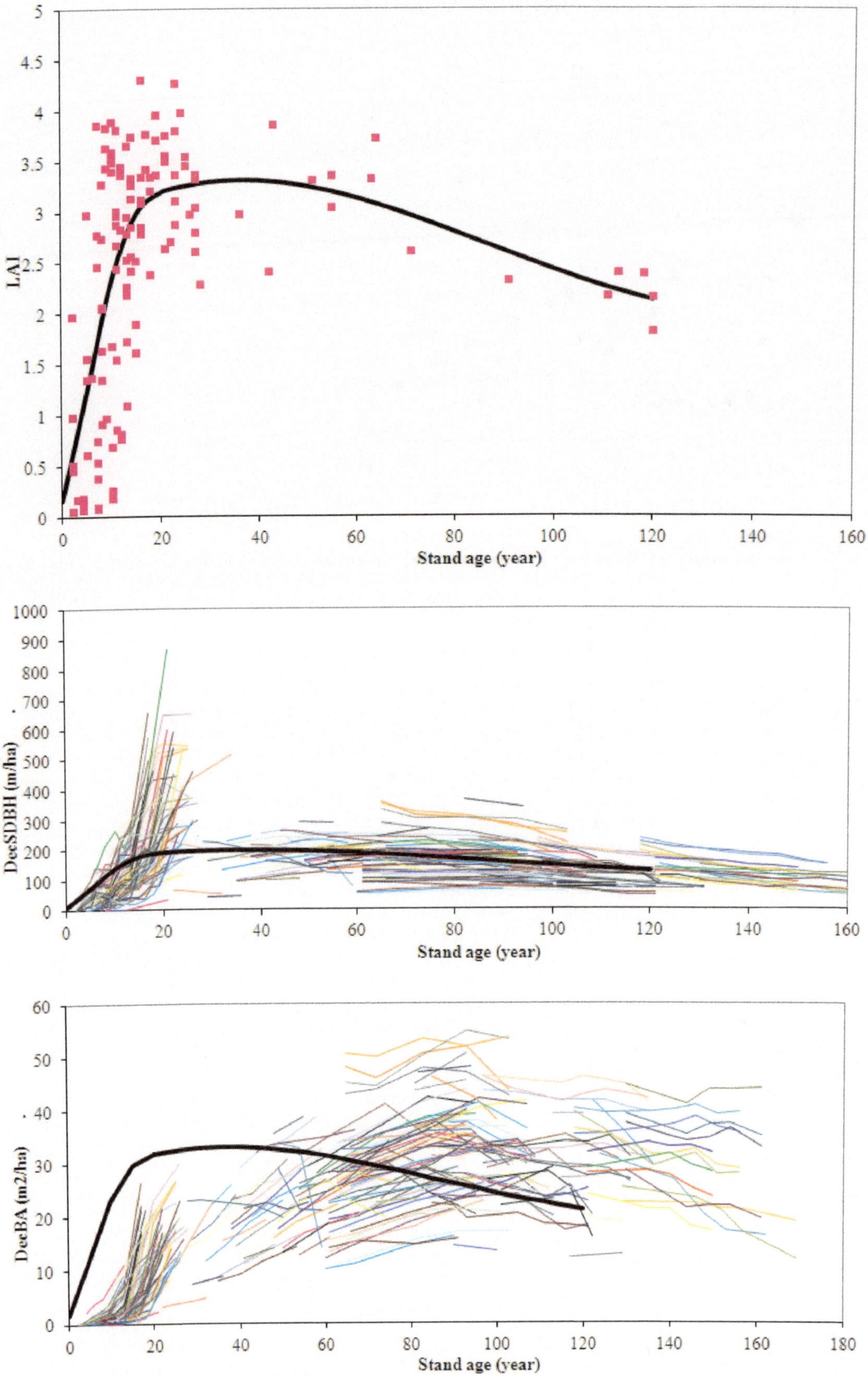

Figure 3. A. Leaf area index plotted against stand age for a chronoseqence of pure aspen stands [5] The dotted line is a natural cubic spline fit to this data. B. Sum of deciduous DBH per hectare (SDBH) for aspen-dominated permanent sample plots (PSPs) plotted against stand age.

Each line represents the SDBH of a PSP through each of its re-measurements. These PSPs are another random sample of the same population (selected only for accessibility). The LAI vs. age trajectory is shown here too as the same dotted line, scaled to approximately fit through the middle of the PSP data. C. Basal area of deciduous trees per hectare plotted against stand age for the same PSPs as in B. The LAI vs. age dotted line is also overlaid on this data with appropriate scaling. Though it may peak more strongly in young stands, SDBH better represents the general trend of LAI vs. time for aspen stands.

deciduous and coniferous competition indices, and natural subregions (Equation 6). Basal area growth of a subject tree is given by

$$G_{m,n} = G_{\max} \times \exp\left[(b_d + b_{dn}NSR)C'_d + (b_s + b_{sn}NSR)C'_s\right] \quad (6)$$

where $G_{m,n}$ is the basal area increment of subject tree n in the mth

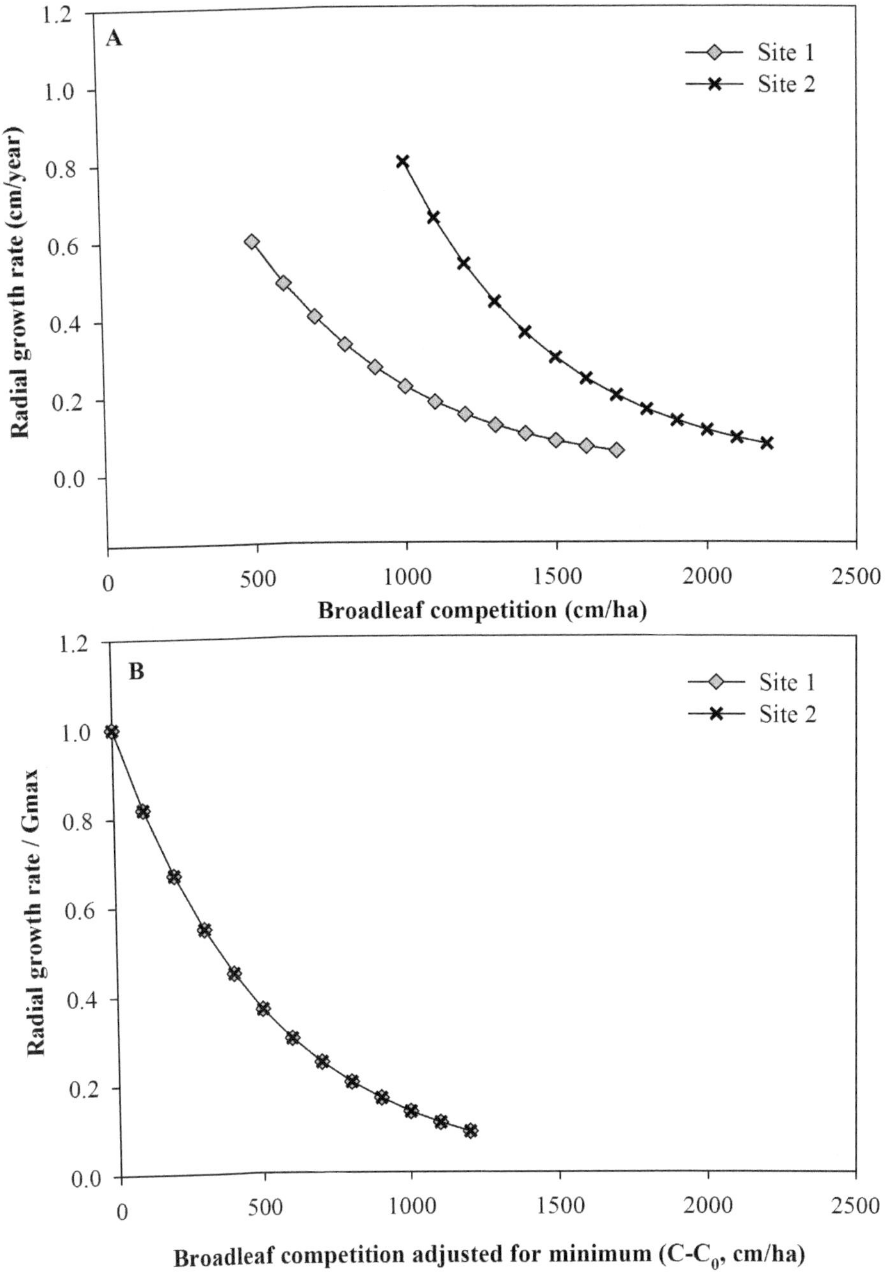

Figure 4. A) Hypothetical growth – competition relationship for two sites. Site 2 has a higher growth potential than site 1 but has no dominant large spruce. B) Adding the maximum growth of dominant trees to the growth response makes the growth response of both sites similar (shifts the growth response to a common maximum of one; and shifting the competition index by subtracting the competition level for the large trees found in the site moves the competition response to the left, bringing all sites onto the same growth – competition response (see Methods).

transect over the last 5, 10, 15, 20, 25, 30, or 35 years, G_{max} is the maximum potential basal area growth rate as above (Equation 3), C'_d is the adjusted deciduous competition index and C'_s is the adjusted coniferous competition index for the transect (Equation 5), *NSR* is a dummy variable (0, 1) for the two natural subregion groups, b_d and b_s are the parameters for deciduous and coniferous competition indices, respectively, and b_{dn} (deciduous) and b_{sn} (conifers) are the parameters corresponding to natural subregion effects that result in an additive shift to b_{dn} and b_{sn}. We fitted two of these models: one for aspen and one for spruce. Note that on Figures 4 and 5, the y-variable is $G_{m,n}/G_{max}$, a re-arrangement of Equation (6) which allows the transects to share a common maximum of one.

Statistical Fitting

Since the data have a nested error structure (multiple subject trees within transects), we used nonlinear mixed modeling to fit Equation (6). Correlation within transects was modeled as a random parameter to capture this restriction on model variance. The minimum Akaike Information Criterion (AIC) [42] was used to choose the best competition index and evaluate the effects of natural subregion. For varying integration periods (from 5 years to 35 years), different dependent variables are used and therefore AIC is not a valid way to compare between models. To compare between models with differing integration periods, we instead used adjusted R^2, a commonly used measure of goodness of fit. All analyses were conducted using the SAS nonlinear mixed modeling macro (%nlinmixed, [43]).

Results

Maximum Growth Model and Diameter at Stump Height: DBH Relationships

To link the maximum potential basal area increment to tree size and develop a link between dominant and subdominant tree growth potential, we developed a relationship between growth rate and size. Figure 2 demonstrates the recent 5 year basal area increment as a function of tree size (DSH) for a selection of large trees. The oscillations in basal area growth reflect the tree's response to longer-term climate fluctuations and other factors; however there was an overall increasing and concave downward trend with increasing size. Site differences were marked, as shown by considerable variation in mean growth rate from site to site. Equation (3) effectively modeled the general increasing trend of

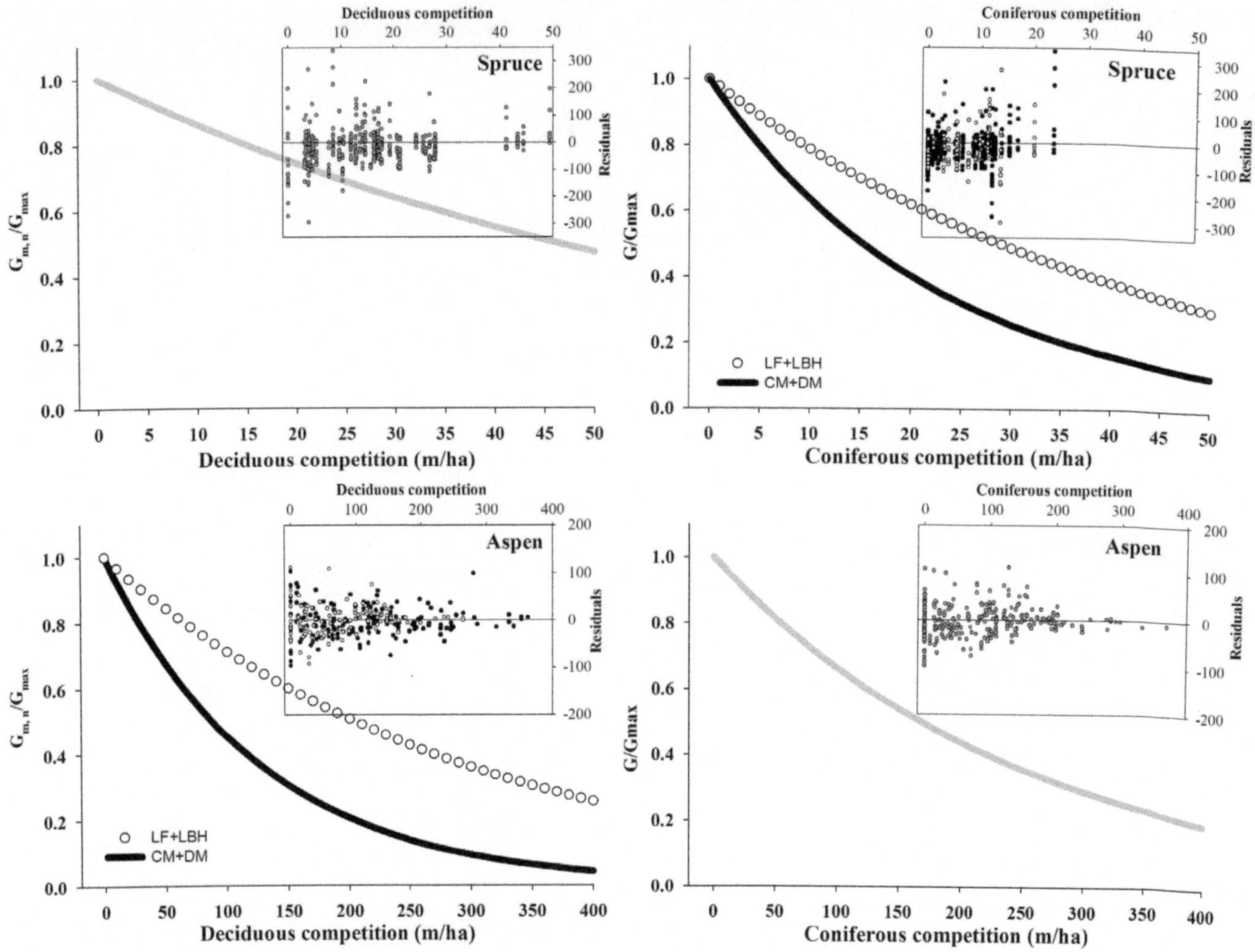

Figure 5. Growth ratio ($G_{m,n}/G_{max}$) vs. competition curves [BAGR for spruce (A, B), and SDBHGR for aspen (C, D)] and the residuals plots from the best nonlinear mixed model for spruce over the most recent 25 years, and for aspen over the recent 15 year (CM + DM: Central Mixedwood and Dry Mixedwood; LF+LBH: Lower Foothills and Lower Boreal Highlands).

maximum growth potential in basal area increment (BAI) vs. size. There was no significant improvement in this model from adding a linear term ($P>0.05$). An exponent of two was also superior to other exponents (1.5 or 2.5). To ensure that the model was biologically reasonable, a refinement was made by fixing the only fitted parameter, b_1, to G_L/DSH_{max}^2, to force the function through zero when diameter of the subject tree (DSH_{sub}) was zero and through the maximum growth when DSH_{sub} was equal to the large tree size (DSH_{max}). This form prevented negative growth in small trees and resulted in very little deterioration of fit. Consequently, this parameter-less version of equation (3) was used for estimating the competition-free growth potential of each subject tree as a function of its size.

We found the relationship between *DSH* and *DBH* for aspen was $DSH = 5.30561 + 1.06508 * DBH$ (Adjusted $R^2 = 0.91$, $P < 0.0001$, $n = 1004$) and that for spruce was $DSH = 0.6099 + 1.18632 * DBH$ (Adjusted $R^2 = 0.98$, $P < 0.0001$, $n = 6913$). These relationships can be used for future modeling when DBH values are required.

Overall Growth-competition Models

When comparing growth-competition models using data integrated over the last 5, 10, 15, 20, 25, 30 and 35 years, we found the models for the past 10 to 25 years of aspen growth had the highest adjusted R^2 values for all competition indices. The 15-year integration period was marginally higher among these, with an adjusted R^2 of 0.667 when using SDBHGR as the predictor (Table 2; integration results with respect to the other indices are not shown). For spruce, we found the models which integrated growth over 25 or more years had the highest adjusted R^2 with a value of 0.743 at 25 years when using BAGR as the predictor (Table 2).

We fixed the integration time at these intervals: the past 15 years for aspen and the past 25 years for spruce, to compare the six

Table 2. Parameters and the statistics of the nonlinear mixed models over the past 5, 10, 15, 20, 25, 30, and 35 years for aspen when using SDBHGR as the predictor, and for spruce when using BAGR as the predictor.

Species	Interval	Parameters ($\times 10^{-2}$)				Adj.R^2
		b_d	b_s	b_{dn}	b_{sn}	
Aspen	5	−0.434	−0.382	−0.343	0	0.573
	10	−0.383	−0.414	−0.435	0	0.665
	15	**−0.341**	**−0.412**	**−0.451**	**0**	**0.667**
	20	−0.323	−0.405	−0.403	0	0.662
	25	−0.300	−0.404	−0.400	0	0.622
	30	−0.299	−0.377	−0.400	0	0.544
	35	−0.263	−0.360	−0.348	0	0.554
Spruce	5	−1.338	−5.511	0	5.442	0.649
	10	−1.530	−5.442	0	4.850	0.682
	15	−1.529	−5.075	0	3.852	0.701
	20	−1.534	−4.751	0	3.013	0.721
	25	**−1.495**	**−4.549**	**0**	**2.160**	**0.743**
	30	−1.680	−4.051	0	0	0.742
	35	−1.824	−4.039	0	0	0.745

Note: See definitions for parameters in the text; Parameters that are insignificant were set to zero. Bold text shows the selected model.

competition indices (Table 3). We found the models for aspen growth consistently perform much better in terms of AIC and adjusted R^2 when NGR and SDBHGR were used vs. the other competition indices. Among the models for spruce growth, those which consistently performed best in terms of AIC and adjusted R^2 used BAGR or SDBHGR as the predictor. Competition indices with all-sized trees considered (BA, SDBH, N) yielded much poorer fits than the indices which included only thicker trees.

When we evaluated asymmetric vs. symmetric competition by testing models which simultaneously incorporated both competition indices for thicker trees and indices for all trees (e.g. BAGR and BA, SDBHGR and SDBH, or NGR and N), we found a significant but very minor improvement in fit over models with only the index in thicker trees. Under likelihood ratio tests, the additional competition index was significant for conspecifics only (i.e. SDBH of deciduous added to the aspen model, BA of conifers improved the spruce model). For aspen BAI, adjusted R^2 increased 0.022 with both SDBH and SDBHGR in the model. For spruce BAI the increase in adjusted R^2 from adding competitor BA to a model with competitor BAGR already in the model, was even less (0.006).

Competition effects on growth differed among natural subregion (NSR) groups. As noted in the Methods, we found the data from the Central and Dry Mixedwood (CM+DM) subregions behaved similarly, as did data from the Lower Foothills and Lower Boreal Highlands (LF+LBH). We accordingly assigned data to two natural subregion groups. Coniferous competitor effects on aspen growth did not significantly change in the two different natural subregion groups ($b_{sn} = 0$); but deciduous effects on aspen growth were different ($b_{dn} \neq 0$) (Tables 2 and 3), Conversely, the effect of conifers on spruce growth was different in the two NSR groups (parameter $b_{sn} \neq 0$), but the effect of deciduous competition on spruce growth did not change with subregion ($b_{dn} = 0$). The effect of these NSR coefficients is that, in the Central and Dry Mixedwoods, the effect of coniferous competitors, $b_s + b_{sn}$, was consistently greater compared to the effect of deciduous competitors, $b_d + b_{dn}$, on the growth of both aspen and spruce, for growth integrated over all but the shortest (5 year) integration period for aspen (Table 2). However, in the Lower Foothills and Lower Boreal Highlands, deciduous competitors had a consistently stronger effect than coniferous competitors on aspen and spruce growth ($b_d + b_{dn} > b_s + b_{sn}$).

As shown in Fig. 5, the effect of intra- and inter-species competition on aspen or spruce relative growth were effectively modeled as a negative exponential, with the residuals evenly distributed around zero, even for some extreme competition values. The spruce growth-coniferous competition curve is steeper in Central Mixedwood and Dry Mixedwood forests than in Lower Foothills and Lower Boreal Highlands forests (Fig. 5). For aspen, the growth response to deciduous competition is steeper in the Central and Dry Mixedwood than the Lower Foothills and Lower Boreal Highlands, whereas the aspen growth response to coniferous competition was not different between these two natural region groups.

Discussion

In western boreal mixedwoods, modeling the long-term relationship between growth and competition has posed a challenge due to the complexity of species interactions, the high variability in growth rates, a lack of data at the critical stem exclusion phase, and the set of competition metrics which have been used. In this study, we developed a nonlinear mixed model to

Table 3. Parameters and the statistics of the nonlinear mixed models for aspen over recent 15 years and for spruce over recent 25 years when using different competition indices including sum of basal area (BA), sum of diameter at breast height (SDBH), and stem density (N), as well as similar indices for trees thicker than each subject (SDBHGR, NGR, and BAGR) as the predictor.

Species	Compet. indices	Parameters ($\times 10^{-2}$)				Adj.R^2	−2 Res log likelihood	AIC
		b_d	b_s	b_{dn}	b_{sn}			
Aspen	BA	−3.737	0	0	0	0.404	3165.3	3169.3
	SDBH	−0.205	−0.064	−0.161	0	0.443	3178.5	3182.5
	N	−0.014	−0.007	−0.014	0	0.411	3206.0	3210.0
	BAGR	−3.709	−1.977	−3.381	0	0.547	2963.2	2967.2
	SDBHGR	**−0.341**	**−0.412**	**−0.451**	**0**	**0.667**	**2920.0**	**2924.0**
	NGR	−0.042	−0.085	−0.078	0	0.743	2905.7	2909.7
Spruce	BA	−0.901	−0.365	0	0	0.636	5815.7	5819.7
	SDBH	−0.066	−0.118	0	0.061	0.649	5828.7	5832.7
	N	0	−0.007	0	0.006	0.632	5831.7	5835.7
	BAGR	**−1.495**	**−4.549**	**0**	**2.160**	**0.743**	**5598.5**	**5602.5**
	SDBHGR	−0.123	−0.761	0	0.395	0.720	5636.5	5640.5
	NGR	−0.012	−0.108	0	0.060	0.691	5689.2	5693.2

Note: See definitions for parameters in the text; Parameters that are not significant were set to zero.

quantify this relationship and tested a series of simple, distance-independent competition indices. Our models demonstrated a remarkable capability to predict stem growth as a function of stem growth of the nearby dominant trees, the size of each tree relative to these dominants, intra- and interspecific competition, and ecoregion. One of the merits of our model is that we were able to model this complex growth vs site, size and competition relation through readily obtainable data, combining dendrochronological (tree-ring) sampling with simple competition indices. The model captured a large degree of the growth variance, as reflected by adjusted coefficient of determination (R^2) values in excess of 0.67. This comprehensive model captures most of the easily described factors driving tree growth: the genetics of the population and site-level environmental factors are captured by sampling nearby dominant trees; this rate is then adjusted for tree size, then reduced by competition on an ecoregion-specific basis. Only tree-level genetic variation and damage from herbivores, pathogens and physical agents (abrasion) are not captured.

Tree-ring analysis was particularly useful for documenting the growth pattern of dominant trees. Using cross-dated basal area increment histories of the largest trees in each plot, we modeled the basal area increment as a function of initial size. As many others [16,31] have observed, the increment of dominant trees initially increased rapidly with tree size, presumably as the trees built their crown and leaf area, then the increment increased less quickly presumably as their leaf area approached a maximum and the maintenance cost of the structural and conductive tissues increased at a greater rate than the photosynthetic supply (Fig. 2). The pattern of BAI vs. initial DSH did not exhibit the bell-shaped pattern Canham et al. [31] observed for radial increment in a sample including much larger cold-temperate trees or Huang and Titus [16] found with older boreal spruce. It is likely that BAI will also decline with size in large boreal trees; we cannot speculate since our sampling frame was young to near-mature stands. Very large trees are rare in the fire-prone boreal region.

The BAI response of dominant trees to initial size was modeled effectively by a simple function of initial DSH^2 with fixed points at the origin and the current BAI at the most recent initial DSH (Fig. 2). Although we tested several functions with fitted parameters, this parameter-free function described the growth response of a leading dominant tree from a juvenile to a near-mature tree, with similar effectiveness. This function represents the maximum growth potential of each tree on that site over time, given the local climate in these years, physiography, soil and population-level genetics. Given that tree size may provide a surrogate measurement of tree leaf area and indirectly reflects the amount of light being absorbed and used by the tree [44], previous studies recommended including initial size of the subject tree as an important explainable variable for better explaining growth-competition relation [45,46]. Filipescu and Comeau [35] also found a significant improvement (33%) in predictive ability of the model when initial size of the subject tree was introduced. Our model of dominant tree growth adjusted for the difference in size of the subject and dominant trees, accounted for the majority of the modeled variation in the final model, i.e., G_{max}, determined by equation (3), was strongly correlated with the observed growth (BAI) of our aspen and of white spruce (Pearson's $r>0.75$), accounting for 82–85% of the total variation (R^2) captured in the final model (Equation 6).

It is generally assumed that height growth is a better indicator of site productivity than diameter or basal area growth, since the former is less sensitive to competition [47]. However, our work indicated that a minor amount of additional variation in dominant tree diameter growth is accounted for by stand density or basal area when fit in addition to competition indices which only consider thicker trees. This result implies that the diameter of leading trees in mid-aged aspen – spruce mixedwoods has little response to overall stand competition. This runs counter to accepted wisdom; however most studies of competition effects on height vs diameter growth have reported on the strong response of diameter (compared to height) for the average tree or with decreasing social status, not simply the response of the dominant trees. For example, the basal area increment function of Yang et al. [17] for aspen includes a competition term; however, for dominant trees (achieved by setting the social status indicators to the maximum), the growth response to a 30% shift in stand basal area is <10% change in basal area growth. Huang and Titus' [16] spruce diameter growth function is similarly affected by a 30% change in stem density (~10% change in growth), but nearly unaffected (<3% response) by a 30% stand basal area change. The longer crowns maintained by the dominant trees (especially spruce) likely maintain diameter growth in these primarily natural stands.

Many forest growth models are driven by site index (SI). Height vs. age and site index (the height of the dominant trees at a reference age) curves provide a simple method to model potential height growth, which is then reduced for sub-dominant trees in some individual tree models [20,48]. The link between SI and diameter growth is less obvious; a height vs. diameter curve or allocation function [16,49] may be used to translate height increment into a diameter increment. We did not make an explicit link to SI in this study for several reasons. SI could be obtained for leading aspen; however, current dominant white spruce were likely not dominant during much of their early growth, rendering the apparent spruce SI of questionable value as an index of spruce potential [50]. Site potential is also demonstrated in diameter growth of dominants, particularly since we found little evidence of diameter suppression in dominant trees due to stand density. By linking sub-dominant growth to leading dominant growth, our model captures the site effect without relying on SI. The competition effects which our model estimates can still be used in a SI-based model, by using Equations (4) and (6) to reduce the modeled diameter growth of the leading dominants [48].

A problem encountered in our region is that white spruce trees are usually overtopped by deciduous hardwoods early in stand development [1,51], except in recently-established (since the 1980's), heavily tended stands. We sought dominant spruce near our sites, but it is very likely these were overtopped earlier in their development. Thus, we cannot fully estimate the growth potential at all sizes and ages. However, when we did not find dominant spruce, we offset our white spruce response by the competition experienced by the largest white spruce sampled in or near our stands, as shown in Fig. 2. This adjustment provided mathematical consistency in the spruce growth response to competition across site differences and despite the lack of true spruce dominants on six sites. This adjustment assumes the trees respond to competition similarly and proportionally to the maximum potential (G_{max}) across all stands (though we did allow for differences in competition response between ecoregions). Since it is almost universal in mixed forests that the potential competition-free growth of the slower-growing species is unknown, but that competition levels and growth rates can be measured, the reconciliation via Equation (5), and illustrated in Fig. 2, is a significant contribution to modeling competition in mixed forests.

Development, discussion and comparisons of indices for competition have been the subject of considerable work [22]. Burton [39] noted a failure of competition indices is that they are generally static snapshots for one stand type at one age. Since we intend this work to support a long-term forest growth model for pure and mixed stands [48], it was important to find an index that could capture the change in competitive dynamics with stand age. We argue in our Methods, that an index summing diameter (SDBH and SDBHGR) should be more effective than one which sums DBH^2, such as the more commonly used basal area (BA or BAGR). We note that SDBH approximates a chronosequence of aspen stand leaf area much more effectively than basal area (Fig. 3). As an index for the 20–80 year-old stands sampled in this study, SDBH and SDBHGR were both more effective (in terms of improved adjusted R^2 and reduced AIC) than BA and BAGR in a growth vs. competition model for aspen. For spruce, SDBHGR was also very effective, though slightly inferior to BAGR. Aspen leaf area (and SDBH) have peaked and are declining in the age range we sampled, while these spruce variables have not, which may be why SDBH and BA or SDBHGR and BAGR show similar effectiveness for spruce. The smaller size of the spruce relative to the aspen may also be a factor.

Interestingly, the density of thicker trees (NGR) was a superior index for aspen growth. Newsome et al. [52] also reported the effectiveness of NGR in pine - aspen stands in British Columbia. However, further analysis [53] has indicated that NGR was less effective for describing variation in height and diameter growth of lodgepole pine than other competition indices which incorporate competitor diameter. During the stem exclusion and understory reinitiation stages in these aspen dominated stands, size and density are strongly coupled through self-thinning dynamics. Therefore, it may not be necessary to include both in a competition index applied to these natural stands. However, there are many silvicultural treatments being applied in the western boreal which strongly alter aspen density, and would lead to uncoupling of the size – density relationship. Since this model will be applied to such stands, we feel SDBHGR or BAGR are more robust indices, but nonetheless report all models (Table 3).

In terms of species' competitive effects, our modeling results showed that spruce competition had a greater effect than deciduous competition on aspen and white spruce growth in Central and Dry Mixedwood ecoregions, as indicated by b_s being more negative than b_d. This is in agreement with Stadt et al. [22] who compared the competitive effect of the five dominant boreal species using PSP data and found white spruce caused larger growth reductions than trembling aspen competition. However, this relation was reversed in the Lower Foothills and Lower Boreal Highlands, with aspen having a stronger competitive effect than spruce. This may be a consequence of the lack of large spruce in our stands in these ecoregions; we may not have obtained sites with enough large spruce in these natural stands to show any strong effect on aspen growth. However, regional differences are common. Wright et al. [10] and Filipescu and Comeau [33], for example, found different growth vs. light competition relationships among regions. This appears to be a complex interaction: regional differences are primarily climatic, and climate effects on spruce growth can be exacerbated by deciduous competition [54].

We acknowledge that while our growth rate sample is retrospective, our competition sample is current. Metasaranta and Lieffers [15] reconstructed past competition levels in small pure pine plots by dendrochronological analysis on all trees (live and dead) within a plot. We were unable to achieve this in our plots due to the rapid decomposition of the broadleaf species, preventing cross-dating of their past size and death year.

Another important finding is that current estimates of competition can be linked to long-term growth, i.e. 10–25 years for aspen and >25 years for spruce. Commonly, growth is determined for a short period, related to the PSP measurement interval or a few recent ring widths [10,15,33,35]. Our study demonstrates that current competition is more strongly related to stem growth over the past 10 to 25 years vs. shorter periods. Our use of stable competition indices may be part of the reason. Longer periods may smooth out noise in the growth of both subject trees and competitors, especially if the competitive position of subdominant trees remains the same throughout this time.

Conclusion

We used tree ring analysis as a substitute for permanent sample plots to obtain data for quantifying the effect of inter- and intra-species competition on stem growth of the two predominant species, white spruce and trembling aspen in the boreal mixed-wood forest of western Canada. Through a nonlinear mixed modeling approach, we established a powerful model to predict basal area growth as the function of the growth of dominant trees, the difference in size between the nearby dominant trees and of subject trees, deciduous and coniferous competition levels, and ecoregions. We found the competition index using the sum of diameter at breast height of thicker trees is a reasonable and ecologically meaningful predictor for predicting aspen and white spruce growth, though other indices also fit well.

Acknowledgments

We thank our research assistants A. Bill, E. Elliot, Y. Jin, N. Luchanski, H.A. Yan, and L.H. Zhai for help with the field and lab work.

Author Contributions

Conceived and designed the experiments: JGH KS PC. Performed the experiments: JGH AD KS. Analyzed the data: JGH KS. Contributed reagents/materials/analysis tools: JGH KS AD PC. Wrote the paper: JGH KS AD PC.

References

1. Lieffers VJ, Stadt KJ, Navratil S (1996) Age structure and growth of understory white spruce under aspen. Can J For Res 26: 1002−1007.
2. PetersVS, Macdonald SE, Dale MRT (2005) The interaction between masting and fire is key to white spruce regeneration. Ecology 86: 1744−1750.
3. MacDonald GB (1995) The case for boreal mixedwood management: an Ontario perspective. For Chron 71: 725−734.
4. Chen HYH, Popadiouk V (2002) Dynamics of North American boreal mixedwoods. Environ Rev 10: 137−166.
5. Peterson EB, Peterson NM (1992) Ecology, management, and use of aspen and balsam poplar in the Prairie Provinces, Canada. For Can North For Cent Spec Rep. 1.
6. Lieffers VJ, Pinno BD, Stadt KJ (2002) Light dynamics and free-to-grow standards in aspen-dominated mixedwood forests. For Chron 78: 137−145.
7. Comeau PG, Wang JR, Letchford T (2003) Influences of paper birch competition on growth of understory white spruce and subalpine fir following spacing. Can J For Res 33: 1962−1973.
8. Bokalo M, Comeau PG, Titus SJ (2007) Early development of tended mixtures of aspen and spruce in western Canadian boreal forests. For Ecol Manag 242: 175–184 doi:10.1016/j.foreco.2007.01.038.
9. Lieffers VJ, Stadt KJ (1994) Growth of understory *Picea glauca*, *Calamagrostis canadensis*, and *Epilobium angustifolium* in relation to overstory light transmission. Can J For Res 24: 1193−1198.
10. Wright EF, Dave Coates K, Canham CD, Bartemucci P (1998) Species variability in growth response to light across climatic regions in northwestern British Columbia. Can J For Res 28: 871−886.
11. Comeau PG, Gendron F, Letchford T (1998) A comparison of several methods for estimating light under paper birch mixedwood stands. Can J For Res 28: 1842−1850.
12. Man R, Lieffers VJ (1999) Are mixtures of aspen and white spruce more productive than single species stands? For Chron 75: 505−513.
13. MacPherson DM, Lieffers VJ, Blenis PV (2001) Productivity of aspen stands with and without a spruce understory in Alberta's boreal mixedwood forests. For Chron 77: 351−356.
14. Vanclay JK (1994) Modelling forest growth and yield. Applications to mixed tropical forests. CAB International, Wallingford UK, ISBN 0851989136.
15. Metsaranta JM, Lieffers V (2009) Using dendrochronology to obtain annual data for modeling stand development: a supplement to permanent sample plots. Forestry 00: 163–173. doi:10.1093/forestry/cpn051.
16. Huang S, Titus SJ (1995) An individual tree diameter increment model for white spruce in Alberta. Can J For Res 25: 1455−1465.
17. Yang Y, Huang S, Meng SX, Trincado G, VanderSchaaf CL (2009). A multilevel individual tree basal area increment model for aspen in boreal mixedwood stands. Can J For Res 39: 2203−2214.
18. Fritts HC (1976) Tree Rings and Climate. Academic Press, New York, NY, USA.
19. Biondi F (1999) Comparing tree-ring chronologies and repeated timber inventories as forest monitoring tools. Ecol Appl 9: 216−227.
20. Biging GS, Dobbertin M (1995) Evaluation of competition indices in individual tree growth models. For Sci 41: 360–377.
21. Cumming SG, Schmiegelow FKA, Burton PJ (2000) Gap dynamics in boreal aspen stands: is the forest older than we think? Ecol Appl 10: 744−759.
22. Stadt KJ, Huston C, Coates KD, Feng ZL, Dale MRT, et al. (2007) Evaluation of competition and light estimation indices for predicting diameter growth in mature boreal mixed forests. Ann For Sci 64: 477−490.
23. Beckingham JD, Corns IGW, Archibald JH (1996) Field guide to ecosites of west-central Alberta. Natural Resource of Canada, Canadian Forest Service, Northwest Reg., North. Forest Centre, Edmonton, Alberta. Special Report 9.
24. Beckingham JD, Archibald J H (1996) Field guide to ecosites of northern Alberta. Natural Resource of Canada, Canadian Forest Service, Northwest Reg., North Forest Centre, Edmonton, Alberta. Special Report. 5.
25. Environment Canada (2012) Canadian climate normal or averages 1971−2000. Canadian Climate Program. Environment Canada. Atmospheric Environment Service, Downsview, Ontario, Canada. Available from: http://www.climate.weatheroffice.gc.ca/climate_normals/index_e.html [access on April 30, 2012].
26. AESRD (2012) Phase 3 Forest Inventory. Alberta Environment and Sustainable Resource Development, Available from: http://srd.alberta.ca/MapsPhotosPublications/Maps/ResourceDataProductCatalogue/ForestVegetationInventories.aspx [Access on July 8, 2013].
27. ASRD (2006) Alberta Forest Management Planning Standard. Alberta Sustainable Resource Development Public Lands and Forests Division Forest Management Branch version 4.2.
28. Holmes RL (1983) Computer-assisted quality control in tree-ring dating and measurement. Tree-Ring Bull 43: 69−78.
29. Gutsell SL, Johnson EA (2002) Accurately ageing trees and examining their height-growth rates: implications for interpreting forest dynamics. J Ecol 90: 153−166.
30. Huang S (1994) Ecologically-based individual tree volume estimation for major Alberta tree species. Report # 1. Individual Tree Volume Estimation Procedures for Alberta: Methods of Formulation and Statistical Foundations. Alberta Environmental Protection. Edmonton.
31. Canham CD, LePage PT, Coates KD (2004) A neighborhood analysis of canopy tree competition: effects of shading versus crowding. Can J For Res 34: 778−787.
32. Friend AD (1993) The prediction and physiological significance of tree height. In: Schugart HH, Solomon AM (eds), Vegetation dynamics modeling and global change. Chapman Hall, London, UK.
33. Filipescu C, Comeau PG (2007a) Aspen competition affects light and white spruce growth across several boreal sites in western Canada. Can J For Res 37: 1701−1713.
34. Lorimer CG (1983) Tests of age-independent competition indices for individual trees in natural hardwood stands. For Ecol Manag 6: 343−360.
35. Filipescu C, Comeau PG (2007b) Competitive interactions between aspen and white spruce vary with stand age in boreal mixedwoods. For Ecol Manag 247: 175−184.
36. Wimberly M, Bare BB (1996) Distance-dependent and distance-independent models of Douglas-fir and western hemlock basal area growth following silvicultural treatment. For Ecol Manag 89: 1−11.
37. Pukkala T (1989) Predicting diameter growth in even-aged Scots pine stands with a spatial and non-spatial model. Silva Fennica 23: 101−116.
38. Feng Z, Stadt KJ, Lieffers VJ (2006) Linking juvenile white spruce density, dispersion, stocking, and mortality to future yield. Can J For Res 36: 3173−3182.
39. Burton PJ (1993) Some limitations inherent to static indices of plant competition. Can J For Res 23(10): 2141−2152.
40. Waring RH, Schroeder PE, Oren R (1982) Application of the pipe model theory to predict canopy leaf area. Can J For Res 12: 556−560.
41. Reid DEB, Silins U, Lieffers VJ (2003) Stem sapwood permeability in relation to crown dominance and site quality in self-thinning fire-origin lodgepole pine stands. Tree Physi 23 (12): 833–840. doi: 10.1093/treephys/23.12.833.
42. Akaike H (1974) A new look at the statistical model identification. IEEE Transactions on Automatic Control 19: 716−723.
43. Littell RC, Milliken GA, Stroup WW, Wolfinger ED, Schabenberger O (2006) SAS for Mixed Models, SecondEdtion. SAS Institute Inc. Cary, NC, U.S.A.
44. Pritchard JM, Comeau PG (2004) Effects of opening size and stand characteristics on light transmittance and temperature under young trembling aspen stands. For Ecol Manag 200(1-3): 119−128, doi 10.1016/j.foreco.2004.06.002.
45. Morris DM, MacDonald GB (1991) Development of a competition index for young conifer plantations established on boreal mixedwood sites. For Chron 67: 403−410.
46. MacFarlane DW, Kobe RK (2006) Selecting models for capturing tree-size effects on growth-resource relationships. Can J For Res 36: 1695−1704.
47. Lanner BM (1985) On the insensitivity of height growth to spacing. For Ecol Manag 13: 143−148.
48. Bokalo M, Stadt KJ, Comeau PG, Titus SJ (2013) The validation of the Mixedwood Growth Model (MGM) for use in forest management decision making. Forests 4: 1–27.
49. Nunifu T (2009) Compatible diameter and height increment models for lodgepole pine, trembling aspen, and white spruce. Can J For Res 39: 180–192, 10.1139/X08-168.

50. Osika DEB, Stadt KJ, Comeau PJ, MacIsaac DM (2013) Sixty-year effects of. deciduous removal on white spruce height growth and site index in the Western Boreal. Can J For Res 43: 139–148, 10.1139/cjfr-2012-0169.
51. Comeau PG, Harper GJ (2009) Effects of vegetation control treatments for release of Engelmann spruce from a mixed-shrub community in southern British Columbia - year 15 results. For Chron 85: 583−592.
52. Newsome TA, Heineman JL, Nemec AFL (2006) Early effects of manipulating aspen density and spatial arrangement on lodgepole pine performance, aspen sucker production, and stand development in an 11-year-old stand in the SBPS subzone of south-central British Columbia. B.C. Min. For., Res. Br.,Victoria, B.C. Tech. Rep. 029.
53. Newsome TA, Heineman JL, Nemec AFL (2010) A comparison of lodgepole pine responses to varying levels of trembling aspen removal in two dry south-central British Columbia ecosystems. For Ecol Manag 259: 1170−1180.
54. Cortini F, Comeau PG, Bokalo M (2012) Trembling aspen competition and climate effects on white spruce growth in boreal mixtures of western Canada. For Ecol Manag 277: 67−73.

15

Ecological Structure of Recent and Last Glacial Mammalian Faunas in Northern Eurasia: The Case of Altai-Sayan Refugium

Věra Pavelková Řičánková*, Jan Robovský, Jan Riegert

Department of Zoology, Faculty of Science, University of South Bohemia, České Budějovice, Czech Republic

Abstract

Pleistocene mammalian communities display unique features which differ from present-day faunas. The paleocommunities were characterized by the extraordinarily large body size of herbivores and predators and by their unique structure consisting of species now inhabiting geographically and ecologically distinct natural zones. These features were probably the result of the unique environmental conditions of ice age ecosystems. To analyze the ecological structure of Last Glacial and Recent mammal communities we classified the species into biome and trophic-size categories, using Principal Component analysis. We found a marked similarity in ecological structure between Recent eastern Altai-Sayan mammalian assemblages and comparable Pleistocene faunas. The composition of Last Glacial and Recent eastern Altai-Sayan assemblages were characterized by the occurrence of large herbivore and predator species associated with steppe, desert and alpine biomes. These three modern biomes harbor most of the surviving Pleistocene mammals. None of the analyzed Palearctic Last Glacial faunas showed affinity to the temperate forest, taiga, or tundra biome. The Eastern part of the Altai-Sayan region could be considered a refugium of the Last Glacial-like mammalian assemblages. Glacial fauna seems to persist up to present in those areas where the forest belt does not separate alpine vegetation from the steppes and deserts.

Editor: Michael Hofreiter, University of York, United Kingdom

Funding: The study was supported by Czech Ministry of Education (MSM 6007665801) http://www.msmt.cz/index.php?lang=2 and Czech Science Foundation (# P504/11/0454) http://www.gacr.cz/en/. The funders had no role in study design, data collection and analysis, decision to publish, or preparation of the manuscript.

Competing Interests: The authors have declared that no competing interests exist.

* E-mail: ricankova@seznam.cz

Introduction

The unique structure of Pleistocene mammalian communities has drawn the attention of scientists for many decades [1–7]. The extraordinary large body size of some Pleistocene mammals (e.g., mammoths, giant deer, or cave bear) and composition of the ice age communities have no analogies in the present-day faunas [8–9]. The Pleistocene communities consisted of species which now inhabit geographically and ecologically distinct natural zones (tundra, forest, steppe, savanna) [5,8]. Arctic reindeer and musk-ox thus lived in sympatry with e.g. steppe horse and bison and/or with forest elk and roe deer [4,10].

The reason for such a peculiar structure of Last Glacial (corresponding with Weichselian Glaciation) mammalian assemblages is probably associable with the unique environmental conditions of ice age ecosystems [8,10,11]. The non-taxonomic, ecological structure of a mammalian community (~ its guild composition according to e.g. Simberloff & Dayan [12]) is determined mainly by environmental factors such as climate, type of biome or vegetation heterogeneity [13–15]. Each type of ecosystem (e.g. woodland or arid ones) is characterized by a specific trophic-size structure of its mammalian community [5,13,16]. Mammalian communities from areas with similar regional climates tend to converge to similar community structures [15]. Historical factors also play an important role in forming community structure, but they operate on a different, probably much longer timescale than environmental factors [14–15].

The non-analogue Last Glacial communities evolved in a cold and dry continental climate, which supported highly heterogeneous vegetation and landscape structure, usually described as tundra-steppe or mammoth steppe [4,8,10,17].

The Pleistocene tundra-steppe ecosystem was quite heterogeneous locally but displayed a relatively high degree of homogeneity on the continental scale. This ecosystem covered wide areas of the northern part of the globe, thrived for approximately 100,000 years without major changes, and then suddenly went extinct about 12,000 years ago [4,10].

Surprisingly, environmental conditions similar to the Last Glacial period have been found in the Central Eurasian Altai-Sayan mountains [18]. This climatic analogue has recently been supported by biological data. Recent findings of the paleo-biome reconstruction [19–20] and pollen-analytical research [21–23] suggest that present-day Altai-Sayan landscapes could be considered the closest modern analogy to the Last Glacial environments. The area is currently inhabited by mollusc assemblages that were characteristic of full-glacial environments across large areas in Eurasia but went extinct in the regions that experienced considerable climatic change, namely in Europe [24]. Simulated paleovegetation maps based on paleoclimatic models and plant functional types have also suggested considerable stability in

Central Eurasia over the last 40,000 years [25,26]. Detailed analysis of the Altai Late Pleistocene assemblages of small mammals revealed that no significant changes occurred between the cold phase of the Pleistocene and the Holocene [27]. The environment of this region can thus be considered as conservative and stable.

In this study, we compared the ecological structure of Recent Altai-Sayan mammalian assemblages to the ecological structure of Last Glacial fauna of Altai-Sayan and several adjacent regions as well as to Recent mammalian communities from various natural zones of northern Eurasia. In order to examine the most important structural characteristics of Pleistocene assemblages, we assigned individual species according to their biome and the trophic-size categories. Given the analysis of Willis et al. [28] and Rodrigues et al.[7,29] showing that glacial vegetation consisted of steppe, tundra and forest and mammalian fauna was characterized by large herbivores and predators, we would expect that glacial communities are characterized (and differ from Recent communities) by the co-occurrence of steppe, tundra, and forest species [28], and by high a proportion of large herbivores and predators [29]. We hypothesize that, given the environmental stability of central Eurasia [18], Altai-Sayan Recent assemblages will be more similar to the glacial communities than to any other Recent community [22–23].

Materials and Methods

Regions and localities

To compare the ecological compositions of Recent and Last Glacial faunas, lists of mammalian species for 14 Recent and seven Last Glacial localities were collected (Fig. 1, see Table S1). The areas were selected in order to cover most of the Palearctic Realm above 35° N, to include well documented Last Glacial localities, and to be compatible with the WWF eco-regions (e.g., Altai-Sayan, Caucasus, Carpathian Mountains). Taking into account (i) the scarcity of paleontological localities, (ii) the general incompleteness of the fossil record, (iii) and relative homogeneity of the glacial fauna, larger regions were used as units for the analysis of Last Glacial faunas. This grouping of data, therefore, helps to average taphonomic biases [29]. The experimental mixture of modern communities tends to increase the taxonomic richness but does not significantly modify the overall ecological diversity [30].

For the analysis of Recent mammalian faunas we used smaller, ecologically homogeneous areas (see Table S1). The Recent fauna of Altai-Sayan region has been assigned to 12 areas, covering most of the region's heterogeneity (Fig. 2; Table S2).

Species

The Recent mammalian communities refer to the interval from now to approximately the 16th century AD in order to respect the IUCN definition of "recent extinction" [31] and to eliminate taxonomical uncertainties. The Last Glacial refers here to the last glacial period of the Late Pleistocene, corresponding with the Weichselian Glaciation. The Last Glacial communities were dated from approximately 125,000 to 12,000 BP, i.e. they included the time interval from MIS 5d to MIS 2 [32] which is well defined in the geological/fossil record [33,34]. The early-Holocene fauna was not included in our analyses.

Presence/absence of 379 mammalian species in Recent and Last Glacial regions were recorded.. The Recent species include extant (or extinct in the historical period) autochthonous elements but not allochthonous elements associated with human unintended or intentional activities (see Notes S1). The marine mammals (Cetacea, Pinnipedia) and island endemics (though geographically associated with the analyzed regions) were excluded from analyses. Bats (Chiroptera) have been removed from the list of examined species for trophic-size categories, following Rodríguez [7]. Domesticated mammals were not considered.

Some Recent species are not diagnosable in the fossil record owing to the lack of diagnostic morphological, cytogenetic, and/or molecular characters. We therefore fused some closely related mammalian species into single operational taxonomic units (for details see Notes S1). No permits were required for the described study, which complied with all relevant regulations.

Biome and trophic-size structure of the faunas

The ecological structure of a community is defined as the number of species in different ecological groups or categories, i.e. species with similar trophic habits, body size and ecological requirements [7]. The species were classified according to two schemes:

(i) The species' affinity to a particular biome (a group of terrestrial ecosystems with similar climates and vegetation structure): we recognized (1) tundra, (2) taiga, (3) deciduous forest, (4) steppe, (5) alpine grassland, and (6) desert species, following Duff & Lawson [35] and Wilson & Reeder [36].

(ii) Trophic and body-size categories combining information about trophic habits, locomotor abilities, microhabitat, and body size of the species examined cf. [7]): (1) aquatic predator (e.g. Eurasian otter *Lutra lutra*); (2) small terrestrial predator (e.g. red fox *Vulpes vulpes*); (3) large terrestrial predator (e.g. wolf *Canis lupus*); (6) aquatic predator of invertebrates (e.g. Russian desman *Desmana moschata*); (7) subterranean predator of invertebrates (e.g. Siberian mole *Talpa altaica*); (9) small terrestrial predator of invertebrates (e.g. Eurasian shrew *Sorex araneus*); (11) small terrestrial omnivore (e.g. Altai birch mouse *Sicista napaea*); (12) large terrestrial omnivore (e.g. brown bear *Ursus arctos*); (13) arboreal omnivore (e.g. forest dormouse *Dryomys nitedula*); (14) small terrestrial herbivore (e.g. common vole *Microtus arvalis*); (15) small sized foregut fermenter (e.g. goitered gazelle *Gazella subgutturosa*); (16) medium sized foregut fermenter (e.g. reindeer *Rangifer tarandus*); (17) large sized foregut fermenter (e.g. red deer *Cervus elaphus*); (18) small sized hindgut fermenter (e.g. mountain hare *Lepus timidus*); (20) large sized hindgut fermenter (e.g. Asiatic wild ass *Equus hemionus*); (21) subterranean herbivore (e.g. Siberian zokor *Myospalax myospalax*); (22) arboreal herbivore (e.g. Eurasian red squirrel *Sciurus vulgaris*); (23) aquatic herbivore (e.g. European water vole *Arvicola amphibius*).

Data analyses

To visualize the overall similarity of Recent and Last Glacial areas according to presence/absence of the mammalian species, we used all 21 regions/localities as "samples" and ecological categories as "species" for Principal Component Analysis (PCA). We also performed analyses complementary to PCA using non-metric multidimensional scaling (NMDS) with Bray-Curtis dissimilarity indices (CANOCO for Windows software [37]). We expressed the proportion of species associated with each biome and trophic-size category as percentages in the input data matrix (see Table S3). The percentages were log-transformed and standardized by species. The sensitivity of PCA was controlled by analyzing datasets after removing rare categories (biomes: tundra and desert, trophic-size: aquatic predator, aquatic predator of invertebrates, subterranean predator of invertebrates, arboreal omnivore, small-sized foregut fermenter, subterranean herbivore).

Figure 1. Location map of Recent and Pleistocene Palearctic localities used in the analyses. Abbreviations: B. R. - Biosphere Reserve; P. A. - Protected Area; Mts. - mountains

Results

(1) Faunas classified according to species-biome associations

Composition of Last Glacial faunas was characterized by the co-occurrence of steppe, desert, alpine, and tundra species (Fig. 3). The first axis of the ordination space was determined by the steppe/desert-to-taiga gradient, whereas presence/absence of the alpine faunas was strongly correlated with the second axis. The first two axes explain 89.9% of variance (see Table S4 for detailed results and comparison with NMDS analyses). The Last Glacial faunas were scattered in the area between the steppe/desert and

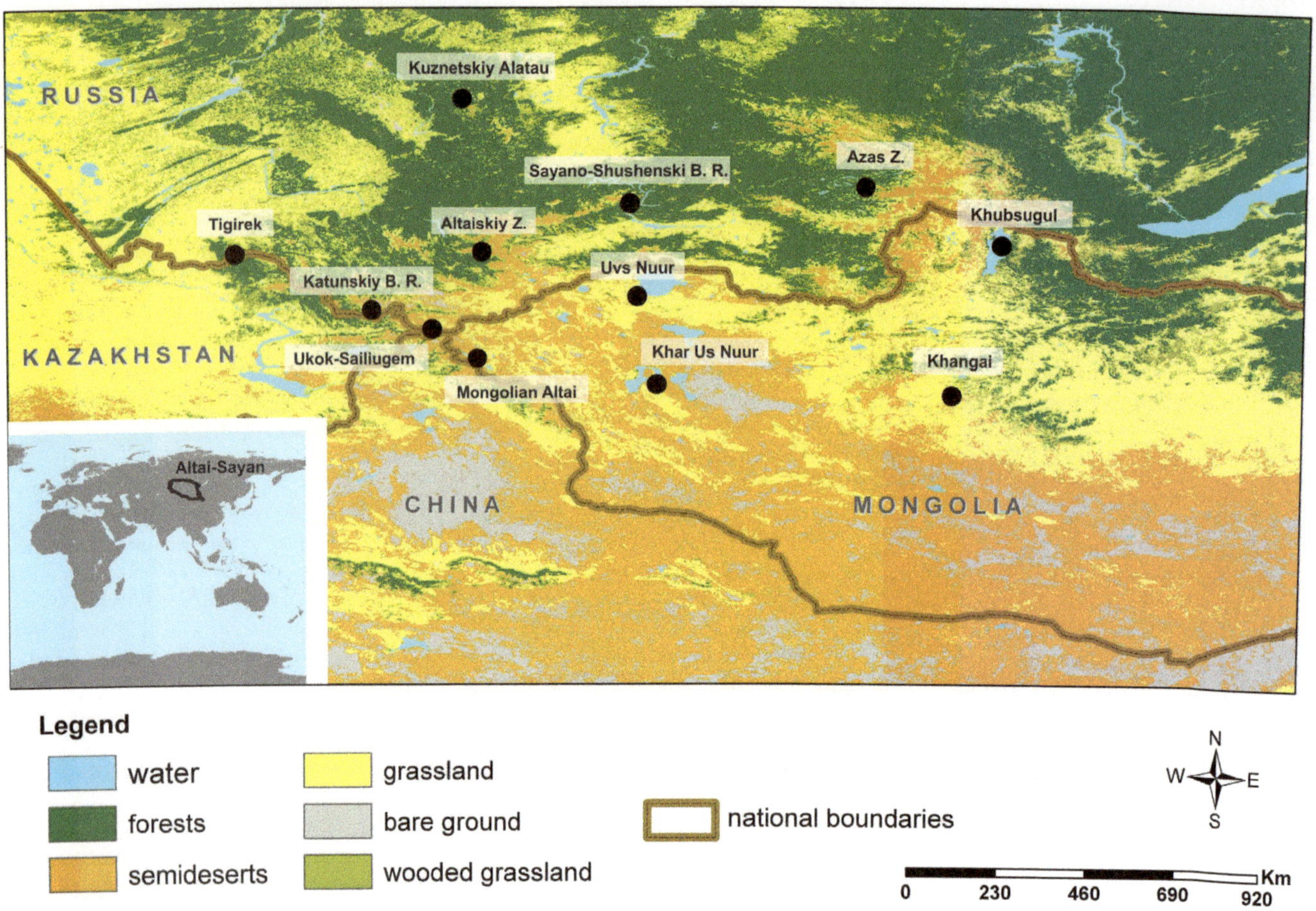

Figure 2. Location map of examined localities in the Altai-Sayan region. Abbreviations: B. R. - Biosphere Reserve; Z. – Zapovednik (National Park). Dashed line: state borders, solid line: Altai-Sayan region borders according to WWF ecoregion.

alpine faunas, with no affinities to the tundra and taiga ones. The Last Glacial faunas of North Yakutia and North Ural were characterized by the presence of alpine faunal elements, and did not resemble any of the Recent localities. The faunas of Recent eastern Altai-Sayan localities (with high proportion of grasslands) were characterized by the co-occurrence of steppe, desert, and alpine species (as well as their Last Glacial counterparts), and were more similar to some Last Glacial faunas than to faunas of any other Recent areas. The NMDS analysis and PCA without rare biomes showed similar results to PCA (see Fig. S1 and Fig. S2).

Comparison of the proportion of species assigned to biome categories between Altai-Sayan and North Yakutia regions (Fig. 4a) revealed that the main difference between the regions consisted of a high proportion of forest and taiga species among Holocene immigrant species in North Yakutia and desert species in Altai-Sayan.

(2) Faunas classified according to trophic-size structure

The gradient between Last Glacial and Recent faunas along the first axis (Fig. 5) was determined by the ratio of insectivores and aquatic predators to large ungulates and predators. The Last Glacial faunas were characterized by a high proportion of the large-sized terrestrial predators (e.g., lion, wolf), large-sized foregut fermenters (e.g., steppe bison, deer, camel), and large-sized hindgut fermenters (e.g., mammoth, horse). Recent communities were characterized by the presence of small terrestrial predators of invertebrates, aquatic predators, subterranean predators of invertebrates, and aquatic predators of invertebrates (Fig. 5a). The occurrence of arboreal and small terrestrial omnivores and subterranean herbivores was positively correlated, while the occurrence of small-sized foregut fermenters was correlated negatively to the above mentioned cluster along the second axis (Fig. 5a). The first two axes explain 63.4% of variance. (see Table S4 for detailed results and comparison with NMDS analyses).

All Last Glacial localities were situated on the right side of the diagram (Fig. 5b), together with the Recent faunas from Khar Us Nuur NP and Ukok-Sailiugem localities (both E Altai). The Khar Us Nuur NP is more similar to the Last Glacial Altai-Sayan fauna than to any Recent locality. The Recent fauna of Tigirek (SW Altai) differs from the other Altai-Sayan localities, as it is situated closer to the European temperate-forest localities (Fig. 5b) characterized by a larger proportion of small terrestrial and arboreal omnivores (Fig. 5a). The Last Glacial assemblage of the Carpathian Mountains was more similar to the Recent community of Central Kazakhstan than to any other assemblage. The NMDS analysis and PCA without rare trophic-sizes showed similar results to PCA (see Fig. S3 and Fig. S4).

The main difference between the proportion of species assigned to trophic-size categories in Altai-Sayan and North Yakutia regions (Fig. 4b) consisted of a high proportion of arboreal

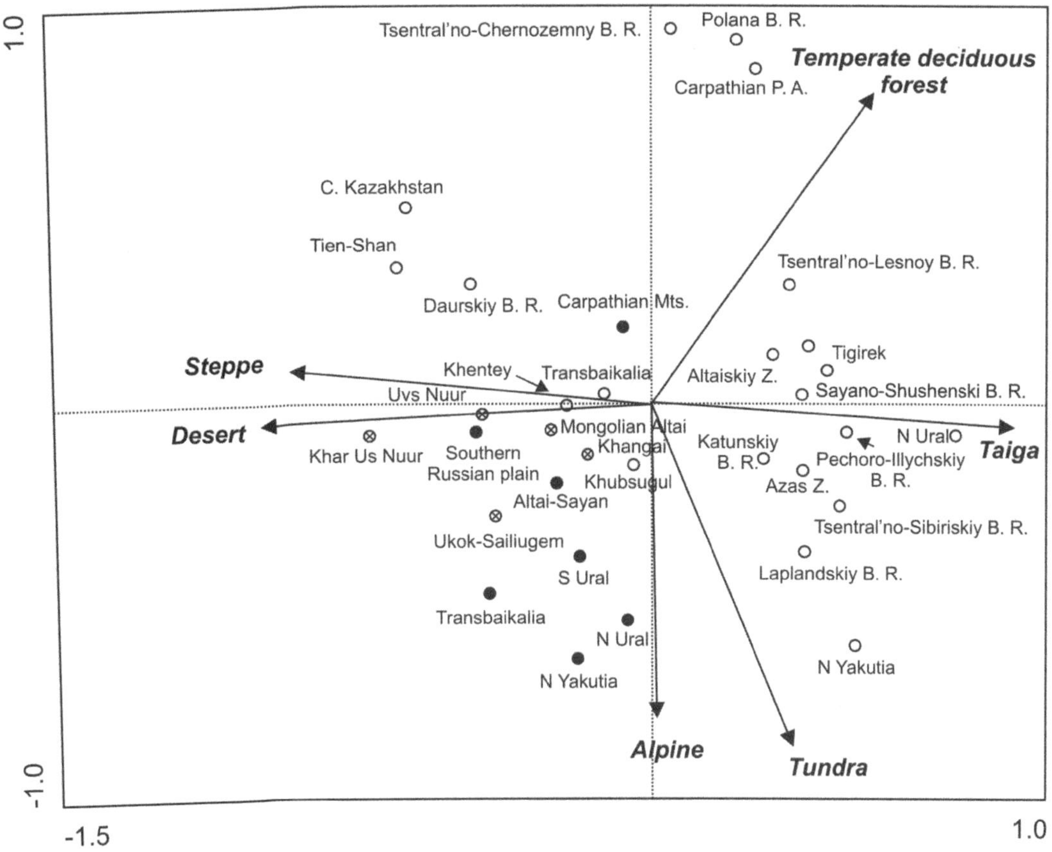

Figure 3. Projection scores of mammalian communities for studied localities (PCA). Species were classified according to their presence in a particular biome. The first two axes describe 89.8% of variance. Open circles – Recent assemblages; crossed circles – Recent eastern Altai assemblages; full circles – Last Glacial assemblages

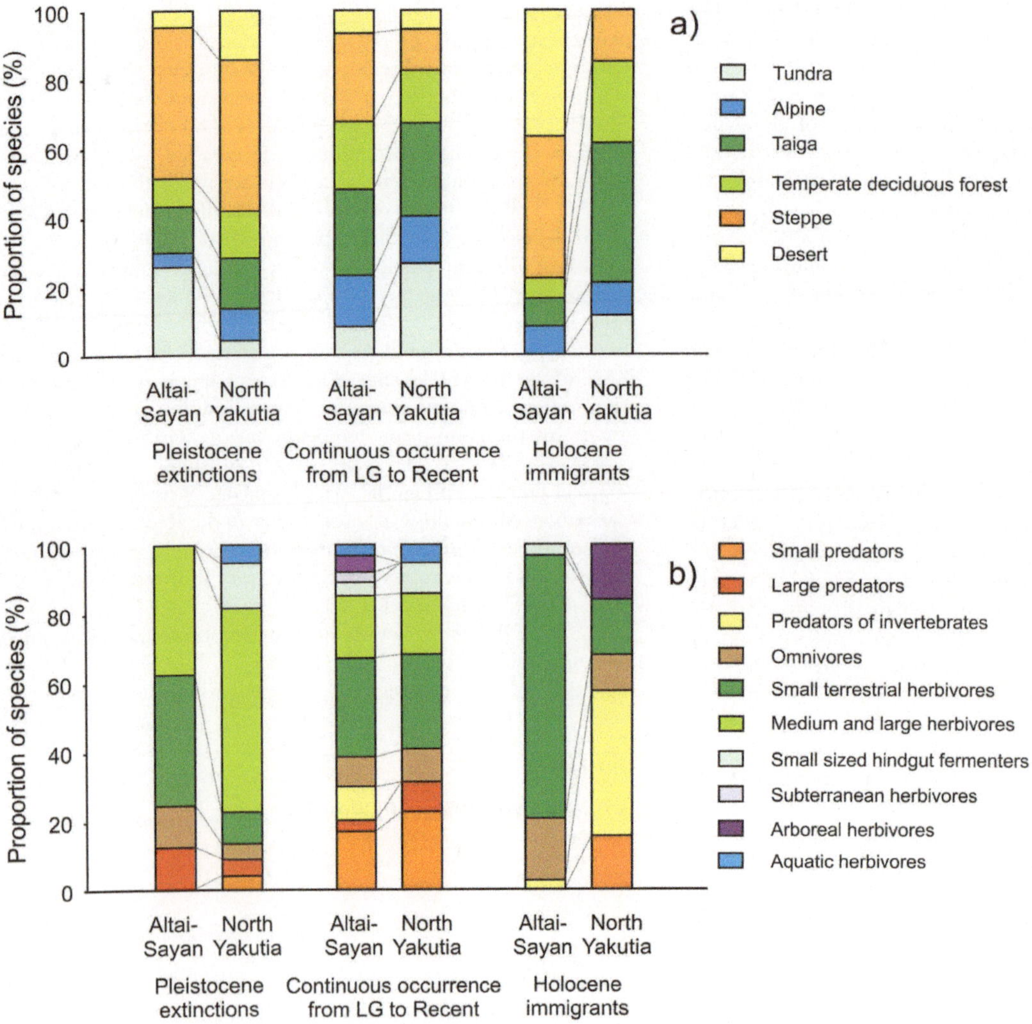

Figure 4. The differences in ecological structure between Altai-Sayan and North Yakutia faunas. Comparison of the proportion of species assigned to biome (a) and trophic-size (b) categories between the regions.

herbivores and predators of invertebrates among Holocene immigrant species in North Yakutia.

Discussion

We found a marked similarity in ecological structure between Recent Altai-Sayan mammalian assemblages and Last Glacial paleocommunities. The ecological structure of mammalian communities confirmed the possible persistence of the Last Glacial-like fauna in the present-day Altai-Sayan region, as the Recent assemblages of eastern Altai-Sayan (i.e. Ukok-Sailiugem, Khar Us Nuur, and Uvs Nuur) were more similar to various glacial localities than to the communities of any other Recent areas. Glacial communities have close modern analogues in the three eastern Altai-Sayan areas where e.g. reindeer and saiga antelope still live in sympatry [38].

Our results are congruent with other evidence supporting the persistence of Pleistocene biota in the Altai-Sayan region. Rodríguez [7] reported a similarity between the ecological structure of present-day mammalian communities from Central Eurasia (Uvs Nuur and Great Gobi) and Iberian Pleistocene communities. The Altai Mountains are an important refugium for full-glacial snail faunas, as recently documented by Horsák et al. [24] and Hoffmann et al. [39]. Kuneš et al. [22] and Pelánková & Chytrý [23] demonstrated a close similarity between glacial pollen samples from central Europe and modern surface-pollen spectra from the Altai-Sayan region. Fossil pollen spectra from the Altai and adjacent regions indicate little difference between modern biomes of this region and those reconstructed for the Last Glacial Maximum [20]. Similar conditions possibly occur in climatically stable areas of North American cold deserts. Fossil evidence from the Great Basin indicate that Pleistocene plant assemblages are comparable with the modern ones [40].

According to vegetation studies, three major biomes occurred widely in the Pleistocene: steppe, tundra, and taiga [28,41–42]. In contrast, the composition of the Last Glacial (and Recent eastern Altai-Sayan) mammalian faunas was characterized by the co-occurrence of steppe, desert and alpine species. These three modern biomes harbor most of the surviving Last Glacial mammals. The importance of the desert biome was probably more pronounced in the examined region of southern Russian plain and central Asia in comparison to the northern and western part of Eurasia.

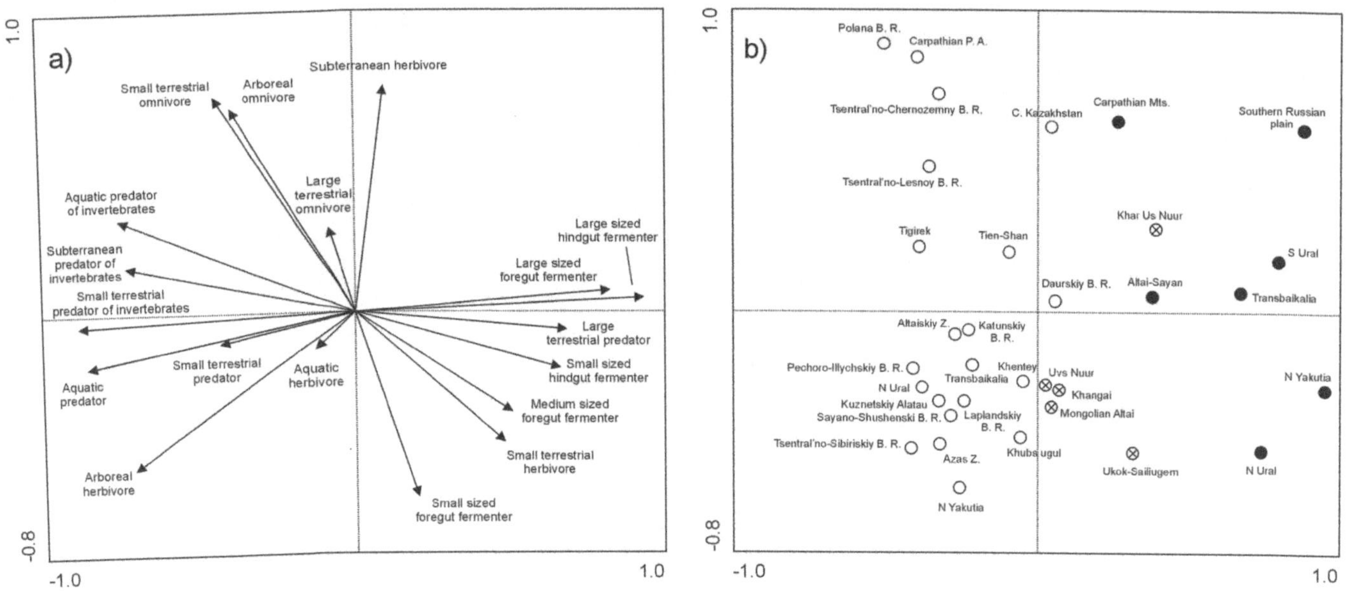

Figure 5. Projection scores of trophic-size community structure (PCA). The first two axes describe 63.4% of variance. Open circles – Recent assemblages; crossed circles – Recent eastern Altai assemblages; full circles – Last Glacial assemblages.

Glacial-like communities still persist in areas where the forest belt does not separate alpine vegetation from the steppes and (semi)deserts. High aridity in eastern Altai-Sayan restricts forests to isolated patches with higher soil moisture [43–44]. Holocene fragmentation of the alpine grasslands to isolated patches surrounded by forests could have lead to the extinction of some large mammals typical of the mammoth steppe. Sharp separation of originally intermixed faunal elements into distinct biomes seems to be the major pattern of the Last Glacial/Holocene transition [8,45].

Most previous attempts to find a modern analogue of the mammoth steppe have been focused on regions of the Arctic tundra, i.e., Yakutia, NE Russia, and Alaska [8,9,46,47]. However, present-day faunas of the Arctic regions have diverged greatly in their ecological structure from Last Glacial ones, showing more affinity to the taiga biome. The divergent ecological structure of modern Yakutian fauna is determined mainly by Holocene immigrant species, characterized by a high proportion of arboreal herbivores and predators of invertebrates associated with forest and taiga biomes (Fig. 4).

The modern arctic tundra is characterized by low productivity and relatively high homogeneity [4,25], in contrast to the alpine biome where tundra vegetation occurs in association with forests and steppes, e.g. [48–49]. During the Last Glacial, tundra vegetation was confined to places with higher precipitation or lower evapotranspiration, e.g. in the mountains or at non-glaciated higher latitudes, often in the close vicinity of steppe [20].

Taiga and temperate forest species occurred in all of the examined Last Glacial assemblages; however their percentage was very low, with the exception of the Last Glacial assemblage of the Carpathian Mountains. This assemblage holds a special position as the site resembles Recent rather than any other Last Glacial community. The Carpathian Mountains of eastern Europe possibly represented a glacial refugium of forest vegetation and forest-dwelling animals, as suggested by Sommer & Nadachowski [50], Jankovská and Pokorný [21], Markova et al. [51], Willis et al. [28], and Willis & van Andel [52].

The main difference in trophic-size structure between present-day and fossil assemblages is the higher richness of large mammals and proportionally lower richness of small mammals in the latter. The high proportion of large herbivores observed in the Last Glacial as well as in some present-day communities is generally typical of areas with low tree cover [15] and could be further supported by the year-round availability of high quality food in the glacial steppe [10,53]. Fossil assemblages are probably biased against small mammal species owing to fossil record incompleteness [54,7,29,55–56]. However, the low proportion of small mammals cannot be considered completely artifactual because the present-day assemblages from the eastern Altai-Sayan region show very similar composition to Last Glacial communities. The low species richness of insectivores and aquatic predators in Last Glacial-like Altai-Sayan assemblages (and probably in other Last Glacial assemblages as well) could be due to the dry and cold climate associated with permafrost which strongly limits insectivores' food sources [57, 48 58]. These results were confirmed using two different ecological classifications and using two independent statistical methods (NMDS, PCA). Moreover, our fossil datasets are based on large areas and a long time period to avoid taphonomic biases in species occurrence. Therefore, we suggest that potential bias in the Pleistocene data subset cannot significantly affect our results.

In general, Pleistocene assemblages were characterized by the occurrence of large herbivore and predator species associated with steppe, desert and alpine biomes. Mammalian paleocommunities classified according to biome type are relatively homogeneous, confirming the view of the mammoth steppe as a single Last Glacial biome [10]. In contrast to biome classification, the trophic-size structure of mammalian paleocommunities shows a degree of heterogeneity, comparable to the Recent localities. The trophic-size structure of communities is probably less influenced by historical factors [16]. Historical processes are considered to be the main factor promoting differences between communities from similar environments [15,29].

Our results open new research possibilities for many aspects of Quaternary paleoecology. The Altai-Sayan region offers a possibility to study factors shaping the structure of so called non-analogue communities and explore vegetation and faunal changes associated with Pleistocene/Holocene transition. Research of soil nutrient availability and cycling in glacial-like localities can provide an insight into the carrying capacity of ice age ecosystems supporting numerous large herbivore species. Modeling the impact of climate changes on the glacial-like landscape may elucidate the process of biome diversification in the Holocene. Our result can be confirmed by thorough paleontological research of the as yet unexplored eastern Altai-Sayan region as well as by phylogeographical analyses of typical glacial species (e.g. steppe lemming or pika) including Altai-Sayan populations.

Supporting Information

Figure S1 The projection scores of studied localities according to biome classification (NMDS analysis based on Bray-Curtis dissimilarity indices).

Figure S2 The projection scores of studied localities according to biome classification without rare categories (PCA analysis).

Figure S3 The projection scores of studied localities according to trophic-size classification (NMDS analysis based on Bray-Curtis dissimilarity indices).

Figure S4 The projection scores of studied localities according to trophic-size classification without rare categories (PCA analysis).

Table S1 Paleartic regions used in the analyses with associated references.

Table S2 Examined localities of Altai-Sayan region with associated references.

Table S3 Percentages of biomes and trophic-size categories for regions and localities (dataset).

Table S4 Component loadings of the PCA and NMDS analyses.

Notes S1 Taxonomic notes with associated references.

Acknowledgments

We thank David Storch, Martin Konvička, Daniel Frynta, Ivan Horáček, Vojtěch Novotný, Michal Horsák, Jan Zrzavý, Milan Chytrý, John Stewart and anonymous referee who helped greatly with useful comments. We thank Ingrid Steenbergen for language corrections and Martin Hais for maps.

Author Contributions

Conceived and designed the experiments: VPŘ JRi. Analyzed the data: JRi. Wrote the paper: VPŘ Jro JRi. Performed the literature search and data extraction: JRo VPŘ.

References

1. Romer AS (1933) Pleistocene vertebrates and their bearing on the problem of human antiquity in North America. In: Jenness D, editor. The American Aborigines: Their Origin and Antiquity. Toronto: University of Toronto Press. pp. 47–84.
2. Hibbard CW (1949) Pleistocene vertebrate paleontology in North America. Geol Soc Am Bull 60: 1417–1428.
3. Kurtén B (1968) Pleistocene mammals of Europe. London: Weidenfeld and Nicolson. 352 p.
4. Guthrie RD (1984) Mosaics, allelochemics and nutrients. An ecological theory of Late Pleistocene megafaunal extinctions. In: Martin PS, Martin RG, editors. Quaternary extinctions: A prehistoric revolution. Tuscon: University of Arizona Press. pp. 259–298.
5. Vereshchagin NK, Baryshnikov GF (1992) The ecological structure of the "Mammoth Fauna" in Eurasia. Ann Zool Fennici 28: 253–259.
6. Van Kolfschoten T (1995) On the application of fossil mammals to the reconstruction of the palaeoenviroment of northwestern Europe. Acta Zool Cracov 38: 73–84.
7. Rodríguez JS (2004) Stability in Pleistocene Mediterranean mammalian communities. Palaeogeogr Palaeoclimatol Palaeoecol 207: 1–22.
8. Guthrie RD (1990) Frozen fauna of the Mammoth steppe: The story of Blue Babe. Chicago: University of Chicago Press. 323 p.
9. Musil R (1999) The enviroment in the last Glacial on the territory of Moravia. Acta Musei Moraviae 84: 161–186.
10. Guthrie RD (2001) Origin and causes of the mammoth steppe: a story of cloud cover, woolly mammal tooth pits, buckles, and inside-out Beringia. Quaternary Review 20: 549–574.
11. Stewart JR (2008). The progressive effect of the individualistic response of species to Quaternary climate change: an analysis of British mammalian faunas. Quat Sci Rev 27: 2499-2508
12. Simberloff D, Dayan T (1991) The guild concept and the structure of ecological communities. Annu Rev Ecol Syst 22: 115–143.
13. Mendoza M, Goodwin B, Criado C (2004) Emergence of community structure in terrestrial mammal-dominated ecosystems. J Theor Biol 230: 203–214.
14. Rodríguez J (2006) Structural continuity and multiple alternative stable States in Middle Pleistocene European mammalian communities. Palaeogeogr Palaeoclimatol Palaeoecol 239: 355–373.
15. Louys J, Meloro C, Elton S, Ditchfield P, Bishop LC (2011) Mammal community structure correlates with arboreal heterogeneity in faunally and geographically diverse habitats: implications for community convergence. Global Ecol Biogeogr 20: 717–729.
16. Mendoza M, Janis CJ, Palmqvist P (2005) Ecological patterns in the trophic-size structure of large mammal communities: a 'taxon-free' characterization Evol Ecol Res 7: 505–530.
17. Blinnikov MS, Gaglioti BV, Walker DA, Wooller MJ, Zazula GD (2011) Pleistocene graminoid-dominated ecosystems in the Arctic. Quat Sci Rev 30: 2906–2929.
18. Frenzel B, Pécsi M, Velichko AA (1992) Atlas of paleoclimates and paleoenvironments of the Northern Hemisphere. JenaNew York: Geographical Research Institute, Hungarian Academy of Sciences Budapest and Gustav Fischer Verlag Stuttgart. 153 p.
19. Tarasov PE, Guiot J, Cheddadi R, Andreev AA, Bezusko LG, et al. (1999) Climate in northern Eurasia 6000 years ago reconstructed from pollen data. Earth Planet Sci Lett 171: 635–645.
20. Tarasov PE, Volkova VS, Webb III T, Guiot J, Andreev AA, et al. (2000) Last glacial maximum biomes reconstructed from pollen and plant macrofossil data from northern Eurasia. J Biogeogr 27: 609–620.
21. Jankovská V, Pokorný P (2008) Forest vegetation of the last full-glacial period in the Western Carpathians (Slovakia and Czech Republic). Preslia 80: 307–324.
22. Kuneš P, Pelánková B, Chytrý M, Jankovská V, Pokorný P, et al.(2008) Interpretation of the last-glacial vegetation of eastern-central Europe using modern analogues from southern Siberia. J Biogeogr 35: 2223–2236.
23. Pelánková B, Chytrý M (2009) Surface pollen-vegetation relationships in the forest-steppe, taiga and tundra landscapes of the Russian Altai Mountains. Rev Palaeobot Palyno 157: 253–265.
24. Horsák M, Chytrý M, Pokryszko BM, Danihelka J, Ermakov N, et al. (2010) Habitats of relict terrestrial snails in southern Siberia: lessons for the reconstruction of palaeoenvironments of full-glacial Europe. J Biogeogr 37: 1450–1462.
25. Allen JRM, Hickler T, Singarayer JS, Sykes MT, Valdes PJ, et al. (2010) Last glacial vegetation of northern Eurasia. Quat Sci Rev 29: 19–20.
26. Huntley B, Allen JRM, Collingham YC, Hickler T, Lister AM, jet al. (2013) Millennial climatic fluctuations are key to the structure of Last Glacial ecosystems. PloS ONE 8: e61963. doi: 10.1371/journal.pone.0061963
27. Agadjanian AK, Serdyuk NV (2005) The history of mammalian communities and paleogeography of the Altai Mountains in the Paleolithic. Paleontological Journal 39: 645–821.
28. Willis KJ, Rudner E, Sumegi P (2000) The full-glacial forests of central and southeastern Europe. Quaternary Research 53: 203–213.
29. Rodríguez J, Hortal J, Nieto M (2006) An evaluation of the influence of environment and biogeography on community structure: the case of Holarctic mammals. J Biogeogr 33: 291–303.
30. Fur SL, Fara E, Vignaud P (2011) Effect of simulated faunal impoverishment and mixture on the ecological structure of modern mammal faunas: Implications for the reconstruction of Mio-Pliocene African palaeoenvironments. Palaeogeogr Palaeoclimatol Palaeoecol 305: 295–309.

31. IUCN Red List Categories and Criteria version 3.1. Available: http://www.iucnredlist.org/technical-documents/categories-and-criteria/2001-categories-criteria. Accessed January 2012.
32. Shackleton NJ (1969) The last interglacial in the marine and terrestrial records. Proc Biol Sci 174:135–154.
33. Williams DF, Thunell RC, Tappa E, Rio R, Raffi I (1988) Chronology of the Pleistocene isotope record 0-1.88 m.y. B.P. Palaeogeogr Palaeoclimatol Palaeoecol 64: 221–240.
34. Horáček I, Ložek V (1988) Palaeozoology and the Mid–European Quaternary past: scope of the approach and selected results. Rozpravy ČSAV, řada matematických a přírodních věd 98: 1–102.
35. Duff A., Lawson A (2004) Mammals of the World: A checklist. London: A & C Black. 312 p.
36. Wilson DE, Reeder D-AM (2005) Mammal species of the World. A taxonomic and geographic reference. Baltimore: Johns Hopkins University Press. 2142 p.
37. Ter Braak CJF, Šmilauer P (2012) Canoco reference manual and user's guide: software for ordination, version 5.0. (Ithaca, NY, Microcomputer Power).
38. Yudin BS, Galkina LI, Potapkina AF (1979) Mammal s of the Altai-Sayan mountainous part. Novosibirsk: Nauka. 296 p. (In Russian).
39. Hoffman MH, Meng S, Kosachev PA, Terechina TA, Silanteva MM (2011) Land snail faunas along an enviromental gradient in the Altai mountains (Russia). J Mollusan Stud 77: 76–86.
40. Wilson JS, Pitts JP (2010) Illuminating the lack of consensus among descriptions of earth history data in the North American deserts: a resource for biologists. Prog in Phys Geogr 34: 419–441.
41. Jankovská V, Chromý P, Nižnianská M (2002) "Šafárka" - first palaeobotanical data on vegetation and landscape character of Upper Pleistocene in West Carpathians (North East Slovakia). Acta Palaeobot 42: 29–52.
42. Simakova AN (2006) The vegetation of the Russian Plains during the second part of the Late Pleistocene (33–18 ka). Quat Int 149: 110–114.
43. Hoffmann MH, Telyatnikov MYu., Ermakov N (2001) Phytogeographical analysis of plant communities along an altitudinal transect through the Kuraiskaya basin (Altai, Russia). Phytocoenologia 31: 401–426.
44. Dulamsuren C, Hauck M, Muhlenberg M (2005) Ground vegetation in the Mongolian taiga forest-steppe ecotone does not offer evidence for the human origin of grasslands. Applied Vegetation Science 8: 149–154.
45. Stewart JR, Lister AM, Barnes I, Dalén L (2010) Refugia revisited: individualistic responses of species in space and time. Proc. R. Soc. B 277: 661-671.
46. Zimov SA, Chuprynin VI, Oreshko AP, Chapin III FS, Reynolds JF, cet al. (1995) Steppe-tundra transition: a herbivore-driven biome shift at the end of the Pleistocene. Am Nat 146: 765–794.
47. Boeskorov GG (2006) Arctic Siberia: refuge of the Mammoth fauna in the Holocene. Quat Int 142–143: 119–123.
48. Smith RL (1986) Elements of Ecology. Second Edition. Harper & Row Publishers 704 p.
49. Wells AG, Rachlow JL, Garton EO, Rice CG (2012) Mapping vegetation communities across home ranges of mountain goats in the North Cascades for conservation and management. Applied Vegetation Science 15: 560–570.
50. Sommer RS, Nadachowski A (2006) Glacial refugia of mammals in Europe: evidence from fossil record. Mamm Rev 36: 251–265.
51. Markova AK, Simakova AN, Puzachenko AYu (2009). Ecosystems of Eastern Europe at the time of maximum cooling of the Valdai glaciation (24–18 kyr BP) inferred from data on plant communities and mammal assemblages. Quat Int 201: 53–59.
52. Willis KJ, van Andel TH (2004). Trees or no trees? The enviroments of central and eastern Europe during the Last Glaciation. Quat Sci Rev 53: 203–213.
53. Faith JT (2011) Late Pleistocene climate change, nutrient cycling, and the megafaunal extinctions in North America. Quat Sci Rev 30: 13–14.
54. Alberdi MT, Caloi L, Dubrovo I, Palombo MR, Tsoukala E (1998) Large mammal faunal complexes and palaeoenviromental changes in the late Middle and Late Pleistocene: a preliminary comparison between the Eastern European and the Mediterranean areas. Geologija 25: 8–19.
55. Brugal J-P, Croitor R (2007) Evolutoin, ecology and biochronology of herbivore associations in Europe during the last 3 million years. Quaternaire 18: 129–151.
56. Meloro C, Clauss M (2012). Predator-prey biomass fluctuations in the Plio-Pleistocene. Palaios 27: 90–96.
57. Smidt S, Oswood MW (2002). Landscape patterns and stream reaches in the Alaskan taiga forest: potential roles of permafrost in differentiating macroinvertebrate communities. Hydrobiologia 468: 1–3.
58. McCain CM (2007) Could temperature and water availability drive elevational species richness patterns? A global case study for bats. Global Ecol Biogeogr 16: 1–13.

16

Discontinuity of Human Presence at Atapuerca during the Early Middle Pleistocene: A Matter of Ecological Competition?

Guillermo Rodríguez-Gómez[1]*, Ana Mateos[1], Jesús Angel Martín-González[2], Ruth Blasco[3], Jordi Rosell[4,5], Jesús Rodríguez[1]

1 Paleofisiología y Ecología Social de homínidos, Centro Nacional de Investigación sobre la Evolución Humana (CENIEH), Burgos, Spain, **2** Departamento de Matemáticas y Computación, Universidad de Burgos, Burgos, Spain, and temporarily assigned to CENIEH, Burgos, Spain, **3** The Gibraltar Museum, Gibraltar, Gibraltar, **4** Àrea de Prehistòria, Universitat Rovira i Virgili (URV), Tarragona, Spain, **5** Institut Català de Paleoecologia Humana i Evolució Social (IPHES), Tarragona, Spain

Abstract

Increasing evidence suggests that the European human settlement is older than 1.2 Ma. However, there is a fierce debate about the continuity or discontinuity of the early human settlement of Europe. In particular, evidence of human presence in the interval 0.7–0.5 Ma is scarce in comparison with evidence for the previous and later periods. Here, we present a case study in which the environmental conditions at Sierra de Atapuerca in the early Middle Pleistocene, a period without evidence of human presence, are compared with the conditions in the previous period, for which a relatively intense human occupation is documented. With this objective in mind, the available resources for a human population and the intensity of competition between secondary consumers during the two periods are compared using a mathematical model. The Gran Dolina site TD8 level, dated to 0.7–0.6 Ma, is taken as representative of the period during which Atapuerca was apparently not occupied by humans. Conditions at TD8 are compared with those of the previous period, represented by the TD6-2 level, which has yielded abundant evidence of intense human occupation. The results show that survival opportunities for a hypothetical human population were lower at TD8 than they were at TD6-2. Increased resource competition between secondary consumers arises as a possible explanation for the absence of human occupation at Atapuerca in the early Middle Pleistocene.

Editor: David Caramelli, University of Florence, Italy

Funding: This research was funded by the MINECO project, CGL2012-38434-C03-02, CGL2012-38434-C03-03, CGL2012-38358 and CGL-BOS-2012-34717. G. Rodríguez-Gómez is the beneficiary of a predoctoral FPI Grant from the Spanish MICINN. The fieldwork at the Atapuerca sites was funded by the Junta de Castilla y León. R. Blasco is a Beatriu de Pinós postdoctoral scholarship recipient from Generalitat de Catalunya and was co-financed by the European Union through Marie Curie Actions, FP7. The funders had no role in study design, data collection and analysis, decision to publish, or preparation of the manuscript.

Competing Interests: The authors have declared that no competing interests exist.

* Email: guillermo.rodriguez@cenieh.es

Introduction

The current paleoanthropological scenario shows an increasingly complex human population dynamic for the Pleistocene period. Some phenomena, such as migrations, replacements of species and abandonment of territories with unfavourable conditions for human colonization, seem to have played significant role in the evolution of the genus *Homo*. From this perspective, one of the most interesting subjects is the continuity or discontinuity between the hominins of the Early and Middle Pleistocene in Western Europe and the origin of the Neandertal lineage [1]. This topic has been treated from different perspectives and by different disciplines. Several authors have proposed the existence of migration waves during the Early Pleistocene taking into account different evidence [2–13]. Bermúdez de Castro and Martinón-Torres [14] suggest a theoretical evolutionary scenario for the Early Pleistocene based on isolation processes and intermittent contacts between the Eurasian populations, as previously proposed by Dennell et al [15] for the Middle Pleistocene. According to their hypotheses, the hominins from Eurasia lost contact with Africa after the first *Out of Africa* because of the formation and evolution of significant biogeographical barriers between the two continents, such as the Negev desert [16,17]. From this point forward, Eurasian populations followed an independent evolutionary process in which the same phenomena of climatic and geographical barriers caused periods of disconnection between some regions [13,18]. Apparently, these frontiers mainly affected the European populations, producing alternating episodes of isolation with population fluxes and refluxes. This dynamic seems to have also occurred during the Middle Pleistocene and may help to explain some significant changes that occurred in Western Europe during this period. An alternative explanation relates human dispersal in Europe to climate change [12] and links the permanence of humans Oldowan technology in Europe to faunal continuity, and specifically to the presence of the sabertooth *Megantereon whitei*, a hypercarnivorous predator which, presumably, provided hominins with large amounts of scavengable carcasses [13]. Arribas and Palmqvist [13] suggest that there was isolation between Europe and Africa between 1.8 Ma until 0.6 Ma, when Achelean populations expanded from Africa into Western Europe.

Currently, human settlement during the Early Pleistocene is represented in Western Europe by a short list of sites, such as Fuente Nueva 3 [19,20], Barranco León [21], Vallparadís [22,23] but see also [24], Sima del Elefante and Gran Dolina in Spain [25], Pirro Nord [26] and Ca'Belvedere di Monte Poggiolo [27] in Italy, Le Vallonet [28], Saint-Hilaire-la Gravelle [29,30] and Pont de Lavaud [31] in France, Happisburgh [32] in England and Untermassfeld [33] and Dorn-Dürkheim 3 [34] in Germany. The site of Pakefield (England), which dates to the beginning of the Middle Pleistocene (0.7 Ma) [35], can also be considered in this list. Lézignan le Cèbe, in southern France, has been proposed as one of the oldest sites documenting human occupation in Europe [36], but this interpretation is controversial. Technologically, all of these archaeological locations can be classified as Oldowan Technological Complexes or Mode 1 [11,37]. *Homo antecessor*, described from level TD6 of Gran Dolina, seems to have inhabited this area at least between the Jaramillo Subchron and the Matuyama-Brunhes Boundary [38]. Human remains older than the Jaramillo Subchron have been found at Sima del Elefante (1.2 Ma) [39] and Barranco León (1.4 Ma) [21], but see also Muttoni et al. [40]. Unfortunately, these fossil remains do not meet enough taxonomic criteria to establish possible relationships with other species such as *H. antecessor* and to describe the human scenario at this early time. In the absence of other data, these fossils have been classified as *Homo* sp. [21,38].

Mosquera et al. [11] postulated a general continuity in the hominin occupation of Sierra de Atapuerca with the exception of a gap between 0.9 and 0.5 Ma at the Gran Dolina site, based on the archaeological evidence. These authors suggest that this depopulation was not a local phenomenon but a European event. During this period, evidence of human presence is scarce in Western Europe. Although no human remains have been recovered between OIS18 and OIS16, some isolated lithic assemblages have been dated to the beginning of this period, such as those found at the Middle Loire River Basin [41] and at Caune de l'Arago [42] in France. In contrast, the number of archaeological sites increases significantly in approximately 0.6−0.5 Ma, revealing new palaeoanthropological and cultural features that mark the European Middle Pleistocene [12,43,44]. New lithic assemblages with early Acheulean industries seem to appear in Western Europe at the end of the Early Pleistocene ([11] and reference therein) although the dates of some sites are debatable [12]. In any case, Acheulean technology arrived in Europe around or before 0.6−0.5. Ma. The new settlers, with full Acheulean Mode 2 technology, occupied the European landscapes intensely at this time, as shown by the increasing number of sites. The sparsely populated northern latitudes began to be systematically occupied. Examples of this phenomenon can be found at Boxgrove in England [45], and at Bilzingsleben [46] and Schöningen [47] in Germany. Interestingly, the oldest European hearths have been dated to approximately 400 ka (Terra Amata, Beeches Pit, Schöningen or Bilzingsleben) [11,48]. The hominin fossil evidence is currently the focus of an intense debate concerning the evolutionary relationships of the Early and Middle Pleistocene European hominins [1,15,49–53] and references therein.

Given the shortage of sites and the sparse anthropic evidence from the beginning of the European Middle Pleistocene, three possible scenarios may be considered. In the first scenario, a discontinuity occurred between the Early and Middle Pleistocene human populations, and thus, Europe was almost depopulated at this time. Alternatively, if continuity between the Early and Middle Pleistocene human populations occurred, the scarcity of evidence might be explained by a contraction of the human populations as a response to adverse environmental conditions. In this latter case, the Early Pleistocene hominins would have survived in some refuges where the environment remained more consistent and eventually evolved into the new Middle Pleistocene forms, both anatomical and culturally. A third possible scenario is that the Early Pleistocene human populations survived in some refuges from OIS18 to OIS14 but were eventually substituted by the Acheulean immigrants approximately 0.5 Ma. Alternatively, it could be argued that the scarcity of evidence of human presence is just a consequence of the low completeness of the fossil record. Although this possibility may not be definitively ruled out, the scarcity of lithic record in this period, in comparison to the previous one, is striking, since stone tools are abundantly produced by humans and they have a high probability of being preserved in the archaeological record.

Here, we focus on the critical period 0.7−0.5 Ma to test whether the lack of evidence of human presence is related to adverse ecological conditions during this period. Resource availability and competition with carnivores have been repeatedly proposed as key limiting factors for Early Pleistocene European hominin populations [4,13,54–59]. Thus, a key question is whether resource availability was lower in the period between OIS18 and OIS12 than at the end of the Early Pleistocene. Rodríguez-Gómez et al. [60] presented a model to study predator-prey relationships and estimate resource availability on a local scale. This model is used here to test the hypothesis that the absence of human settlement was coincident with a period of low resource availability.

The Gran Dolina site in the Sierra de Atapuerca (in northern Spain) provides a unique opportunity to test this hypothesis on a local scale. This site has a long stratigraphic sequence that dates from the Jaramillo Subchron to the end of the Middle Pleistocene. Evidence of human presence at Sierra de Atapuerca before and after the critical period 0.7−0.5 Ma has been registered at different sites and stratigraphic levels. Human settlement in the 1.2−0.7 Ma period is documented at the TE9 level at Sima del Elefante, and the TD3-TD4 and TD6 levels at Gran Dolina, while the GII, GIIIa, GIIIb units of Galería, the TD10-2 and TD10-1 levels of Gran Dolina and the Sima de los Huesos site document human settlement during the period 0.5−0.25 Ma [61–64]. The interval between OIS18 and OIS16 (roughly from 0.7 to 0.5 Ma) is represented at Atapuerca by the TD7, TD8 and TD9 levels of Gran Dolina, all of which lack evidence of human presence in the Atapuerca area [25]. This archaeological gap in Atapuerca has been discussed taking as a perspective the technological evidence [11,43]. The TD8 level provides the richer faunal assemblage of this period at Atapuerca [62,63]; thus, it has been selected for the study of this critical period.

The aim of this study is to compare competition intensity and resource availability at Atapuerca during the period for which evidence of human presence is lacking, represented by the TD8 assemblage, with the conditions at the end of the Early Pleistocene, represented by the TD6-2 assemblage, a period of intense human occupation at Atapuerca. We introduce in this analysis an index of intraguild carnivore competition to compare the changes in the local conditions between these two periods. These data will provide us which information about the main features of the palaeoecosystem and the survival opportunities for a hypothetical human population at the Sierra de Atapuerca during this period.

Materials and Methods

The Atapuerca Gran Dolina TD8 Level

The Sierra de Atapuerca is located 15 km east of Burgos, in the north of the Iberian Peninsula. This Cretaceous limestone massif has various cavities filled with well-stratified Pleistocene sediments, known from south to north as Sima del Elefante (TE), Galería (G) and Gran Dolina (TD). These sites have been excavated since 1981 by Atapuerca Team with permission of the Consejería de Cultura y Turismo de la Junta de Castilla y León. Gran Dolina (TD) (3°31′08 W, 42°21′09 N; UTM coordinates: X = 457279, Y = 4689172) is a cavity approximately 18 m in height that was filled with Lower and Middle Pleistocene sediments. Its stratigraphic sequence was initially divided into 11 stratigraphic levels (TD1-TD11 from bottom to top) and revised in subsequent studies; e.g., [25].

The TD8 level was formed by a succession of brecciated flows of red lutites with boulders and gravels [65]. Its sedimentation is composed of overlapping cones with main vertexes at the north of the site. The vertical positions of these vertexes and their proximity to the roof suggest a small entrance into the cave during the TD8 formation. The ESR and U-series dates taken from the middle part of the sedimentary deposit suggest an average age of 600 kyr (602±52 kyr) [63]. The result obtained by TL from one sample collected at the base of TD8 correlates with the Matuyama-Brunhes boundary when the range of error is taken into account (820±140 kyr) [62]. From an archaeological perspective, evidence of human presence was not documented at TD8, as noted above. It might be argued that the small entrance and the general conditions (humidity and darkness) of the cave prevented a continuous access by a hypothetical human population, although the evidence available is not conclusive on this respect.

Taking into account the species of carnivores recovered at the TD8 level (see below) and the taphonomic characteristics identified in the faunal assemblage (e.g., tooth-mark dimensions), hyena is proposed as the main agent responsible for the ungulate accumulations [66]. However, a certain degree of variability with respect to those features traditionally used to define carnivore dens can be observed at TD8 (e.g., an absence of immature carnivore remains, a low proportion of coprolites, few marks related to the end stages of carnivore consumption, the absence of an attritional age profile and large quantities of whole bones and epiphyses) (see [67]). According to Blasco et al. [66], this variability seems to be the result of a combination of several types of dwellings (dens and refuges) and the occasional access of other carnivores during the formation of the TD8 deposit.

The TD8 Faunal assemblage

The faunal list for TD8 was obtained from Rodríguez et al. [25], Blasco et al. [66], and van der Made [68]. We restricted our analysis to mammal species of weighing more than 10 kg because these include the main sources of meat and fat, as well as the main potencial competitors and predators of a hunter-gatherer population [69,70]. This size interval includes small to medium-sized predators such as the lynx that may prey on small ungulates that are potentially important in a human diet. Rodríguez [71] provides estimations for the body mass of the mammal species identified at Gran Dolina based on the few fossils available at that time and using allometric equations. However, excavation of Gran Dolina in subsequent years increased the fossil sample from TD8. Gran Dolina TD8 site contains 899 remains that belong to mammal species of more than 10 kg. No specific permits are required for this study, which complied with all relevant regulations. The current regulation is Decreto 37/2007, 25th of April 2007, Junta de Castilla y León Cultural Heritage (BOCyL 79, 25/05/2007). The material revised is temporary housed at Institut Català de Paleoecologia Humana i Evolució Social (IPHES) at Tarragona (Spain). This research is under the frame of the Research Project CGL2012-38434-C03-02 from Spanish MINECO. The specimen numbers are provided in (Table S1).

Thus, we analysed the entire TD8 fossil assemblage and reviewed the body size estimations provided by Rodriguez [71], recalculating the body masses with allometric equations [72] when new data were available. There are eleven primary consumer species in TD8 weighing more than 10 kg: *Bison voigtstedtensis, Cervus elaphus* (red deer), *Dama vallonetensis, Equus altidens, Eucladoceros giulii, Hippopotamus* sp., *Macaca* sp., *Megaloceros solilhacus, Stephanorhinus etruscus, Sus scrofa* (wild boar) and *Ursus* sp. The bear, *Ursus* sp., is considered to be an omnivore, and in this study, it was included as both a potential prey and a potential predator. Five large predators were identified in the TD8 assemblage: *Canis mosbachensis, Crocuta crocuta* (spotted hyaena), *Hyaena* sp. (considered analogous to striped hyaena), *Lynx* sp., *Panthera gombaszoegensis and Ursus* sp. The small canid *Vulpes* sp. was not included in the analyses because its diet was presumed to be composed of small mammals, as discussed by Rodríguez-Gómez et al. [60]. The families Viverridae and Mustelidae were also excluded because their diets are also based mainly on small mammals. The TD8 fossil assemblage lacks adequate elements to estimate the body masses of *Hippopotamus* sp., *Hyaena* sp., *Macaca* sp., *Megaloceros solilhacus*, *Panthera gombaszoegensis* and *Sus scrofa* using allometric equations. Thus, for these species, we used the body masses provided by Rodríguez [71] and Blasco et al [66], except for *Hippopotamus* sp., for which the body mass for *Hippopotamus* gr. *H. antiquus* provided by Mazza and Bertini [73] was used. The body masses used are shown in Table 1.

Although we lack evidence of human presence at TD8, *Homo* sp. has been included in the analyses as a hunter-gatherer to estimate competition and resource availability for a hypothetical human population living in Atapuerca at that time. We assume for *Homo* sp. the same trophic behaviour inferred for *Homo antecessor* in the TD6-2 assemblage. Hunting has been interpreted as the main food procurement strategy at TD6-2 [74,75], although other Early Pleistocene sites like Fuente Nueva-3 evidence a predominantly scavenging behaviour of early *Homo* [55].

Thus, we analysed the distribution of resources between secondary consumers for two different scenarios and examined the effects of these different configurations of the TD8 paleoecommunity: 1) a TD8 assemblage and 2) a TD8 assemblage + *Homo* sp.

Rodríguez Gómez et al. [60] used a mathematical model to quantify the resources available for a human population in the ecosystem represented at TD6-2. TD6-2 is another level at the Gran Dolina site, lately dating to approximately 0.9 Ma [61]. This level has yielded abundant faunal remains [25,76,77] and Mode 1 stone tools, together with a large assemblage of human fossils attributed to *Homo antecessor* [8]. The study by Rodríguez-Gómez et al. [60] on the TD6-2 assemblage, suggests the existence of a rich environment, abundant in trophic resources for a hominin population, at Atapuerca at the end of the Early Pleistocene.

To make the results for TD6-2 and TD8 directly comparable we also ran the model for TD6-2, including *Homo antecessor* as a secondary consumer, because it was not included in [60]. Two different levels of animal food in the diet of *Homo* were tested to represent either a diet with a low (30%, H_{min}) or high (60%, H_{max}) meat component. In addition, we included two scenarios for TD6-2, adding a large felid to its carnivore guild, although none was recorded in the fossil assemblage. Thus, we repeated our analyses,

Table 1. Estimated body masses of the species in the TD8 assemblage.

Species	Estimated Body Mass (kg)
Bison voigtstedtensis	397
Cervus elaphus	163
Dama vallonetensis	84
Equus altidens	324
Eucladoceros giulii	360
Hippopotamus sp.	2,225
Megaloceros solilhacus	383
Macaca sylvanus	18
Stephanorhinus struscus	1,400
Sus scrofa	85
Ursus sp.	282
Canis mosbachensis	12
Crocuta crocuta	75
Hyaena sp.	50
Lynx sp.	10
Panthera gombaszoegensis	90

Weights of *Hyaena* sp., *Macaca* sp., *Megaloceros solilhacus*, *Panthera gombaszoegensis* and *Sus scrofa* were taken from Rodríguez [71] and Blasco et al [66]. To *Hippopotamus* sp. weight was taken from [73].

including *Homotherium latidens* as a member of the TD6-2 predatory guild, to evaluate the effect of its possible presence on our results, as explained in Rodríguez-Gómez [60].

The model

We investigated the distribution of resources, primary consumer biomass, among secondary consumers in different scenarios. On the one hand, it is necessary to estimate resource availability, i.e., the biomass of primary consumers available to secondary consumers or total available biomass (TAB). On the other hand, the requirements of secondary consumers (TBD) should also be estimated (Figure 1) [78]. A summary description of the model components is provided below; for a detailed formal description of this model see Rodríguez-Gómez et al. [60]. The model was written and executed in Matlab R2009b.

Total Available Biomass

Our model was developed on the basis of the assumption that all of the variations in population size and composition may be taken as oscillations around a mean value that is constant through time-an assumption that is widely accepted in population dynamics studies [79]. We represented the average long-term condition of every population using a Leslie Matrix [80,81]. Leslie Matrices are used in population dynamics to represent a population structure at different times and to describe its oscillations (Figures 1 and 2A). We sought the Leslie Matrix that represented the average structure over time by introducing two additional conditions:

1. - The population should be stable (i.e., population size should be constant from year to year).
2. - The population should be stationary (i.e., the age structure should be constant from year to year).

Input data are physical and physiological variables (adult body mass, body mass at birth, litter size, breeding interval, age at reproductive maturity, growth rate and lifespan) that are species specific. For species with living representatives, such as *Sus scrofa* and *Cervus elaphus*, the values of these physiological variables were taken directly from the literature. For species without living representatives, the values reported in the literature for closely related living species of similar body mass were used (e.g., *Equus zebra* was used as an analogue for *Equus altidens*). For species defined to genus level such as *Macaca* sp., we used values for species of similar size, as *Macaca sylvanus*. In the cases of *Bison voigtstedtensis*, *Dama vallonetensis*, *Eucladoceros giulii*, *Hippopotamus* sp., *Stephanorhinus etruscus* and *Ursus* sp., which lack closely related living species of similar size, we computed a least square regression equation for each physiological variable with respect to mean body weight using data for species in the same family; see further details in Supplementary Material, Table S3 in [60].

The population profiles obtained from this model for every primary consumer population provide estimates of the average sustainable biomass output by age classes, which were eventually translated into body size intervals. Biomass output by age interval was obtained from the annual mortality rates obtained from the Leslie Matrix (Figure 2B). Each dead individual of a primary consumer species was assigned to one of six size categories according to its average body mass at the age of death: 10–45 kg, 45–90 kg, 90–180 kg, 180–360 kg, 360–1,000 kg or >1,000 kg (see [57]). The biomass made available for secondary consumers by each single primary consumer population was obtained as the sum of the biomass of all dead individuals (Figure 2C).Total biomass output (TBO) was obtained as the sum of the biomass outputs in each size category from each primary consumer population (Figure 1). The distribution of sustainable TBO by size category is an important feature of the model, because prey body size is a main selection factor for predators [82–84]. With this treatment of the data it is then possible to represent features such as the hunting of sub-adults in species with very low adult mortality rates and very low predation at the adult stage, such as rhinos and hippos. This model may overestimate TBO because it

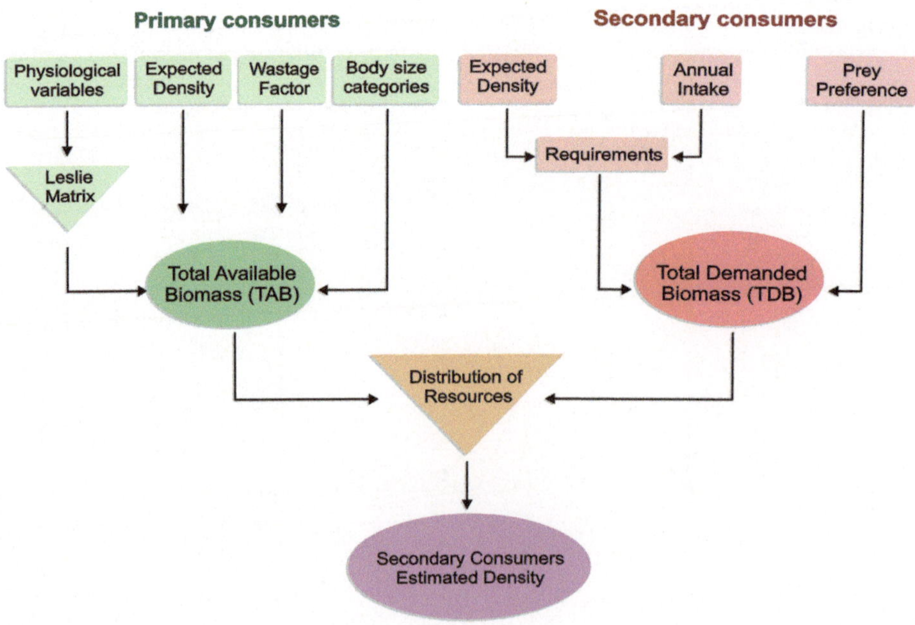

Figure 1. Flow diagram showing the components of the model used to evaluate trophic resource availability and intraguild competition (modified from [78]**).**

considers that all the available biomass was consumed exclusively by mammals. The role of other secondary consumers, like vultures is not been considered in the model, although we acknowledge its relevance in actual ecosystems. There is not way to estimating the amount of TBO consumed by non-mammalian secondary consumers in the past, but it is reasonable to assume that it was similar in both levels (TD6-2 and TD8), an assumption that strengthens the null hypothesis of the absence of ecological differences between both assemblages.

Only female individuals are represented in a Leslie Matrix. Thus, we assumed the sex ratio was equal to 1:1, that the population profile was the same for males and females and that the survival rate was equal for both sexes. A further assumption is that the sub-adult survival rate should be lower than the adult survival rate [60].

The model solutions are not dependent on population size: thus, an estimate of population density is needed to estimate the sustainable biomass output. Mediterranean taxa dominate the TD8 pollen spectrum [25]; thus, we used the equation provided by Damuth [85] to estimate the density of primary consumers in a European mixed temperate forest:

$$\log(D) = -0.79 * \log(m) + 4.33$$

where D is the population density in number of individuals per square kilometre and m is the body mass in grams.

Combining the mortality profiles obtained from the Leslie Matrix with the mean body size per age class and the estimated population density of the species, the sustainable biomass output can be computed. Because a carcass includes a variable amount of non-edible tissues (horns, bones, hide, etc.), this sustainable biomass output cannot be fully used by secondary consumers. The percentage of non-edible TBO is represented in the model by a size-specific "wastage factor" [86]. Once this percentage has been subtracted, the final amount of biomass available to secondary consumers or total available biomass (TAB) is obtained. TAB is also distributed by body mass classes (Figure 1).

Our model yields several population profiles for each species, corresponding to different mortality rates. We selected extreme values with maximum and minimum pressure on sub-adults (or maximum and minimum mortality rates) that produce minimum and maximum TAB levels, respectively (TAB-m and TAB-M, respectively). We limited the results to those solutions for which sub-adult mortality rates were higher than adult mortality rates because this is the pattern usually observed in natural populations [87].

Total Demanded Biomass (TDB)

Carnivore-demanded resources should be estimated as a first step in evaluating resource distribution among secondary consumers. The carnivore intake rate was estimated using the equation reported by Farlow [88]:

$$\log I = 0.69686 \log(m) + 0.27747;$$

$$r^2 = 0.96758;\ F_{1;100} = 2{,}984.773;\ p < .01$$

where I is the intake rate in kcal per day and m is the body mass in grams. Some adjustments were made for each secondary consumer according to its inferred dietary preferences. Taking the coyote as an analogue, we estimated that large mammal flesh represented only 20% of the energetic requirements of *Canis mosbachensis* and that it was primarily consumed as carrion [57]. The spotted hyena, *Crocuta crocuta*, is able to exploit bone marrow thanks to its bone-crashing abilities. We assumed that approximately 2% of the total energetic requirements of the spotted hyaena were satisfied by bone marrow, and consequently, its energetic requirements were reduced by a factor of 0.02 [89,90]. In the case of *Hyena* sp., by analogy with the striped

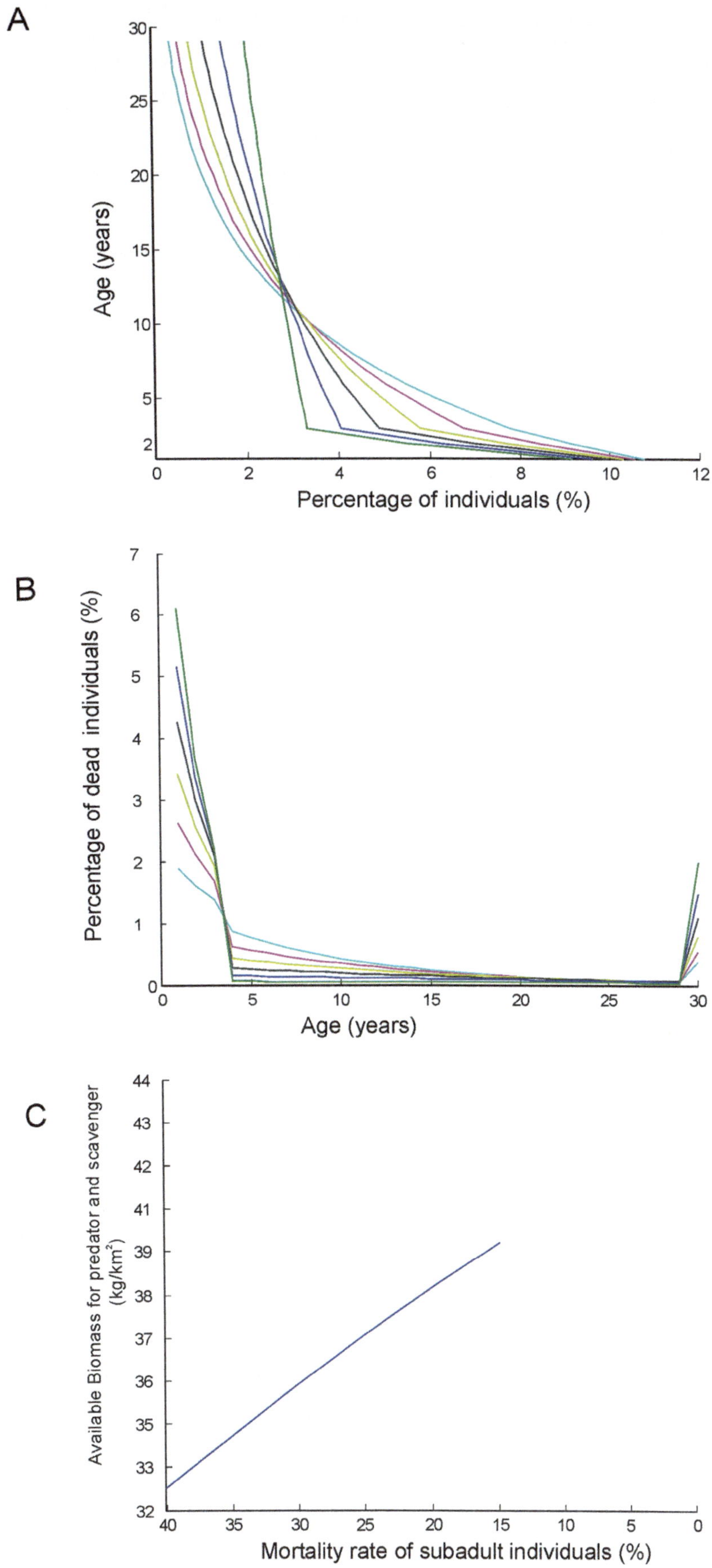
A
30
25
20
15
10
5
2
Age (years)
0
2
4
6
8
10
12
Percentage of individuals (%)
B
7
6
5
4
3
2
1
0
Percentage of dead individuals (%)
0
5
10
15
20
25
30
Age (years)
C
44
43
42
41
39
38
37
36
35
33
32
Available Biomass for predator and scavenger (kg/km²)
40
35
30
25
20
15
10
5
0
Mortality rate of subadult individuals (%)

Figure 2. Graphical representation of population profiles (A), mortality profiles (B) and biomass output available for secondary consumers (C) at different sub-adult mortality rates for *Equus altidens*. Each line represents the solution of the model for a different level of sub-adult mortality. The percentage of dead individuals in the mortality profiles represents the percentage of individuals in the population that died at age *i*.

hyaena, carrion of large mammals would represent approximately 75% of its diet; the remainder could be invertebrates, vegetable food and small vertebrates [91]. The diet of the TD8 lynx (*Lynx* sp.), was presumed to be similar to that of the recent Iberian Lynx (*Lynx pardina*) [57]. Thus, it was assumed that species between 10 and 90 kg represented only 5% of the lynx diet [92–94], lagomomorphs and other small mammals making up the bulk of its diet. Because *Ursus* sp. was likely to have been highly omnivorous, it was assumed that meat represented only 10% of its energetic requirements as in recent European brown bears (*Ursus arctos*) [95,96]. Likely, fish was also an important component of its diet as suggested by the nitrogen-isotope values obtained for the Venta Micena bears [97].

The annual energetic requirements of a carnivore population by km^2 are obtained multiplying the individual annual intake by the population density (Figure 1). The equation provided by Damuth [98] for African flesh-eaters was used to estimate the typical carnivore density:

$$\log(D) = -0.64 * \log(m) + 2.33$$

where D is the population density in number of individuals per square kilometre and m is the body mass in grams. In the case of *Ursus* sp., the equation for carnivores was used instead of the equation for primary consumers because it predicts the population densities of recent bear species more accurately [99–101].

The requirements of *Homo* sp. were estimated to be similar to those of recent hunter-gatherer populations, with a mean daily requirement of 3,000 kcal per individual [102]. The average body mass was considered to be 76 kg, based on pelvis and femur estimates for *Homo heidelbergensis* from the Sima de los Huesos site [103,104]. Studies of recent hunter-gatherer populations have shown that animal resource consumption represents between 30% and 60% of their nutritional intake [105,106]. Thus, we modelled *Homo* sp. as a species that meets 30% (H_{min}) or 60% (H_{max}) of its energetic requirements (3,000 kcal/day) by consuming the meat of large mammals. Population densities and diet compositions observed for recent hunter-gatherers populations are shown for comparison in Table 2 (extracted from [107]).

As in the case of TAB, TDB was distributed over the same six body size categories based on the inferred prey size preferences of each predator, on the basis of the behaviour of their living relatives (Table 3 and Figure 1) [57]. The preference of a predator for a body size category is represented by the percentage of predation (PD) that this size category was presumed to represent in its diet. If a predator was presumed to be unable to kill prey in a given size category and to not consume carrion, a PD of 0 was assigned to the predator in that size category.

Distribution of TAB between secondary consumers

The distribution of TAB between secondary consumers is based on the proportional predation pressure (PPP_{ij}) of each species in each body size category. PPP_{ij} represents the relative amount of biomass demanded by the *jth* carnivore species from the *ith* prey body size category and is calculated as the proportion of the total amount of biomass demanded from a prey body size category by all carnivores that corresponds to the requirements of a single carnivore species. PPP_{ij} incorporates intraguild competition in the model because the resources obtained by a carnivore species depends on the number and characteristics of its competitors. See a detailed formal description of resource distribution computation in Rodríguez-Gómez et al. [60].

To compare the different scenarios considered in our analysis, we generated two indices (SCI and GCI) that relate the estimated and expected (estimated from allometric equations [98]) secondary consumer densities.

Species competition index (SCI):

$$SCI_i = 1 - (Ds_i / Dx_i)$$

where Dx_i is the expected density for species i obtained from the allometric equation in Damuth [98]. Ds_i is the estimated density for species i and is obtained from our model.

Global competition index (GCI):

$$GCI = 1 - (\sum Ds / \sum Dx)$$

where $\sum Ds$ is the summation of estimated densities for all species and $\sum Dx$ is the summation of expected densities for all species.

These indices provide information about competition intensity in the ecosystem with regard to an ideal condition in which all species reach optimal densities. We assume that if the densities estimated from our model approach these thresholds densities, all species would fulfil needs with the have plenty of resources available to them. The SCI index shows to what degree a species fulfils its requirements in a given environment. GCI performs similarly but at the scale of the whole guild of secondary consumers guild. In both cases, however, if the estimated densities were closer to the expected densities or optimal ones, we considered that resources were abundant when compared with requirements (i.e. that competition is low). Conversely, if our model results indicate that the secondary consumers were able to coexist only at low densities, this would be so because there was a scarcity of resources in relation to requirements, which implies a high competition.

Results

Total Available Biomass

The minimum TAB, corresponding to the maximum sub-adult mortality rate, was 496,814 kcal/km^2*year (TAB-m), and the maximum TAB, corresponding to the minimum sub-adult mortality rate, was 643,968 kcal/km^2*year (TAB-M). TAB-m is 23% lower than TAB-M. In addition, the distributions of TAB-M and TAB-m by size category are different (Table 4 and Figure 3), mainly because TAB-m has more biomass in the first category (10–45 kg) than TAB-M. The reason for the different distribution is that both adults and sub-adults of *Macaca* sp. are present in this category and with a low sub-adult mortality rate, the biomass output of macaques is higher, increasing TAB in this size category. For the rest of species considered, adult and sub-adult individuals belong to different body mass categories, and biomass increases in larger body size categories as sub-adult mortality rate decreases. These results demonstrate the importance of including the age structure of primary consumer populations in the model, because the distribution of TAB in size categories is not homogeneous, and

Table 2. Population densities and food resource use observed in recent hunter-gatherer populations [107].

Group	Area	Density ind./km^2	Gathering%	Hunting%	Fishing%
Punan	29.6	0.118	65.00	30	5
Siberian Eskimo	274.9	0.047	1.00	30	60
Guayaki-Ache	28.7	0.0348	30.00	62	10
Efe	47.0	0.1596	88.20	11	0,8
Hadza	25.0	0.240	60.00	40	0
!Kung	110.0	0.066	67.00	33	0
G/Wi	180.0	0.0293	55.00	45	0
Mardudjara	226.0	0.0075	70.00	30	0
Kiowa	280.0	0.014	20.00	80	0
Caribou Inuit	2365.0	0.003	0.10	55	44.9
Nunamiut Inuit	249.0	0.0096	0.10	89	10.9
Polar Inuit	731.0	0.0041	0.01	30	69.99

prey body size is a key factor in prey selection by most predators. Thus, the TAB of a prey size category is not equally available for all secondary consumers.

Requirements of the secondary consumers

The energy necessary to maintain the species of the TD8 carnivore guild (first scenario) at their ecological densities is 693,725 kcal/km^2*year (Table 5). This energy requirement exists if all species fulfil all of their expected requirements. The species with higher energetic requirements are *Crocuta crocuta* and *Panthera gombaszoegensis*, approximately 222,000 kcal/km^2*year each, or 32% of total requirements. Next is *Hyaena* sp., which accounts for approximately 160,000 kcal/km^2*year or 23% of the total. *Canis mosbachensis* accounts for approximately 40,000 kcal/km^2*year or 6% of the total. *Lynx* sp. and *Ursus* sp. account for approximately 20,000 kcal/km^2*year. However, these are estimates, and to establish whether all of the requirements of the secondary consumers are fulfilled, the distribution of TAB by size category and the prey preferences of the carnivores should be taken into consideration.

Comparing different scenarios and conditions

Two scenarios were considered for TD8: only the species in the TD8 assemblage and the TD8 assemblage plus *Homo* sp. In addition, we considered two other factors: TAB (TAB-m or TAB-M) and *Homo* sp. requirements (H_{min} or H_{max}). Thus, we had six different scenarios for TD8 (Table S2 and Figure 4). Comparing the scenarios, we found the worst case to be the hypothetical assemblage of TD8 with *Homo* sp. meeting 60% of its energetic requirements from meat (H_{max}) and TAB at a minimum (TAB-m). The scenario with less competition between secondary consumers occurs when TAB is at a maximum (TAB-M). As might be expected, the parameter with the greatest effect is TAB (Table 6 and Figure 4).

Distribution of resources

The distribution of resources in all scenarios yield viable ecosystems with regard to the minimun viable population density (MVPD) of the species involved, estimated using equations provided by Silva and Downing [108] (Table 6 and Figure 4). Nevertheless, it should be noted that the MVPD corresponds to

Table 3. Percentage of predation (PD) of the TD8 carnivores plus *Homotherium latidens* and *Homo* sp. by body mass category.

Species	Body size range (kg)					
	10–45	45–90	90–180	180–360	360–1,000	>1,000
Canis mosbachensis	17	17	17	17	17	17
Crocuta crocuta	21	32	26	11	5	5
Homo sp.	24	29	19	14	10	5
Homotherium latidens	6	6	28	33	22	6
Hyaena sp.	17	17	17	17	17	17
Lynx sp.	75	25	0	0	0	0
Panthera gombaszoegensis	6	6	31	38	19	0
Ursus sp.	17	17	17	17	17	17

The scores represent the relative importance that animals in each body size category had in the carnivore's diet. The sum of the scores in a row is 1. A score of 0 indicates no consumption of mammals in that body size category.

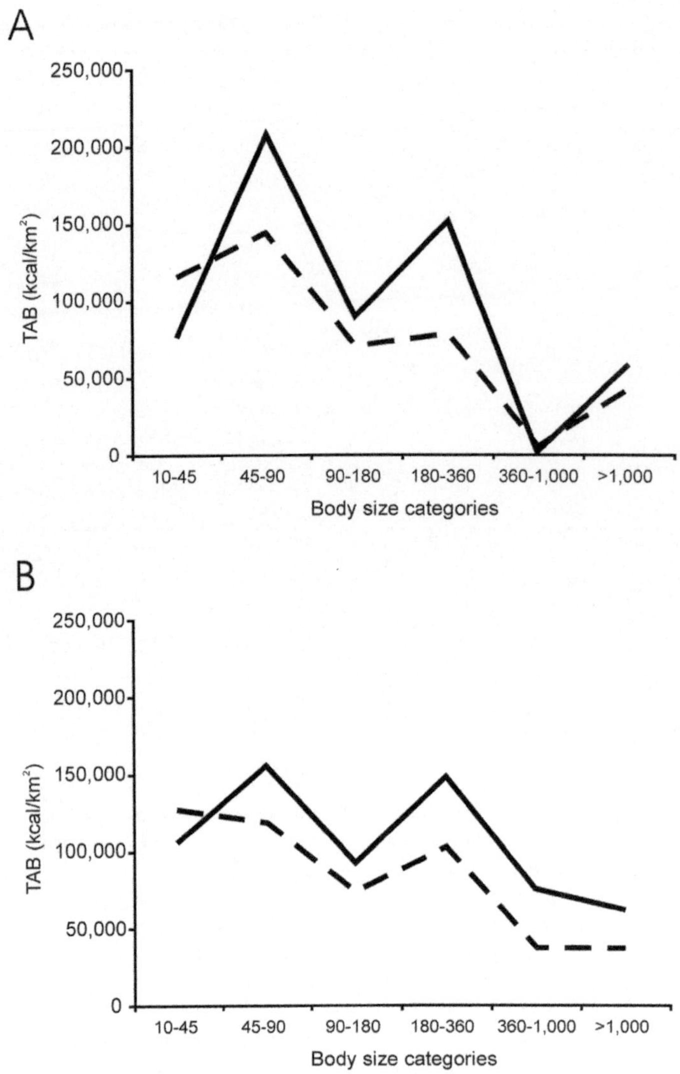

Figure 3. Biomass output distributed among six body size categories (see text). Dashed line: maximum sub-adult mortality rate for all primary consumer species or TAB-m; solid line: minimum sub-adult mortality rate (greater than or equal to the adult mortality rate) or TAB-M. A: TD6-2 available biomass; B: TD8 available biomass.

populations living in extreme conditions and prone to extinction because MVPD is estimated on the basis of populations classified as "endangered", "vulnerable", or "close to extinction" by the IUCN [109].

With respect to the energy requirements met, mainly the species that satisfies a smaller percentage of its energy requirement in all cases is *Panthera gombaszoegensis* while *Lynx* sp. is the least affected by the competition for resources (Table S2). Even with the inclusion of *Homo* sp. all species reach viable densities, slightly higher when *Homo* sp. has minor requirements. We used as reference density for *Homo* sp. 0.24 individuals per square kilometre, the highest density observed in recent hunter-gatherers (Table 2). In the worst case, scenario with *Homo* sp. and TAB-m condition, *Panthera gombaszoegensis* met 49% of their requirements with 0.06 individuals per square kilometre; *Canis mosbachensis*, *Ursus* sp. and *Hyaena* sp. met 71% of their requirements with 0.24, 0.03 and 0.09 individuals per square kilometre, respectively; *Crocuta crocuta* met 61% of its requirements with a density of 0.08 individuals per square kilometre; and *Lynx* sp. met 81% of its requirements with a density of 0.38 individuals per square kilometre. *Homo* sp. met 60% of its requirements with a density of 0.15 individuals per square kilometre for TAB-m (see SCI in Table S2).

Comparison between TD8 and TD6-2

We take the conditions for the human groups inhabiting Atapuerca 900,000 years ago, represented by the TD6-2 assemblage, as a reference against which resource availability and competition between carnivores in TD8 may be compared (Figure 4). TAB varies in TD6-2 from a minimum of 457,693 kcal/km^2*year to a maximum of 583,638 kcal/km^2*year (see Table 3 in [57]), in both of which are below the respective values observed for TD8 (496,659 and 643,201 kcal/km^2*year, respectively). Moreover, in the TD6-2 assemblage, the maximum available resources (TAB-M) are higher than the maximum consumer requirements (TDB-M = 461,028 kcal/km2*year when

Table 4. Maximum and minimum total available biomass (TAB) in kcal/km^2*yr, distributed over the six body size categories for the TD8 assemblage.

Body size range (kg)	10–45	45–90	90–180	180–360	360–1,000	>1,000	Total
TAB-M	106,512	155,722	92,136	149,351	76,849	62,630	643,201
TAB-m	126,928	118,768	72,589	102,942	38,128	37,304	496,659

The maximum TAB (TAB-M) is obtained when the sub-adult mortality rate is 0, while the minimum TAB (TAB-m) is obtained when all populations sustain the maximum possible sub-adult mortality rate while simultaneously satisfying the premises of stability and stationarity.

Homo antecessor consumes more meat) and thus, all species are able to fulfil their requirements. Meat resources were high enough at TD6-2 to sustain the expected densities, even in excess (Table S3), except in the case of the most unfavourable scenario with TAB-m and the diet of *Homo antecessor* including a high amount of meat. Conversely, in the case of the TD8 assemblage, this does not occur, and TAB-M (643,201 kcal/km^2*year) is insufficient to satisfy the secondary consumers' requirements (TBD is equal to 689,279 kcal/km^2*year). Despite that TAB being lower in TD6-2 than in TD8, TD6-2-TAB meets the carnivores' requirements and TD8-TAB does not. This implies a higher degree of competition between secondary consumers in TD8 than in TD6-2. When the global competition indices are compared (Figure 4), we found that TD8 had a higher GCI than TD6-2, whether *Homo antecessor* is assumed to be highly dependent on animal resources (H_{max}) or not (H_{min}) for TAB-M condition but not for H_{max} with TAB-m. In addition, we evaluated two hypothetical scenarios: the TD6-2 assemblage with a large felid such as *Homotherium latidens* [60] and TD8 with *Homo* sp. In the hypothetical scenario of the TD6-2 assemblage with *Homotherium latidens* and TAB-m, TD8-GCI is similar to TD6-2-GCI with *Homo antecessor* consuming H_{min}. When *Homo antecessor* is considered to be more dependent on meat resources (H_{max}), TD6-2-GCI is slightly higher than TD8-GCI. However, if *Homo* sp. is included in the TD8 assemblage, both for H_{min} and H_{max} scenarios, competition increases to a level higher than observed in TD6-2 with *Homotherium latidens* (Figure 4 and Table 6). As it might be expected, the highest competition is observed when *Homo* is included in the TD8 assemblage with high meat consumption requirements (H_{max}). Under these conditions, GCI is 0.35 for TAB-m and greater than 0.2 for TAB-M (Figure 4 and Table 6). It is worth noting that the inclusion of *Homo* sp. and *Homotherium latidens* in TD8 and TD6-2 levels softens the differences between both macromammal guilds.

Discussion

Evidence of human presence in Europe during the period 0.5–0.7 Ma is very scarce. Although several European sites provide abundant evidence of human presence before and after this period, the European continent was apparently depopulated during the 0.5–0.7 Ma interval. The results presented here suggest that during this period, the environment was more hostile to a hominin population than it was previously. The Atapuerca environment was able to sustain the expected population densities of a diverse carnivore guild, including *Homo antecessor*, 0.9 Ma, as shown by the application of the model to the TD6-2 assemblage, even with the inclusion of a large felid not recorded in the fossil assemblage. In contrast, the higher values of the global competition index (GCI) obtained for TD8 in comparison to TD6-2 suggest an environment with a higher intraguild competition for resources at Atapuerca 0.6 Ma. Consequently, survival opportunities for *Homo* at Atapuerca would have been higher in the late Early Pleistocene than in the early Middle Pleistocene. Certainly the value of the population density obtained for *Homo* at TD8 is higher than the minimum viable population size (MVPS) for these species, according to the equation provided by Silva and Downing [108]. Nevertheless, the value of MVPD should be considered an absolute minimum and not an average for a sustainable population. For this reason, we focus our interpretation on the relative values of GCI in concluding that the conditions for a human population were worse at TD8 than at TD6-2 and that the Atapuerca environment was less suitable for a permanent human settlement at 0.6 Ma than at 0.9 Ma. These results support the

Table 5. Requirements of the secondary consumers in the TD8 assemblage plus *Homo* sp.

Species	W (kg)	D (ind/km^2)	Requirements (kcal/km^2/year)	Percentage requirement (%)	Percentage requirement with a less *Homo* sp. hunter (%)	Percentage requirement with a more *Homo* sp. hunter (%)
Canis mosbachensis	12	0.42	40,062	0.06	0.05	0.05
Crocuta crocuta	75	0.13	217,864	0.32	0.28	0.26
Lynx sp,	10	0.47	19,825	0.03	0.03	0.02
Hyaena sp,	50	0.17	162,932	0.24	0.21	0.19
Panthera gombaszoegensis	90	0.11	224,627	0.33	0.29	0.27
Ursus dolinensis	282	0.06	23,970	0.03	0.03	0.03
Homo sp. (30%)	76	0.24	78,840	-	0.10	-
Homo sp. (60%)	76	0.24	157,680	-	-	0.19

Two different levels of meat consumption were tested for *Homo* sp. In the first case, meat represents 30% of the energy intake, while in the second case, meat represents up to 60% of energy intake.

hypothesis that the human absence from TD8 was related to a more hostile environment during this period at Atapuerca, characterised by a higher competition and less access to resources than in the previous period. These results support the interpretation of Blain et al. [110] that the Atapuerca area was not necessarily always a suitable place for human settlement.

The differences between TD6-2 and TD8 are due to different community structures. Cuenca-Bescós and García [111] and Cuenca-Bescós et al. [112] differentiate several faunal units (FU) in Gran Dolina and separate TD6-2 and TD8 assemblages into two different faunal units FU4 and FU5, respectively. FU4 is characterised by the presence of *Homo antecessor* and the large red-toothed shrew *Dolinasorex glyphodon*, and FU5 is considered the local range zone of *Microtus ratticepoides* and *Blanus cinereus*

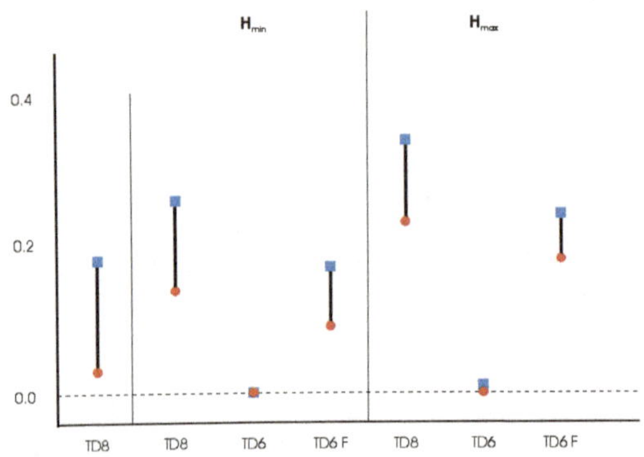

Figure 4. Graphical representation of the global competition index (GCI) for different scenarios. H_{min} denotes the scenarios that include human presence and assume low animal resource requirements for the human population, while H_{max} denotes the scenarios that include human presence and assume high animal resource requirements. TD8: TD8 assemblage; TD6: TD6-2 assemblage. TD6 F: TD6-2 assemblage plus *Homotherium latidens*. GCI takes values from 0 to 1, 0 being minimum competition and 1 being maximum competition. The solid circles represent the condition of maximum total available biomass (TAB-M). The solid squares represent the condition of minimum total available biomass (TAB-m). The black bars represent the range of values between TAB-m and TAB-M.

[25,111,112]. With respect to large primary consumer mammals, it can be observed that the differences between TD6-2 and TD8 are reflected in *Castor fiber* and *Mammuthus* sp. which do not appear in the TD8 assemblage, and *Macaca* sp., *Megaloceros solilhacus* and *Hippopotamus* sp., which do not appear in TD6-2 assemblage. The rest of the species are equal in both assemblages. As *Castor fiber* and *Mammuthus* sp. are approximately equivalent in terms of biomass to *Macaca* sp. and *Hippopotamus* sp., respectively (because these species belong to the same body mass categories) the main change between both assemblages is the presence of *Megaloceros solilhacus*. This species provides biomass to the categories of 45 to 90 kg, 180 to 360 kg and 360 to 1,000 kg. As shown in Figures 3A and 3B, the patterns of biomass output distribution are different in both assemblages. In the TD8 assemblage, the fourth (180–360 kg) and fifth (360–1,000 kg) categories contain higher proportion of biomass than in TD6-2. TD8 presents a more balanced biomass distribution than TD6-2, with a trapezoidal form rather than a triangular form.

The differences between TD6-2 and TD8 are larger for the secondary consumer group. *Canis mosbachensis*, *Crocuta crocuta*, *Lynx* sp. and *Ursus* sp. (*Ursus dolinensis* to TD6-2 level) are present in both assemblages. The TD8 level marks the last occurrence of *Canis mosbachensis* and *Crocuta crocuta* in the Gran Dolina sequence [112]. The main difference between TD8 and TD6-2 is the absence of *Homo antecessor* and the presence of *Panthera gombaszoegensis* and *Hyaena* sp. in the younger assemblage. *Panthera gombaszoegensis* is recorded at Atapuerca in the lower levels of Gran Dolina in TD3-TD4 and TD5 and also in Sima del Elefante in TE9 and TE12, all of which are older than 0.7 Ma [25]. If *Homo antecessor* were considered an effective hunter, the main difference between the TD6-2 and TD8 palaeoecosystems would be the presence of the scavenger *Hyaena* sp. In addition, it may be assumed that Middle Pleistocene human groups also used scavenging as a feeding strategy and were competitors of *Hyaena* sp., which might also be considered a partial ecological equivalent of *Homo*. In this interpretation, *Homo* would be replaced in the TD8 assemblage by two competitors: *Hyaena* sp. and *Panthera gombaszoegensis*. Alternatively, the absence of *Homo* may be interpreted as a key factor in determining the structure of the TD8 carnivore guild, allowing the presence of certain species that would be competitively displaced if a human population, absent for reasons other than competition, were

Table 6. Global competition index (GCI) for TD8 and TD6-2 assemblages and several hypothetical scenarios.

	TAB-m	TAB-M
TD8	0.19	0.04
TD8 H_{min}	0.27	0.14
TD6 H_{min}	0.00	0.00
TD6 F H_{min}	0.17	0.09
TD8 H_{max}	0.35	0.23
TD6 H_{max}	0.01	0.00
TD6 F H_{max}	0.24	0.18

TAB-m: minimum total available biomass; TAB-M: maximum total available biomass; H_{min}: meat represents 30% of the energy intake; H_{max}: meat represents 60% of the energy intake; F: scenario with a large felid such as *Homotherium latidens*.

present in the area. In this alternative interpretation, the absence of human settlement at Atapuerca 0.6 Ma would not be the consequence but the cause of the community structure observed in the TD6-2 assemblage.

Although the results presented here suggest more adverse conditions in TD8 than in TD6-2 other possible explanations for the absence of hominins at TD8 should be considered. The Gran Dolina cave and its sedimentation dynamics constituted a favourable environment for the establishment of carnivore dens and shelters 0.6 Ma [66], but may not have been favourable for human access, unlike TD6-2, because the entrance into the cave was smaller during the deposition of TD8 [66]. The possible effect of an accumulation bias in the TD8 fossil assemblage against certain species should also be considered because the main accumulation agents were hyaenas [66]. Proboscideans were a common faunal element in the Iberian Peninsula during the late Early, Middle and Late Pleistocene [113], but their remains are rarely transported to dens and carnivore shelters, except maybe in the case of *Homotherium* [114,115]. However, because the secondary consumers present at TD8 were not able to prey on proboscideans, their presence at Atapuerca 0.6 Ma would have had little effect on the intensity of competition and thus would not affect our results. The hypothetical presence of the sabertooth *Homotherium latidens*, a large predator that has been recorded at Gran Dolina in level TD5 (Early Pleistocene) and in level TD10-3 (Middle Pleistocene) [25] would have significantly increased intraguild competition. Given that saberthooths were likely able to kill juvenile, and very old proboscideans [55,57], if both *Homotherium latidens* and a proboscidean species were present at Atapuerca 0.9 Ma, biomass output would have been increased in the heavier body mass categories, but little of this additional biomass could have been consumed by predators. The presence of a proboscidean would increase resource availability carrion, which would have reduced competition among scavengers. Conversely, competition between predators would be increased by the presence of a top predator like *Homotherium*. Thus, the presence of these species at Atapuerca in the early Middle Pleistocene would change the values of the results presented here, but not their meaning nor our interpretation.

With respect to the debate about the continuity or discontinuity of the peopling of Europe in the early Middle Pleistocene, the results presented here show that a discontinuity in hominin occupation existed at Gran Dolina that coincided with an increase in carnivore competition. According to this interpretation, at approximately 0.6 Ma the TD8 palaeoecosystem was not as favourable to maintaining a hominin settlement as it was at 0.9 Ma, and Atapuerca had no conditions that made it favourable as refuge for human populations, thus representing a "bottleneck" [49] to the human population at the Sierra de Atapuerca. Finlayson et al. [116] propose that *Homo* preferred mosaic landscapes and that homogeneous habitats were ignored by humans throughout the Palaeolithic. Although the Palaeolithic human groups were able to adapt to a wide range of environments [117], they required heterogeneous habitats [116]. If the Atapuerca palaeoecosystem were more homogenous 0.6 than 0.9 Ma, the absence of hominins at TD8 could be explained by their inability to settle homogeneous environments. With respect to vegetation, and according to Rodríguez et al. [25], TD8 presents a high dominance of Mediterranean taxa, while in TD6-2, Mediterranean taxa occur together with cold and dry adapted species, suggesting the existence of a mosaic habitat with areas of steppe vegetation. The preference of *Homo* for heterogeneous environments could indicate its inability to directly compete against specialist species, like open habitat cursorial predators [118], and its need for high environmental diversity to obtain resources using a generalist strategy.

It is likely, environmental conditions for *Homo* were better 0.9 Ma than in later periods, due to higher resource availability and less competition with other carnivores. A faunal turnover occurred in Europe at the beginning of the Middle Pleistocene [4,54,58], marked by the appearance of new carnivores and increased herbivore richness [4,57], that changed the structure of mammalian communities throughout Europe.

This dramatic change in the ecological scenario at the beginning of the Middle Pleistocene might be linked to the appearance in Europe of behavioral and technological improvements and innovations that increased the hunting abilities and survival opportunities of hominins. The use of fire for food processing improved the palatability and edibility of foods and increased energy gain [119]. Interestingly, both technology and hominin populations were completely different before and after the critical interval of the alleged European depopulation. Nevertheless, the subsistence strategies of human groups were successful in both the Early and Middle Pleistocene [120,121]. Roebroeks [67] suggests a demographic population increase and a human expansion in the Middle Pleistocene on the basis of the technological improvements that occurred during this time.

It is difficult to point towards a unique explanation for the absence of humans from TD8, but what can be said is that survival opportunities for a human population were worse at TD8 than at

TD6-2 because intraguild competition was higher. Nevertheless, it is necessary to take into account that the excavation of TD8 affected an area of only 24 m^2 [122]. It is expected that future excavation of a larger area of this level will provide more faunal remains and other evidence to shed light on the hominin population continuity or discontinuity debate.

The structure of the mammalian palaeocommunity and the intensity of competition in the predatory guild likely constrained human ecodynamics in the Pleistocene. This study provides a new avenue for evaluating the relative suitability of a palaeoenvironment for a human hunter-gatherer population, taking into consideration resource availability and competition intensity. Fossil assemblages with and without evidence of human presence may be compared providing a different way to approach the study of variations in human presence in the Palaeolithic. Although this approach does not provide information about the mechanisms that may have produced changes from one community structure to another, it quantifies changes in trophic relationships using the global competition index (GCI). Investigation of more European sites using a similar approach will help to test whether the apparent discontinuity in the peopling of Europe is related to more adverse environmental conditions or decreased survival opportunities at the beginning of the Middle Pleistocene (0.7−0.5 Ma), as suggested by the Atapuerca record, or whether other mechanisms are implicated.

Conclusions

The reconstruction of trophic relationships in palaeocommunities and the comparison of the intensities of intraguild competition, using the global competition index (GCI) presented here make it possible to evaluate the relative suitability of past ecosystems for hominins. Increased intraguild competition made conditions at Atapuerca more adverse for secondary consumers at 0.6 than 0.9 Ma. This increased competition is a possible cause of the apparent depopulation of the Atapuerca area in the early Middle Pleistocene. Nevertheless, more sites of this period should be studied with a similar approach to determine whether increased intraguild competition for resources played a role in changes in the distribution of *Homo* in Europe in the 0.5–0.7 Ma interval. Both sites with and without evidence of human presence should be compared to shed light on the fiercely debated question of the continuity *vs.* discontinuity of human occupation in Europe at the beginning of the Middle Pleistocene. This study demonstrates that mathematical modeling is a helpful tool in addressing this topic.

Supporting Information

Table S1 Specimen Number.

Table S2 Sustainable densities of the TD8 secondary consumers for six different scenarios (see text) for maximum and minimum total available biomass (TAB). Estimated density of carnivores (individuals per square kilometre), nutritional requirements (kilocalories per year), total intake (kilocalories per year), unsatisfied requirements (kilocalories per year), sustainable density (individuals per square kilometre), Species Competition Index (SCI). Total intake is defined as the biomass (in kcal) consumed by the species after dividing TAB among the secondary consumers, taking into account the distribution of TAB by body size category (Table 4) and the carnivore preferences (Table 3).

Table S3 Sustainable densities of TD6-2 secondary consumers for eight different scenarios (see text) for maximum and minimum total available biomass (TAB). Estimated density of carnivores (individuals per square kilometre), nutritional requirements (kilocalories per year), total intake (kilocalories per year), unsatisfied requirements (kilocalories per year), sustainable density (individuals per square kilometre), Species Competition Index (SCI), Total intake is defined as the biomass (in kcal) consumed by the species after dividing TAB among the secondary consumers, taking into account the distribution of TAB by body size category (Table 4) and the carnivore preferences (Table 3).

Acknowledgments

We are grateful to P. Palmqvist and the Academic Editor for their useful comments and suggestions on a previous version of this manuscript. An anonymous editor from Elsevier's Language services improved the English of the original manuscript. We also thank Estefanía Pérez Fernández for her suggestions and support in the revision of the manuscript. We are very grateful to the participants in the fieldwork and all of the members of the Atapuerca Research Team.

Author Contributions

Conceived and designed the experiments: AM J. Rodríguez. Performed the experiments: GRG. Analyzed the data: GRG AM J. Rodríguez. Contributed reagents/materials/analysis tools: JAMG. Wrote the paper: GRG AM JAMG RB J. Rosell J. Rodríguez. Discussion: GRG AM JAMG RB J. Rosell J. Rodríguez.

References

1. Gómez-Robles A, Bermúdez de Castro JM, Arsuaga J-L, Carbonell E, Polly PD (2014) No known hominin species matches the expected dental morphology of the last common ancestor of Neanderthals and modern humans. Proc Natl Acad Sci. In press.
2. Bermúdez de Castro JM, Martinón-Torres M, Blasco R, Rosell J, Carbonell E (2013) Continuity or discontinuity in the European Early Pleistocene human settlement: the Atapuerca evidence. Quat Sci Rev 76: 53–65.
3. Muttoni G, Scardia G, Kent DV (2010) Human migration into Europe during the late Early Pleistocene climate transition. Palaeogeogr Palaeoclimatol Palaeoecol 296: 79–93.
4. Palombo MR (2010) A scenario of human dispersal in the northwestern Mediterranean throughout the Early to Middle Pleistocene. Quatern Int 223–224: 179–194.
5. O'Regan HJ (2008) The Iberian Peninsula – corridor or cul-de-sac? Mammalian faunal change and possible routes of dispersal in the last 2 million years. Quat Sci Rev 27: 2136–2144.
6. O'Regan HJ, Turner A, Bishop LC, Lamb AL (2011) Hominins without fellow travellers? First appearances and inferred dispersals of Afro-Eurasian large-mammals in the Plio-Pleistocene. Quat Sci Rev 30: 1343–1352.
7. Made Jvd, Mateos A (2010) Longstanding biogeographic patterns and the dispersal of early Homo out of Africa and into Europe. Quatern Int 223–224: 195–200.
8. Carbonell E, Sala Ramos R, Rodríguez XP, Mosquera M, Ollé A, et al. (2010) Early hominid dispersals: A technological hypothesis for "out of Africa". Quatern Int 223–224: 36–44.
9. Palombo MR (2013) What about causal mechanisms promoting early hominin dispersal in Eurasia? A research agenda for answering a hotly debated question. Quatern Int 295: 13–27.
10. Agustí J, Blain H-A, Cuenca-Bescós G, Bailon S (2009) Climate forcing of first hominid dispersal in Western Europe. J Hum Evol 57: 815–821.
11. Mosquera M, Ollé A, Rodríguez XP (2013) From Atapuerca to Europe: Tracing the earliest peopling of Europe. Quatern Int 295: 130–137.
12. Jiménez-Arenas JM, Santonja M, Botella M, Palmqvist P (2011) The oldest handaxes in Europe: fact or artefact? J Archaeol Sci 38.
13. Arribas A, Palmqvist P (1999) On the Ecological Connection Between Sabretooths and Hominids: Faunal Dispersal Events in the Lower Pleistocene and a Review of the Evidence for the First Human Arrival in Europe. J Archaeol Sci 26: 571–585.

14. Bermúdez de Castro JM, Martinón-Torres M (2013) A new model for the evolution of the human Pleistocene populations of Europe. Quatern Int 295: 102–112.
15. Dennell RW, Martinón-Torres M, Bermúdez de Castro JM (2011) Hominin variability, climatic instability and population demography in Middle Pleistocene Europe. Quat Sci Rev 30: 1511–1524.
16. Almogi-Labin A (2011) The paleoclimate of the Eastern Mediterranean during the transition from early to mid Pleistocene (900 to 700 ka) based on marine and non-marine records: An integrated overview. J Hum Evol 60: 428–436.
17. Amit R, Enzel Y, Sharon D (2006) Permanent Quaternary hyperaridity in the Negev, Israel, resulting from regional tectonics blocking Mediterranean frontal systems. Geology 34: 509–512.
18. Rodríguez J, Martín-González JA, Goikoetxea I, Rodríguez-Gómez G, Mateos A (2013) Mammalian paleobiogeography and the distribution of Homo in early Pleistocene Europe. Quatern Int 295: 48–58.
19. Duval M, Falguères C, Bahain J-J (2012) Age of the oldest hominin settlements in Spain: Contribution of the combined U-series/ESR dating method applied to fossil teeth. Quat Geochronol 10: 412–417.
20. Duval M, Falguères C, Bahain J-J, Grün R, Shao Q, et al. (2012) On the limits of using combined U-series/ESR method to date fossil teeth from two Early Pleistocene archaeological sites of the Orce area (Guadix-Baza basin, Spain). Quatern Res 77: 482–491.
21. Toro-Moyano I, Martínez-Navarro B, Agustí J, Souday C, Bermúdez de Castro JM, et al. (2013) The oldest human fossil in Europe dated to ca. 1.4 Ma at Orce (Spain). J Hum Evol 65: 1–9.
22. Martínez K, Garcia J, Carbonell E, Agustí J, Bahain J-J, et al. (2010) A new Lower Pleistocene archeological site in Europe (Vallparadís, Barcelona, Spain). Proc Natl Acad Sci 107: 5762–5767.
23. Duval M, Moreno D, Shao Q, Voinchet P, Falguères C, et al. (2011) Datación por ESR del yacimiento arqueológico del Pleistoceno inferior de Vallparadís (Terrassa, Cataluña, España). Trabajos de Prehistoria 68: 7–24.
24. Madurell-Malapeira J, Alba DM, Minwer-Barakat R, Aurell-Garrido J, Moyà-Solà S (2012) Early human dispersals into the Iberian Peninsula: A comment on Martínez et al. (2010) and Garcia et al. (2011). J Hum Evol 62: 169–173.
25. Rodríguez J, Burjachs F, Cuenca-Bescós G, García N, Made Jvd, et al. (2011) One million years of cultural evolution in a stable environment at Atapuerca (Burgos, Spain). Quat Sci Rev 30: 1396–1412.
26. Arzarello M, Marcolini F, Pavia G, Pavia M, Petronio C, et al. (2007) Evidence of earliest human occurrence in Europe: the site of Pirro Nord (Southern Italy). Naturwissenschaften 94: 107–112.
27. Bahain J-J, Falguères C, Voinchet P, Duval M, Dolo J-M, et al. (2007) Electron Spin Resonance (ESR) dating of some European Late Lower Pleistocene Sites. Quaternaire 18: 175–186.
28. de Lumley H, Kahlke H-D, Moigne A-M, Moulle P-E (1988) Les faunes de grands mammifères de la grotte du Vallonnet, Roquebrune-Cap-Martin, Alpes-Maritimes. L' Anthropologie 92: 465–495.
29. Voinchet P (2002) Datations par résonance paramagnétique électronique (RPE) de quartz blanchis extraits de sédiments fluviatiles pléistocènes: contribution méthodologique et application aux systèmes de la Creuse, du Loir et de l'Yonne. Paris.
30. Despriée J, Gageonnet R (2003) La très haute nappe alluviale d'âge pléistocène inférieur de la vallée de la Creuse à Eguzon (Indre): figures de cryoturbations, habitats préhistoriques et datations absolues. Bulletin de la Société géologique de France 174: 383–400.
31. Despriée J, Gageonnet R, Voinchet P, Bahain J-J, Falguères C, et al. (2006) Une occupation humaine au Pléistocène inférieur sur la bordure nord du Massif central. C R Palevol 5: 821–828.
32. Parfitt SA, Coope GR, Field MH, Peglar SM, Preece RC, et al. (2010) Middle Pleistocene biota of the early Anglian 'Arctic Fresh-water Bed' at Ostend, Norfolk, UK. Proc Geol Assoc 121: 55–65.
33. Parés JM, Duval M, Arnold L (2013) New views on an old move: Hominin migration into Eurasia. Quatern Int 295: 5–12.
34. Fiedler L, Franzen JL (2002) Artefakte vom altpleistozänen Fundplatz "Dorn-Dürkheim 3" am nördlichen Oberrhein. Germania 80: 421–440.
35. Stuart AJ, Lister AM (2001) The mammalian faunas of Pakefield/Kessingland and Corton, Suffolk, UK: evidence for a new temperate episode in the British early Middle Pleistocene. Quat Sci Rev 20: 1677–1692.
36. Crochet J-Y, Welcomme J-L, Ivorra J, Ruffet G, Boulbes N, et al. (2009) Une nouvelle faune de vertébrés continentaux, associée à des artefacts dans le Pléistocène inférieur de l'Hérault (Sud de la France). C R Palevol 8: 725–736.
37. Barsky D, Garcia J, Martínez K, Sala R, Zaidner Y, et al. (2013) Flake modification in European Early and EarlyeMiddle Pleistocene stone tool assemblages. Quatern Int 316: 140–154.
38. Bermúdez de Castro JM, Martinón-Torres M, Gómez-Robles A, Prado-Simón L, Martín-Francés L, et al. (2011) Early Pleistocene human mandible from Sima del Elefante (TE) cave site in Sierra de Atapuerca (Spain): A comparative morphological study. J Hum Evol 61: 12–25.
39. Carbonell E, Bermúdez de Castro JM, Parés JM, Pérez-González A, Cuenca-Bescós G, et al. (2008) The first hominin of Europe. Nature 452.
40. Muttoni G, Scardia G, Kent DV (2013) A critique of evidence for human occupation of Europe older than the Jaramillo subchron (w1 Ma): Comment on 'The oldest human fossil in Europe from Orce (Spain)' by Toro-Moyano et al. (2013). J Hum Evol 65: 746–749.
41. Despriée J, Voinchet P, Tissoux H, Bahain J-J, Falguères C, et al. (2011) Lower and Middle Pleistocene human settlements recorded in fluvial deposits of the middle Loire River Basin, Centre Region, France. Quat Sci Rev 30: 1474–1485.
42. Barsky D, de Lumley H (2010) Early European Mode 2 and the stone industry from the Caune de l'Arago's archeostratigraphical levels "P". Quatern Int 223–224: 71–86.
43. Ollé A, Mosquera M, Rodríguez XP, Lombera-Hermida Ad, García-Antón MD, et al. (2013) The Early and Middle Pleistocene technological record from Sierra de Atapuerca (Burgos, Spain). Quatern Int 295: 138–167.
44. Doronichev V, Golovanova L (2010) Beyond the Acheulean: A view on the Lower Paleolithic occupation of Western Eurasia. Quatern Int 223–224: 327–344.
45. Roberts MB, Parfitt SA (2000) Boxgrove: A Middle Pleistocene Hominid Site at Eartham Quarry, Boxgrove, West Sussex. Geoarchaeology: An International Journal, 15: 819–822.
46. Mania D, Vlcek E (1999) The Bilzingsleben site: Homo erectus, his culture and his ecosphere. In: Ullrich H, editor. Hominid Evolution: Lifestyles and Survival Strategies: Archaea. pp. 293–314.
47. Thieme H (1997) Lower Palaeolithic hunting spears from Schöningen, Germany. Nature 358: 807–810.
48. Roebroeks W, Villa P (2011) On the earliest evidence for habitual use of fire in Europe. Proc Natl Acad Sci 108: 5209–5214.
49. Bennett KD, Provan J (2008) What do we mean by 'refugia'? Quat Sci Rev 27: 2449–2455.
50. Tattersall I (2011) Before the Neanderthals: Hominid Evolution in Middle Pleistocene Europe. In: Condemi S, Weniger G-C, editors. Continuity and Discontinuity in the Peopling of Europe: One Hundrer Fifty Years of Neanderthal Study: Vertebrate Paleobiology and Paleoanthropology - Springer.
51. Hublin JJ (2009) The origin of Neandertals. Proc Natl Acad Sci 106: 16022–16027.
52. Mallegni F (2011) The Earliest European Human Peopling After the Recent Discoveries: Early Neanderthals or Different Lineages? In: Condemi S, Weniger G-C, editors. Continuity and Discontinuity in the Peopling of Europe: One Hundrer Fifty Years of Neanderthal Study: Vertebrate Paleobiology and Paleoanthropology - Springer.
53. Stringer C (2012) The Status of Homo heidelbergensis (Schoetensack 1908). Evol Anthropol 21: 101–107.
54. Turner A (1992) Large carnivores and earliest European hominids: changing determinants of resource availability during the Lower and Middle Pleistocene. J Hum Evol 22: 109–126.
55. Espigares MP, Martínez-Navarro B, Palmqvist P, Ros-Montoya S, Toro I, et al. (2013) *Homo* vs. *Pachycrocuta*: Earliest evidence of competition for an elephant carcass between scavengers at Fuente Nueva-3 (Orce, Spain). Quatern Int 295: 113–125.
56. Palmqvist P, Martínez-Navarro B, Pérez-Claros JA, Torregrosa V, Figueirido B, et al. (2011) The giant hyena Pachycrocuta brevirostris: Modelling the bone-cracking behavior of an extinct carnivore. Quatern Int 243: 61–79.
57. Rodríguez J, Rodríguez-Gómez G, Martín-González JA, Goikoetxea I, Mateo A (2012) Predator–prey relationships and the role of *Homo* in Early Pleistocene food webs in Southern Europe. Palaeogeogr Palaeoclimatol Palaeoecol 365–366: 99–114.
58. Palombo MR (2007) What is the Boundary for the Quaternary Period and Pelistocene Epoch? The Contribution of Turnover Patterns in Large Mammalian Complexes from North-Western Mediterranean to the Debate. Quaternaire 18: 35–53.
59. Martínez-Navarro B, Palmqvist P (1996) Presence of the African Sabertoothed Felid *Megantereon whitei* (Broom, 1937) Mammalia, Carnivora, Machair-odontinae in Apollonia-1 (Mygdonia Basin, Macedonia, Greece). J Archaeol Sci 23: 869–872.
60. Rodríguez-Gómez G, Rodríguez J, Martín-González JA, Goikoetxea I, Mateos A (2013) Modeling trophic resource availability for the first human settlers of Europe: The case of Atapuerca-TD6. J Hum Evol 64: 645–657.
61. Parés JM, Arnold L, Duval M, Demuro M, Pérez-González A, et al. (2013) Reassessing the age of Atapuerca-TD6 (Spain): New paleomagnetic results. J Archaeol Sci 40: 4586–4595.
62. Berger GW, Pérez-González A, Carbonell E, Arsuaga JL, Bermúdez de Castro JM, et al. (2008) Luminescence chronology of cave sediments at the Atapuerca paleoanthropological site, Spain. J Hum Evol 55: 300–311.
63. Falguères C, Bahain J-J, Yokoyama Y, Arsuaga JL, Bermúdez de Castro JM, et al. (1999) Earliest humans in Europe: the age of TD6 Gran Dolina, Atapuerca, Spain. J Hum Evol 37: 343–352.
64. Bischoff JL, Williams RW, Rosenbauer RJ, Aramburu A, Arsuaga JL, et al. (2007) High-resolution U-series dates from the Sima de los Huesos hominids yields 600 kyrs: implications for the evolution of the early Neanderthal lineage. J Archaeol Sci 34: 763–770.
65. Parés JM, Pérez-González A (1999) Magnetochronology and stratigraphy at Gran Dolina section, Atapuerca (Burgos, Spain). J Hum Evol 37: 325–342.
66. Blasco R, Rosell J, Made Jvd, Rodríguez J, Campeny G, et al. (2011) Hiding to eat: the role of carnivores in the early Middle Pleistocene from the TD8 level of Gran Dolina (Sierra de Atapuerca, Burgos, Spain). J Archaeol Sci 38: 3373–3386.

67. Palmiqvist P, Arribas A (2001) Taphonomic decoding of the paleobiological information locked in a lower Pleistocene assemblage of large mammals. Paleobiology 27: 512–530.
68. Made Jvd (2013) First description of the large mammals from the locality of Penal, and updated faunal lists for the Atapuerca ungulates Equus altidens, Bison; and human dispersal into Western Europe. Quatern Int 295: 36–47.
69. Owen-Smith N, Mills MGL (2008) Predator–prey size relationships in an African large-mammal food web. Journal of Animal Ecology 77: 173–183.
70. Roebroeks W (2001) Hominid behaviour and the earliest occupation of Europe: an exploration. J Hum Evol 41: 437–461.
71. Rodríguez J (1997) Análisis de la estructura de las comunidades de mamíferos del Pleistoceno de la Sierra de Atapuerca. Revisión de metodologías. Madrid: Universidad Autónoma de Madrid.
72. Damuth JD, MacFadden BJ (1990) Body Size in Mammalian Paleobiology: estimations and Biological Implications: Cambridge University Press.
73. Mazza PPA, Bertini A (2013) Were Pleistocene hippopotamuses exposed to climate-driven body size changes? Boreas 42: 194–209.
74. Díez JC, Fernández-Jalvo Y, Rosell J, Cáceres I (1999) Zooarchaeology and taphonomy of Aurora Stratum (Gran Dolina, Sierra de Atapuerca, Spain). J Hum Evol 37: 623–652.
75. Saladié P, Huguet R, Díez C, Rodríguez-Hidalgo A, Cáceres I, et al. (2011) Carcass transport decisions in *Homo antecessor* subsistence strategies. J Hum Evol 61: 425–446.
76. Carbonell E, Bermúdez de Castro JM, Arsuaga JL, Allue E, Bastir M, et al. (2005) An Early Pleistocene hominin mandible from Atapuerca-TD6, Spain. Proc Natl Acad Sci 102: 5674–5678.
77. Carbonell E, Bermúdez de Castro JM, Arsuaga JL, Díez JC, Rosas A, et al. (1995) Lower Pleistocene Hominids and Artifacts from Atapuerca-TD6 (Spain). Science 269: 826–830.
78. Rodríguez-Gómez G, Martín-González JA, Goikoetxea I, Mateos A, Rodríguez J (2014) A New Mathematical Approach to Model Trophic Dynamics of Mammalian Palaeocommunities. The Case of Atapuerca-TD6. In: Pardo-Igúzquiza E, Guardiola-Albert C, Heredia J, Moreno-Merino L, Durán JJ et al., editors. 15th Annual Conference of the International Association for Mathematical Geosciences. Madrid: Lectures Notes in Earth System Sciences - Springer. pp. 739–745.
79. Owen-Smith N (2010) Dynamics of Large Herbivore Populations in Changing Environments: Towards Appropriate Models. Malaysia: Blackwell Publishing Ltd.
80. Leslie PH (1945) On the use of matrices in certain population mathematics. Biometrika 33: 183–212.
81. Leslie PH (1948) Some further notes on the use of matrices in population mathematics. Biometrika 35: 213–245.
82. Levinton JS (1982) The body size-prey size hypothesis: the adequacy of body size as a vehicle for character displacement. Ecology 63: 869–872.
83. Carbone C, Mace GM, Roberts SC, Macdonald DW (1999) Energetic constraints on the diet of terrestrial carnivores. Nature 402: 286–288.
84. Radloff FGT, Toit JTD (2004) Large predators and their prey in a southern African savanna: a predator's size determines its prey size range. Journal of Animal Ecology 73: 410–423.
85. Damuth J (1981) Population density and body size in mammals. Nature 290: 699–700.
86. Viljoen PC (1993) The effects of changes in prey availability on lion predation in a natural ecosystem in northern. Zoological symposium 65: 193–213.
87. Gaillard J-M, Festa-Bianchet M, Yoccoz NG, Loison A, Töigo C (2000) Temporal variation in fitness components and population dynamics of large herbivores. Annual Review of Ecology Systems 31: 367–393.
88. Farlow JO (1976) A Consideration of the Trophic Dynamics of a Late Cretaceous Large-Dinosaur Community (Oldman Formation). Ecology 57: 841–857.
89. Blumenschine RJ, Madrigal TC (1993) Variability in Long Bone Marrow Yields of East African Ungulates and its Zooarchaeological Implications. J Archaeol Sci 20: 555–587.
90. Outram A, Rowley-Conwy P (1998) Meat and Marrow Utility Indices for Horse (Equus). J Archaeol Sci 25: 839–849.
91. Holekamp KE, Kolowski JM (2009) Family Hyaenidae (Hyaenas). In: Wilson DE, Mittermeier RA, editors. Handbook of the mammals of the world. Barcelona: Lynx Ediciones. pp. 234–262.
92. Beltrán JF, Sanjosé C, Delibes M, Braza F (1985) An analysis of the Iberian lynx predation upon fallow deer in the coto Doñana, SW Spain. XVIIth Congress of the International Union of Game Biologist: 961–967.
93. Delibes M (1980) El lince ibérico. Ecologia y comportamientos alimenticios en el Coto de Doñana, Huelva. Doñana, Acta Vertebrata 7: 1–128.
94. Rodríguez A (2008) Lince ibérico – Lynx pardinus (Temminck, 1827). Enciclopedia virtual de los vertebrados españoles.
95. Bocherens H, Argant A, Argant J, Billiou D, Crégut-Bonnoure E, et al. (2004) Diet reconstruction of ancient brown bears (Ursus arctos) from Mont Ventoux (France) using bone collagen stable isotope biogeochemistry (13C, 15N). Can J Zool 82: 576–586.
96. Parde J-M, Camarra J-J (1992) Encyclopédie des Carnivores de France, Part 5: L'Ours: (Ursus arctos, Linnaeus, 1758): Société Française Pour L'Etude et la Protection des Mammifères.
97. Palmqvist P, Pérez-Claros JA, Janis CM, Figueirido B, Torregrosa V, et al. (2008) Biogeochemical and ecomorphological inferences on prey selection and resource partitioning among mammalian carnivores in an Early Pleistocene community. PALAIOS 23: 724–737.
98. Damuth J (1993) Copes rule, the island rule and the scaling of mammalian population density. Nature 365: 748–750.
99. DeMaster DP, Stirling I (1981) Ursus maritimus. Mammalian Species 145: 1–7.
100. Larivière S (2001) Ursus americanus. Mammalian Species 647: 1–11.
101. Pasitschniak-Arts M (1993) Ursus arctos. Mammalian Species 439: 1–10.
102. Eaton SB, Eaton III SB, Konner MJ (1997) Paleolithic nutrition revisited: A twelve-year retrospective on its nature and implications. Nature 51: 207–216.
103. Carretero JM, Arsuaga JL, Martínez I, Quam R, Lorenzo C, et al. (2002) Los humanos de la Sima de los Huesos (Sierra de Atapuerca) y la evolución del cuerpo en el género Homo. In: Baquedano E, Rubio Jara S, editors. Miscelánea Homenaje a Emiliano Aguirre. Madrid: Paleoantropología. Museo Arqueológico Regional. pp. 120–134.
104. Bonmatí A, Arsuaga JL (2005) Inventory and preliminary description of Middle Pleistocene pelvis remains from the site of Sima de los Huesos, Atapuerca (Spain). Am J Phys Anthropol Res 126 (Suppl 40): 76.
105. Jenike MR (2001) Nutritional ecology: diet, physical activity and body size. In: Panter-Brick C, Layton RH, Rowley-Conwy P, editors. Hunter-gatherers An interdisciplinary perspective: Cambridge University Press.
106. Leonard WR, Robertson ML, Snodgrass JJ (2007) Energetic Models of Human Nutritional Evolution. In: Ungar PS, editor. Evolution of the human diet The Known, the Unknown, and the Unknowable: Oxford University Press. pp. 344–359.
107. Binford LR (2001) Constructing frames of reference. An analytical method for archaeological theory building using ethnographic and environmental data sets: University of California Press. 118–129 p.
108. Silva M, Downing JA (1994) Allometric Scaling of Minimal Mammal Densities. Conserv Biol 8: 732–743.
109. Thornback J, Jenkins M (1982) The IUCN mammal red data book. Gland, Switzerland: International Union for the Conservation of Nature.
110. Blain H-A, Cuenca-Bescós G, Burjachs F, López-García JM, Lozano-Fernández I, et al. (2013) Early Pleistocene palaeoenvironments at the time of the Homo antecessor settlement in the Gran Dolina cave (Atapuerca, Spain). J Quaternary Sci 28: 311–319.
111. Cuenca-Bescós G, García N (2007) Biostratigraphic succession of the Early and Middle Pleistocene mammal faunas of the Atapuerca cave sites (Burgos, Spain). Cour Forsch-Inst Senckenberg 255: 99–110.
112. Cuenca-Bescós G, Rofes J, López-García JM, Blain H-A, De Marfá RJ, et al. (2010) Biochronology of Spanish Quaternary small vertebrate faunas. Quatern Int 212: 109–119.
113. Made Jvd, Mazo AV (2001) Spanish Pleistocene Proboscidean diversity as a function of climate. The World of Elephants: Proceedings of the 1st International Congress Rome. pp. 214–218.
114. Antón M, Galobart A, Turner A (2005) Co-existence of scimitar-toothed cats, lions and hominins in the European Pleistocene. Implications of the post-cranial anatomy of Homotherium latidens (Owen) for comparative palaeoecology. Quat Sci Rev 24: 1287–1301.
115. Meade GE (1961) The saber toothed cat, *Dinoblastis serus*. Texas Memorial Museum 3: 23–60.
116. Finlayson C, Carrión J, Browna K, Finlayson G, Sánchez-Marco A, et al. (2011) The Homo habitat niche: Using the avian fossil record to depict ecological characteristics of Palaeolithic Eurasian hominins. Quat Sci Rev 30: 1525–1532.
117. Carrión JS, Rose J, Stringer CB (2011) Early Human Evolution in the Western Palaearctic: Ecological Scenarios. Quat Sci Rev 30: 1281–1295.
118. Saladié P, Rodríguez-Hidalgo A, Huguet R, Cáceres I, Díez C, et al. (2014) The role of carnivores and their relationship to hominin settlements in the TD6-2 level from Gran Dolina (Sierra de Atapuerca, Spain). Quat Sci Rev 93: 47–66.
119. Carmody RN, Weintraub GS, Wrangham RW (2011) Energetic consequences of thermal and nonthermal food processing. Proc Natl Acad Sci 108: 19119–19203.
120. Huguet R, Saladié P, Cáceres I, Díez C, Rosell J, et al. (2013) Successful subsistence strategies of the first humans in south-western Europe. Quatern Int 295: 168–182.
121. Blasco R, Rosell J, Fernández Peris J, Arsuaga JL, Bermúdez de Castro JM, et al. (2013) Environmental availability, behavioural diversity and diet: a zooarchaeological approach from the TD10-1 sublevel of Gran Dolina (Sierra de Atapuerca, Burgos, Spain) and Bolomor Cave (Valencia, Spain). Quat Sci Rev 70: 124–144.
122. Rosell J, Blasco R, Made Jvd, Cáceres I, Campeny G, et al. (2010) Acumulaciones producidas por carnívoros en el Pleistoceno medio inicial: una reconstrucción a partir del conjunto faunístico del nivel TD8 de Gran Dolina (Sierra de Atapuerca, Burgos). Zona arqueológica 13: 158–172.

Available Phosphorus in Forest Soil Increases with Soil Nitrogen but Not Total Phosphorus: Evidence from Subtropical Forests and a Pot Experiment

Xingzhao Liu[1]◔, Wei Meng[2], Guohua Liang[3]◔, Kun Li[1], Weiqiang Xu[1], Liujing Huang[4], Junhua Yan[1]*

1 South China Botanical Garden, Chinese Academy of Sciences, Guangzhou, China, 2 Paulownia Research and Development Centre, China Forestry Administration, Zhengzhou, China, 3 State Key Laboratory of Conservation and Utilization of Subtropical Agro-Bioresources, South China Agricultural University, Guangzhou, China, 4 Guangdong Provincial Academy of Environmental Science, Guangzhou, China

Abstract

This paper aims to establish evidence for available phosphorous (AP) binding with total nitrogen (N) in subtropical forest soils. Soil organic carbon (SOC), total N, total phosphorous (P) and AP concentration were measured for three contrasting forest types in southern China: Masson pine forest (MPF), coniferous and broadleaved mixed forest (CBMF) and monsoon evergreen broadleaved forest (MEBF). A pot experiment with N addition was conducted to confirm the dominant factor to affect on soil AP concentration. The results showed that mean soil total N concentration in 0–10 cm soil layer was 440±50 for MPF, 840±80 for CBMF and 1020±50 mg kg^{-1} for MEBF, respectively. The mean soil AP concentration in 0–10 cm soil layer was 2.67±0.87 for MPF, 2.65±0.58 for CBMF, 4.10±0.29 mg kg^{-1} for MEBF, respectively. The soil total N concentration could explain about 70% of the variations in soil AP concentration in the top 20 cm soil layers in the three forest types. A pot experiment with N addition also showed an increase of AP concentration from 2.56 to 5.63 mg kg^{-1}, when N addition increased from 5 g to 17 g NH_4NO_3. Our results therefore suggested that N addition significantly increased soil AP concentration, which might be beneficial for stabilizing the net primary production of subtropical forests that were limited by soil AP. This finding may provide a theory basis for tropical and subtropical forests management.

Editor: Vishal Shah, Dowling College, United States of America

Funding: This study was funded by the National Natural Science Foundation of China (No. 30872012 and 41301058) and National Basic Research Program of China (2011CB403202). The funders had no role in study design, data collection and analysis, decision to publish, or preparation of the manuscript.

Competing Interests: The authors have declared that no competing interests exist.

* E-mail: jhyan@scib.ac.cn

◔ These authors contributed equally to this work.

Introduction

Soil nitrogen (N) and phosphorus (P) are the most common macronutrients limiting plant growth under natural conditions. Inventories of global soil P found that soil P amount is lowest in the tropical and subtropical regions [1,2] that account for about 40% of the global gross primary production and net carbon uptake over the past two decades [3,4]. It is generally accepted that the total soil P gradually decreases as the result of weathering [5], ecosystems may decline at their advanced stage [6], which resulting in a decrease of biomass and diversity due to soil P limitation [7,8]. Previous studies also found that available soil P reduced the responses of tropical forests to increasing CO_2 concentrations [9,10]. Under global increasing atmospheric CO_2 and N deposition in the future, P limitation of net primary production (NPP) in terrestrial ecosystems, particularly tropical and subtropical forests, will likely exacerbate. However, a number of continuous eddy covariance measurements and long-term biomass inventories showed that old-growth forests can be significant carbon sinks, including tropical and subtropical forests [11–15].

The Dinghushan Biosphere Reserve (DBR) is located in the southern China experiencing a typical subtropical monsoon climate. The monsoon evergreen broadleaved forest (MEBF) is more than 400 years old [16], and is dominated by *Castanopsis chinensis*, *Schima superba*, *Cryptocarya chinensis*, *C. concinna*, and *Machilus chinensis*, which are typical species of forests at the advanced stage of succession in the subtropical area [17]. The prevailing theory of ecosystem biogeochemistry suggests that soil P availability would decrease in the advanced forest. A recent study also showed that soil P was relatively low in this region and P was possibly one of the limiting factors to tree growth in MEBF at the DBR [18]. However, long-term forest inventories and eddy covariance measurements showed that the NPP in MEBF was quite similar to that in the pioneer forest (Masson pine forest, MPF), or transition forest (coniferous and broadleaved mixed forest, CBMF) [19,20].The reasons of such high ecosystem production in the P-limited, late successional subtropical forest are still unclear.

Remarkably, the advanced forest, MEBF, was saturated with N [21], as a result of very high input from atmospheric deposition (>30 kg N ha^{-1} yr^{-1}) for much of the last two decades [22]. Accumulation of soil N can also affect soil P availability by changing soil ion balance and soil phosphatase production. Fertilization experiments demonstrated that N addition increased P cycle rate [23,24]. Previous studies also suggested that N addition (mainly N deposition and N fertilizer applications) can significantly increase the availability of soil N and reduce P limitation in forest ecosystems [10,25]. Some of those studies focused on the effects and mechanisms of soil N or P alone or N-P together, but very few studies examined the mechanisms of interaction between P limitation and N cycle, or whether N addition will alter P availability [10]. In tropical and subtropical regions where productivity is considered to be strongly P limited, a large carbon uptake above forest canopy [14] and accumulated in forest soil [16] have been found, but the mechanism of those observed large carbon uptake has not been understood yet. In this study, we hypothesize that N addition will increase available soil P production and reduce the affect of P limitation.

Materials and Methods

Ethics statement

The study site is maintained by South China Botanical Garden, Chinese Academy of Sciences. This site is within Dinghushan Forest Ecosystem Research Station. All necessary permits were obtained for the field study. The field study did not involve endangered or protected species. Data will be made available upon request.

Site description

Our study site is located at the DBR (112.10° E and 23.10° N) in southern China. The reserve was established in 1950s to protect the natural forests, with an area of 1 133 ha and an elevation ranging from 10 to 1000 m above sea level. The local climate is subtropical monsoon climate, with a distinctive wet season (April-September) and dry season (October-March) within a year. The annual average rainfall is 1 700 mm, of which more than 80% falls in wet seasons. The mean annual temperature is 21.0°C, with average temperatures of 12.6°C and 28.0°C in the coolest month (January) and the hottest month (July) respectively. The mean annual relative humidity is 82%. Bedrocks are classified as Devonian sandstone and shale [26]. The lateritic red-earth at elevations below 400–500 m and yellow earth at higher elevations are major soil types in this region [27].

In the subtropical region, stand composition generally changes from coniferous to coniferous and broad-leaved mixed to broad-leaved in the process of forest succession [17]. During natural succession, heliophytes (*e.g. S. superba* and *C. chinensis*) gradually invade coniferous forest to form mixed forest, and mesophytes (*e.g. C. concinna, C. chinensis*) subsequently invade mixed forest and eventually transform them into broadleaved forest. MPF, CBMF and MEBF mentioned above represent the forests in pioneer, transition and advanced stage, respectively [17]. Characteristics of these three forests are listed in Table 1.

Natural forest soil sampling

To quantify how soil nutrient varies with forest succession in DBR, soil samples were collected at depths of 0–10, 10–20, 20–40 and 40–60 cm using a 30 mm diameter auger, after removing organic layer of 1–2.5 cm above the soil surface. In each forest, three plots (10 m × 10 m) were established. In each plot, 3 randomly located subplots were selected. In each subplot, four augers of soil in different layers were taken and mixed to make up a composite sample, resulting in 9 composite samples for each soil layer for each forest type. All samplings were collected in September-October 2005.

Pot-based N addition experiment

Seedlings of four species were selected for the pot experiment, among which *P. massoniana* is the pioneer species, while *S. superb, C. chinensis* are dominant native tree species. *Ormosia pinnata* was chosen as the main legume tree species to evaluate the influences of N-fixing on soil nutrient availability. Clay pots with 33 cm in diameter and 30 cm in depth were used for cultivating tree seedlings. The soil was collected from the transitional forest at DHR. About 18.6 kg mixed soil (fresh weight) was put into one pot. Seedlings of different species with similar height were collected from the forests. One seedling was planted into one pot.

To simulate the soil N and P conditions in subtropical forests of China, we employed seven levels of N addition (Control, 5, 8, 11, 13, 15, 17 g NH_4NO_3 for each time), four replicates for each level. For each pot, NH_4NO_3 was dissolved with 1 L spring water and then added to the pot soil, with exceptions in the control treatment where no NH_4NO_3 was added but only the equal volume of spring water. A total of 10 times of N addition were conducted during the experiment with a frequency of every 20 days.

The amount of N addition was designed based on the range of soil N/P ratios of six subtropical forests in China (unpublished data, Fig. 1) and the background data of the pot soil in this experiment (Table 2). The background N/P ratio of pot soil was about 5. We considered that it is unnecessary to simulate soil N/P ratios less than 5 because soil in the DHR is rich in N but poor in P and this trend is continuing [33].

Measurement

Soil was sampled with a soil auger (diameter 2.5 cm) from surface to bottom in each pot after the tenth N addition. The samples were dried by free air and sieved (2 mm) to remove plant

Table 1. Stand characteristics of three forests in DBR.

	MPF	CBMF	MEBF	References
Successional stage	**Pioneer**	**Transition**	**Advanced**	**[17]**
Leaf area index	4.3	6.5	7.8	[28]
Biomass (kg C m^{-2})	6.13	8.21	14.52	[29]
Litter input (kg C m^{-2} $year^{-1}$)	0.18	0.43	0.42	[30]
Soil bulk density (g cm^{-3})	1.49	1.22	1.09	[31]
Soil pH	4.02	3.92	3.80	[31]
Microbial biomass (10^6 g^{-1} dry soil)	1.2	1.4	2.1	[32]
Fine root biomass in top soil (Mg C ha^{-1})	1.9	2.8	4.9	[31]

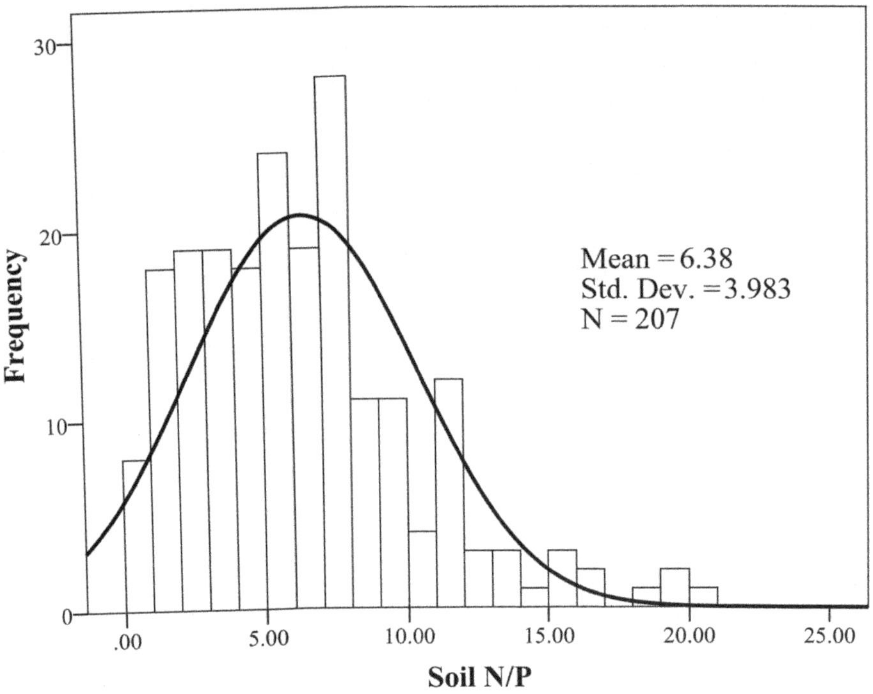

Figure 1. Frequency histograms of soil N/P ratio in six subtropical forests in China.

residues, and mixed thoroughly by hand. To reduce the analytical error, the average value of two replicates per sample was used and all data were expressed on an oven dry weight basis. Soil organic carbon (SOC), total N, total P and available P (AP) concentration were measured by standard element analyzer. The determination was in accordance with standard methods for observation and analysis in Chinese Ecosystem Research Network. SOC concentration was determined by potassium dichromate oxidation [34]. Semi-micro Macro Kjeldahl method was adopted to analyze soil total N concentration [35]. Soil total P concentration was measured by acid dissolve molybdenum and antimony anti-spectrophotometry (Spectrum lab 24, Shanghai, China). Soil AP was extracted with ammonium fluoride (NH_4F, 0.03 mol L^{-1}) and hydrochloric acid (HCl, 0.025 mol L^{-1}) [36], and measured by UV-Vis spectrophotometer (Spectrum lab 24, Shanghai, China).

Data analysis

Data were tested for normality and equality of variance. All analyses were conducted with the SPSS statistical package for Windows (16.0, SPSS, Inc, Chicago, IL). Differences of SOC, total N, total P and AP concentration among the three forests were tested by One-Way ANOVA. The effects of treatments on pot soil total N, total P, AP concentration and other soil parameters were analyzed using one way ANOVA with N addition as the fixed factor. Correlations between SOC, total N, total P and AP concentration were analyzed using Pearson correlation tests.

Table 2. The properties of pot soil before N addition treatment.

	SOC (mg kg^{-1})	N (mg kg^{-1})	P (mg kg^{-1})	AP (mg kg^{-1})	N/P
Mean	32230	1080	220	3.061	4.960
n	50	50	50	5	50
s.d.	3661	140	30	1.25	0.54

Results

Patterns of SOC and soil total N in three successional forests

Fig. 2A showed that SOC concentration in the three forests decreased with soil depth. SOC concentration in 0–10 cm soil layer was about 2–5 times higher than that in 40–60 cm soil layer for the three forests (Fig. 2A). Mean SOC concentration in 0–10 cm soil layer increased with the forest succession, being 14868±2987 for MPF, 28508±3175 for CBMF and 31406±3572 mg kg^{-1} for MEBF. This increase was significant between pioneer (MPF) and transitional (CBMF) forests ($P<0.05$), while no significant difference between transitional (CBMF) and advanced forests was found ($P>0.05$).

Fig. 2B showed that soil total N concentration in the three forests also decreased with soil depth. In each forest, soil total N concentration in 0–10 cm soil layer was significantly greater than that in each of the other three soil layers ($P<0.05$). There was no significant difference in soil total N concentration among soil layers of 10–20 cm, 20–40 cm and 40–60 cm ($P>0.05$). Mean soil total N concentration in 0–10 cm soil layer increased with the forest succession, being 440±50 for MPF, 840±80 for CBMF to 1020±50 mg kg^{-1} for MEBF, respectively. The difference in soil total N concentration between MPF and CBMF or MEBF was significant ($p<0.05$), and insignificant ($P>0.05$) between CBMF and MEBF.

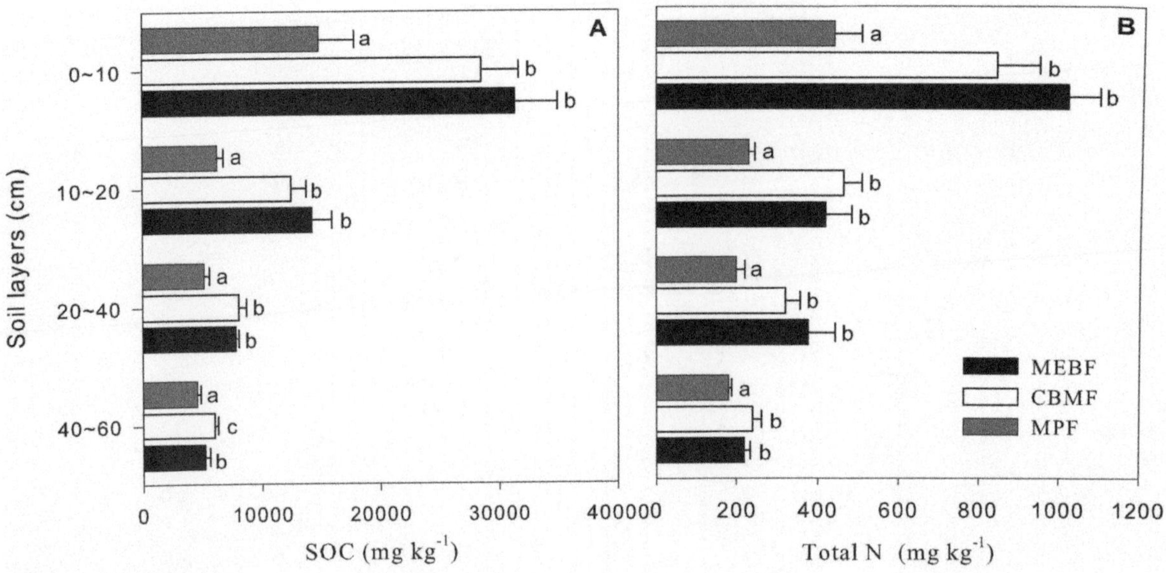

Figure 2. SOC and total N concentrations among three forest types at different succession stages in DBR. Error bars indicate one standard deviation. Different letters indicate significant differences at the confidence level of $P<0.05$ in the same soil layer among the three forests. (OneWay ANOVA, $P<0.05$).

Patterns of soil total P and soil AP in three successional forests

Fig. 3A showed that the variations in soil total P concentration with soil depth were quite different among the three forests. In the pioneer forest, soil total P concentration increased with soil depth and this increase was only significant between 20–40 cm and 40–60 cm soil layers ($P<0.05$). In the transitional forest, soil total P concentration was not significant difference among the different soil layers ($P>0.05$). In the advanced forest, soil total P concentration decreased with soil depth and this decrease was only significant between 0–10 cm and 10–20 cm soil layers ($P<0.05$).

Soil total P concentration was significantly greater in the transitional forest when compared to the pioneer or advanced forest ($P<0.05$). Mean soil total P concentration from 0 to 60 cm was 209.2 for the pioneer, 309.3 for the transition and 208.7 mg kg^{-1} for the advanced forest, respectively. In the 0–10 cm soil layer, soil total P concentrations was 190±10 for MPF, 340±20 for CBMF and 280±10 mg kg^{-1} for MEBF, respectively (Fig. 3A).

However, soil AP concentration in 0–10 cm soil layer was the least in the transitional forest, and the highest in the advanced

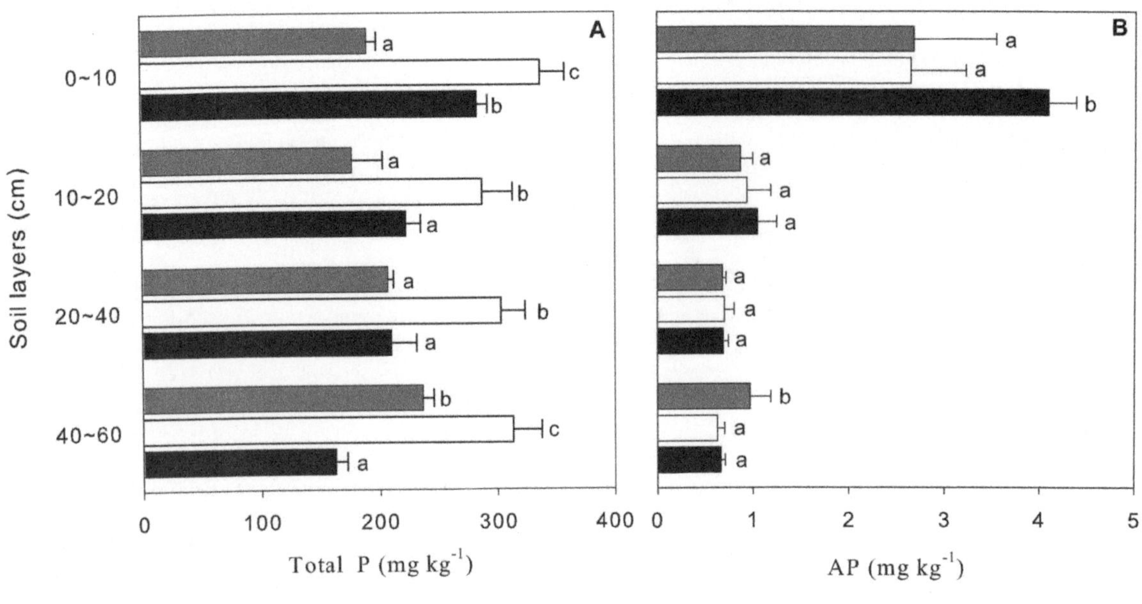

Figure 3. Soil total P and AP concentrations among three forest types at different succession stages in DBR. Error bars indicate one standard deviation. Different letters indicate significant differences at the confidence level of $P<0.05$ in the same soil layer among the three forests. (OneWay ANOVA, $P<0.05$).

forest. The mean soil AP concentration in 0–10 cm soil layer was 2.67±0.87 for MPF, 2.65±0.58 for CBMF, 4.10±0.29 mg kg^{-1} for MEBF, respectively. In each of the three forests, soil AP concentration in 0–10 cm soil layer was significantly greater than that in each of the other soil layers ($P<0.05$). There was no significant difference in soil AP concentration among soil layers of 10–20 cm, 20–40 cm and 40–60 cm ($P>0.05$) (Fig. 3B).

Correlations of soil AP with SOC, total N and total P

As shown in Fig. 3, the patterns (distribution along soil depth and variation with forest succession) of soil AP concentration were quite different with soil total P concentration. Therefore, the variation in soil AP concentration was not consistent with the variation in soil total P concentration. As a test of this result, we examined the relation between soil total P concentration and soil AP concentration (Fig. 4A). A weak positive relationship ($R^2=0.096$) was found between soil total P concentration and soil AP concentration. However, the degree of scatter in the plotted points ($P=0.920$) suggests that the trends were as expected.

However, the distribution of soil AP concentration along soil depth were quite similar to the distribution of soil total N concentration or SOC concentration in each of the three forests. We plotted soil AP concentration against soil total N or SOC concentration in the soil layers of 0–10 cm and 10–20 cm in the three forests together (Fig. 4B and 4C). A linear regression fitted well between soil AP concentration and soil total N or SOC concentration. Soil total N or SOC concentration could explain about 70% of variations in soil AP concentration.

Effects of soil total N or SOC on soil AP concentration

As mentioned above, each of soil total N and SOC could explain much of the variations in soil AP concentration. Both of soil total N and SOC increased with the forest succession and showed co-variation significantly (Fig. 5). Therefore, it was difficult to distinguish which was the dominant factor to drive the variations in soil AP concentration based on the field measurements. To confirm the dominant factor on soil AP concentration, a pot experiment with N addition was employed. The results from this experiment showed that soil total N concentration increased with N addition level (Fig. 6A). This increase was significant under the different N addition levels ($P<0.05$) (Fig. 6). SOC concentration also increased with N addition level (Fig. 6B), but this increase was not significant under the different N addition levels ($P>0.05$) (Fig. 6). The variations in soil AP concentration with N addition level were similar to soil N concentration (Fig. 6C). There was significant difference in soil AP concentration under different N addition levels ($P<0.05$) (Fig. 6).

Soil total N concentration increased from 960.7 (control) to 1212.3 mg kg^{-1} (highest N addition level) (Fig. 5). Soil AP concentration increased from 3.5 (control) to 5.7 mg kg^{-1} (highest N addition level) (Fig. 5). We found a strong relationship ($AP=0.006$ N-2.443, $R^2=0.578$, $P<0.001$) between soil AP concentration and total N concentration. Soil total N could explain about 60% of the variations in soil AP concentration (Fig. 7). However, the linear correlation of soil AP concentration with SOC was much weaker than that with soil total N concentration ($AP=0.0007$ SOC-1.577, $R^2=0.359$, $P<0.001$). SOC only explained about 36% of variations in soil AP concentration (Fig. 7).

Discussion

Why can N addition increase AP in subtropical forest soil?

Soil AP mainly came from the cleavage of soil organic P via biochemical P mineralization or decomposition of soil organic matter via biological mineralization. Both plant roots and soil micro-organism produced phosphatase for biochemical P mineralization [37]. Secreted by plants or microbes, extracellular phosphatases hydrolyzed phosphodiester-bonded P to phosphate ions that were available for uptake by plant roots [37]. It was reported that phosphodiester-bonded P accounted for 20% to 80% of soil organic P [38].

Phosphatase came from the physiological activity of soil microorganisms. Meanwhile, phosphatase was a constitutively N-rich (15% N, [24]) class of enzymes. The addition of N could provide the necessary element for the production of phosphatase [23,24]. The soil in MEBF in the DBR was relatively richer in N, therefore it produced more AP than other two forests.

P limitation and C allocation in forests

N addition can accelerate soil P cycle by increasing the production of extracellular phosphatase enzyme. However, N was not the only factor which affects P cycle. The physiological activity of soil microorganisms, in which extracellular phosphatase enzyme

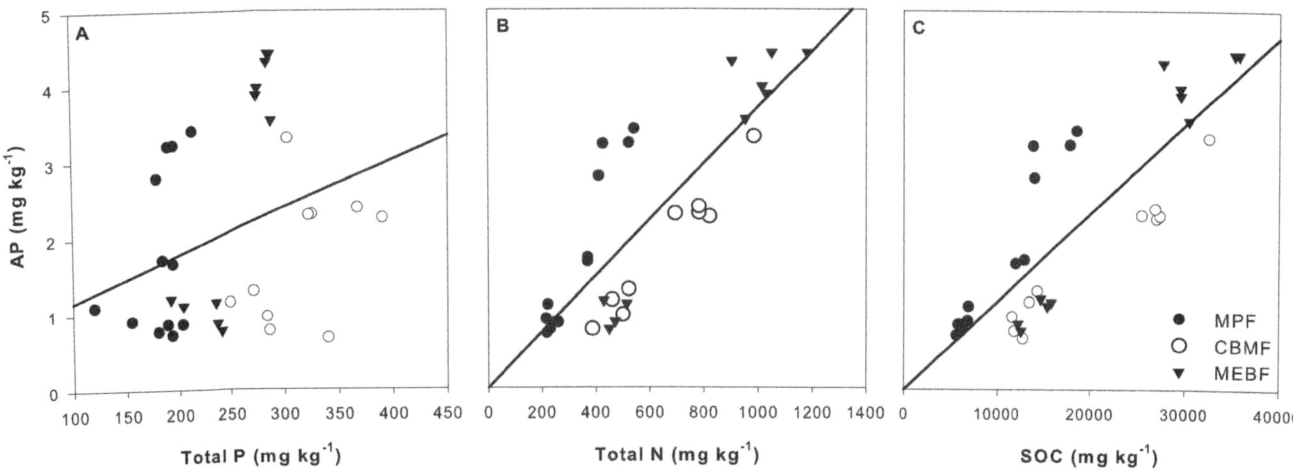

Figure 4. The correlation between soil total P (a), total N (b), SOC (c) and AP concentrations in 0–20 cm soil of three forests in DBR. (a): $AP=0.0062$ P+0.536, $R^2=0.096$, $P=0.920$; (b) $AP=0.003784$ N-0.0003, $R^2=0.699$, $P<0.0001$; (c): $AP=0.0001$ SOC-0.0016, $R^2=0.713$, $P<0.0001$.

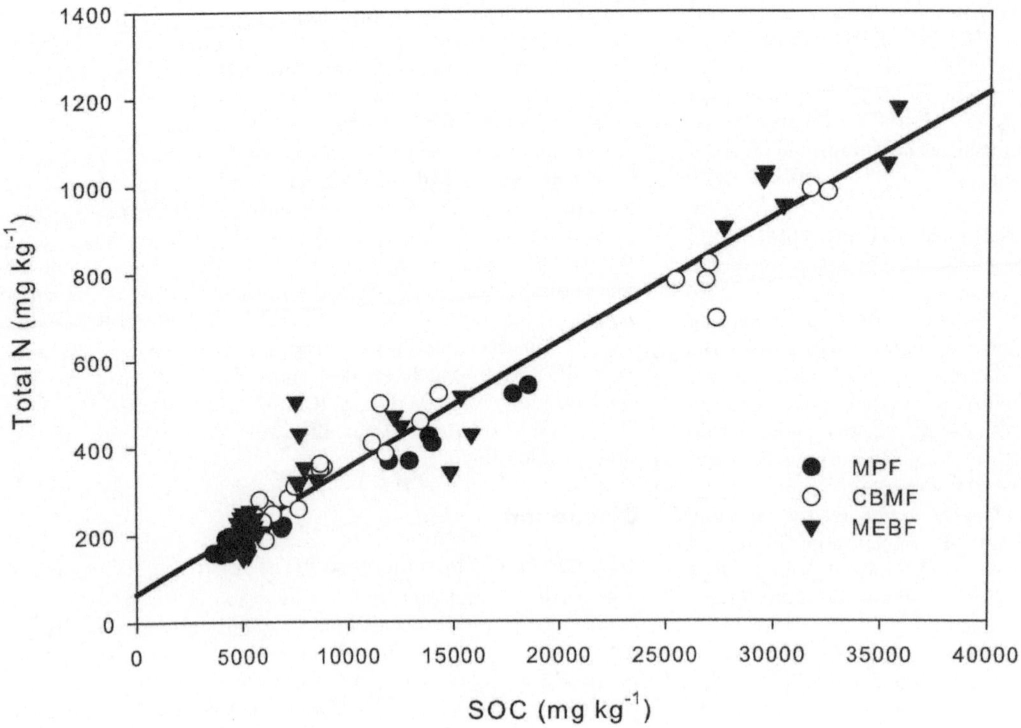

Figure 5. The correlation between soil total N and SOC concentrations in 0–20 cm soil of three forests in DBR. N = 0.029 SOC-66.627, R^2 = 0.954, P<0.0001.

was produced, depends on available C for growth and energy. Consequently, about 36–71% of variation of AP in the soil could be explained by SOC (see Fig. 4 and 7A). Therefore, C was another factor influencing AP in soil. The root exudates and dead root can be used by soil microbe to produce enzymes, which could increase soil AP [23,39].

In turn, P limitation can impact C cycle. P limitation was very common in old-growth forests, especially at low latitudes [6]. In this case, plants should develop strategies for adapting to low P soil, and make necessary physiological adjustments to alleviate P limitation. The activation of insoluble phosphate by the interaction of root and soil microbe was an effective mechanism. However, this strategy required high C input. Plants allocated new biomass C preferentially to roots for the growth of root and exudates, both of which could increase the absorption of P. The allocation of C to belowground, to some extent, would decrease the aboveground biomass, even the total biomass if the proportion of exudate was too high. Actually, the exudates accounted for a significant proportion of annual net primary production. Some studies demonstrated that the annual rate of exudates production in rhizosphere can account for 15–23% of total net primary production [40,41]. Marschner (1995) indicated that the amount of organic matter released from root to rhizosphere accounted for 1–30% of the photosynthetic carbon assimilation [42].

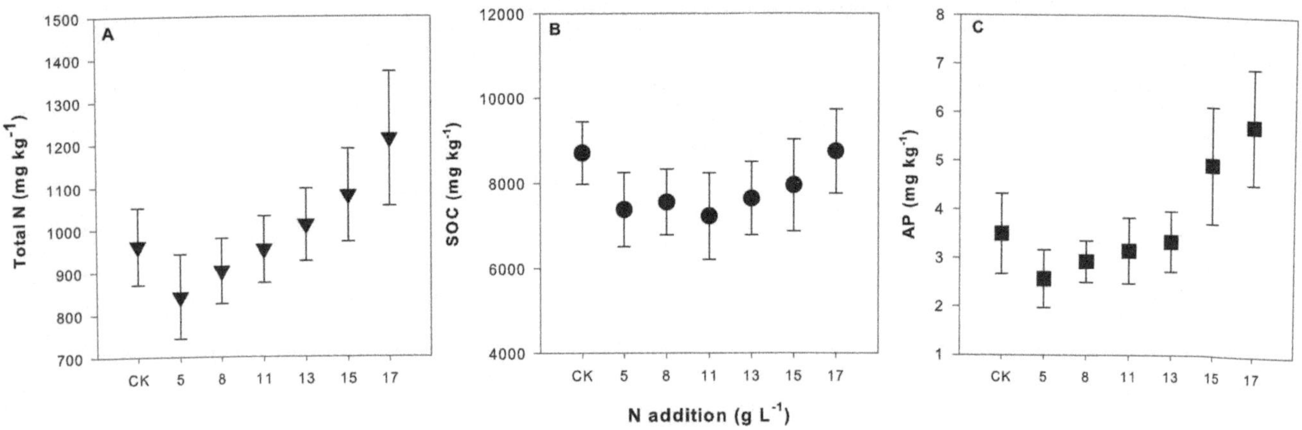

Figure 6. Variation of pot soil total N (A), SOC (B) and AP (C) concentrations with the rate of N addition. Error bar represents one standard deviation. (A): r^2 = 0.307, P = 0.000; (B): r^2 = 0.029, P = 0.072; (C): r^2 = 0.222, P = 0.000.

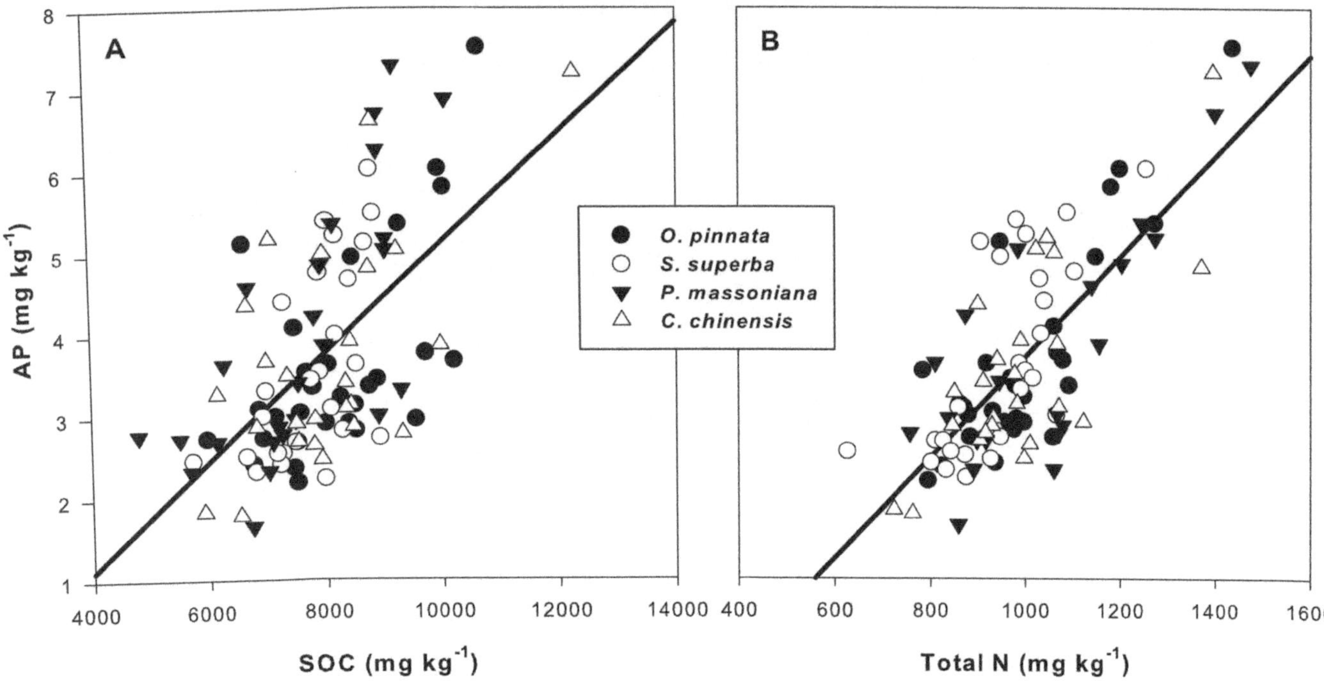

Figure 7. The correlation between pot soil C (A), N (B) and AP concentrations. (A): AP = 0.00067 SOC-1.577, $R^2 = 0.359$, $P < 0.0001$; (B): AP = 0.006 N-2.443, $R^2 = 0.578$, $P < 0.0001$

The changes of C allocation in P-limitation forests would bring a great amount of C into soil. MEBF in DBR, older than 400 years, was considered as a P-deficiency and N-adequacy ecosystem [43]. However, it has accumulated C with a steady rate during the last 23 years [20]. Furthermore, the standing biomass and annual litter decreased [20,43] and soil C increased [16,43,44]. Carbon invested into belowground will contribute to the accumulation of soil C. A shift in C allocation toward belowground was the physiological adjustments to alleviate P limitation in MEBF. As a result, the AP content in MEBF was highest among the three forests in DBR. Our measurements showed that annual NPP in MEBF had been relatively stable for at least twenty years. Consequently, C allocation change was an effective adaptation strategy under P limitation in old-growth forests.

The mechanism of P limitation in subtropical forest ecosystems

Available soil P changed along a forest successional gradient [2,6]. As forest canopy developed, the microenvironment below the forest became wetter and shadier, which will increase chemical weathering of parent material, and therefore soil P lost [5]. As a result, P became limiting to plant growth in the old-growth forests [6,33].

Because of input and output of P to soil depends on many other geological and ecological processes as well as plants themselves, the mechanisms of P limitation were variable. Vitousek *et al.* concluded six mechanisms that can cause P limitation to terrestrial ecosystems: P depletion, soil barriers, transactional, low-P parent materials, P sinks, and anthropogenic forcings [10]. These mechanisms were not exclusive. It was likely that P limitation in any ecosystem will be more or less controlled by more than one mechanism.

Perhaps the most likely cause of P limitation to old-growth tropical forest ecosystems was depletion-driven P limitation, the progressive loss of P that occurs during long term soil and ecosystem development [10].The regional bedrock in the DBR was thick sandstone and sand shale from Devonian [26] that has undergone a very long period of weathering. At the same time, locating in low-subtropical made their parent material P impossible to be rejuvenated or subsidized by glacial or periglacial processes. Except for the topsoil which was strongly affected by biology, depletion of P was evident in the deep soil (40–60 cm) along the forest succession in the DBR. Much of P may be lost via leaching of dissolved organic P [45].

The other possible mechanism of P limitation in the DBR was anthropogenic P limitation. The DBR locates in East Asia, where human-enhanced N deposition was one of highest in the world [46]. The annual N wet deposition amounted to 38.4 kg N $ha^{-1} \cdot yr^{-1}$ in the DBR [22] and soil was saturated with N in the old-growth forest [21]. The pot fertilization experiments further demonstrated that N addition can accelerate soil P cycle by increasing the production of extracellular phosphatase enzyme [23,24]. However, the increase in P cycling was insufficient to balance the increased rate of N inputs, hence, P often became limiting.

Conclusions

The field surveys of forest soil total N, SOC, total P, AP concentration and the N-addition experiments in Dinghushan, Southern China, allowed us to estimate the relationship between soil total N, SOC, total P and AP. The main findings of the present study are:

1. Comparing to the two younger forests (MPF and CBMF), the advanced forest (MEBF) had greater AP concentration but less total P concentration in soil.
2. The concentration of soil total N or SOC increased with the forest succession and showed a similar pattern with soil AP concentration. A significant correlation was found between soil AP concentration and soil total N concentration ($R^2 = 0.713$,

$P<0.001$) or SOC concentration ($R^2 = 0.699$, $P<0.001$) at top 20 cm soil layer. However, there was very weak correlation ($R^2 = 0.096$, $P = 0.920$) between soil P concentration and AP concentration.

3. Results from the pot experiment with N addition showed that soil AP concentration increased with N addition. Soil total N concentration could explain about 60% of the variations in soil AP concentration.

Acknowledgments

We are grateful to Mr. Ze Meng, Mr. Chuanyin Xiang and Mr. Dingsheng Mo for their assistance in field work. We would like to thank Xuli Tang, Guomin Huang and Honglang Duan for their constructive comments.

Author Contributions

Conceived and designed the experiments: JY XL. Performed the experiments: XL MW GL KL WX. Analyzed the data: XL GL MW WX KL. Contributed reagents/materials/analysis tools: XL WM GL. Wrote the paper: XL GL MW WX KL LH JY.

References

1. Taylor SR (1964) Abundance of chemical elements in the continental crust: A new table. Geochimica et Cosmochimica Acta 28: 1273–1285.
2. Zhang C, Tian HQ, Liu JY, Wang SQ, Liu ML, et al. (2005) Pools and distributions of soil phosphorus in China. Global biogeochemical cycles 19: doi: 1010.1029/2004GB002296.
3. Beer C, Reichstein M, Tomelleri E, Ciais P, Jung M, et al. (2010) Terrestrial gross carbon dioxide uptake: global distribution and covariation with climate. Science 329: 834–838.
4. Pan Y, Birdsey RA, Fang J, Houghton R, Kauppi PE, et al. (2011) A large and persistent carbon sink in the world's forests. Science 333: 988–993.
5. Walker TW, Syers JK (http://www.sciencedirect.com/science/article/pii/0016706176900665-AFF11976) The fate of phosphorus during pedogenesis. Geoderma 15: 1–19.
6. Wardle DA, Walker LR, Bardgett RD (2004) Ecosystem properties and forest decline in contrasting long-term chronosequences. Science 305: 509–513.
7. Hooper DU, Vitousek PM (1998) Effects of plant composition and diversity on nutrient cycling. Ecological monographs 68: 121–149.
8. Wassen MJ, Venterink HO, Lapshina ED, Tanneberger F (2005) Endangered plants persist under phosphorus limitation. Nature 437: 547–550.
9. Lloyd J, Bird MI, Veenendaal E, Kruijt B (2001) Should phosphorus availability be constraining moist tropical forest responses to increasing CO2 concentrations? In Global biogeochemical cycles in the climate system. Schulze ED, Harrison SP, Heimann M, Holland EA, Lloyd J, et alEds. San Diego, CA: Academic Press: 96–114.
10. Vitousek PM, Porder S, Houlton BZ, Chadwick OA (2010) Terrestrial phosphorus limitation: mechanisms, implications, and nitrogen-phosphorus interactions. Ecological applications 20: 5–15.
11. Phillips OL, Malhi Y, Higuchi N, Laurance WF, Núñez PV, et al. (1998) Changes in the carbon balance of tropical forests: evidence from long-term plots. Science 282: 439–442.
12. Knohl A, Schulze ED, Kolle O, Buchmann N (2003) Large carbon uptake by an unmanaged 250-year-old deciduous forest in Central Germany. Agricultural and Forest Meteorology 118: 151–167.
13. Luyssaert S, Schulze ED, Börner A, Knohl A, Hessenmöller D, et al. (2008) Old-growth forests as global carbon sinks. Nature 455: 213–215.
14. Lewis SL, Lopez-Gonzalez G, Sonké B, Affum-Baffoe K, Baker TR, et al. (2009) Increasing carbon storage in intact African tropical forests. Nature 457: 1003–1006.
15. Yan JH, Zhang YP, Yu GR, Zhou GY, Zhang LM, et al. (2013) Seasonal and inter-annual variations in net ecosystem exchange of two old-growth forests in southern China. Agricultural and Forest Meteorology 182-183: 257–265.
16. Zhou GY, Liu SG, Li ZA, Zhang DQ, Tang XL, et al. (2006) Old-growth forests can accumulate carbon in soils. Science 314: 1417.
17. Peng SL, Wang BS (1995) Forest succession at Dinghushan, Guangdong, China. Chinese Journal of Botany 7: 75–80 (in Chinese with English abstract)
18. Mo JM, Zhang DQ, Huang ZL, Yu QF, Kong GH (2000) Distribution pattern of nutrient elements in plants of Dinghushan lower subtropical evergreen broad-leaved forest. Journal of Tropical and Subtropical Botany 8: 198–206 (in Chinese with English abstract).
19. Li YL, Zhou GY, Zhang DQ, Wenigmann KO, Otieno D, et al. (2012) Quantification of ecosystem carbon exchange characteristics in a dominant subtropical evergreen forest ecosystem. Asia-Pacific Journal of Atmospheric Sciences 48: 1–10.
20. Tang XL, Wang YP, Zhou GY, Zhang DQ, Liu S, et al. (2011) Different patterns of ecosystem carbon accumulation between a young and an old-growth subtropical forest in Southern China. Plant Ecology 212: 1385–1395.
21. Mo JM, Xue JH, Fang YT (2004) Litter decomposition and its responses to simulated N deposition for the major plants of Dinghushan forests in subtropical China. Acta Eclogica Sinica 24: 1413–1420 (in Chinese with English abstract).
22. Zhou GY, Yan JH (2001) The influences of regional atmospheric precipitation characteristic sand its element inputs on the existence and development of Dinghushan forest ecosystems. Acta Ecologica Sinica 21: 2002–2012 (in Chinese with English abstract).
23. Olander LP, Vitousek PM (2000) Regulation of soil phosphatase and chitinase activity by N and P availability. Biogeochemistry 49: 175–190.
24. Treseder KK, Vitousek PM (2001) Effects of soil nutrient availability on investment in acquisition of N and P in Hawaiian rain forests. Ecology 82: 946–954.
25. Gress SE, Nichols TD, Northcraft CC, Peterjohn WT (2007) Nutrient limitation in soils exhibiting differing nitrogen availabilities: what lies beyond nitrogen saturation? Ecology 88: 119–30.
26. Wu HS, Deng HZ, Chen HT, Zheng LW, Liu YT (1982) Physico-geographical features of Ding Hu Shan and their dynamic analyses. Tropical Subtropical Forest Ecosystem 1: 1–10 (in Chinese with English abstract).
27. Ding MM, Brown S, Lugo AE (2001) A continental subtropical forest in China compared with an insular subtropical forest in the Caribbean. General Technical Report II TF-17, USDA (United States Department of Agriculture) Forest Service, International Institute of Tropical Forestry, Río Piedras, Puerto Rico.
28. Yan JH, Wang YP, Zhou GY, Zhang DQ (2006) Estimates of soil respiration and net primary production of three forests at different succession stages in South China. Global Change Biology 12: 810–821.
29. Liu S, Luo Y, Huang YH, Zhou GY (2007) Studies on the community biomass and its allocations of five forest types in Dinghushan Nature Reserve. Ecological Science 26: 387–393 (in Chinese with English abstract).
30. Huang YH, Li YL, Xiao Y, Wenigmann KO, Zhou GY, et al. (2011) Controls of litter quality on the carbon sink in soils through partitioning the products of decomposing litter in a succession series in South China. Forest Ecology and Management 261: 1170–1177.
31. Tang XL, Liu SG, Zhou GY, Zhang DQ, Zhou CY (2006) Soil-atmospheric exchange of CO_2, CH_4, and N_2O in three subtropical forest ecosystems in southern China. Global Change Biology 12: 546–560.
32. Zhou LX, Yi WM, Yi ZG, Ding MM (2002) Soil microbial characteristics of several vegetations at different elevation in Dinghushan Bio-sphere Reserve. Tropical and Subtropical Forest Ecosystem Research 9: 169–174 (in Chinese with English abstract).
33. Liu XZ, Zhou GY, Zhang DQ, Liu SZ, Chu GW, et al. (2010) N and P stoichiometry of plant and soil in lower subtropical forest successional series in southern China. Chinese Journal of Plant Ecology 34: 64–71 (in Chinese with English abstract).
34. Nelson DW, Sommers LE (1982) Methods of soil analysis, Part 2: chemical and microbiological properties (2nd ed) Page AL, Miller RH, Keeney DR (eds). American Society of Agronomy, and Soil Science Society of America, Madison 539–579.
35. Bremner JM, Mulvaney CS (1982) Methods of soil analysis, Part 2: chemical and microbiological properties (2nd ed.). Page AL, Miller RH, Keeney DR (eds). American Society of Agronomy, and Soil Science Society of America, Madison 595–624.
36. Liu GS, Jiang NH, Zhang LD (1996) Soil physical and chemical analysis and description of soil profiles. Standards Press of China, Beijing(in Chinese).
37. Duff SMG, Sarath G, Plaxton WC (1994) The role of acid phosphatases in plant phosphorus metabolism. Physiol Plant 90: 791–800.
38. McGill WB, Cole CV (1981) Comparative aspects of cycling of organic C, N, S and P through soil organic matter. Geoderma 26: 267–286.
39. Hocking PJ (2001) Organic acids exuded from roots in phosphorus uptake and aluminum tolerance of plants in acid soils. Advances in Agronomy 74: 63–97.
40. Gardner WK, Barber DA, Parbery DG (1983) The acquisition of phosphorus by *Lupinus albus L.* III. The probable mechanism by which phosphorus movement in the soil/root interface is enhanced. Plant and Soil 70: 107–124.
41. Dinkelaker B, Römheld V, Marschner H (1989) Citric acid exudation and precipitation of calcium citrate in the rhizosphere of white lupin (*Lupinus albus L.*). Plant Cell and Environment 12: 265–292.
42. Marschner H (1995) Mineral nutrition in higher plants. Academic Press, London, UK.
43. Huang WJ, Zhou GY, Liu JX (2011) Nitrogen and phosphorus status and their influence on aboveground production under increasing nitrogen deposition in three successional forests. Acta Oecologica-International Journal of Ecology 44: 20–27.
44. Yan JH, Zhou GY, Zhang DQ, Chu GW (2007) Changes of soil water, organic matter, and exchangeable cations along a forest successional gradient in southern China. Pedosphere 17: 397–405.
45. Hedin LO, Vitousek PM, Matson PA (2003) Nutrient losses over four million years of tropical forest development. Ecology 84: 2231–2255.
46. Galloway JN, Cowling EB (2002) Reactive nitrogen and the world: 200 years of change. Ambio 31: 64–71.

Shifts in Species Interactions Due to the Evolution of Functional Differences between Endemics and Non-Endemics: An Endemic Syndrome Hypothesis

Courtney E. Gorman[1*], Brad M. Potts[2], Jennifer A. Schweitzer[1], Joseph K. Bailey[1,2]

1 Dept. of Ecology and Evolutionary Biology, University of Tennessee, Knoxville, Tennessee, United States of America, **2** School of Plant Science, University of Tasmania, Hobart, Tasmania, Australia

Abstract

Species ranges have been shifting since the Pleistocene, whereby fragmentation, isolation, and the subsequent reduction in gene flow have resulted in local adaptation of novel genotypes and the repeated evolution of endemic species. While there is a wide body of literature focused on understanding endemic species, very few studies empirically test whether or not the evolution of endemics results in unique function or ecological differences relative to their widespread congeners; in particular while controlling for environmental variation. Using a common garden composed of 15 *Eucalyptus* species within the subgenus *Symphyomyrtus* (9 endemic to Tasmania, 6 non-endemic), here we hypothesize and show that endemic species are functionally and ecologically different from non-endemics. Compared to non-endemics, endemic *Eucalyptus* species have a unique suite of functional plant traits that have extended effects on herbivores. We found that while endemics occupy many diverse habitats, they share similar functional traits potentially resulting in an endemic syndrome of traits. This study provides one of the first empirical datasets analyzing the functional differences between endemics and non-endemics in a common garden setting, and establishes a foundation for additional studies of endemic/non-endemic dynamics that will be essential for understanding global biodiversity in the midst of rapid species extinctions and range shifts as a consequence of global change.

Editor: Paul Adam, University of New South Wales, Australia

Funding: The Australian Research Council funded the Future Fellowship (to JKB), http://www.arc.gov.au. The funders had no role in study design, data collection and analysis, decision to publish, or preparation of the manuscript.

Competing Interests: The authors have declared that no competing interests exist.

* Email: cgorman5@utk.edu

Introduction

Species ranges have been shifting since the Pleistocene [1], whereby fragmentation, isolation, and the subsequent reduction in gene flow have resulted in local adaptation of novel genotypes and the repeated evolution of endemic species. Endemic species have long been valued for their novelty by both the general and scientific communities, which has resulted in a vast body of evolutionary and natural history research [2–4]. However, in the midst of a biodiversity crisis where species extinction rates are 100 to 1000 times greater than the background geological rate [5], understanding the biology of endemic species has become a priority rather than a pursuit of novelty, as these species are often the ones most at risk [6]. Studies have investigated the causes and consequences of endemism [2–4], the geography, risks, and prospects of endemic species [7–10], as well as the genetic differences between endemic (or rare/narrowly distributed) vs. widespread species [11–13]. The literature is generally lacking, however, in studies that attempt to investigate the ecological significance of endemic species. Because the formation of relict populations and the evolution of endemic species is thought to be a major consequence of species range shifts due to climate change [14], identifying whether endemics are functionally different and support unique species interactions may place even greater conservation value on these populations and species.

Linking evolutionary history to contemporary ecological interactions is a burgeoning field that is bringing with it many new insights into the relationship between biodiversity, species interactions, and ecosystem function [15]. Despite studies on their evolutionary novelty, few studies have experimentally investigated the ecological differences between endemic species and their non-endemic congeners or how endemicity may influence species interactions; particularly while controlling for environmental variation. A 2003 study measured net photosynthesis, leaf nitrogen content, and specific leaf area of 78 crop, endemic, and non-endemic plant species [16]. They noted variation between endemic and non-endemic species, however did not find statistical significance for the observed differences between any of the measured parameters in the field. Additionally, a recent study compared traits of 20 congeneric pairs of endemic and widespread plant species and while they found that endemics were smaller and produced fewer flowers, they found no differences in traits related to resource acquisition, resource conservation, and patterns of herbivory [17]. While these studies provide a valuable basis for understanding the ecological differences between endemic and

Table 1. Mixed model analysis of plant functional traits.

Response Variable	$F_{1,13}$	p
% Insect Herbivory	4.446	0.039* (0.078)
% Insect Herbivory DB	9.932	0.002* (0.008*)
% Mammal Herbivory	0.454	0.502 (0.502)
Specific Leaf Area (cm^2/g)	41.919	<0.001* (0.001*)
Height (cm)	5.588	0.021* (0.063)
Internode Length (mm)	45.066	<0.001* (0.001*)
Leaf Thickness (mm)	49.318	<0.001* (0.001*)

Summary of mixed model analysis using Restricted Maximum Likelihood (REML) of the differences between samples of endemic ($n = 9$) and non-endemic ($n = 6$) eucalypt species of the subgenus *Symphyomyrtus* growing on the island of Tasmania. Holm- Bonferroni corrected *p*-values are given in parentheses.

non-endemic species, they are limited in the inferences that they can make, since traits were measured *in situ* rather than in an experimental common garden and are thus influenced by a range of environmental variables. Common garden experiments provide an opportunity to more accurately partition genetic and environmental components of trait variation when attempting to characterize the ecology of a set of species [18] and provide a powerful tool for linking evolutionary history to contemporary ecological interactions.

Endemic species have frequently been characterized based on generalizations of their perceived commonalities, such as low genetic diversity [19], [11–13] and limited reproduction and dispersal abilities [20], [6]. For example, a 2000 study summarized the generalizations that are often made regarding the reproductive biology of endemic species as an increased tendency for self-compatibility, lower investment in reproduction, poorer dispersal abilities, and shorter generation times in comparison to common species [6]. Although attempts have been made to characterize endemic species based on their shared traits, the extent of this convergence on an endemic syndrome of traits remains unclear, along with how these shared differences may influence species interactions differently than those of common species. Here we hypothesize that endemic species are a homogenous group that can be characterized based on commonalities that result from isolation and lead to an 'endemic syndrome'.

The genus *Eucalyptus* in Tasmania provides an ideal natural system for examining an endemic syndrome among congeners, as the island has 29 native eucalypts from two subgenera, 17 of which are endemic to the island of Tasmania, while the others also occur on the Australian mainland [21]. We used a common garden with 15 *Eucalyptus* species (9 endemic, 6 non-endemic) to test the hypothesis that functional plant traits and associated patterns of herbivory of endemic species differ from those traits in closely related non-endemic species. To our knowledge this is the first endemic/non-endemic comparative study to use an experimental common garden design to separate differential environmental conditions as explanatory variables. Here we show that endemic plant species are ecologically different than non-endemics. We show that these differences include functional plant traits with extended effects across trophic levels. Furthermore, we found that while endemics occupy many diverse habitats (from loamy sites near sea-level to alpine scrub), they share similar functional traits potentially resulting in an endemic syndrome of traits.

Materials and Methods

Common Garden

In order to test whether endemic species are ecologically different than non-endemics without the constraints of environmental/habitat variation, we used a common garden experiment. The common garden was part of a forestry trial established by The Cooperative Research Centre for Forestry (CRC). This experimental forest trial was established in 2009 with 15 species of closely related Eucalypts native to Tasmania that occur in the subgenus *Symphyomyrtus* [21]. Nine of these species are endemic to Tasmania, while the other 6 are native non-endemics that also occur on the Australian mainland. Both groups of species exhibit a widespread distribution within Tasmania and co-occur throughout the state. Non-endemic species included in the trial were *E. dalrympleana*, *E. rubida*, and *E. viminalis*, *E. brookeriana*, and *E. ovata* and *E. perriniana*. Endemic species included in the trial were *E. johnstonii*, *E. subcrenulata*, and *E. vernicosa*, *E. archeri*, *E. cordata*, *E. gunnii*, *E. morrisbyi*, and *E. barberi* and *E. rodwayi*. The endemic species included in the trial occupy a diverse variety of habitats ranging from loamy sites near sea-level (*E. cordata* and *E. morrisbyi*), poorly drained montane forest (*E. johnstonii*), well-drained subalpine rainforest (*E. subcrenulata*), and alpine scrub (*E. vernicosa*) [21]. Each species was represented by an average of four open-pollinated families collected from native trees in Tasmania with between 1 and 17 plants per family. Individuals were planted in rows that were 36 trees long. Plant positions within a row were allocated randomly, and the total sample size was 412 trees. Both mammalian and insect herbivores had unrestricted access to the garden. No specific permissions were required to carry out this study and field studies did not involve endangered or protected species.

Plant Measurements

To quantify differences between endemics and their closely related non-endemic species, common plant functional traits (height, internode length, leaf thickness, and specific leaf area (SLA)) and herbivory were measured in 2011 on 4 year-old plants. Total tree height was measured to the nearest cm. Two random

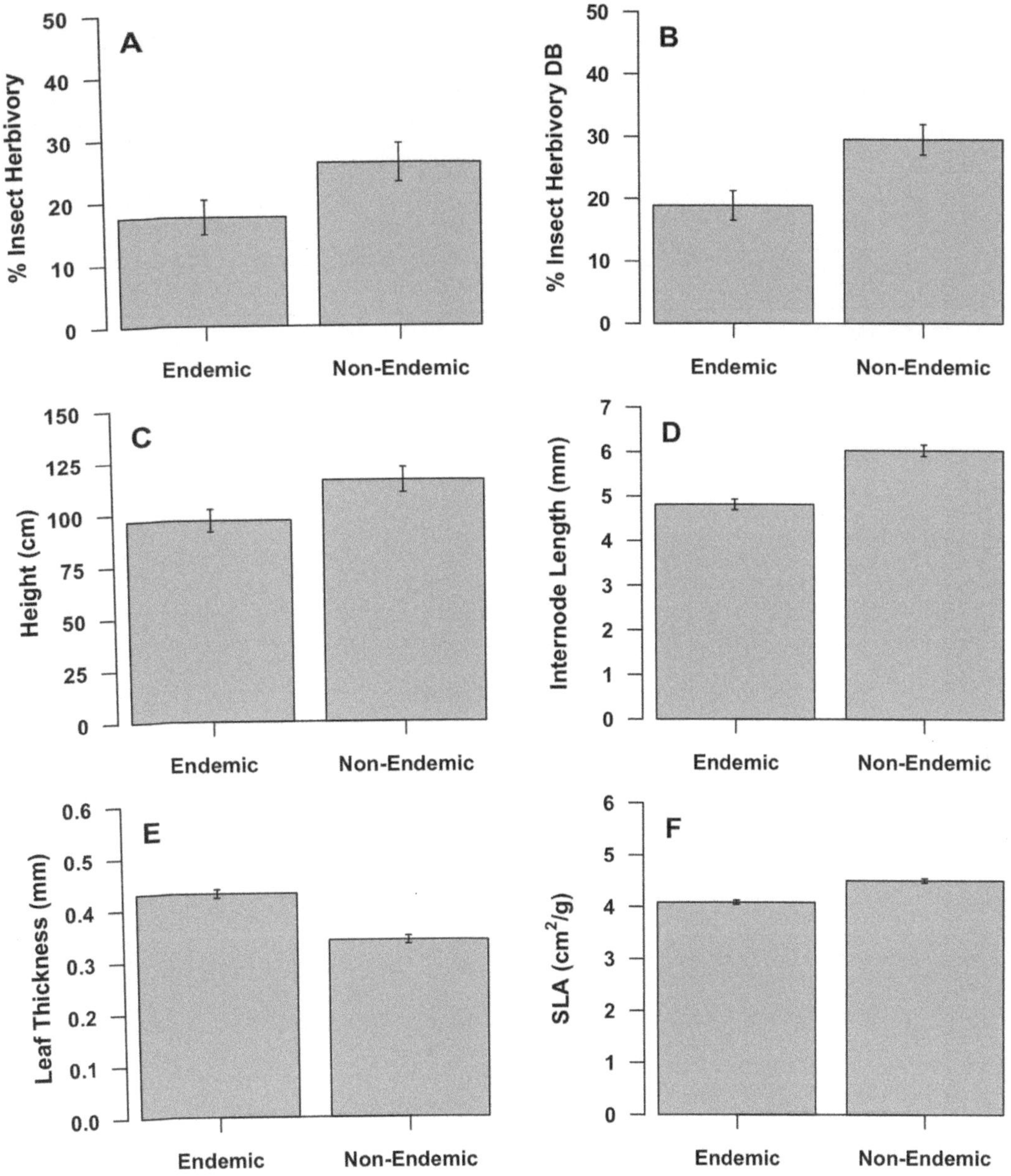

Figure 1. Functional traits and patterns of herbivory differ between endemic *Eucalyptus* species and their non-endemic congeners. Variation in plant functional traits and insect herbivory relative to level of endemism: (A) total foliar herbivory (B) foliar herbivory on the most damaged branch (DB), (C) height, (D) internode length, (E) leaf thickness, (F) specific leaf area (SLA). Total sample size was 412 trees. Error bars represent the standard error.

shoots and two fully expanded leaves were collected from the terminal stems of each tree (juvenile foliage) for measurements of shoot and leaf functional traits. Internode lengths (in mm) were measured on these shoots as the length between the first two fully expanded leaves; typically the 4th and 5th plastochron. Leaf thickness (in mm) was measured with digital calipers. Leaf length, width and area were estimated from the leaf samples using the imaging program ImageJ. Leaves were oven-dried at 70°C for 48 h. Specific leaf area (SLA) was calculated as the average leaf area/average dry weight (cm^2/g).

To understand how potential functional differences between the endemic and non-endemic species might influence the response of interacting species, we quantified herbivory by common mammals and arthropods. Herbivory was estimated in three ways: total insect folivory on the whole tree, insect folivory on the most damaged branch, and total mammal browsing damage. Total insect folivory was visually surveyed and characterized as percent

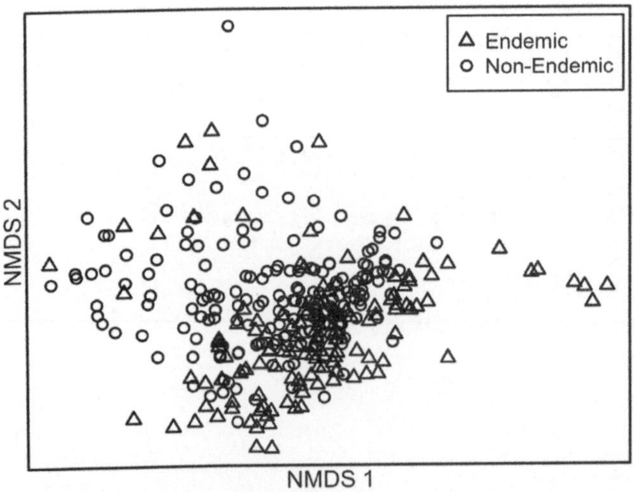

Figure 2. Endemic *Eucalyptus* species contribute a unique suite of functional traits to the landscape. Non-metric multidimensional scaling (NMDS) plot using functional trait and herbivory data showing separation between suites of functional traits and patterns of herbivory of endemic species versus non-endemics. Open circle symbols represent non-endemic species and open triangle symbols represent endemic species.

foliar tissue removed from 1–100 percent (i.e., 0, 1, 2, 3, 5, 10, 20, continuing in 10% increments). Because herbivory is often not uniform across an individual tree, a second survey was conducted on the most damaged branch of each tree using the same methodology. Characteristic shoot clipping by mammal browsing [22], typically by *Trichosurus vulpecula* (common brushtail possum) and *Thylogale billardierii* (red-bellied pademelon), was estimated on each tree as a total damage score. Scores were characterized as the percentage of shoot tips clipped from each tree (using the same scale as insect survey's described above).

Statistical Analyses

The data were analyzed using mixed effect models and Restricted Maximum Likelihood (REML) using the statistical program JMP 10. We tested for quantitative differences in several plant functional traits (height, internode length, leaf thickness, and specific leaf area (SLA)), as well as herbivory between endemic species and non-endemics. We used a conservative approach and constructed a mixed model that included seed family nested within tree species and row as random effects to account for variance explained by these factors that would otherwise contribute to differences between endemics and non-endemics. Endemism/non-endemism and tree species nested within endemism were included as fixed effects. Additionally, to account for multiple comparisons of traits between endemic and non-endemic species, we used the function 'p.adjust' in R (2.15.3) [23] to apply a Holm-Bonferroni correction to estimates of significance.

Because the divergence of *Eucalyptus* species in the subgenus *Symphyomyrtus* has been relatively recent, resolution of phylogenetic relationships at the species level has not been possible with standard DNA sequence markers [24–26]. This makes it impossible to use advanced comparative methods such as phylogenetically independent contrasts (PIC) to account for phylogenetic dependency of our data. In an attempt to account for phylogenetic dependency, we took a conservative approach and constructed a mixed model identical to the one above but that also included clade as a fixed effect. Clade was defined by taxonomic series (Ovatae or Viminales; [21]). Species included in the series Ovatae were *E. brookeriana, E. barberi, E. ovata,* and *E. rodwayi*, and species included in the series Viminales were *E. archeri, E. cordata, E. dalrympleana, E. gunnii, E. johnstonii, E. morrisbyi, E. perriniana, E. rubida, E. subcrenulata, E. vernicosa,* and *E. viminalis*.

Additionally, regression analyses were used to examine the relationships between plant functional traits (height, internode length, leaf thickness, and SLA) and percent foliar herbivory. Because variation in functional traits can reflect niche differentiation, we compared suites of functional traits that differed between endemic and non-endemic species using a two-dimensional ordination of multivariate data using Nonmetric Multidimensional Scaling (NMDS) (R 2.15.3, vegan package) [27]. A distance matrix was constructed using Euclidean distances based on the values of functional plant traits and patterns of herbivory, which were standardized by maximum resemblance for all individuals in the study. Differences were quantified using ANOSIM (analysis of similarity) (R 2.15.3, vegan package) [27], a non-parametric method for determining if there is significant variation between groups of samples based on a Euclidean distance [28].

Results

Functional trait differences between endemic and non-endemic species

The endemic species differed from non-endemic species in functional plant traits. Height, internode length, leaf thickness, and specific leaf area varied significantly between endemic and non-endemic *Eucalyptus* species (**Table 1**, **Figure 1**). The endemic species had 22% thicker leaves (Figure 1E) and 9% lower average SLA (Figure 1F) than the non-endemic congeners. The endemics also had 23% shorter internodes (Figure 1D) and were 18% shorter in height (Figure 1C) than their non-endemic congeners (Table 1). The Holm corrected estimates of significance generally supported our inferences from uncorrected p-values (Table 1). When the functional traits were combined in a multivariate framework, there were significant differences between the endemic and non-endemic species (**Figure 2**, ANOSIM: $R = 0.119$, $p < 0.001$) providing evidence of an endemic syndrome of traits.

Endemic species also exhibited less herbivory than non-endemic species. The endemic species had 40% less total insect folivory (Figure 1A) and 44% less herbivory on the most damaged branch (Figure 1B) than the non-endemics (Table 1). Additionally, the response of insect herbivores was correlated with plant functional traits (height, internode length, leaf thickness, and specific leaf area) (**Table 2**). Significant differences in mammal browsing were not detected ($p = 0.502$); all trees experienced ~10.5% of mammal damage.

It is also possible that shared evolutionary history could influence the differences between endemic and non-endemic species in functional traits. When evolutionary history was accounted for in the mixed model, the levels of significance of endemism as a fixed effect did not change among internode length, leaf thickness, and specific leaf area (**Table 3**), suggesting that shared evolutionary history was not driving the differences in functional traits or patterns of herbivory.

Table 2. Multiple regression model results for leaf traits on herbivore response.

	Coefficient	Standard Error	p	R^2
Height (cm)	0.211	2.779	<0.001*	0.152
Internode Length (mm)	4.195	1.041	<0.001*	0.048
Leaf Thickness (mm)	−62.654	13.018	<0.001*	0.067
SLA (cm^2/g)	7.610	3.222	0.018*	0.017

Summary of regression analysis of the correlation between plant functional traits and total foliar herbivory ($n = 412$).

Discussion

This study demonstrates that despite having evolved from sea-level to tree-line, and under a broad range of selective pressures, endemic *Eucalyptus* species are functionally different from closely related non-endemic congeners. These results support a general hypothesis of convergence on an endemic syndrome of traits. Specifically, we found that endemics have more stress tolerant resource acquisition traits, such as lower SLA, thicker leaves, shorter internodes, and slower growth than widespread, non-endemic species. Although studies involving more species are required to fully understand the driving forces behind these differences, we believe that convergent evolution in response to an environmental gradient (such as elevation or harsh soil conditions) is likely playing a substantial role in the differences in functional traits that we found. Regardless of the environmental conditions driving this convergence, such functional differences in plant traits between endemics and non-endemics reflect differences in nutritional quality and palatability of these species, which in turn likely impacted the response of insect herbivores.

In general, the functional plant traits associated with the endemic species reflect a poorer quality resource for herbivores. For example, we found endemics to have lower SLA than non-endemics, a trait correlated with water use, leaf life span, and leaf nitrogen content [29]. We also found that the endemic species experienced less insect herbivory than non-endemics. This is consistent with the resource availability hypothesis [30] that suggests that the local environment heavily influences anti-herbivore defenses, and that plants with traits such as slow growth rates and long leaf lifespans generally invest more in anti-herbivore defense. Additionally, the response of herbivores was correlated with internode length, leaf thickness, and SLA. While this result suggests that endemics represent a poorer quality resource for herbivores than non-endemics, the alternate hypothesis, that the herbivores specialized for the endemic or non-endemic species were absent from the common garden, cannot be dismissed. Additionally, significant differences in mammal browsing were not detected. However, this result is inconsistent with those from a 2002 study of eucalypt susceptibility to marsupial damage that found that the endemic species *E. gunnii* and *E. morrisbyi* are significantly more susceptible to possum browsing than the two non-endemic species used in the study (*E. globulus* and *E. ovata*) [31]. More studies should examine both insect and mammalian herbivory to determine if there are general differences between endemic and non-endemic species in this ecologically important interaction.

Endemic species are highly valued from a biodiversity standpoint, as the scientific community has made preventing extinctions an urgent priority [32]. Our research shows that endemic *Eucalyptus* species contribute a novel syndrome of traits, with extended consequences across trophic levels (i.e., endemic species experienced less herbivore damage). These results contribute to a growing body of research that suggests genetically based plant traits can have direct and indirect effects on communities [33–35], that can in turn influence ecosystem processes [36–37]. For example, a recent study showed that variation in species interactions has major consequences for community composition and ecosystem processes, such as energy flow, that increase across levels of organization [37]. This has important implications for the conservation of biodiversity, as the loss of endemics as a group might also represents the loss of novel ecological interactions.

Endemic plant species generally evolve in response to a broad range of environmental conditions, including edaphic factors, altitude, geographic isolation, and several other ecological conditions. In the context of climate change where species ranges have been shifting since the Pleistocene [1], fragmentation, isolation, and the subsequent reduction in gene flow have resulted in local adaptation of novel genotypes and the evolution of

Table 3. Mixed model analysis of functional trait measures including clade as a fixed effect.

	Endemism		Clade	
Response Variable	$F_{1,13}$	p	$F_{1,13}$	p
% Insect Herbivory	1.572	0.232	0.987	0.337
% Insect Herbivory DB	2.768	0.121	0.016	0.901
Height (cm)	1.518	0.248	0.741	0.411
Internode Length (mm)	5.671	0.038*	0.065	0.805
Leaf Thickness (mm)	7.633	0.017*	0.904	0.361
Specific Leaf Area (cm^2/g)	11.794	0.005*	1.267	0.282

Summary of mixed model analysis using Restricted Maximum Likelihood (REML) for the difference between samples of endemic ($n = 9$) and non-endemic ($n = 6$) eucalypt species of the subgenus *Symphyomyrtus* growing on the island of Tasmania when evolutionary history is accounted for.

endemics [38–41], [14]. It remains to be seen if endemics across gradients are commonly different from closely related non-endemics, but our results provide a testable hypothesis for endemic syndromes that is worthy of future attention across plant systems. Much more research is needed to elucidate the causes and consequences of the evolution of endemism and to understand whether the conservation of endemics also preserves a unique suite of species interactions.

Acknowledgments

We would like to thank Mark Genung for assistance with data interpretation and support. Thanks to Justin Bloomfield, Camilla Bloomfield, and Paul Tilyard for field and lab help. Thanks to the CRC for Forestry for supporting the trial establishment and Forestry Tasmania for providing the trial site. Comments from Dan Simberloff, Tom Whitham, the Bailey-Schweitzer lab group, and anonymous reviewers greatly improved the manuscript.

Author Contributions

Conceived and designed the experiments: BMP JAS JKB. Performed the experiments: BMP JAS JKB. Analyzed the data: CEG BMP JAS JKB. Contributed to the writing of the manuscript: CEG BMP JAS JKB.

References

1. Davis MB, Shaw RG (2001) Range shifts and adaptive responses to Quaternary climate change. Science (New York, NY) 292: 673–679.
2. Kruckeberg A, Rabinowitz D (1985) Biological aspects of endemism in higher plants. Annu Rev Ecol Syst 16: 447–479.
3. Kunin WE, Gaston KJ (1993) The biology of rarity: Patterns, causes and consequences. Trends Ecol Evol 8: 298–301.
4. Ferreira PM, Boldrini II (2011) Potential reflection of distinct ecological units in plant endemism categories. Conserv Biol 25: 672–679.
5. Pimm S, Russell G (1995) The future of biodiversity. Science(New York) 269: 347–347.
6. Cowling RM (2000) Endemism. In: (Encyclopedia of biodiversity), Levin, SA, editor. Academic Press, San Diego, California. pp. 497–507.
7. Myers N, Mittermeier RA, Mittermeier CG, da Fonseca GA, Kent J (2000) Biodiversity hotspots for conservation priorities. Nature 403: 853–858.
8. Dirzo R, Raven PH (2003) Global state of biodiversity and loss. Annu Rev Environ Resour 28: 137–167.
9. Kier G, Kreft H, Lee TM, Jetz W, Ibisch PL, et al. (2009) A global assessment of endemism and species richness across island and mainland regions. Proc Natl Acad Sci U.S.A 106: 9322–9327.
10. Caujapé-Castells J, Tye A, Crawford DJ, Santos-Guerra A, Sakai A, et al. (2010) Conservation of oceanic island floras: Present and future global challenges. Perspect Plant Ecol Evol Syst 12: 107–129.
11. Karron J (1987) A comparison of levels of genetic polymorphism and self-compatibility in geographically restricted and widespread plant congeners. Evol Ecol 1: 47–58.
12. Gitzendanner M, Soltis P (2000) Patterns of genetic variation in rare and widespread plant congeners. Am J Bot 87: 783–792.
13. Cole CT (2003) Genetic variation in rare and common plants. Annu Rev Ecol Evol Syst 34: 213–237.
14. Hampe A, Petit R (2005) Conserving biodiversity under climate change: the rear edge matters. Ecol Lett 8: 461–467.
15. Cadotte MW, Jonathan Davies T, Regetz J, Kembel SW, Cleland E, et al. (2010) Phylogenetic diversity metrics for ecological communities: integrating species richness, abundance and evolutionary history. Ecol Lett 13: 96–105.
16. Gulias J, Flexas J, Mus M, Cifre J, Lefi E, et al. (2003) Relationship between maximum leaf photosynthesis, nitrogen content and specific leaf area in balearic endemic and non-endemic mediterranean species. Ann Bot 92: 215–222.
17. Lavergne S, Thompson JD, Garnier E, Debussche M (2004) The biology and ecology of narrow endemic and widespread plants: a comparative study of trait variation in 20 congeneric pairs. Oikos 107: 505–518.
18. Reich PB, Oleksyn J, Modrzynski J, Mrozinski P, Hobbie SE, et al. (2005) Linking litter calcium, earthworms and soil properties: a common garden test with 14 tree species. Ecol Lett 8: 811–818.
19. Stebbins GL (1942) The genetic approach to problems of rare and endemic species. Madroño 6: 241–272.
20. Gottlieb LD (1979) The origin of phenotype in a recently evolved species. In: Solbrig OT, Jain S, Johnson GB, Raven, PH, editors. Topics in plant population biology. Columbia University Press, New York. 264–286.
21. Williams K, Potts B (1996) The natural distribution of Eucalyptus species in Tasmania. Tasforests 2: 39–165.
22. O'Reilly-Wapstra J, McArthur C, Potts B (2002) Genetic variation in resistance of Eucalyptus globulus to marsupial browsers. Oecologia 130: 289–296.
23. R Core Team (2013). R: A language and environment for statistical computing.R Foundation for Statistical Computing, Vienna, Austria. ISBN 3-900051-07-0, URL http://www.R-project.org/.
24. Steane DA, McKinnon, Vaillancourt RE, Potts BM (1999) ITS sequence data resolve higher level relationships among the eucalypts. Mol Phylogenet Evol 12: 215–223.
25. Steane DA, Nicolle D, McKinnon GE, Vaillancourt RE, Potts BM (2002) Higher-level relationships among the eucalypts are resolved by ITS-sequence data. Aust Syst Bot 15: 49–62.
26. McKinnon GE, Vaillancourt RE, Steane DA, Potts BM (2008) An AFLP marker approach to lower-level systematics in Eucalyptus (Myrtaceae). Am J Bot 95: 368–380.
27. Oksanen J, Blanchet FG, Kindt R, Legendre P, O'Hara RB, Simpson GL, Solymos P, Stevens MHH, Wagner H. (2013) Vegan: community ecology package. R package version 2.0–7.
28. Clarke K (1993) Non-parametric multivariate analyses of changes in community structure. Aust J Ecol 18: 117–143.
29. McIntyre S, Lavorel S (1999) Disturbance response in vegetation: Towards a global perspective on functional traits. J Veg Sci 10: 621–630.
30. Coley P, Bryant J, III FC (1985) Resource availability and plant antiherbivore defense. Science(Washington) 230: 895–899.
31. Dungey H, Potts B. (2001) Susceptibility of some Eucalyptus species and their hybrids to possum damage. Aust For 65: 23–30.
32. Ricketts TH, Dinerstein E, Boucher T, Brooks TM, Butchart SHM, et al. (2005) Pinpointing and preventing imminent extinctions. Proc Natl Acad Sci U.S.A 102: 18497–18501.
33. Whitham T, Young W, Martinsen G (2003) Community and ecosystem genetics: A consequence of the extended phenotype. Ecology 84: 559–573.
34. Wimp GM, Young WP, Woolbright SA, Martinsen GD, Keim P, et al. (2004) Conserving plant genetic diversity for dependent animal communities. Ecol Lett 7: 776–780.
35. Johnson MTJ, Agrawal A (2005) Plant genotype and environment interact to shape a diverse arthropod community on Evening Primrose (Oenothera Biennis). Ecology 86: 874–885.
36. Schweitzer JA, Bailey JK, Rehill BJ, Martinsen GD, Hart SC, et al. (2004) Genetically based trait in a dominant tree affects ecosystem processes. Ecol Lett 7: 127–134.
37. Bailey JK, Wooley SC, Lindroth RL, Whitham TG (2006) Importance of species interactions to community heritability: a genetic basis to trophic-level interactions. Ecol Lett 9: 78–85.
38. Dynesius M, Jansson R (2000) Evolutionary consequences of changes in species' geographical distributions driven by Milankovitch climate oscillations. Proc Natl Acad Sci U.S.A 97: 9115–9120.
39. Jansson R, Dynesius M (2002) The fate of clades in a world of recurrent climatic change: Milankovitch Oscillations and Evolution. Annu Rev Ecol Syst 33: 741–777.
40. Jansson R (2003) Global patterns in endemism explained by past climatic change. Proc R Soc B 270: 583–590.
41. Jump AS, Peñuelas J (2005) Running to stand still: adaptation and the response of plants to rapid climate change. Ecol Lett 8: 1010–1020.

19

Utility of Marine Benthic Associations as a Multivariate Proxy of Paleobathymetry: A Direct Test from Recent Coastal Ecosystems of North Carolina

Carrie L. Tyler*, Michał Kowalewski

Florida Museum of Natural History, University of Florida, Gainesville, Florida, United States of America

Abstract

Benthic marine fossil associations have been used in paleontological studies as multivariate environmental proxies, with particular focus on their utility as water depth estimators. To test this approach directly, we evaluated modern marine invertebrate communities along an onshore-offshore gradient to determine the relationship between community composition and bathymetry, compare the performance of various ordination techniques, and assess whether restricting community datasets to preservable taxa (a proxy for paleontological data) and finer spatial scales diminishes the applicability of multivariate community data as an environmental proxy. Different indirect (unconstrained) ordination techniques (PCoA, CA, DCA, and NMDS) yielded consistent outcomes: locality Axis 1 scores correlated with actual locality depths, and taxon Axis 1 scores correlated with actual preferred taxon depths, indicating that changes in faunal associations primarily reflect bathymetry, or its environmental correlatives. For datasets restricted to taxa with preservable hard parts, heavily biomineralized mollusks, open ocean habitats, and a single onshore-offshore gradient, the significant correlation between water depth and Axis 1 was still observed. However, for these restricted datasets, the correlation between Axis 1 and bathymetry was reduced and, in most cases, notably weaker than estimates produced by subsampling models. Consistent with multiple paleontological studies, the direct tests carried out here for a modern habitat using known bathymetry suggests that multivariate proxies derived from marine benthic associations may serve as a viable proxy of water depth. The general applicability of multivariate paleocommunity data as an indirect proxy of bathymetry is dependent on habitat type, intrinsic ecological characteristics of dominant faunas, taxonomic scope, and spatial and temporal scales of analysis, highlighting the need for continued testing in present-day depositional settings.

Editor: Kerrie Swadling, University of Tasmania, Australia

Funding: This research was funded by the National Science Foundation (EAR-1243484). The funders had no role in study design, data collection and analysis, decision to publish, or preparation of the manuscript.

Competing Interests: The authors have declared that no competing interests exist.

* E-mail: ctyler@flmnh.ufl.edu

Introduction

Relative species abundance can frequently be described as a function of measured environmental variables (direct gradient analysis), as community composition varies along environmental gradients. Conversely, environmental gradients may be inferred by detecting patterns of variation in community composition (indirect gradient analysis) [1]. The latter approach is frequently employed in paleoecological analyses (e.g., [2,3,4]), and consists of arranging community samples along axes of variation based on their composition, followed by interpretation of the axes in terms of environmental gradients [5]. Indirect gradient analysis is typically performed using multivariate ordination techniques applied to community abundance data [1]. These techniques allow for plotting samples in ordination space, to capture the major directions of variation in faunal composition. Ordination axes are then interpreted *post-hoc* in terms of putative eco-environmental gradients controlling species composition and sample distribution. Present-day settings allow for a direct comparison of ordination scores derived from community data against actual environmental variables. This study employs modern benthic invertebrate communities to directly test the reliability of depth estimates derived from indirect ordinations of quantitative community data.

In present-day ecosystems changes in environmental conditions associated with water depth, such as decrease in light intensity, decrease in wave energy, changes in ambient temperature, and changes in salinity (related to distance from shore and precipitation), are often reflected by fundamental differences in taxonomic and ecological composition of marine benthic communities [6,7,8,9,10,11]. In the fossil record faunal composition is also likely to change notably with water depth (e.g., [12,13,14,15,16]), a view often supported by indirect multivariate analyses of fossil associations evaluated against independent lithological and/or ecological proxies (e.g., [4,13,17,18,19,20,21,22,23]). Many modern ecological studies, however, identify factors other than depth as primary controls (e.g., [24,25,26,27,28]). Furthermore, most paleontological studies are limited to one, or at most a few, higher taxa (but see [13,23,29]), particularly heavily biomineralized organisms such as brachiopods and mollusks (e.g., [4,22,30]). Because studies examining the effect of preservation biases on community ordination patterns are lacking, it remains unclear whether ordinations based on subsets of communities, confined to

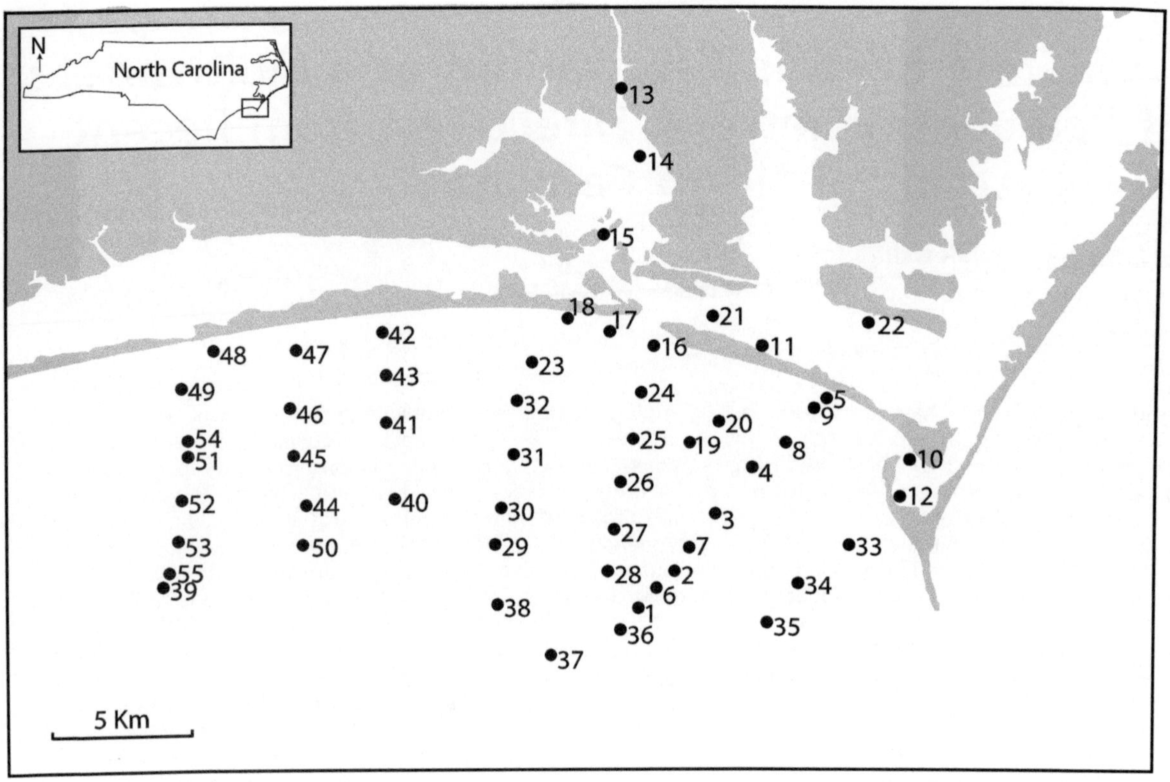

Figure 1. Map of sample locations. Inset box in top left corner shows location of the field area in North Carolina. Each locality (numbered 1–55) was sampled a minimum of three times. In the restricted data analyses, open ocean localities are all localities except 10–15 and 21–22, and the small grid of samples consists of localities 39–55.

biomineralized organisms, accurately detect ecological gradients observed for the entire community.

In this study, dredge samples collected along an onshore-offshore gradient off the coast of North Carolina, were used to assess the reliability and fidelity of multivariate community proxies of bathymetry. Using the resulting benthic invertebrate community data, we evaluated four research questions:

First, we assessed the hypothesis that depth estimates based on faunal composition provided reliable measures of actual bathymetry (and related environmental parameters). Under this hypothesis, indirect ordination scores derived solely from faunal composition data are expected to correlate with locality water depth (known *a priori*, and independent of faunal data). This hypothesis predicts that indirect ordination scores of species should correlate with the actual preferred species depth, which can be estimated directly in modern settings via direct ordination methods (weighted averaging *sensu* [31]).

Second, by deriving ordinations using various subsets of the community, we assess the multivariate fidelity of the marine benthic associations from a paleontological perspective; specifically, comparing the performance of indirect ordinations for all taxa, preservable taxa, and heavily biomineralized mollusks. These three datasets, which represent neontological data (all taxa) and two paleontological proxies (preservable taxa and heavily biomineralized mollusks), respectively, are used here to evaluate their mutual consistency and compare their relative effectiveness in capturing environmental information.

Third, we compare the results of Principal Coordinates Analysis, Correspondence Analysis, Detrended Correspondence Analysis, and Non-Metric Multidimensional Scaling, in terms of their consistency and their effectiveness in deriving robust environmental proxies from multivariate community datasets.

Finally, we examine how geographic scope of the analysis alters the strength of the bathymetric signal, as the relationship between community composition and depth is expected to scale spatially and temporally [10,20,32]. Results are compared for the entire study area (~952 km^2), open ocean habitats (~578 km^2), and a confined onshore-offshore gradient (~180 km^2).

Study Area Description

The coast of North Carolina is protected from the open ocean by barrier islands and sandbars. Storm activity is concentrated in the fall and winter (September-February), with waves arriving predominantly from the northeast during winter months and from the southeast during summer months. The area behind these islands and bars, referred to as the back-sound, is typically more lagoonal/estuarine and somewhat sheltered from swells and storms. The average salinity of the open marine waters in the region is 36 ppt, with the warm Gulf Stream flowing from the south. Inner shelf waters vary seasonally in temperature (>28°C in summer, 12–14°C in winter) [33]. Nearshore average salinity is 34 ppt, and estuarine waters vary in salinity levels dependent on precipitation. Water depth is relatively shallow on the continental shelf, and increases gradually to ~70 m with increasing distance from shore to the edge of the continental shelf [33]. The shelf break (~120 km off the coast of Onslow Bay) marks a sudden dramatic increase in depth. Sediments include fine and medium to coarse sands and gravel [33].

Table 1. Sample location data.

Locality	Samples	Richness	Abundance	Depth (m)	GPS
1	3	3	47	19.0	N34° 34.383′ W76° 40.474′
2	3	12	47	18.8	N34° 35.044′ W76° 39.650′
3	6	18	105	18.3	N34° 36.506′ W76° 38.372′
4	6	29	256	16.7	N34° 37.691′ W76° 37.404′
5	3	25	409	11.7	N34° 39.516′ W76° 34.956′
6	3	9	63	18.7	N34° 34.863′ W76° 39.714′
7	3	20	72	18.7	N34° 35.888′ W76° 38.913′
8	6	36	408	15.5	N34° 38.363′ W76° 36.237′
9	5	22	247	8.0	N34° 39.473′ W76° 34.576′
10	3	21	119	8.5	N34° 38.258′ W76° 32.752′
11	6	55	562	9.9	N34° 40.698′ W76° 36.834′
12	4	31	537	8.8	N34° 37.449′ W76° 32.854′
13	6	23	435	5.2	N34° 48.006′ W76° 41.216′
14	8	40	398	5.4	N34° 45.751′ W76° 40.562′
15	8	45	593	5.6	N34° 43.445′ W76° 41.622′
16	8	25	1253	13.6	N34° 40.877′ W76° 39.929′
17	6	12	73	6.7	N34° 41.307′ W76° 41.185′
18	7	18	283	7.9	N34° 41.460′ W76° 42.554′
19	5	30	189	9.6	N34° 40.499′ W76° 38.665′
20	4	17	208	15.5	N34° 39.028′ W76° 38.063′
21	7	16	235	8.3	N34° 41.528′ W76° 38.142′
22	5	29	213	4.9	N34° 41.374′ W76° 33.769′
23	3	22	179	16.0	N34° 40.375′ W76° 43.591′
24	3	31	779	15.0	N34° 39.755′ W76° 40.362′
25	3	21	116	16.3	N34° 38.595′ W76° 40.642′
26	3	16	87	17.0	N34° 37.585′ W76° 40.868′
27	3	13	56	18.0	N34° 36.316′ W76° 41.138′
28	3	15	65	18.3	N34° 35.403′ W76° 41.331′
29	6	14	62	18.0	N34° 35.952′ W76° 44.557′
30	3	14	91	17.5	N34° 36.819′ W76° 44.487′
31	3	18	134	16.0	N34° 38.218′ W76° 44.358′
32	3	22	128	17.0	N34° 39.459′ W76° 44.176′
33	3	19	217	16.0	N34° 35.740′ W76° 34.505′
34	3	9	60	18.0	N34° 34.928′ W76° 35.792′
35	3	9	55	18.0	N34° 34.009′ W76° 36.660′
36	3	8	59	19.0	N34° 33.921′ W76° 40.914′
37	3	7	64	19.0	N34° 33.230′ W76° 43.016′
38	3	13	82	18.7	N34° 34.442′ W76° 44.632′
39	3	8	106	18.3	N34° 34.939′ W76° 54.526′
40	3	15	82	16.9	N34° 37.281′ W76° 47.791′
41	6	21	143	16.8	N34° 38.852′ W76° 48.196′
42	4	21	137	13.3	N34° 41.149′ W76° 48.010′
43	4	19	207	16.0	N34° 40.118′ W76° 48.143′
44	4	33	356	17.2	N34° 36.891′ W76° 50.267′
45	3	20	319	17.0	N34° 38.122′ W76° 50.702′
46	3	22	442	16.0	N34° 39.306′ W76° 50.736′
47	3	21	205	13.0	N34° 40.761′ W76° 50.570′
48	3	16	257	11.3	N34° 40.739′ W76° 53.045′
49	3	20	251	16.0	N34° 40.015′ W76° 54.028′
50	3	15	100	16.0	N34° 35.938′ W76° 50.392′

Table 1. Cont.

Locality	Samples	Richness	Abundance	Depth (m)	GPS
51	6	20	129	16.3	N34° 38.005′ W76° 53.658′
52	3	12	94	17.0	N34° 37.060′ W76° 54.003′
53	3	12	126	17.0	N34° 36.077′ W76° 54.068′
54	3	12	546	17.0	N34° 34.772′ W76° 54.699′
55	3	12	143	18.0	N34° 35.938′ W76° 50.392′

Number of samples collected at each locality is listed under "Samples".

Ethics Statement

All necessary permits were obtained for the described study, which complied with all relevant regulations. No permits are required for general dredging in the study area. As all dredging was conducted using the Duke University Marine Lab (DUML) facilities and equipment, field collections fell under the DUML invertebrate collections permits (Duke University Marine Lab Scientific or Education Permit 707075 for 2011, 2012, and 2013). With the exception of species identification voucher specimens, all individuals were released *in situ* after counting and identification. No protected species were identified in the sampled material. Data and R script to perform all analyses are included in supplemental materials.

Materials and Methods

Samples were collected in a series of dredges, near the city of Beaufort, North Carolina, U.S.A. (Figure 1). Sampling was completed in four field seasons (June 2011, November 2011, May of 2012, and April 2013) to capture seasonal variation in community composition. Repeat visits to localities served to minimize seasonal variation in sample composition and reduce the magnitude to which richness and relative abundances in the living population may be underestimated. Day et al. [12] conducted similar surveys of the benthic invertebrate fauna repeated in different seasons within a year, and determined that seasonal effects on faunal composition were negligible in the study area.

Dredging was conducted at 55 localities, resulting in a total of 221 dredge samples collected from a variety of habitats, depths, and distances from shore (Table 1). At each locality a minimum of three samples were collected, utilizing the following types of equipment (minimum one sample each): a benthic sled, a dredge basket, and a van Veen grab. The benthic sled was lined with 1 mm wire mesh to ensure representative sampling of smaller species and juveniles, and van Veen samples were wet sieved (1 mm mesh). In the case of benthic sled and basket samples, the entire sample was carefully examined, and all invertebrates identifiable without the aid of a microscope, with the exception of small encrusting species (such as bryozoans and sponges), were counted and identified to the lowest taxonomic level possible (typically species). The resulting live samples consist of multiple higher taxonomic groups (e.g., bivalves, gastropods, arthropods, echinoderms, annelids, etc.). Although surficial dredging of this nature provides limited coverage of the sea floor both in terms of surface area covered, and depth below sediment-water interface, infaunal organisms were adequately represented; 27% of live species surveyed and 36% of individuals were infaunal.

Sampling resulted in an abundance matrix consisting of 11248 individuals, 231 species, and 220 samples from 55 localities. Localities with less than 30 individuals, species with less than 10 individuals, and taxa occurring only at a single locality were removed from the data matrix (see below for justification). The resulting matrix, consisting of 9505 individuals (Table S1), 69 species (Table S2), and 49 localities (Table S3), was used as the initial dataset in all subsequent analyses. Data were analyzed using common multivariate ordination techniques: Principal Coordinates Analysis (PCoA or Classical Multidimensional Scaling), Correspondence Analysis (CA), Detrended Correspondence Analysis (DCA), and Non-Metric Multidimensional Scaling (NMDS). The bivariate relationship between community composition and bathymetry was evaluated via a reduced major axis regression of scores from Axis 1 of a given ordination against the appropriate water depth (m) estimates (locality water depths or weighted species occurrence depths). Locality water depth was recorded when sampling localities, using the onboard depth sounder ($\pm$0.3 m).

Weighted species occurrence depths were obtained for each species by weighted averaging of a species across all localities in which that species was present. For example, Species A found in two localities only, with Locality X at 10 m including $n=5$ specimens and Locality Y at 8.5 m including $n=100$ specimens, would have a preferred water depth calculated as $(10*5+8.5*100)/(100+5)=8.57$ m. Note that this approach represents a direct ordination method based on direct measurements of water depth.

Multiple indirect ordination strategies have been developed for community and other compositional data, including four widely used methods. PCoA represents relationships between objects in multidimensional space, and involves translation of dissimilarities between objects into Euclidian distances [34,35,36]. Ordinations produced in PCoA, however, can distort ordination gradients [37], especially for sparse data matrices, forcing long gradients into curved patterns (the horseshoe effect). CA involves the repeated averaging of scores, maximizing the correspondence between species and locality scores [38,39], and can also suffer from curvilinear distortions (the arch effect) and compression of gradient extremes. DCA was developed to counter these distortions [40]. DCA divides the first axis into segments and centering each segment on zero of Axis 2 (detrending) removing any systematic relationship between scores, and shifting the positions of localities along the first ordination axis by rescaling the segments [40]. This straightens out the arch generated by correspondence analysis, and the ends of the gradient tend to be less compressed. This rescaling may not, however be desirable in all cases [37,41]. NMDS is an iterative technique that optimizes the placement of samples into a low-dimensional space minimizing the mismatch between rank-order of multivariate ecological dissimilarity and Euclidean distances in NMDS space [42,43,44,45]. NMDS is also potentially affected by arch effect and, in addition, this non-eigenvector method does not maximize the variability associated with individual axes [46]. These indirect ordination methods (especially

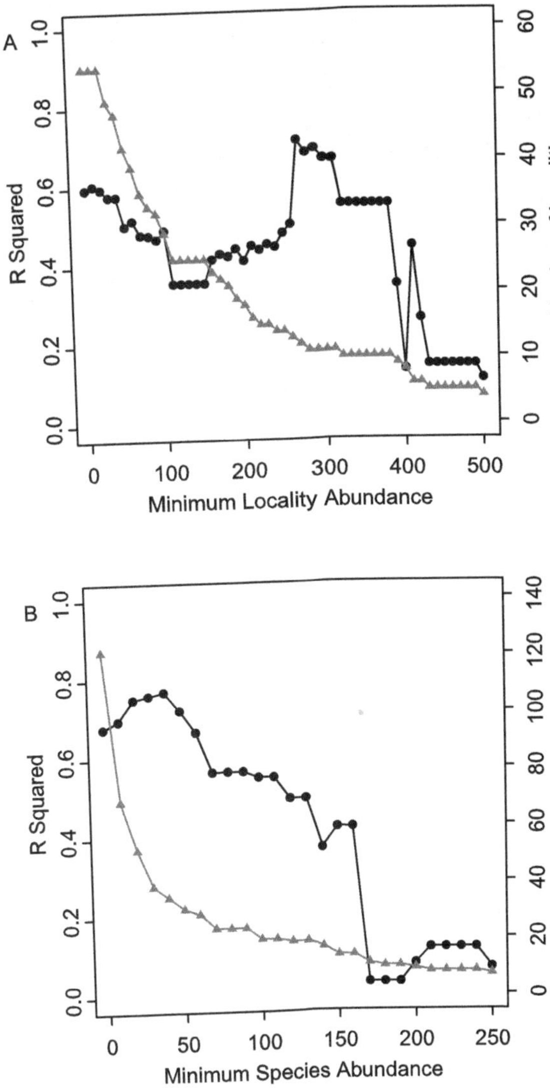

Figure 2. Locality and species abundance filtering. (A) Dashed grey line indicates number of localities retained at a given minimum acceptable sample size. Solid black line indicates adjusted R^2 values from regression of Axis 1 scores with locality depths. (B) Grey line with triangles indicates number of species retained at a given minimum acceptable sample size (i.e., a minimum acceptable number of specimens per species). Solid black line with circles indicates adjusted R^2 values from regression of Axis 1 scores with weighted species occurrence depths. In both figures, each R^2 value represents a new ordination. Note that a minimum species abundance of 10 and minimum locality abundance of 30 was used for all analyses, which allowed us to retain the bulk of the data and R^2 consistent with values observed for larger sample sizes.

CA, NMDS, and DCA) are widely employed in ecology and paleoecology, but numerous controversies surround their relative effectiveness and validity (e.g., [20,31,41,46]). Analyses were therefore performed using all four types of ordination techniques to both facilitate comparison with other studies of both modern and ancient communities, and to compare their relative performance in the specific case of the data analyzed here.

To evaluate the effect of selective restriction of community data (which is inevitable in the case of paleontological data and, in practice, also affects ecological community analysis), ordinations were performed using all taxa, preservable taxa, and heavily biomineralized mollusks to imitate common types of paleontological data. Preservable taxa were defined herein as species with biomineralized skeletal components, which therefore could potentially be preserved in the fossil record (e.g., arthropods, echinoderms, mollusks). This category excluded organisms with no macroscopic hard-parts, such as polychaetes, sponges, and soft corals. Heavily biomineralized mollusks consisted of species with thick shells, having a relatively high preservation potential. Note that the heavily biomineralized mollusk category restricted the analysis to one higher taxon and also eliminated mollusk species that have thin, fragile shells (e.g., *Anomia simplex*) or lack shells (e.g., nudibranchs).

To investigate the effect of spatial scaling on the relationship between bathymetry and community composition, results were compared for the study area (all localities, area ~952 km^2, depth range 4–20 m, and variable salinity), open ocean habitats (all localities excluding 10–15 and 21–22, area ~578 km^2, depth range 7–19 m, and relatively invariant salinity), and a small grid of samples along an onshore-offshore gradient (localities 39–55, area ~180 km^2, depth range 11–18 m, and relatively invariant salinity). At finer spatio-temporal scales, samples may not ordinate along bathymetric gradients [30], and consequently, the strength of the relationship between bathymetry and community composition in ordination space may be expected to deteriorate.

The above datasets, selectively restricted by taphonomic, taxonomic, and geographic criteria, are subsets of the entire dataset. The ordinations based on reduced datasets may be expected to perform less effectively in regression models owing to loss of information associated with loss of taxa, localities, and specimens, resulting in lower values of R^2, merely as an artifact of reduced dimensions of the data matrix and smaller numbers of specimens per sample and/or per taxon. Random subsampling without replacement was therefore used to generate replicate random subsets of the total dataset while mimicking dimensions and sample sizes of the restricted datasets analyzed above. NMDS was performed for each randomly resampled dataset, and only NMDS ordinations with a stress of 0.2 or lower were retained. Stress values >0.2 are generally considered poor and potentially uninterpretable [31,47]. Although such generalizations regarding the interpretability of stress values are arguably oversimplified as stress values vary based on the number of samples and species [47,48], the 0.2 cutoff value is still relatively lenient, allowing us to include many more outcomes in resampling simulations. Outcomes of individual simulation runs are thus more variable, producing more conservative estimates of standard errors. In addition, NMDS analyses performed in three dimensions (which inevitably notably reduced stress values) produced ordinations consistent with their two-dimensional counterparts (only the latter results are reported here). For all NMDS runs with stress <0.2, R^2 values were computed for Axis 1 scores and the independent depth estimates. 1000 iterations were run for each subset. The distribution of the resampled R^2 values were compared with the actual values observed for the entire dataset and for the actual restricted datasets. Separate simulations were performed for each of the four restricted dataset (preservable, robust mollusks, open ocean, and onshore-offshore gradient). The offset between the mean R^2 values obtained in simulation and the R^2 value for the total dataset provide an estimate of the bias due solely to the effect of data restriction. Note that if the R^2 value for a given restricted dataset approximates the mean R^2 value in the corresponding simulation, the poorer performance of the restricted dataset can be attributed to sampling information loss rather than lower informative value of the specific non-random subset of data targeted in a given restricted analyses. Conversely, departures

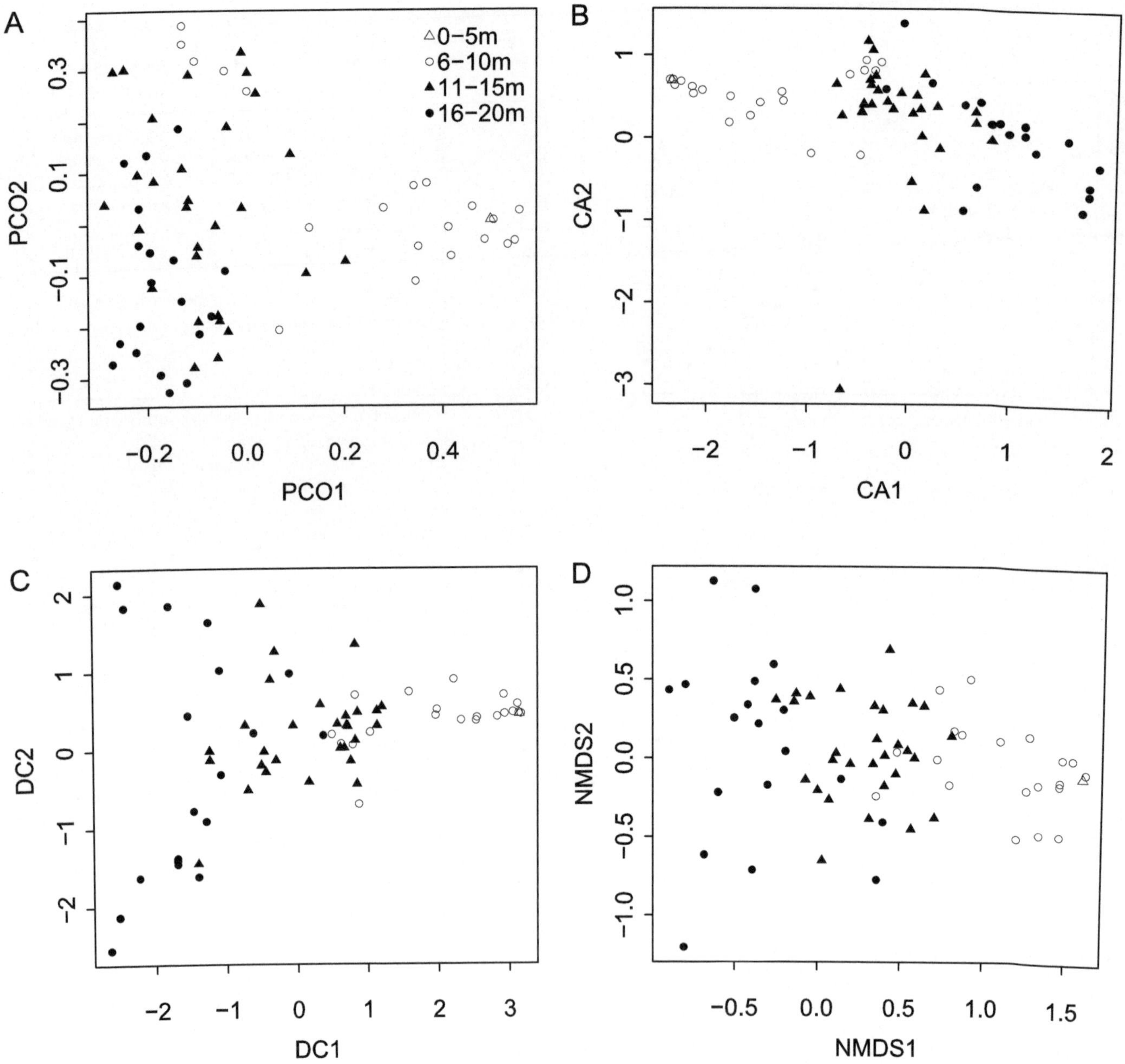

Figure 3. Ordination species score plots for all four ordinations types. (A) PCoA (Axis 1 = 18%, Axis 2 = 12%), (B) CA (principal inertias CA1 = 51%, CA2 = 49%), (C) DCA, (D) NMDS (stress 0.19). Symbols in (A), top right, denote depth ranges.

from the model prediction would suggest that the restricted dataset is less (or more) informative than would be expected for a random subset derived from the entire dataset.

Minimum acceptable sample size per locality was determined by performing NMDS, and correlating Axis 1 scores with species and locality depths at 10 specimen increments. Iterative removal of localities demonstrated that R^2 values were relatively stable even when localities with $n<30$ were included, and only became volatile above the threshold value of $n = 280$ (Figure 2A). In contrast, for species, R^2 values were relatively stable regardless of the threshold. Consequently, a relatively small threshold value of $n = 10$ specimens per species was used, which allowed the retention of the majority of species and a substantial fraction of localities in the final analysis.

All ordinations and analyses were performed using R (version 2.15.1; Tables S4-S7). PCoA (Bray-Curtis distance) and DCA (Chi-square distance) were performed using the "Vegan" package [49]. The default setting of 26 segments was used in all DCA analyses. NMDS (Bray-Curtis distance) and CA (Chi-squared distance) were performed using the "MASS" package [50]. NMDS was performed using two and three dimensions (see above).

Reduced major axis regression (RMA), also known as Standard Major Axis Regression, was used to develop linear models relating water depth and Axis 1 ordination scores. This method is particularly applicable here because the compared variables (ordination scores and water depth) are of intrinsically different types and thus require standardization prior to analysis (e.g., [46]).

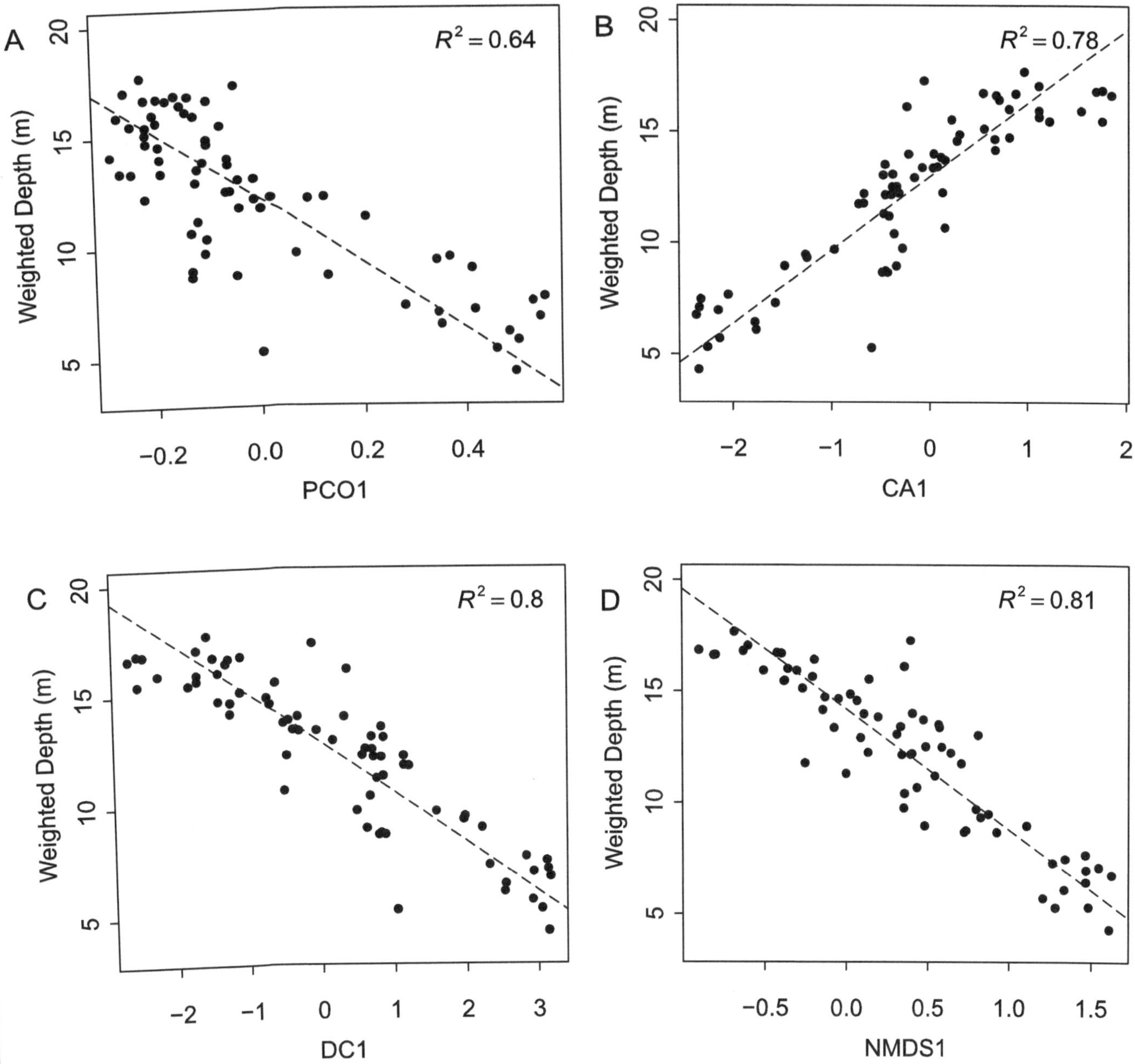

Figure 4. Reduced major axis regression of weighted species depth estimates and Axis 1 species scores for four types of ordinations. (A) Principal Coordinates Analysis, (B) Correspondence Analysis, (C) Detrended Correspondence Analysis, (D) Non-Metric Multidimensional Scaling.

Furthermore, as substantial errors potentially affect both compared variables, a Type II regression model is more appropriate (e.g., [51]). RMA models were based on ordination scores regressed against depth values (either locality depth or weighted species occurrence depths).

All R^2 values reported herein are the adjusted R^2 values. Whereas adjusted R^2 is most commonly applied to multiple regression problems (to compensate for adding additional effects to the model), it is also appropriate for simple linear regression when data represent a sample, rather than exhaustive data for entire statistical populations [46]. The adjusted R^2 is always lower than unadjusted R^2 and provides a more conservative estimate of amount of variance accounted for by the independent effect variable. Because our samples are generally large and all models are 1-parameter models, the differences between the adjusted R^2 and non-adjusted R^2 are trivial (often non-observable when R^2 value is rounded to 2 decimal places). Pearson's correlation coefficient (*r*) and 95% confidence intervals were calculated for each data subset using both standard parametric approximations and bootstrap resampling. Each bootstrap estimate was based on 1000 replicate samples obtained by sampling pairs of observations with replacement.

Results

When species ordination scores are plotted by depth range (Figure 3), species separate by depth along the first axis for all four

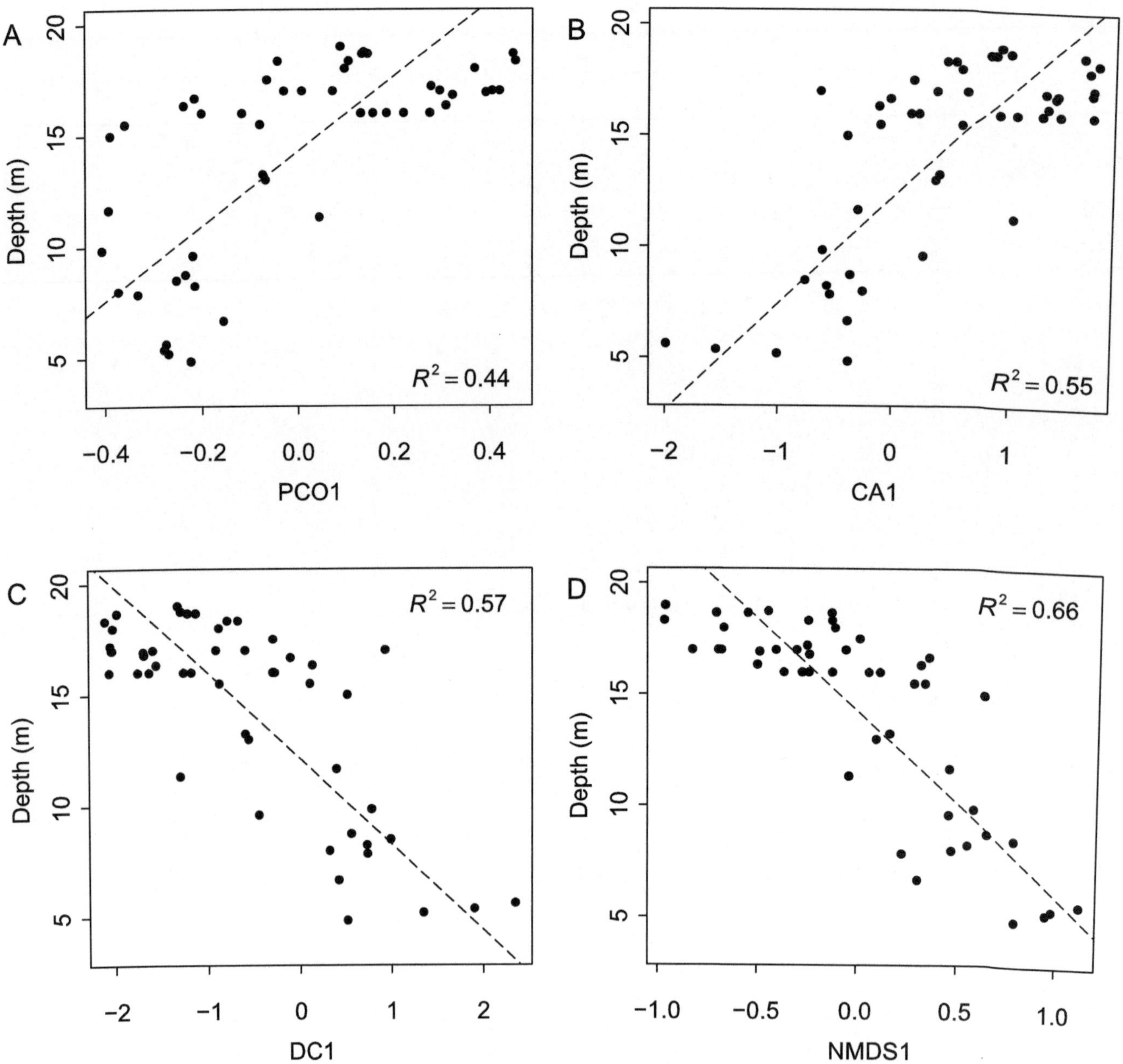

Figure 5. Reduced major axis regression of locality depth and Axis 1 locality scores for four types of ordinations. (A) Principal Coordinates Analysis, (B) Correspondence Analysis, (C) Detrended Correspondence Analysis, (D) Non-Metric Multidimensional Scaling.

ordination techniques. Axis 1 scores are a significant predictor of weighted species occurrence depths for all four ordination techniques (Figure 4). Adjusted R^2 values are comparable across ordination methods, with Axes 1 yielding the highest values for NMDS and DCA (0.81 and 0.80 respectively), and lowest for PCoA (0.64). Although regressions between Axis 1 scores and locality depth are also all significant ($p << 0.0001$), R^2 values are lower (Figure 5). Again, NMDS yielded the highest (0.66) and PCoA the lowest (0.44) R^2 value. As all four ordination techniques produced similar regression results, all subsequent analyses focused on the NMDS results. The depth gradient is clearly apparent in the NMDS ordination plot (Figure 6), with species and locality depths increasing to the left on Axis 1 (stress = 0.2). Axis 2 species and samples do not align or group according to habitat type or distance from the coastline. Whereas scores for the first axes are highly consistent across different ordination types with absolute value of r exceed 0.7 for all comparisons (Table 2), the scores for the second axes are poorly correlated, with most absolute values of r below 0.2 (Table 2).

When data are restricted to preservable taxa only (46 localities, 61 species, 7545 individuals, Table S8; Figure 7), the regression for species scores is still significant, although not as strong (Figure 8 C; $R^2 = 0.79$, $p << 0.0001$). When only heavily biomineralized mollusks are retained (28 localities, 18 species, 3042 individuals, Table S9) the relationship weakens considerably (Figure 8 E; $R^2 = 0.25$, $p = 0.003$). Locality scores performed similarly; when preservable taxa are retained the results are nearly identical to those obtained for all taxa (Figure 8 D; $R^2 = 0.67$, $p << 0.0001$),

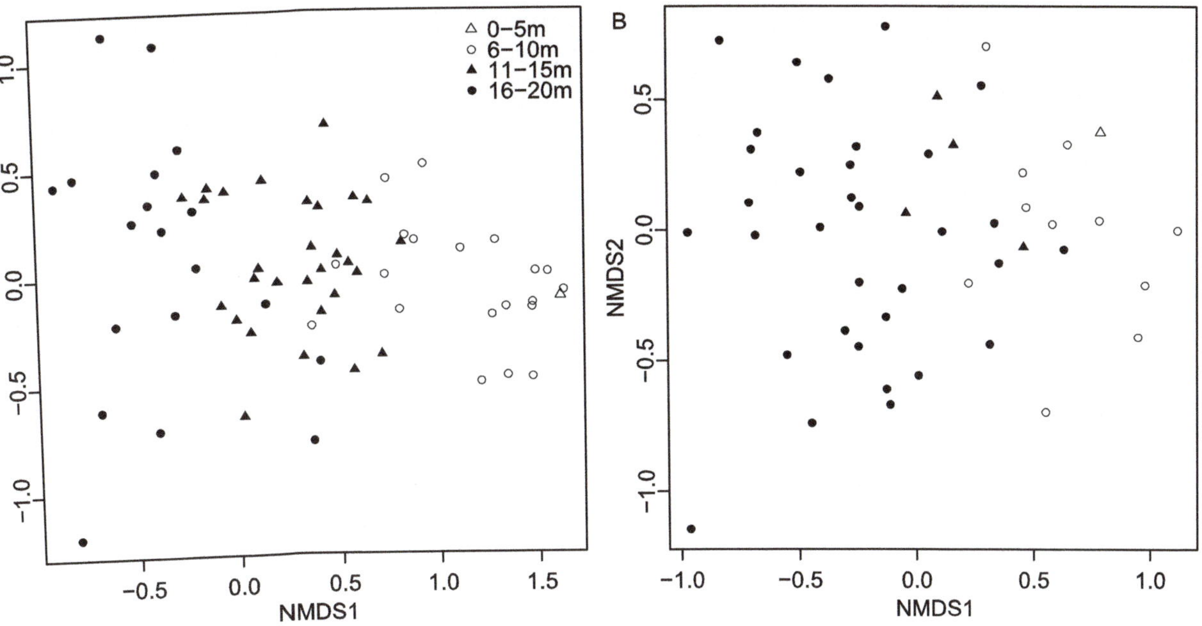

Figure 6. NMDS plots of locality and species scores. (A) plot by species, (B) plot by localities. Symbols denote depth ranges of species (A) or localities (B).

out the association between depth and locality scores weakens considerably when only heavily biomineralized mollusks are retained (Figure 8 F; $R^2 = 0.40$, $p = 0.04$). It should be noted that organisms which vary in preservational potential (i.e., heavily biomineralized species, species with fragile but skeletonized hard parts, species with some hard parts such as chitin, and species with soft tissues only) are present along the entire length of Axis 1 (Figure 7). Thus, the differences observed across the three compared datasets (all taxa, preservable taxa, and heavily biomineralized mollusks) are unlikely to represent an artifact of a non-random clustering of organisms with high (versus low) preservation potential along the first ordination axis.

Table 2. Axis score correlation coefficients.

Method	*r* (L)	*r* (Sp)
CA1 vs. DC1	−1.00*	−1.00*
CA1 vs. MDS1	−0.81*	−0.91*
CA1 vs. PCO1	0.91*	−0.84*
PCO1 vs. MDS1	−0.88*	0.74*
DC1 vs. MDS1	0.81*	0.91*
DC1 vs. PCO1	−0.90*	0.84*
CA2 vs. DC2	0.09	0.11
CA2 vs. MDS2	−0.06	−0.09
CA2 vs. PCO2	0.21	0.54*
PCO2 vs. MDS2	0.56*	0.08
DC2 vs. MDS2	−0.19	−0.09
DC2 vs. PCO2	−0.48*	0.11

Spearman's Rank correlation coefficients (*r*) of Axis 1 and Axis 2 locality (L) and species (Sp) scores for each ordination method.
*Denotes significant correlations ($\alpha = 0.05$).

NMDS ordinations performed at successively smaller spatial scales yielded qualitatively consistent results. The dimensions of the resultant restricted datasets are as follows: open ocean localities consists of 42 localities, 52 species, 7120 individuals (Table S10), and the onshore-offshore gradient consists of 17 localities, 50 species, 2830 individuals (Table S11). Localities separate by depth along the first axis for all localities (Figure 9 A), open ocean localities (Figure 9 C), and for the onshore-offshore gradient

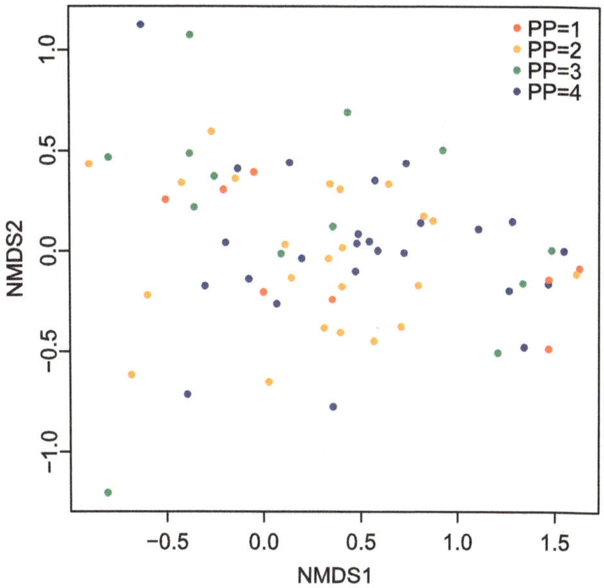

Figure 7. NMDS plot of species scores and preservation potential. Symbols denote relative preservation potential (4 = heavily biomineralized, 3 = fragile but skeletonized hard parts, 2 = some hard parts such as chitin, 1 = all soft parts). Note that species with various ranks of preservation potentials are distributed along the entire range of Axis 1.

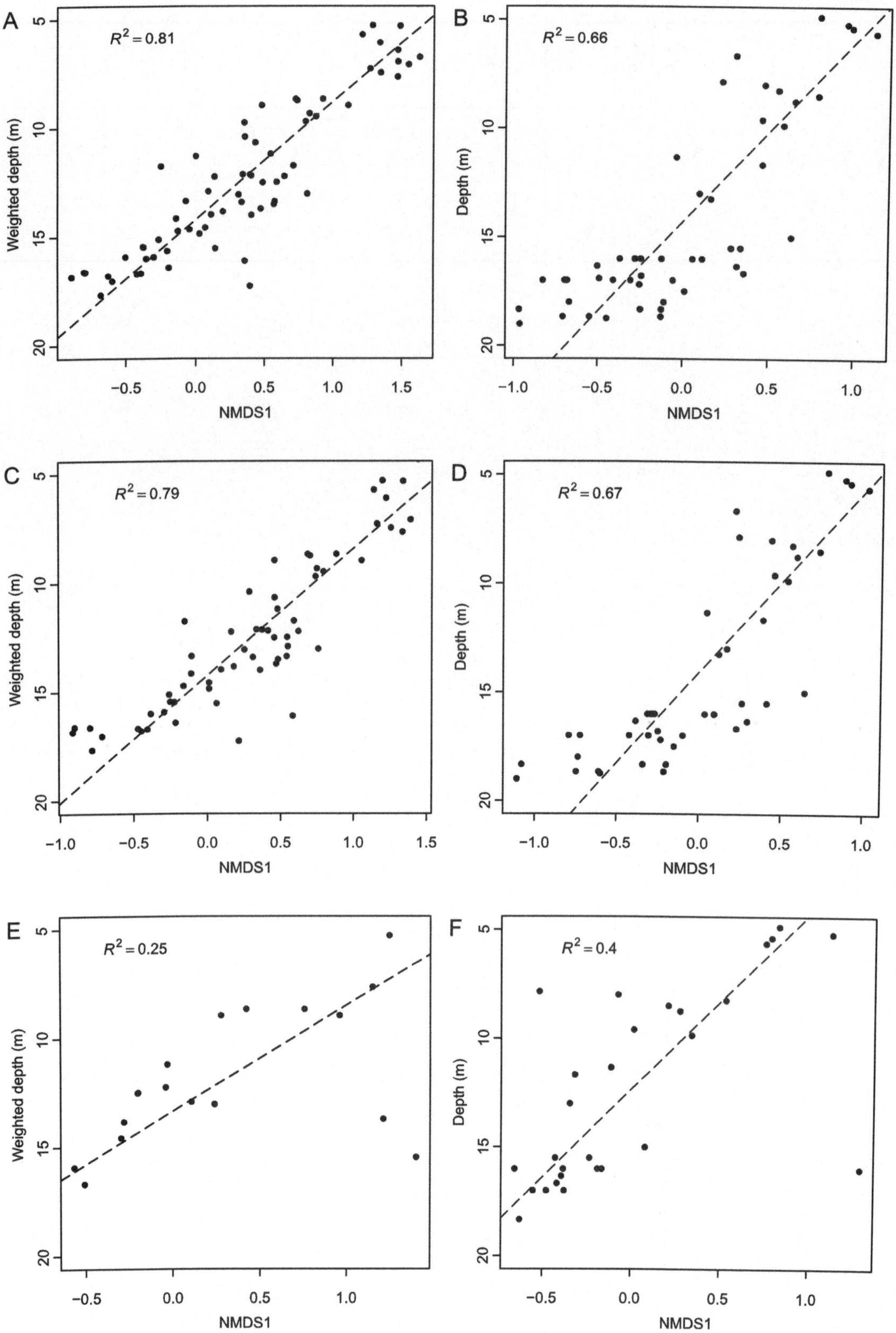

Figure 8. Effects of taphonomic biases and taxonomic scaling. Reduced major axis regressions between Axis 1 scores and depth for: (A–B) all taxa, (C–D) preservable taxa (stress 0.2), and (E–F) heavily biomineralized mollusks (stress 0.2). Plots on the left utilize species scores, plots on the right use locality scores. Note the presence of two notable outliers in 8E (bottom right quadrant, bivalves *Spisula solidissima* and *Ensis directus*), and one in 8F (bottom right quadrant, locality 49), without which the relationship between species or sample depth and Axis 1 scores would greatly improve.

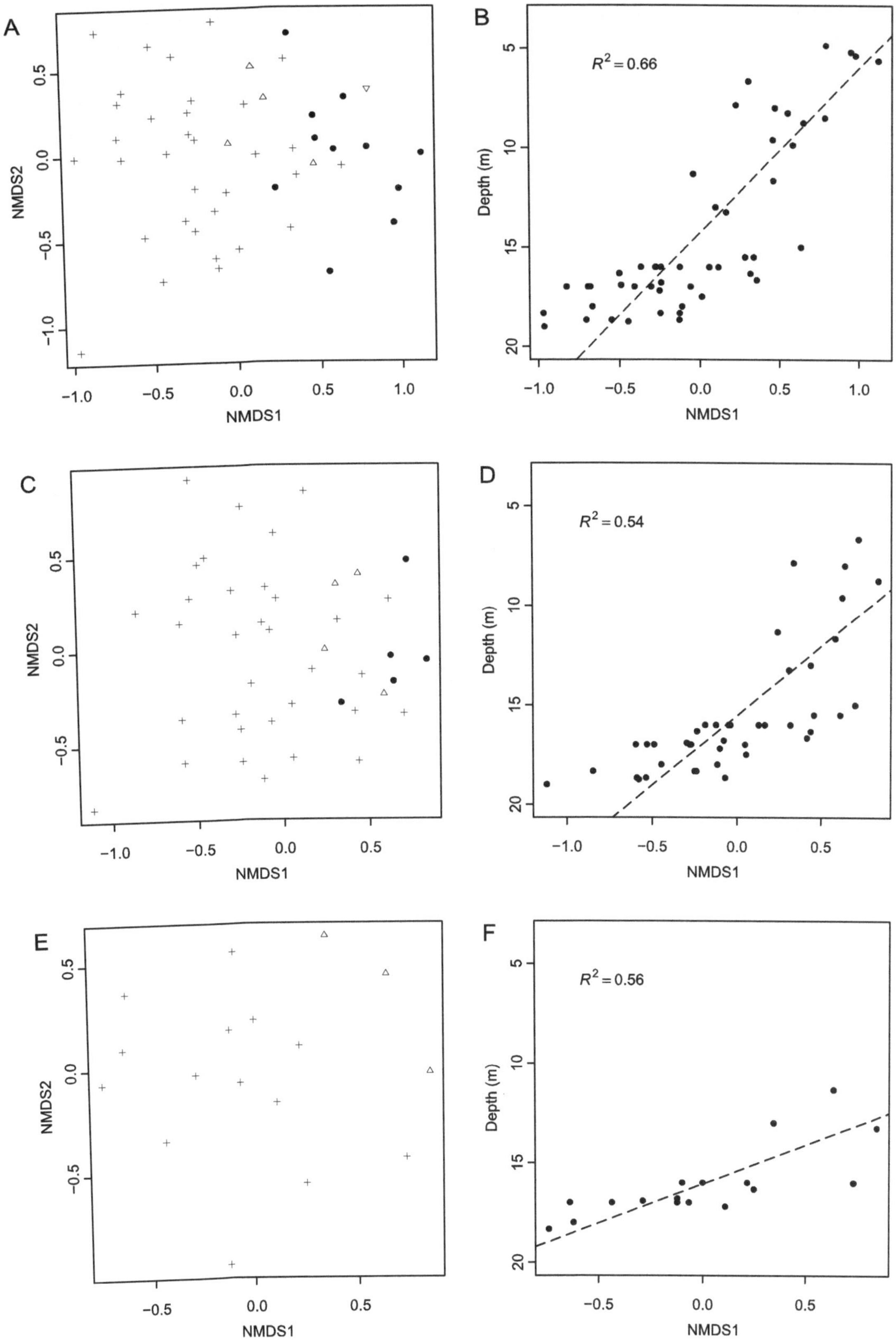

Figure 9. Effects of geographic scaling. NMDS locality score plots and corresponding reduced major axis regressions between Axis 1 locality scores and depth for: (A–B) the entire study area (stress 0.2), (C–D) open ocean localities only (stress 0.2), (E–F) a small onshore-offshore gradient (stress 0.17). Symbols in NMDS plots denote depth ranges (reversed triangle = 0–5 m, filled circle = 6–10 m, inverted triangle = 11–15 m, cross = 16–20 m).

Table 3. Confidence intervals for correlation coefficients.

Species Scores	n	r	p	Lower CI	Upper CI	Lower CI (bootstrap)	Upper CI (bootstrap)
Entire Dataset	69	−0.90	<<0.0001	−0.94	−0.85	−0.94	−0.85
Preservable	61	−0.89	<<0.0001	−0.93	−0.83	−0.94	−0.83
Robust Mollusks	18	−0.54	0.02	−0.80	−0.10	−0.94	−0.03
Open Ocean	52	−0.72	<<0.0001	−0.83	−0.55	−0.83	−0.57
Grid Transect	50	−0.45	0.001	−0.65	−0.20	−0.64	−0.21
Locality Scores							
Entire Dataset	49	−0.82	<<0.0001	−0.89	−0.70	−0.89	−0.72
Preservable	46	−0.82	<0.0001	−0.90	−0.70	−0.90	−0.73
Robust Mollusks	28	−0.65	0.002	−0.83	−0.37	−0.93	−0.28
Open Ocean	42	−0.74	<<0.0001	−0.85	−0.57	−0.84	−0.64
Grid Transect	17	−0.77	0.0003	−0.91	−0.46	−0.91	−0.59

Pearson correlation coefficients, parametric *p*-values, and 95% confidence intervals (CI) are shown for the entire dataset, and each data subset.

(Figure 9 E). Axis 1 scores are a significant predictor of locality depths for open ocean localities (Figure 9 D; $R^2 = 0.56$, $p << 0.0001$), and the onshore-offshore gradient (Figure 9 F; $R^2 = 0.54$, $p < 0.001$), although the strength of the relationship is weaker, in both cases, when compared to the results for all localities (Figure 9 B; $R^2 = 0.66$, $p << 0.0001$).

When compared to the observed r^2 value for the total dataset, the mean R^2 values estimated by random subsampling are notably lower for heavily biomineralized mollusks, open ocean localities, and the onshore-offshore gradient (Figure 10), but comparable for preservable taxa. Note that the simulations for preservable taxa and open ocean localities result in the removal of a small number of taxa and localities from the dataset. For three out of the four restricted datasets (robust mollusks, open ocean localities, and the onshore-offshore gradient), the observed R^2 values for the subsets of data are lower than the mean R^2 values produced by random subsampling. Simulated R^2 values for heavily biomineralized mollusks and the onshore-offshore gradient, are widely dispersed. Pearson's correlation coefficients are all negative and statistically significant (Table 3). The absolute values of coefficients are high (mean = −0.73) and 95% confidence intervals are relatively narrow, with the exception of the two datasets with smaller sample sizes (robust mollusks with 18 species and 28 localities, and the onshore-offshore gradient with 50 species and 17 localities).

Discussion

A strong relationship between bathymetry and faunal composition of marine benthic communities was observed along the studied onshore-offshore gradient suggesting that environmental factors associated with increasing water depth were the primary factor controlling faunal composition of the marine benthos in the study area, despite substantial habitat heterogeneity. Thus, although multiple gradients likely affected community composition in this region, water depth appears to have been a strong correlative of primary processes that control community composition. Altering the minimum acceptable sample size required for retention of localities and species did not obscure the observed relationships between ordination outcomes and the bathymetry, which appeared relatively strong even when rare taxa and localities with low abundance were included. Regressions between the known depth and Axis 1 ordination scores indicated that all four ordination techniques identified bathymetry (and its implied environmental correlatives) as the greatest source of variation in the data, with NMDS scores yielding the highest R^2 values. NMDS produced ordinations highly consistent with other methods despite relatively high stress values (cutoff value of 0.2). These results are consistent with previous studies concluding that the results of DCA and NMDS are often comparable [20]. The second axis could not be interpreted within the context of the available environmental information. Moreover, the ordinations of samples and species are highly inconsistent across different ordination techniques suggesting strong and variable distortions along second axes.

When only preservable taxa were retained in the analysis, the bathymetric signal was still detectable, although somewhat weaker relative to ordinations using the entire community. Preservable taxa therefore served as a reasonable proxy for all taxa. Because these taxa could potentially be preserved in the fossil record, the results suggest that indirect ordinations of fossil communities may provide viable proxies for bathymetric gradients in cases when such gradients were ecologically important. Although the use of restricted datasets limited to preservable species should provide a proxy for fossil communities, such data do not incorporate the

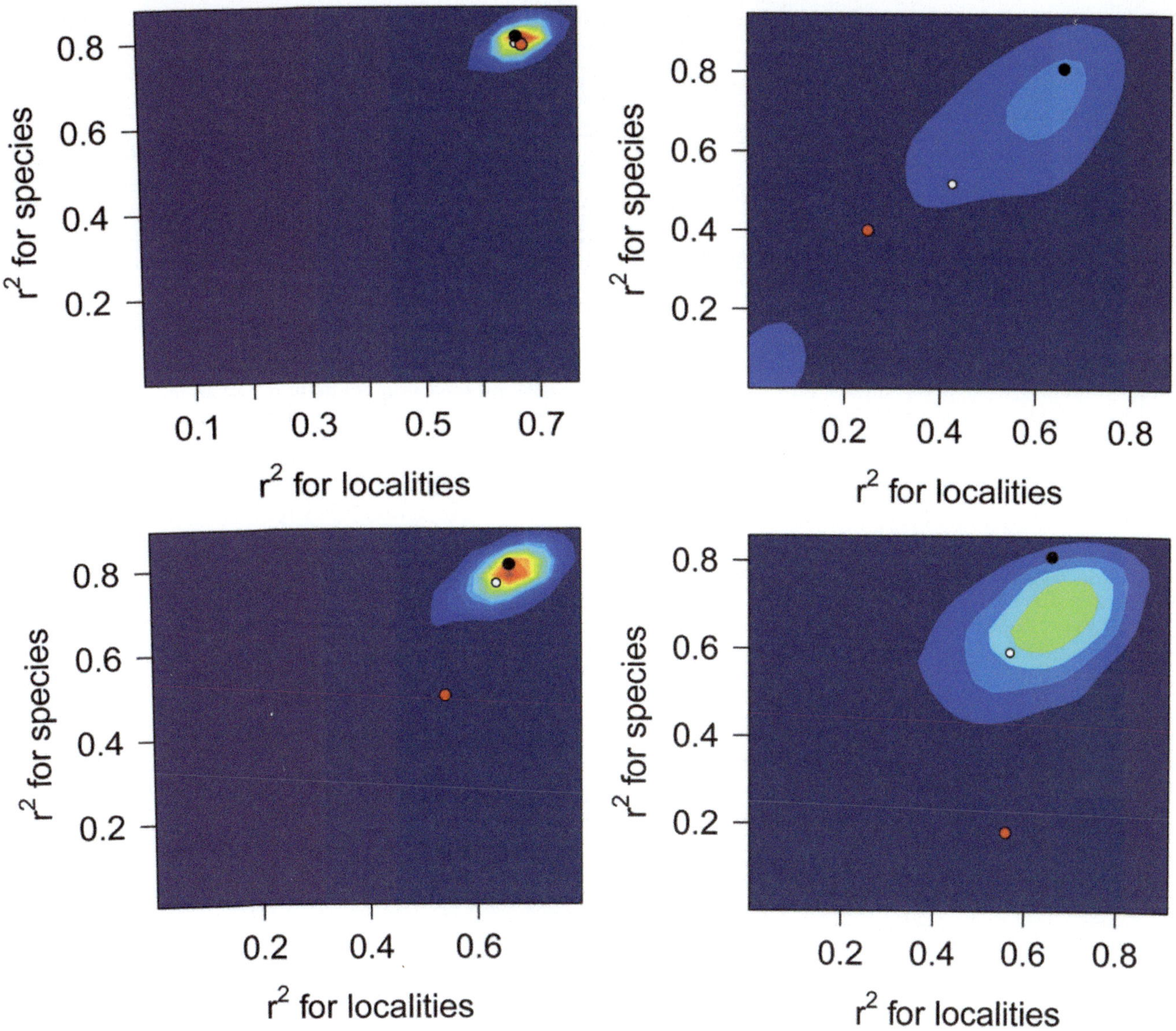

Figure 10. Distributions of r^2 values resulting from random resampling. (A) Preservable Taxa, (B) Heavily Biomineralized Mollusks, (C) Ocean localities, and (D) Small Grid. White dot denotes mean of simulations, red dot denotes actual observed value, and black dot denotes value for complete data set. Color contours within plots denote density from low (dark blue) to high (red).

potential effects of substantial time-averaging that affects fossil samples [52,53,54,55,56,57]. However, time-averaging tends to average seasonal variations in faunal compositions and enhance alpha diversity [58,59,60], while exerting only small loss of information on beta diversity [56,58]. Consequently, it is likely that time averaged samples would perform adequately in cases when the preservable part of the community offers an informative proxy for environmental gradients. Finally, the dataset restricted to robust mollusks was less effective in detecting the bathymetric gradient suggesting that, for our study area, the restriction of data to robust mollusks would hamper paleontological analyses aimed at detecting environmental gradients. While the poor performance of mollusks cannot be readily evaluated from the existing data, the underperformance of robust mollusks may be due to lower diversity/high evenness (only 13 species total) and, possibly, broad bathymetric distributions of some of the most abundant species (e.g., *Mulinia lateralis* is known to occur in depths up to 119 m, *Spisula solidissima* up to 276 m, and *Chione cancellata* up to 108 m).

When analyses are restricted spatially to a single onshore-offshore gradient reducing the geographic area from ~952 to 180 km^2 and the depth range from 16 to 7 m, the bathymetric gradient is poorly manifested in ordination patterns. This outcome is not inconsistent with the hypothesis proposed by Redman et al. [30], that multivariate ordinations at fine spatial scales may be unable to detect bathymetric gradients. At finer spatial scales, water depth may no longer be the strongest control on the distribution of taxa, and factors that were secondary at larger spatial scales may become dominant (e.g., life mode and grain size).

Indirect ordination methods applied to marine benthic communities spanning multiple habitats along a depth gradient appear suitable for detecting environmental gradients even when restricted to non-random subsets of organisms that can be preserved in the fossil record. The results reported here reinforce paleontological studies that successfully employed such analytical strategies in the fossil record. However, restricting data to one group of organisms or smaller spatial scales, where water depth may no longer be the primary control, may reduce the effectiveness of multivariate ordination methods to delineate bathymetric gradients even when strong gradients can be detected when multiple groups of benthic organisms are analyzed simultaneously.

Supporting Information

Table S1 Filtered dataset used for all initial analyses. Species by locality abundance matrix, rows are localities and columns are species.

Table S2 Species list. Species number listed in column one, weighted species depth (m) in column two, preservation potential in column 3 (1 being no hard parts, 3 being robust, see text for full description of categories), and range of depths in column 4 (1 = 0–4.9 m, 2 = 5–9.9 m, 3 = 10–14.9 m, 4 = 15–20 m). To be used with R script.

Table S3 Locality list. Locality number listed in column one, depth (m) in column two, general habitat type in column 3 (1 = harbor, 2 = back-sound, 3 = 0–3 miles offshore, 4 = 6–4 miles offshore, 5 = river), anthropogenic effects in column 4 (1 = least disturbed, 2 = moderately impacted, 3 = highly impacted), depth range (1 = 0–4.9 m, 2 = 5–9.9 m, 3 = 10–14.9 m, 4 = 15–20 m). To be used with R script.

Table S4 R Script. Script to perform all analyses, associated data files must be downloaded.

Table S5 Weighted species depths.

Table S6 Locality depths.

Table S7 Filtered dataset used for all initial analyses for resampling analysis. Species by locality abundance matrix, rows are localities and columns are species, row and column headings have been removed.

Table S8 Filtered dataset with preservable taxa. Species by locality abundance matrix, rows are localities and columns are species. Taxa with no macroscopic hard parts have been removed.

Table S9 Filtered dataset with robust mollusks. Species by locality abundance matrix, rows are localities and columns are species. Only robust mollusks have been retained.

Table S10 Filtered dataset with open ocean habits. Species by locality abundance matrix, rows are localities and columns are species. Only open marine localities have been retained.

Table S11 Filtered dataset with an onshore-offshore transect. Species by locality abundance matrix, rows are localities and columns are species. Only localities form a small set of transects have been retained.

Acknowledgments

We wish to thank the Duke University Marine Lab for providing facilities and equipment, with special thanks to John T. Wilson, and to the following people for assistance in the lab and field: A.J. Giuffre, A. Webb, T.A. Dexter, S. Casebolt, K. O'Donnell, A. Hendy, A. Tucker, M. Meyer, J. Sliko, J. Broce, D. Hawkins, H. McGettigan, K. Mack, S. Paskovitch, A. Flemming, I. Sàtiro, and P. Riggs.

Author Contributions

Conceived and designed the experiments: CLT MK. Performed the experiments: CLT MK. Analyzed the data: CLT MK. Contributed reagents/materials/analysis tools: CLT MK. Wrote the paper: CLT MK.

References

1. Ter Braak CJ, Prentice IC (1988) A theory of gradient analysis. In: Begon M, Fitter AH, Ford ED, MacFadyen A, editors. Advances in Ecological Research. San Diego: Academic Press Inc. 271–317.
2. Springer DA, Bambach RK (1985) Gradient versus cluster analysis of fossil assemblages: a comparison from the Ordovician of southwestern Virginia. Lethaia 18: 181–198.
3. Holland SM (2005) The signatures of patches and gradients in ecological ordinations. Palaios 20: 573–580.
4. Olszewski TD, Patzkowsky ME (2001) Measuring recurrence of marine biotic gradients: A case study from the Pennsylvanian-Permian midcontinent. Palaios 16: 444–460.
5. Whittaker RH (1967) Gradient analysis of vegitation. Biological Reviews 42: 207–264.
6. Walton WR (1955) Ecology of living benthonic foraminifera, Todos Santos Bay, Baja California. Journal of Paleontology 29: 952–1018.
7. Franz DR, Worley EK, Merrill AS (1981) Distribution patterns of common seastars of the middle Atlantic continental shelf of the northwest Atlantic (Gulf of Maine to Cape Hatteras). The Biological Bulletin 160: 394–418.
8. Carney RS (2005) Zonation of deep biota on continental margins. Oceanography and Marine Biology: An Annual Review 43: 211–278.
9. Olabarria C (2006) Faunal change and bathymetric diversity gradient in deep-sea prosobranchs from Northeastern Atlantic. Biodiversity & Conservation 15: 3685–3702.
10. Konar B, Iken K, Edwards M (2009) Depth-stratified community zonation patterns on Gulf of Alaska rocky shores. Marine Ecology 30: 63–73.
11. Smale DA (2008) Continuous benthic community change along a depth gradient in Antarctic shallows: evidence of patchiness but not zonation. Polar Biology 31: 189–198.
12. Day JH, Field JG, Montgome.Mp (1971) The use of numerical methods to determine the distribution of the benthic fauna across the continental shelf of North Carolina. Journal of Animal Ecology 40: 93–125.
13. Holland SM, Miller AI, Meyer DL, Dattilo BF (2001) The detection and importance of subtle biofacies within a single lithofacies: The Upper Ordovician Kope Formation of the Cincinnati, Ohio region. Palaios 16: 205–217.
14. Bilyard GR, Carey AG (1979) Distribution of western Beaufort sea Polychaetous Annelids. Marine Biology 54: 329–339.
15. Webber AJ (2002) High-resolution faunal gradient analysis and an assesment of the causes of meter-scale cyclicity in the type Cincinnatian Series (Upper Ordovician). Palaios 17: 545–555.
16. Scarponi D, Angeletti L (2008) Integration of palaeontological patterns in the sequence stratigraphy paradigm: a case study from Holocene deposits of the Po Plain (Italy). GeoActa 7: 1–13.
17. Scarponi D, Kowalewski M (2004) Stratigraphic paleoecology: Bathymetric signatures and sequence overprint of mollusk associations from upper Quaternary sequences of the Po Plain, Italy. Geology 32: 989–992.
18. Amati L, Westrop SR (2006) Sedimentary facies and trilobite biofacies along an Ordovician shelf to basin gradient, Viola Group, south-central Oklahoma. Palaios 21: 516–529.
19. Kowalewski M, Gurs K, Nebelsick JH, Oschmann W, Piller WE, et al. (2002) Multivariate hierarchical analyses of Miocene mollusk assemblages of Europe: Paleogeographic, paleoecological, & biostratigraphic implications. Geological Society of America Bulletin 114: 239–256.
20. Patzkowsky ME, Holland SM (2012) Stratigraphic Paleobiology: Understanding the Distribution of Fossil Taxa in Time and Space. Chicago: University of Chicago Press.
21. Ziegler AM, Cocks RM, Bambach RK (1968) The composition and structure of Lower Silurian marine communities. Lethaia 1: 1–27.
22. Hendy AJW (2013) Spatial and stratigraphic variation of marine paleoenvironments in the middle-upper Miocene Gutan Formation, Isthmus of Panama. Palaios 28: 210–227.

23. Miller AI, Holland SM, Meyer DL, Datillo BF (2001) The use of faunal gradient analysis for intraregional correlation and assessment of changes in sea-floor toporgraphy in the type Cincinnatian. The Journal of Geology 109: 603–613.
24. Feder HM, Naidu AS, Jewett SC, Hameedi JM, Johnson WR, et al. (1994) The northeastern Chukchi Sea: benthos-environmental interactions. Marine ecology progress series 111: 171–190.
25. Holmes NJ, Harriott VJ, Banks SA (1997) Latitudinal variation in patterns of colonisation of cryptic calcareous marine organisms. Marine Ecology Progress Series 155: 103–113.
26. Jorgensen LL, Pearson TH, Anisimova NA, Gulliksen B, Dahle S, et al. (1999) Environmental influences on benthic fauna associations of the Kara Sea (Arctic Russia). Polar Biology 22: 395–416.
27. Leliaert F, Anderson RJ, Bolton JJ, Coppejans E (2000) Subtidal understorey algal community structure in kelp beds around the Cape Peninsula (Western Cape, South Africa). Botanica Marina 43: 359–366.
28. McKinney FK, Hageman SJ (2006) Paleozoic to modern marine ecological shift displayed in the northern Adriatic Sea. Geology 34: 881–884.
29. Holland SM, Patzkowsky ME (2007) Gradient ecology of a biotic invasion: biofacies of the type Cincinnatian Series (Upper Ordovician), Cincinnati, Ohio Region, USA. Palaios 22: 392–407.
30. Redman CM, Leighton LR, Schellenberg SA, Gale CN, Nielsen JL, et al. (2007) Influence of spatiotemporal scale on the interpretation of paleocommunity structure: Lateral variation in the imperial formation of california. Palaios 22: 630–641.
31. McCune B, Grace JB (2002) Analysis of Ecological Communities. Gleneden Beach: MjM Software Design.
32. Bandy OL, Arnal RE (1960) Concepts of foraminiferal paleoecology. AAPG Bulletin 44: 1921–1932.
33. Menzel DW (1993) Ocean Processes: US Southeast Continental Shelf. Oak Ridge: US Department of Energy.
34. Gower JC (1966) Some distance properties of latent root and vector methods used in multivariate analysis. Biometrika 53: 325–338.
35. Kenkel NC, Bradfield GE (1981) Ordination of epiphytic bryophyte communities in a wet-temperate coniferous forest, South-Coastal British Columbia. Vegetatio 45: 147–154.
36. Williamson MH (1978) The ordination of incidence data. Journal of Ecology 66: 911–920.
37. Minchin PR (1987) An evaluation of the relative robustness of techniques for ecological ordination. Vegetatio 69: 89–107.
38. Hatheway WH (1971) Contingency-table analysis of rain-forest vegetation. In: Patil GP, Pielou EC, Waters WE, editors. Statistical Ecology. University Park: Pennsylvania State University Press.
39. Hill MO (1973) Reciprocal averaging: an eigenvector method of ordination. Journal of Ecology 61: 237–249.
40. Hill MO, Gauch HG, Jr. (1980) Detrended correspondence analysis: An improved ordination technique. Vegetatio 42: 47–58.
41. Wartenberg D, Ferson S, Rohlf FJ (1987) Putting Things in Order: A Critique of Detrended Correspondence Analysis. The American Naturalist 129: 434–448.
42. Shepard R (1962) The analysis of proximities: Multidimensional scaling with an unknown distance function. I. Psychometrika 27: 125–140.
43. Shepard R (1962) The analysis of proximities: Multidimensional scaling with an unknown distance function. II. Psychometrika 27: 219–246.
44. Kruskal JB (1964) Multidimensional scaling by optimizing goodness of fit to a nonmetric hypothesis. Psychometrika 29: 1–27.
45. Kruskal JB (1964) Nonmetric multidimensional scaling: A numerical method. Psychometrika 29: 115–129.
46. Legendre P, Legendre L (2012) Numerical Ecology. New York: Elsevier.
47. Clarke KR (1993) Non-parametric multivariate analyses of change in community structure. Australian Journal of Ecology 18: 117–143.
48. Vanderklift MA, Ward TJ, Jacoby CA (1996) Effect of reducing taxonomic resolution on ordinations to detect pollution-induced gradients in macrobenthic infaunal assemblages. Marine Ecology Progress Series 136: 137–145.
49. Oksanen J, Blanchet FG, Kindt R, Legendre P, Minchin PR, et al. (2012) Vegan: Community Ecology Package.
50. Venables WN, Ripley BD (2002) Modern Applied Statistics with S;. editor. New York: Springer.
51. Sokal RR, Rohlf JF (1995) Biometry. New York: W.H. Freeman and Company.
52. Kidwell SM, Bosence WJ (1991) Taphonomy and time-averaging of marine shelly faunas. In: Briggs DEG, Allison PA, editors. Taphonomy: Releasing the Data Locked in the Fossil Record. New York: Plenum Press. 115–209.
53. Flessa KW, Cutler AH, Meldahl KH (1993) Time and Taphonomy: Quantitative Estimates of Time-Averaging and Stratigraphic Disorder in a Shallow Marine Habitat. Paleobiology 19: 266–286.
54. Kowalewski M (1996) Time-averaging, overcompleteness and the geological record. Journal of Geology 104: 317–326.
55. Kowalewski M, Bambach RK (2003) The limits of paleontological resolution. In: Harris PJ, editor. High Resolution Approaches in Stratigraphic Paleontology, Topics in Geobiology. New York: Kluwer Academic/Plenum,. 1–48.
56. Kidwell SM, Tomasovych A (2013) Implications of Time-Averaged Death Assemblages for Ecology and Conservation Biology. Annual Review of Ecology, Evolution, and Systematics 44.
57. Kidwell SM (1997) Time-averaging in the marine fossil record: Overview of strategies and uncertainties. Geobios 30: 977–995.
58. Tomašových A, Kidwell SM (2009) Fidelity of variation in species composition and diversity partitioning by death assemblages: time-averaging transfers diversity from beta to alpha levels. Paleobiology 35: 94–118.
59. Olszewski TD (1999) Taking advantage of time-averaging. Paleobiology 25: 226–238.
60. Olszewski TD, Kidwell SM (2007) The preservational fidelity of evenness in molluscan death assemblages. Paleobiology 33: 1–23.

20

Late Miocene–Pliocene Paleoclimatic Evolution Documented by Terrestrial Mollusk Populations in the Western Chinese Loess Plateau

Fengjiang Li[1]*, Naiqin Wu[1], Denis-Didier Rousseau[2,3], Yajie Dong[1], Dan Zhang[1], Yunpeng Pei[4]

1 Key Laboratory of Cenozoic Geology and Environment, Institute of Geology and Geophysics, Chinese Academy of Sciences, Beijing, China, **2** Laboratoire de Meteorologie Dynamique, UMR INSU-CNRS 8539 & CERES-ERTI, Ecole Normale Superieure, Paris, France, **3** Lamont-Doherty Earth Observatory of Columbia University, Palisades, New York, United States of America, **4** School of the Earth Sciences and Resources, China University of Geosciences, Beijing, China

Abstract

The Neogene eolian deposits in the Chinese Loess Plateau (CLP) are one of the most useful continental deposits for understanding climatic changes. To decipher Late Neogene paleoclimatic changes in the CLP, we present a terrestrial mollusk record spanning the time interval between 7.1 and 3.5 Ma from the western CLP. The results indicate four stages of paleoclimatic evolution: From 7.1 to 6.2 Ma, cold and dry climatic conditions prevailed as evidenced by high values of the total number of cold-aridiphilous (CA) mollusk species and by low values of all of the thermo-humidiphilous (TH) mollusk indices. From 6.2 to 5.4 Ma, the climate remained cold and dry but was not quite as dry as during the preceding phase, as indicated by the dominance of CA mollusks and more TH species and individuals. From 5.4 to 4.4 Ma, a warm and moist climate prevailed, as indicated by high values of the TH species and individuals and by the sparsity of CA species and individuals. From 4.4 to 3.5 Ma, all of the CA indices increased significantly and maintained high values; all of the TH indices exhibit high values from 4.4 to 4.0 Ma, an abrupt decrease from 4.0 Ma and a further increase from 3.7 Ma. The CA species of *Cathaica pulveraticula, Cathaica schensiensis*, and *Pupopsis retrodens* are only identified in this stage, indicating that the CA species were diversified and that the climate was becoming drier. Moreover, the CA mollusk group exhibits considerable diversity from 7.1 to 5.4 Ma when a cold, dry climate prevailed; whereas the diversity of the TH group was high during the relatively warm, wet interval from 5.4 to 4.4 Ma. This indicates that variations in the diversity of the CA and TH mollusk groups were closely related to climatic changes during the Late Miocene to Pliocene.

Editor: Lorenzo Rook, University of Florence, Italy

Funding: This study is supported by the Chinese Academy of Sciences (KZCX2-EW-QN107), the National Basic Research Program of China (No. 2010CB950204), the Strategic Priority Research Program: Climate Change, Carbon Budget and Relevant Issues (XDA05120203), the National Natural Science Foundation of China (Projects 41072130, 41272205, and 41230104), and part of the work was achieved during a stay in Paris through a CNRS-NSFC grant. This is Lamont-Doherty Earth Observatory contribution 7763. The funders had no role in study design, data collection and analysis, decision to publish, or preparation of the manuscript.

Competing Interests: The authors have declared that no competing interests exist.

* E-mail: fengjiangli@mail.iggcas.ac.cn

Introduction

As evidenced by the marine benthic foraminiferal $\delta^{18}O$ record, Earth's climate underwent a gradual cooling trend during the Late Miocene and Pliocene. This interval witnessed the progressive cooling of oceanic deep water, the expansion of permanent ice sheets in Antarctica, the occurrence of ice rafted detritus in Northern Hemisphere (mostly in north Atlantic), and both hemispheres covered by ice sheets after middle Pliocene [1]. Coincident with ice development in both Polar regions are significant ecological, climatic, and tectonic events elsewhere and especially around Asia [1–12], and which demonstrate that the Late Miocene and Pliocene was an important and complex interval that needs to be better understood.

In East Asia, one of the most important climatic changes is the evolution of the East Asian (EA) monsoon. Numerous sedimentological, geochemical, and paleontological studies, including of fossil mammals, mainly from the Chinese Loess Plateau (CLP) and the South China Sea (SCS), have contributed significantly to our understanding of the EA monsoon changes during the Late Miocene to Pliocene. However, the results are inconsistent and some are even in conflict. In the CLP, sedimentological evidence of changes in sediment grain size and sedimentation rate from the Xifeng, Lingtai, Lantian, Xunyi, Luochuan, Jiaxian, Baode, and Jingle (Figure 1) Red Clay deposits indicate the occurrence of a strong winter monsoon and pronounced aridity in the Asian interior during the Late Miocene [8,13–16]. However, the geochemical record of $^{87}Sr/^{86}Sr$ ratios from the Lingtai Red Clay sequence implies a weak EA winter monsoon from 7 to 2.5 Ma [17]; and in addition, this interval can be sub-divided into a large amplitude and high frequency stage from 7–4.2 Ma and a weak, stable stage from 4.2–2.6 Ma, as evidenced by records of Zr/Rb ratio and mean quartz grain size [18]. Furthermore, grain-size records from fluvial deposits in the Linxia Basin (Figure 1) indicate that the EA winter monsoon intensified at 7.4 Ma and 5.3 Ma [19], the latter datum being significantly different to evidence from the Red Clay record.

In the case of the EA summer monsoon, the magnetic susceptibility record from the Xifeng Red Clay sequence indicates a weakened summer monsoon during the Late Miocene (6–5.4

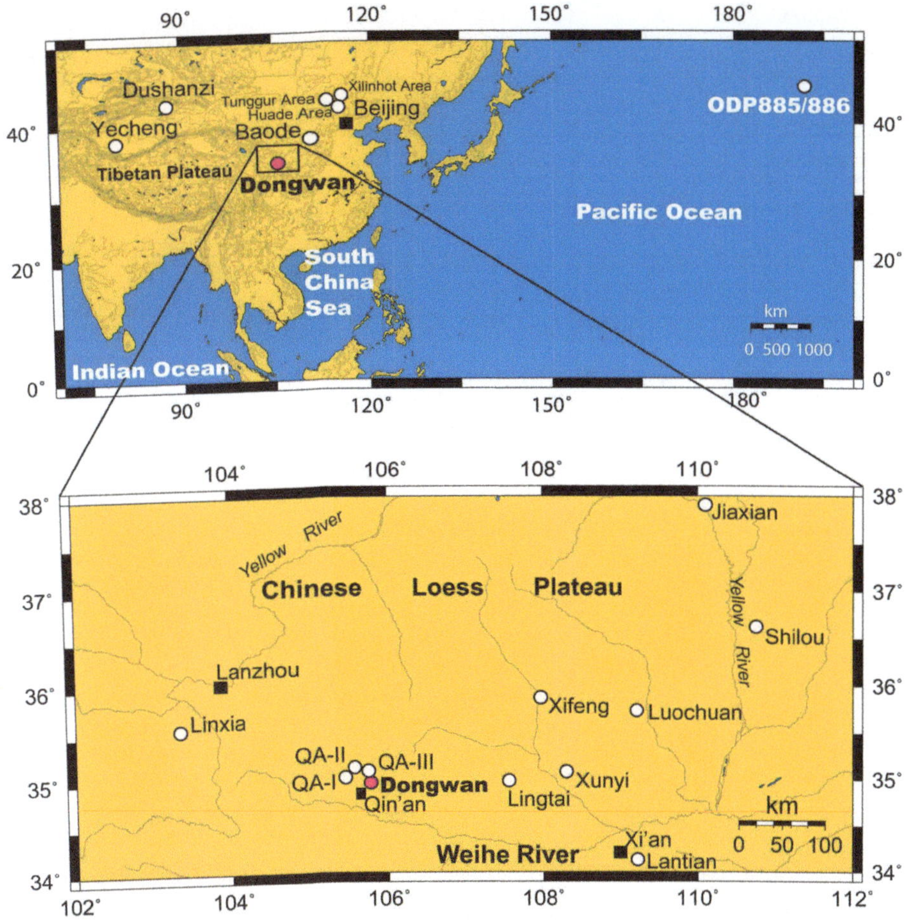

Figure 1. Location of the Chinese Loess Plateau (CLP), the studied Dongwan loess-paleosol sequence and other Neogene sections mentioned in the text. The Dongwan sequence is indicated by a red circle and other sections by white circles. Main cities are indicated by black squares.

Ma) and a strong summer monsoon during the early Pliocene [8]. Additional information about the EA summer monsoon is provided by records of pedostratigraphy and iron geochemistry from the Lingtai Red Clay deposits. These records indicate that the EA summer monsoon was relatively weak from 7.05 to 6.2 Ma, strengthened from 6.2 to 5.5 Ma, was very strong from 5.5 to 3.85 Ma corresponding to the interval of strongest soil development, and weakened significantly over the interval 3.85 to 3.15 Ma [20,21].

However, studies of mammalian fossils from the CLP have yielded very different results: Hypsodonty analysis indicates that northern China became more humid at 7–8 Ma, coincident with the onset of Red Clay deposition in the eastern CLP, and which was interpreted by the authors as representing the onset or intensification of summer monsoonal precipitation [22,23]. A general rise in $\delta^{18}O$ values of soil carbonate from the Lantian fluvial and Red Clay deposits reflects increased summer precipitation related to the onset and/or intensification of the EA monsoon during the Late Miocene to Pliocene [24,25], and which is supported by a mammalian faunal turnover event implying a marked change to more humid and closed habitats [26]. Recent isotopic evidence from fossil mammals and soil carbonates indicates a strengthened EA summer monsoon from 7–4 Ma [27]. However, the $\delta^{13}C$ values of fossil enamel from a diverse group of herbivores and of paleosol carbonate and organic matter from the Linxia Basin indicate that C4 grasses were either absent or insignificant in the Linxia Basin prior to 2–3 Ma, suggesting that the EA summer monsoon was probably not strong enough to affect this part of China throughout much of the Neogene [28]. However, this result conflicts with the interpretations of Fortelius et al. (2002) [22], Kaakinen et al. (2006, 2013) [24,25], Liu et al. (2009) [23] and Passey et al. (2009) [27].

In the SCS, the EA winter monsoon developed progressively from 7 Ma as shown by an increasing trend in black carbon concentration and accumulation rate [29]. In addition the EA summer monsoon weakened after 7.5 Ma, as evidenced by combined planktonic foraminiferal Mg/Ca and $\delta^{18}O$ records [30], consistent with the Red Clay sequences in the CLP. The suggestion of a weakened EA summer monsoon is supported by geochemical records which indicate that chemical weathering intensity decreased from the early Miocene with a rapid decrease centered at 7.2 Ma [31]. This result is generally consistent with records of radiolarian species numbers and individuals, and diversity, from the southern SCS and which suggest a major decrease in summer monsoon intensity after 7.70 Ma [32]. In contrast, records of clay/feldspar ratio, kaolinite/chlorite ratio and biogenic opal MAR from the SCS suggest that the summer monsoon was strong from 7.1–6.2 Ma and remained relatively stable from 6.2–3.5 Ma [33]; however, the authors emphasize that their study is a schematic reconstruction which only outlines the principal stages but neglects the details.

A semi-quantitative reconstruction of the Neogene vegetation in China indicates that Miocene aridification associated with strengthening of the EA winter monsoon is consistent with Neogene global cooling, and that the EA summer monsoon did not weaken during the Pliocene [34], and this result agrees with most of the geological records from the CLP and SCS. However another quantitative reconstruction from plant fossil records yielded contrasting results, indicating that records of both temperature and precipitation from north China exhibit no significant difference between the western and eastern regions during the Miocene, suggesting that the monsoon climate did not commence or intensify at that time [35]. Regional climate model experiments also reveal that during the Late Miocene, from 11–7 Ma, the monsoonal climate may not have been fully established in various Asian regions, including northern China [36]; and this finding is contrary to that of numerous previous studies of the EA monsoon which suggest that it was initiated around the time of the Oligocene/Miocene boundary [37–39].

The foregoing review demonstrates that more work is needed to better understand the evolution of the EA monsoon during the Late Miocene and Pliocene. In particular, higher resolution studies using more sensitive monsoon proxies from key monsoon-dominated regions may be one of the best solutions for resolving the various inconsistencies and even conflicts regarding the process of monsoon development.

As mentioned above, the CLP (Figure 1), located to the northeast of the Tibetan Plateau, is a key continental region for the study of the EA monsoon. Deposition of eolian sediments commenced in the western CLP from 22 Ma, as observed in the QA-I, QA-II, and QA-III (Figure 1) Miocene loess-paleosol sequences which have a basal age of about 22 Ma [37,40], while the upper boundary is dated at about 3.5 Ma in the Dongwan late Miocene-Pliocene loess-paleosol sequence in the western CLP [41]. In contrast, in the eastern CLP, the age of the lower boundary of eolian deposits, the Red Clay sequences, is about 7–8 Ma as evidenced in the Lingtai, Xifeng, Liantian, and Baode sequences [15,20,42,43]. However, results from the recently reported Shilou Red Clay sequence indicate that eolian sediments were deposited from 11 Ma in the eastern CLP [44]. Despite the occurrence of a totally different lower boundary age between the east and west CLP, these deposits have great potential for deciphering in detail the processes of ecological and climatic evolution in the CLP, and by extension in East Asia, during the Miocene and Pliocene. As reviewed above, sedimentological and geochemical studies have so far contributed a great deal to our understanding of climatic changes during the interval of interest [6,8,9,13–18,20,21,24,27,37,40–45]; however, to date there has been only a limited application of a biological approach to analyzing these sequences [22,26,46,47].

Land snails are generally the most common and abundant fossils in Quaternary loess sequences, and this fact, together with the limited degree of success of most of the other paleontological studies of Quaternary loess, makes them especially important paleoenvionmental indicators for loess deposits. Their occurrence in Quaternary loess was first documented in Europe in the early 1820s [48], and since then they have contributed significantly to understanding the origin and paleoclimatic evolution of Quaternary loess deposits in Eurasia and especially in the CLP [46,49–57]. However, terrestrial mollusks preserved in the Neogene sequences in the CLP have not been investigated in detail until recently. Although they have provided crucial paleontological evidence for the wind-blown origin of the Neogene loess sequences in the western CLP [58,59], the ecological and climatic information that terrestrial mollusks may provide has not been well deciphered, with the exception of the record from the Xifeng Red Clay sequence spanning the interval from 6.2 to 2.4 Ma in the eastern CLP [46]. Furthermore, it is unknown how terrestrial mollusk diversity varied in the CLP against the background of climate changes during the Neogene, since the necessary studies have not been performed. In this study, we studied the record of terrestrial mollusks preserved in the Late Miocene to Pliocene Dongwan section in order to investigate the evolution of paleoclimate and terrestrial mollusk diversity in the western CLP during the Late Miocene to Pliocene.

Materials and Methods

The Dongwan loess-paleosol sequence (105°47′E, 34°58′N) [41] is located in the northeastern part of Qin'an County in the western CLP (Figure 1). The current climate in Qin'an County is mainly controlled by the EA monsoon, with mean annual precipitation of about 400–500 mm and mean annual temperature of ~10.4°C. Mean July and January temperatures are ~22.7°C and −3.4°C, respectively. Vegetation in the region corresponds to a semi-arid temperate steppe [60].

The Dongwan sequence, located in the western CLP, is the first counterpart of the Red Clay sequences in the eastern CLP [41]. The section is about 73.7 m thick, and is composed of 84 distinguishable loess-paleosol couplets. The chronology of the sequence has been established by Hao and Guo (2004) [41] using magnetostratigraphic and micromammalian studies (Figure 2). First, Hao and Guo (2004) [41] ascribed the approximate age of the sequence to the Late Miocene to Pliocene using micromammalian fossil teeth sampled from 20 depths of the Dongwan sequence. Second, they defined a series of magnetozones based on 319 oriented samples collected at 20 or 25 cm intervals and correlated them with the Geomagnetic Polarity Timescale [61]. Finally, they established a chronology using paleomagnetic reversals as age controls and interpolated in between them using the magnetic susceptibility age model developed by Kukla et al. (1990) [62]. Their results yield an age duration of 7.1 to 3.5 Ma for the Dongwan loess-palaeosol sequence [41].

A total of 310 mollusk assemblages were collected from the Dongwan sequence using continuous 20-cm-thick samples; however, some intervals were sampled at intervals of 10–50 cm, according to lithological changes [58]. About 30 kg of sediment were obtained for each sample. In the field, we progressively broke each sediment sample into smaller pieces of about 0.5 mm in diameter, at the same time collecting all available shells and visible broken pieces. No necessary permits for the described field investigations were needed. In the laboratory, we attempted to restore any broken shells, and then identified and counted them under a binocular microscope. All of the identifiable mollusk remains were considered in the individual totals using the method of Puisségur (1976) [63]. All of the mollusk shells are stored in the Institute of Geology and Geophysics, Chinese Academy of Sciences.

For each mollusk assemblage, the numbers of species (S) and individuals were counted and a diversity index was calculated for all species, thermo-humidiphilous (TH) species, and cold-aridiphilous (CA) species in order to investigate changes in terrestrial mollusk populations and in different ecological groups. We used the most widely applied Shannon index [64], sometimes referred to as the 'Shannon–Weaver' index and sometimes as the 'Shannon–Wiener' index by researchers, in order to calculate the values of diversity, H (S), of the total, TH, and CA species, as follows:

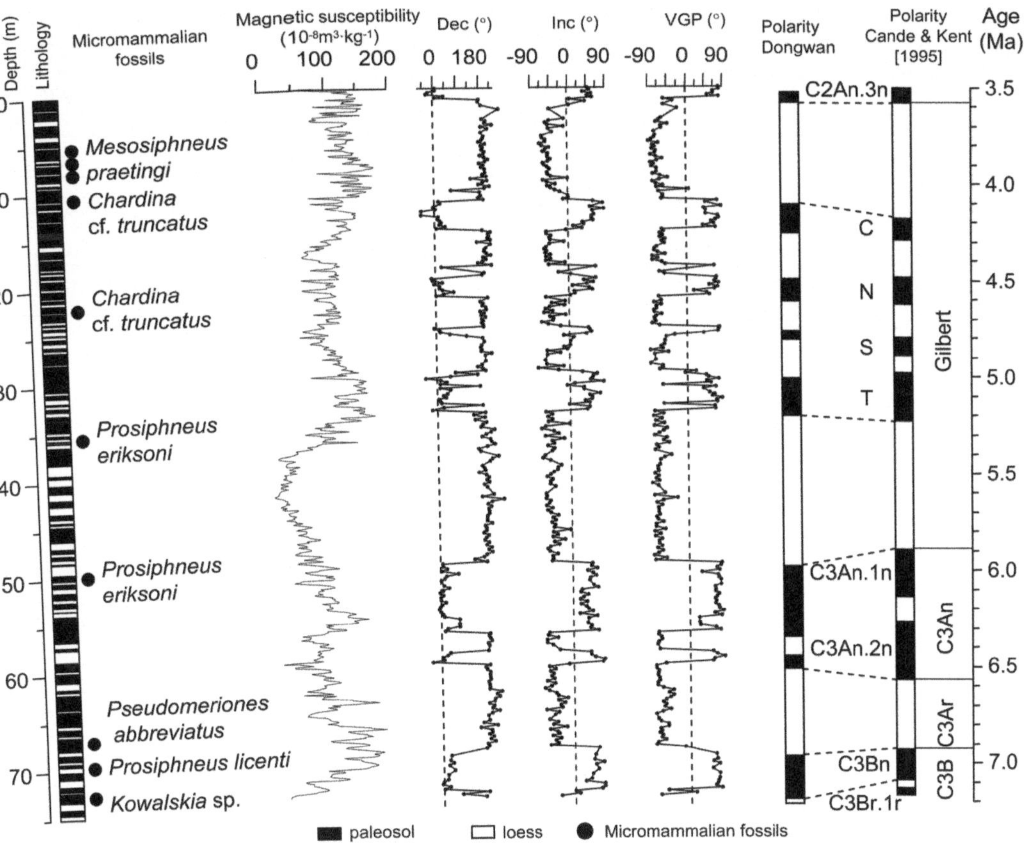

Figure 2. Chronology of the Dongwan loess-paleosol sequence (modified after Hao and Guo, 2004) [41].

$$H = -\sum_{i=1}^{S} \frac{ni}{N} \log_2 \frac{ni}{N}$$

where n_i is the density measure (in this case the mollusk individuals) of the i-th species (I varying between 1 and n); S is the number of species in the sample, and

$$N = -\sum_{i=1}^{S} n_i$$

The theoretical maximum ($H_{\max}$) of diversity in any sample is expressed as

$$H_{\max} = \log_2 S$$

Equitability (or evenness, E) is expressed as

$$E = \frac{H}{H_{\max}}$$

These indices have been applied to European and North American Quaternary terrestrial mollusk assemblages [65–67].

Results

Mollusk fossils are relatively abundant in the Dongwan sequence with significant concentrations at ~2 m, 10 m, 30 m, 50 m, and 70 m depth. Amongst the total of 310 samples, 298 yielded 16439 mollusk individuals and 12 samples were barren. The maximum count reached 1121/30 kg at 54 m depth (Figure 3). Variations in total mollusk individuals parallel fluctuations in magnetic susceptibility [58], indicating that pedogenic processes such as carbonate dissolution did not affect the preserved assemblages [58].

A total of 24 mollusk species were identified in the Dongwan section. They are all terrestrial taxa and consist of *Gastrocopta armigerella* (Reinhardt, 1877), *Gastrocopta* sp., *Punctum orphana* (Heude, 1882), *Punctum* sp., *Metodontia huaiensis* (Crosse, 1882), *Metodontia yantaiensis* (Crosse et Debeaux, 1863), *Metodontia beresowskii* (Moellendorff, 1899), *Metodontia* cf. *huaiensis*, *Metodontia* cf. *yantaiensis*, *Metodontia* cf. *beresowskii*, *Metodontia* sp., *Kaliella* sp., *Macrochlamys* sp., *Opeas* sp., *Cathaica* sp., *Cathaica pulveratrix* (Martens, 1882), *Cathaica pulveraticula* (Martens, 1882), *Cathaica schensiensis* (Hilber, 1883), *Cathaica placenta* (Ping et Yen, 1933), *Pupilla aeoli* (Hilber, 1883), *Pupilla grabaui* (Ping, 1929), *Pupilla* sp., *Vallonia* sp., and *Pupopsis retrodens* (Martens, 1879).

All of these species, except *Pupilla grabaui* and *Pupopsis retrodens*, have been previously identified in the Chinese Quaternary loess-paleosol sequences and most of them have modern representatives. For example, *Cathaica pulveratrix*, *C. pulveraticula*, *C. schensiensis*, and *Pupilla aeoli* are the most common species that prefer living in relatively cold, dry environments, and are presently distributed in northwestern China. They have been regarded as indicative of a

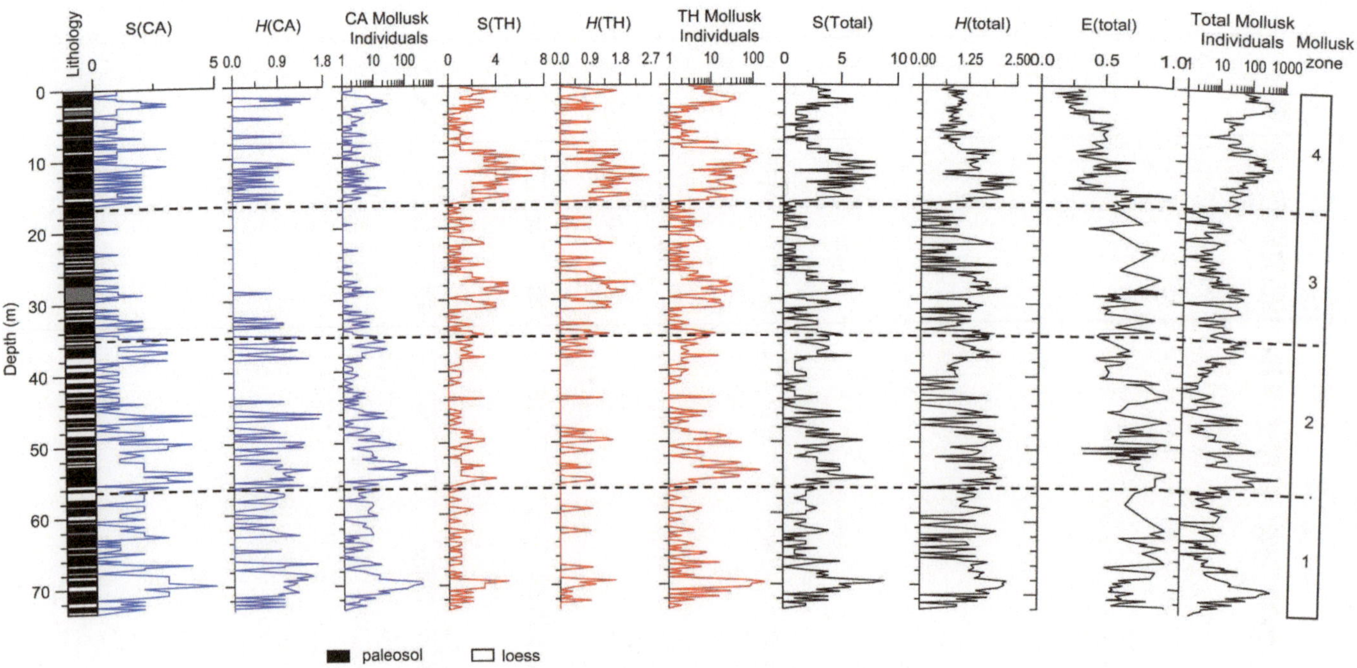

Figure 3. Variations in terrestrial mollusks versus depth in the Dongwan sequence. Lithology and total mollusk individuals are from Li et al., (2006a) [58] and Li et al. (2008) [47]. S(CA)–Total number of species of the cold-aridiphilous (CA) mollusk group. *H*(CA)–Diversity of the CA mollusk group. S(TH)–Total number of species of the thermo-humidiphilous (TH) mollusk group. *H*(TH)–Diversity of the TH mollusk group. S(total)–Total number of species of the total mollusk group. *H*(total)–Diversity of the total mollusk group. E(total)–Equitability of the total mollusk group.

strengthened winter monsoon [49–58]. Conversely, *Metodontia huaiensis, M. yantaiensis, M. beresowskii, Gastrocopta armigerella,* and *Punctum orphana* are species living in a warmer and more humid environment, and are distributed mainly in the southeastern part of the CLP, where the warm, moist summer monsoon brings sufficient precipitation [49–58]. Therefore based on their present requirements of moisture and temperature, as well as their modern geographical distribution, the Dongwan terrestrial mollusks can be grouped into CA and TH ecological groups, as have been previously defined in the Chinese Quaternary loess-paleosol sequences [49–57]. In the Dongwan section, the TH group comprises *Gastrocopta armigerella*, *Gastrocopta* sp., *Punctum orphana*, *Punctum* sp., *Metodontia beresowskii*, *Metodontia huaiensis*, *Metodontia yantaiensis*, *Metodontia* cf. *huaiensis*, *Metodontia* cf. *yantaiensis*, *Metodontia* cf. *beresowskii*, *Metodontia* sp., *Kaliella* sp., *Macrochlamys* sp., and *Opeas* sp.; and the CA group comprises *Cathaica* sp., *Cathaica pulveratrix*, *Cathaica pulveraticula*, *Cathaica schensiensis*, *Cathaica placenta*, *Pupilla aeoli*, *Pupilla grabaui*, *Pupilla* sp., *Vallonia* cf. *pulchella*, *Vallonia* sp., and *Pupopsis retrodens*.

The distributions of these terrestrial mollusks in the Dongwan section have been described previously [58]. *Cathaica* sp. is the most continuously distributed mollusk taxon in the Dongwan section, and *Gastrocopta* sp., *Pupilla* sp., and *Vallonia* sp. are other well represented taxa. The TH species of *Metodontia*, *Punctum*, *Macrochlamys*, and *Opeas* are concentrated in the upper part of the section. In general, the CA species are dominant in the loess layers, while the TH species mainly occur in the paleosols [58].

Variations in the number of species (S), diversity (*H*), equitability (E), and total individuals of all species, CA species, and TH species in the Dongwan sequence are plotted against depth in Figure 3. These data can be found in the supporting information data (Table S1). Variations in these parameters allow definition of four mollusk zones. In Zone 1, from the bottom of the sequence to about 56.8 m depth (~7.1–6.2 Ma), high values of all the CA species (S(CA)) and individuals and diversity (*H*(CA)) dominate over low values of all of the TH species (S(TH)) and individuals and diversity (*H*(TH)). A prominent feature of this zone is that high values of S(CA), *H*(CA), CA mollusk individuals, S(TH) and *H*(TH), as well as total mollusk species and individuals, occur at around 70 m depth, at the very base of the sequence; however, these values then decrease upwards. *H*(CA) and *H*(total) remain at high levels throughout the zone, and the equitability (E) of total species is also higher than in the other zones.

In Zone 2, from 56.8–34.8 m depth (~6.2–5.4 Ma), the total number of CA species and individuals remains at a similarly high level as in the previous zone, except for the interval from 42 to 38 m. In contrast, the total number of TH species and individuals is slightly higher than in the previous zone. The total number of species and individuals of all of the species (CA and TH combined) are generally higher than in the previous zone but with a decreasing trend. The diversity of all species, *H*(total), does not exhibit significant changes compared to Zone 1, and E(total) is somewhat lower than in Zone 1.

Zone 3, from 34.8–16.2 m depth (~5.4–4.4 Ma), is characterized by low numbers of CA species and individuals being almost absent in the upper part, from 24 m to 16.2 m. This pattern of variation is paralleled by a clear increase of all the TH indices, including diversity. The total number of species and individuals of all of the species is lower than in Zone 2, except for the interval from about 32 to 28 m. There are no large magnitude changes in *H*(total) and E(total) within the zone.

Zone 4 corresponds to the depth from 16.2 m to the top of the sequence (~4.4 to 3.5 Ma). The number of CA species and individuals increases markedly and maintains relatively high values throughout the zone. The TH indices exhibit high values from about 16.2 to 9 m and then decrease significantly to very low values up to 4 m, and increase again above about 4 m depth. All of the indices for all species, i.e., S(total), *H*(total), and E(total), increase markedly; however, *H*(total) and E(total) exhibit a generally declining trend, implying that the diversity of the terrestrial mollusk populations decreased and that the distribution of individuals of different species was uneven. The total mollusk individuals in this zone is high and remains relatively stable except for the interval between 8 and 4 m, a pattern which differs from the other three zones where the total number of mollusk individuals is high at the beginning of the zone and then declines thereafter.

Moreover, the variation in diversity of the CA and TH ecological groups from the entire Dongwan sequence, as exhibited in Figure 3, can be differentiated into two major intervals based solely on the Shannon index of the CA and TH groups; and this result probably reflects different ecological population dynamics. First, from the base of the Dongwan sequence to 34.8 m depth, the CA group is dominant, exhibiting high diversity with *H* values varying between 0 and 1.75 (mean of 0.52). In contrast, the diversity of the TH populations exhibits significantly low values, ranging from 0 to 1.58 (mean of 0.18) within this depth interval, although the diversity of all mollusk species, i.e. the sum of CA and TH, does not exhibit any clear changes. Second, from 34.8 m depth to the top of the sequence, the diversity of the CA group declines significantly with *H* values varying between 0 and 1.56 (mean of 0.14); and in contrast the TH group becomes dominant in terms of diversity with *H* values ranging from 0 to 2.64 (mean of 0.64). The diversity and equitability of all mollusk species also exhibit a significant change at about 16.2 m depth within this interval.

Discussion

Late Miocene and Pliocene Paleoclimatic Evolution in the Western CLP

Previous studies of European and North American Quaternary terrestrial mollusks have shown that mollusk assemblages, based not only on the occurrence of characteristic species but also on the statistical dispersion of the assemblages (e.g., diversity, as described by the Shannon index (*H*)), can provide information about climatic conditions and mollusk populations [65–67]. For example, correlation of terrestrial mollusk groups with changes in dust flux, and thus climatic conditions, has previously been observed in the European Quaternary terrestrial mollusk diversity record [65]. Here we use the Dongwan terrestrial mollusk record to extend the statistical analysis of terrestrial mollusk populations and the interpretation of climatic changes to the Late Neogene. As shown in Figure 4, four stages can be recognized according to changes in the mollusk indices used in the present study, and which outline the evolutionary history of ecological and climatic conditions in the CLP from 7.1 to 3.5 Ma.

First, from 7.1 to 6.2 Ma (Zone 1), the total number of CA species (S(CA)) and diversity (*H*(CA)) index are high; however, the number of CA mollusk individuals is generally low apart from a large high peak at around 7 Ma and which indicates diversified CA species. Conversely, all of the indices of the TH species exhibit low values except for the peak at around 7 Ma, indicating low diversity. These features may be related to the occurrence of very cold, dry climatic conditions, the occurrence of which is roughly coincident with the expansion of C4 vegetation in the northeastern CLP [27], Central Inner Mongolia (Tunggur Area, Xilinhot Area, and Huade Area) [68] (Figure 1), and Pakistan [2]. They also indicate the occurrence of seasons with water stress for vegetation and terrestrial mollusks as well as the fact that relatively dry climatic conditions had already appeared in northern China during the Late Miocene, coincident in age with the global transition from C3 to C4 vegetation [4]. The occurrence of C4 plants during the latest Miocene should have been restricted to limited areas within deserts, and the occurrence of these niches could be coincident with, or occurred after, major uplift of the Tibetan Plateau at about 8 Ma [3,8,10,11]. However, C4 vegetation was not yet to develop in the central and southern CLP, demonstrating that C3 plants were still dominant as indicated by $\delta^{13}C$ values of soil carbonate in the Lingtai, Xifeng and Liantian Red Clay sequences in the eastern CLP [6,9,24]. This is also evidenced by the thickness of loess layers in the studied Dongwan sequence and by dust mass accumulation rates in the western Pacific. High dust deposition rates with a gradually increasing trend culminated at about 6.2 Ma in the western Pacific [69], corresponding to thicker loess layers in the Dongwan sequence (Figure 4). Both features indicate drier climatic conditions in the Asian interior, the potential source region of dust deposits in the CLP and western Pacific. While dust accumulation in the western Pacific reached a maximum from about 6.5 to 6.2 Ma, the number of mollusk species and individuals at Dongwan was very low, implying very dry climatic conditions which were unsuitable even for the development of CA species. Pedostratigraphy and iron geochemistry of the Lingtai Red Clay deposits in the CLP indicate that the EA summer monsoon was relatively weak from 7.05 to 6.2 Ma [20,21]. The $\delta^{13}C$ values of fossil enamel from a diverse group of herbivores and of paleosol carbonate and organic matter from the Linxia Basin indicate that C4 grasses were either absent or insignificant in the Linxia Basin prior to ~2–3 Ma, suggesting that the East Asian summer monsoon was probably not strong enough to affect this part of China throughout much of the Neogene [28].

However, during our studied interval there is much evidence for a strong summer monsoon, evidenced mainly from mammal assemblages as well as analyses of mammal tooth and soil carbonate isotopes [22–25,27]. Hypsodonty analysis of fossil mammals indicates that northern China became more humid at 7–8 Ma, coincident with the previously recognized onset of Red Clay deposition in the eastern CLP, and which was interpreted by the authors as reflecting the onset or intensification of summer monsoonal precipitation [22,23]. Soil carbonate isotopes from the Lantian fluvial and Red Clay deposits indicate the absence of any marked change in plant photosynthetic pathway or climate, implying the occurrence of pure C3 vegetation with no indications of any C4 plants during the Late Miocene and Pliocene. A general rise in $\delta^{18}O$ values probably reflects increased summer precipitation related to the onset and/or intensification of the Asian monsoon system [24,25], which is supported by a mammalian faunal turnover event implying a marked change to more humid and closed habitats [26]. A strong summer monsoon at 7.1–6.2 Ma is also supported by the records of clay/feldspar ratio, kaolinite/chlorite ratio and biogenic opal mass accumulation rates (MAR) from the SCS, although the authors emphasize that their study is a schematic reconstruction which only outlines several principal stages and neglects the details [33].

There is also much evidence for a weakening of the summer monsoon and strengthening of the winter monsoon during the Late Miocene. The EA summer monsoon weakened after 7.5 Ma as evidenced by combined planktonic foraminifera Mg/Ca and $\delta^{18}O$ records from Ocean Drilling Program (ODP) Site 1146, northern SCS [30]. Radiolarian species numbers and individuals and diversity from ODP Site 1143, southern SCS, suggest a summer monsoon maximum at 8.24 Ma and a major decline after 7.70 Ma [32]. Geochemical records from the ODP Site 1148,

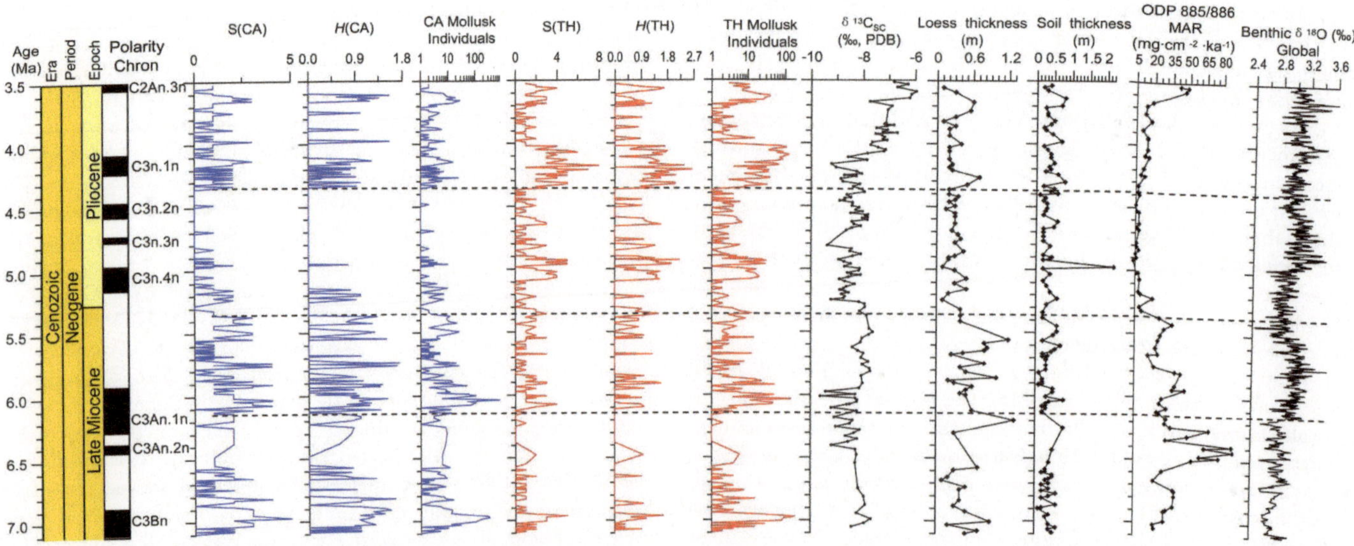

Figure 4. Variations in the terrestrial mollusks of the Dongwan sequence in the western CLP from 7.1 to 3.5 Ma, and comparison with other climate proxies. The proxies are $\delta^{13}C$ of soil carbonate ($\delta^{13}C_{SC}$) in the Lingtai Red Clay sequence [6], loess and soil thickness in the Dongwan sequence (this study), global cooling trend deduced from the marine benthic foraminiferal $\delta^{18}O$ record [1], and continental aridity inferred from the mass accumulation rate (MAR) of eolian dust at ODP Site 885/886 in the western Pacific [69]. Geomagnetic polarity chronology and age of the Dongwan sequence is from Hao and Guo (2004) [41]. S(CA)–Total species of the cold-aridiphilous (CA) mollusk group. *H*(CA)–Diversity of the CA mollusk group. S(TH)–Total number of species of the thermo-humidiphilous (TH) mollusk group. *H*(TH)–Diversity of the TH mollusk group.

northern SCS, indicate that chemical weathering intensity decreased after the early Miocene with a rapid decrease centering at 7.2 Ma, compatible with a weakening of the EA summer monsoon [31]. The trend of increasing black carbon concentration and accumulation rate at ODP Site 1148 suggests the progressive development of the East Asian winter monsoon after about 7 Ma [29]. In addition, grain-size evidence from the Linxia Basin of the CLP indicates that the EA winter monsoon intensified after 7.4 Ma and again at 5.3 Ma [19].

Thus it is noteworthy that the TH mollusks at Dongwan do exhibit peak values at around 7 Ma, suggesting warm, humid climatic conditions consistent with a mammalian faunal turnover event [26]; however, the peak does not extend to the subsequent period, implying that the interval of relatively warm and humid climate may have been a relatively brief event within the context of an overall very cold and arid climate. In addition, a recent synthesis of isotopic and sedimentological analyses, climate modeling and an extensive mesowear analysis of the Baode Red Clay sequence indicates that the climate at 7.5 Ma was more humid than during the youngest interval at 5.7 Ma and that variable climatic conditions occurred at ~6.5 Ma. Thus a significant decrease in the EA summer monsoon strength from 7–5.7 Ma in the Baode region, based on three samples dated at 7.0, 6.5, and 5.7 Ma [70], does not contradict the mollusk results. We suggest that much more definitive results would have been obtained if a higher sampling density had been used. In addition, numerous paleoclimatologists regard the occurrence of Red Clay sequences as reflecting aridity or desertification and an intensified winter monsoon [16,37,42,43,69,71], which contradicts the mammal and isotope record.

We suggest four possible reasons for the discrepancies amongst the extant research results. First, the sensitivity of the proxies used may be one of the main reasons; and indeed, it has long been found in Quaternary loess studies that different climatic proxies can respond differently to climatic changes [55]. Second, as summarized by Wang et al. (2005) [72] and Kaakinen et al. (2006) [24], interpretations of expansions of C4 vegetation are contentious in terms of whether or not they are a summer monsoon proxy [2,4,45]. Third, regional difference in patterns of climate change may have occurred, as suggested by Passy et al. (2009) [27]. Finally, low stratigraphic resolution may have resulted in the failure to resolve environmental changes in sufficient detail, and high resolution mammalian faunal studies (for example using a 20-cm interval) may shed light on the observed differences. In summary, the conflicts cannot be resolved by the present study and future studies employing more sensitive proxies, higher resolution or continuous sampling combined with accurate dating may be necessary.

Second, from 6.2 to 5.4 Ma (Zone 2), the dominant CA terrestrial mollusks indicate that the climate during this interval remained cold and dry, but was not drier than in Zone 1, as evidenced by the slightly greater abundance of TH species and individuals. At the beginning of this zone, from about 6.2 to 5.8 Ma, the number of TH mollusk individuals was higher, and this corresponds to a declining rate of dust deposition as indicated by the thinness of the loess layers in the Dongwan sequence (Figure 4). This supports the finding that decreasing or increasing dust deposition affects the development of terrestrial mollusks by creating either more- or less- favorable environmental conditions [65]. After about 5.8 Ma the number of TH mollusk individuals was reduced, which is probably related to the increase of dust deposition in the CLP, as shown by the increased thickness of the loess represented by this interval (Figure 4). These cold, dry climatic conditions were not solely restricted to the western CLP; in the eastern CLP, mollusks in the Xifeng Red Clay sequence exhibit dominant percentages of the CA group associated with a few meso-xerophilous components and a lesser occurrence of TH species, also indicating a cold, dry climate [46]. This is also reflected by the coarse grain size of the Red Clay sequence in the CLP [16], and by the fact that dust deposition rates in the western Pacific remained high, albeit lower than before, and with a declining trend towards the subsequent time interval from 5.4 and

4.4 Ma [69]. The thickness of the Dongwan loess layers was not higher than before, except at about 6.2 and 5.6 Ma; however, the thickness of the soil layers does not change significantly. The $\delta^{13}C$ values of the soil carbonates in the Lingtai and Xifeng Red Clay sequences were similar to those of the previous time interval from 7.1 to 6.2 Ma, indicating again that C4 plants were not the dominant vegetation in the studied area of the CLP [6,9]. These lines of evidence indicate generally cold, dry climatic conditions in the EA continent but which were not drier than the lower stage. However, the global mean benthic foraminiferal $\delta^{18}O$ record demonstrates that global ice volume was increasing with the highest values occurring at about 5.8–5.7 Ma, followed by a decrease towards the upper stage [1].

There is much evidence supporting the conclusion from the Dongwan mollusk results that the winter monsoon may still have been strong from 6.2–5.4 Ma but that the summer monsoon strengthens than before although may be to a lesser extent, as confirmed by the pedostratigraphy and iron geochemistry of the Lingtai Red Clay deposits in the CLP [20,21]. The coeval pollen record from the Lingtai section exhibits a predominance of Chenopodiaceae and *Artemisia*, indicating a desert or desert-grassland landscape during the Late Miocene [73]. A higher dust accumulation rate and coarser eolian grain size in the CLP and in the North Pacific suggest stronger continental aridity in the Asian interior from ~6.2 to ~5 Ma [14,16,69]. The variation in the U-ratio of the grain size of the Red Clay, reflecting the changing strength of the winter monsoon, indicates a strong winter monsoon between 6.1 and 5.4 Ma [13]. The sedimentation rate across the CLP, including at the Baode Red Clay sequence, indicates that the EA winter monsoon strengthened between 6.26 and 5.4–5.25 Ma [14,15].

Third, from 5.4 to 4.4 Ma (Zone 3), the prominent feature is that all of the CA species and individuals are very few in number and in fact they decrease significantly with almost no occurrence after about 5 Ma. In contrast, there are high values of the TH species numbers and individuals with the maxima occurring at around 5–5.1 Ma when a thick soil layer formed, indicating warm, moist climatic conditions. In the eastern CLP, the mollusk fauna was characterized by maximum abundance of TH species, an absence of CA species, and relatively abundant meso-xerophilous taxa [46]. The Xifeng pollen record indicates a significant increase of temperate forest plants, also implying humid regional climatic conditions during this period [74]. The $\delta^{13}C$ values indicate a slight decrease and thus a shift towards more C3 plants in the CLP in agreement with the pollen results (Figure 4). Ding et al. (1999) identify extremely mature soils in the 5.5–3.85 Ma interval at Lingtai, and interpret these as indicating warm and wet climates, and possibly the strongest summer monsoons in the past 7 Ma [20]. Minimum grain size and sedimentation rate in the CLP [15,16], decreased thickness of loess layers in the Dongwan sequence (Figure 4), and very low dust deposition rates in the western Pacific [69] together suggest very warm, moist climatic conditions in the CLP and less dry conditions in the Asian interior, the potential source area of dust in the CLP and western Pacific. However, this warm climate is not clearly recorded by the global mean benthic foraminiferal $\delta^{18}O$ record compiled by Zachos et al. (2001) [1].

Last, from 4.4 to 3.5 Ma (Zone 4), all of the CA indices increase significantly and remain almost unchanged, while in contrast all of the TH indices exhibit a different pattern of variability with high values from 4.4 to 4.0 Ma, coincident with changes in the CA indices, an abrupt decrease from 4.0 Ma and increased values again from 3.7 Ma. The CA species during this time interval were different from those between 7.1 and 5.4 Ma. Indeed, the CA species of *Cathaica pulveraticula*, *Cathaica schensiensis*, and *Pupopsis retrodens* have only been identified in this zone, as shown by Li et al. (2006a) [58], probably indicating that climatic conditions became drier. The coeval Xifeng land snail record from the eastern CLP is dominated by meso-xerophilous taxa, associated with a significantly reduced abundance of TH species and the paucity of CA taxa during the late period [46]. Pollen data from the Xifeng Red Clay sequence indicate a typical steppe ecosystem during this period [74]. These variations are in good agreement with the expansion of C4 plants in the central CLP as observed at the Lingtai and Xifeng Red Clay sequences [6,9]. In addition, the loess layers in the Dongwan section during this time interval are of moderate thickness, thinner than from 7.1 to 6.2 Ma but thicker than from 5.4 to 4.4 Ma. In contrast the paleosols are somewhat thicker than before (Figure 4), indicating longer pedogenesis under different climate conditions, and probably enhanced seasonality, corresponding to expansions of C4 plants in the central CLP [6,9]. Pedostratigraphy and iron geochemistry of the Lingtai Red Clay deposits in the CLP indicate that the EA summer monsoon weakened significantly over the interval 3.85 to 3.15 Ma [20,21]. Dust deposition increased again in the Pacific in parallel with increased aridity in the dust source areas [69] (Figure 4).

Moreover, the entire Dongwan terrestrial mollusk record provides information about the evolution of mollusk diversity and climatic changes from the Late Miocene to the Pliocene. Previous studies indicate that in most Quaternary loess sequences in Europe and North America generally fewer than 20 mollusk species can be identified and diversity varies with H values between 0 and 4, suggesting that the loess deposits provided few additional ecological niches for land snails to grow and develop [65–67]. In the Neogene Dongwan loess sequence in the CLP, East Asia, 24 mollusk species were identified and diversity varies between 0 and 2.5, which also suggests that fewer niches were provided. The response of the Neogene terrestrial mollusk populations to climatic changes depends on their ecological requirements. Different ecological groups, such as CA and TH, respond differently. During the time interval from 7.1 to 5.4 Ma when a cold, dry climate obtained, as indicated by a high flux of dust in the CLP and western Pacific, the diversity of the CA group was high, indicating that relatively cold, dry climatic conditions may be favorable for the development of CA terrestrial mollusk populations. In contrast, during the relatively warm, moist time interval from 5.4 to 4.4 Ma, corresponding to a reduced loess thickness in the CLP and low dust flux in the western Pacific, the diversity of the TH terrestrial mollusk populations was high.

Possible Causes of Paleoecological and Paleoclimatic Evolution in the Western CLP during the Late Miocene and Pliocene

Dust in the Asian interior, including northwestern China, was emitted and transported in two possible modes, corresponding to patterns of low and high level atmospheric circulation. One mode is that the Asian dust was transported eastwards by high atmospheric circulations (westerlies) and reached northwestern Pacific, as recorded by high values of dust flux to the northwestern Pacific. The other mode is that dust in the Asia interior was transported by low level atmospheric circulation, the EA winter monsoon, to the middle reaches of the Yellow River, leading to the formation of the CLP [37,49,75,76]. Thus loess sequences in the CLP and dust deposits in the western Pacific both relate to climatic changes that impacted the regions to the north of the Tibetan Plateau, i.e., the Asian interior, as has been indicated by the previous studies of Hovan et al., (1989) [77] and Rea et al., (1998) [69].

At approximately 8–7 Ma, the accumulation rate of eolian deposits in the CLP reached high values of about 4 cm/ka [37] and dust deposition in the western Pacific was maximal [69] (Figure 4), indicating that particularly dry climatic conditions must have prevailed in the Asian interior resulting in the mobilization and transport of large amounts of dust. Indeed, a palynological study of the Late Miocene–Pliocene sediments of the Dushanzi section from northwestern China (Figure 1) indicates that steppe taxa (*Artemisia* and Chenopodiaceae) were generally dominant, implying that a dry climate existed in the inland basins of northwestern China since 8.7 Ma, except for a warm and humid phase that lasted from 5.8 to 3.9 Ma [78]. In addition, significant desertification in northwestern China prevailed as early as 7.2–7 Ma, as shown by the development of eolian sand dunes in the Taklimakan Desert [79] and eolian Red Clay deposition around the Lanzhou region [80]. These climatic and environmental conditions probably provided a suitable environment for the growth and development of C4 plants in northern China and promoted the expansion of eolian deposits from the west towards the eastern part of the CLP [42,43] and which constituted suitable environments for terrestrial mollusks. Thus terrestrial mollusks grew and developed in the CLP and were able to record the patterns of environmental evolution as recorded by the Dongwan sequence.

Variations in the Dongwan terrestrial mollusk assemblages indicate that, during the time interval from 7.1 to 3.5 Ma, major paleoecological and paleoclimatic changes occurred at about 6.2, 5.4, and 4.4 Ma. The change at around 6.2 Ma, the boundary between Zones 1 and 2, is less significant than the one at 4.4 Ma, and therefore it is not clearly reflected in terms of a transition in periodicity recorded by the relative abundance of mollusks at this time [47]. Indeed, a relatively high number of CA mollusk species and individuals, consistent with the high relative abundance of CA mollusks [47], prevailed during Zone 1 and Zone 2, indicating that a cold, arid climate prevailed in the study area. These two zones exhibit the same dominant 100 kyr periodicity [47], again suggesting that the 6.2 Ma datum was not particularly significant at Dongwan. However, the TH mollusk individuals and species are somewhat different within these two zones, and in addition the appearance of mollusks in the Xifeng Red Clay sequence and the upper age of the QA-I section imply that 6.2 Ma may be an important datum in the paleoclimatic evolution of the CLP [37,46].

Changes in the CA mollusks of Zones 1 and 2 are paralleled by a global cooling trend [1] (Figure 4), increased ice-rafted detrital flux in the Northern Hemisphere [81–85], and the buildup of the Western Antarctic ice sheet [1]. The CLP is particularly sensitive to changes in high northern latitudes through the EA winter monsoon circulation [76,86]. Extended ice sheets in the Northern Hemisphere reinforce the southward movement of cold air and thereby enhance the Siberian High that controls the EA winter monsoon wind system [86]. Thus, global cooling, especially the expansion of ice deposits in high northern latitudes, could have affected the Siberian High and the EA winter monsoon, thereby expanding habitats for the CA mollusks in the CLP.

There is no evidence for major uplift of the Tibetan Plateau at about 6.2 Ma, and therefore this mechanism is excluded as a cause of the change at this time. However, if the Tibetan Plateau reached a significant height at about 8 Ma it would have accelerated climatic cooling and strengthened the EA winter monsoon [3,8,87–90], causing cold, dry climatic conditions in the western CLP during the Late Miocene, as discussed above.

The shift at about 5.3–5.4 Ma roughly corresponds to the onset of the Pliocene when global climate became warmer than before. However, the climatic drivers contributing to the Pliocene global warming are still highly debated. As summarized by Haywood et al. (2009) [91], possible candidates include paleogeographic changes [92], altered atmospheric trace gas concentrations and water vapor content [93], changes in oceanic circulation [94,95], oceanic heat transport [5,96], thermal structure of the oceans [97–99], and feedbacks generated through altered land cover (including ice sheet extent), surface albedo, cloud cover and temperature [100].

Our results seem to support the suggestion that the warming interval may be related to changes in ocean circulation and ocean heat transport caused by the closures of the Panama and Indonesian seaways [5,94–96]. Closure of these two seaways may have caused changes in heat distribution between the Pacific and Atlantic, causing reorganization of global climatic patterns and changing the pattern of atmospheric moisture flux from latitudinal to meridional, resulting in increased moisture flux to high latitudes [5,95,101] and contributing to climate changes in the CLP. Moreover, both geological records and modeling studies show that closures of the Panama and Indonesian seaways likely played important roles in the strengthening and enlargement of the western Pacific warm pool [5,102–106], providing more moisture and heat to the CLP favorable for the abundant occurrence of TH mollusk species.

The 4.4 Ma shift, representing a climatic transition from early Pliocene warming to late Pliocene cooling, is observed in the Dongwan mollusk diversity and species numbers and individuals; however, it is not clearly shown in the relative abundance of mollusks [47]. It may be related to tectonic events such as the uplift of Tibetan Plateau, since climate models suggest that uplift played a particularly important role in the evolution of the global paleoenvironment [8,87–90], and the effects of which on the EA winter monsoon are more significant than on the summer monsoon [89]. Changes in depositional facies from distal alluvial plains to proximal alluvial fans and an increase in sedimentation rate near Yecheng (Figure 1) in the western Kunlun Mountains indicate the uplift of the northern Tibetan Plateau at about 4.5 Ma [7]. This tectonic activity is thought to trigger not only the enhancement of the EA monsoon, but is also considered to be a driver of general cooling through the consequent increase in the rate of chemical weathering, thus accelerating ice expansion in Northern Hemisphere high latitudes [88,107,108] and further strengthening the Siberian High and the EA winter monsoon and transportation of dust to the CLP. Furthermore, uplift of the Tibetan Plateau could have blocked moisture transport to the Asian interior, contributing to its aridification. Under these climatic conditions, the northern CLP would have commenced a drying trend earlier than the southern CLP, causing significant ecological changes in terrestrial ecosystems including terrestrial mollusks and C4 plants at roughly 4.4 Ma.

Conclusions

The Dongwan terrestrial mollusk record from the western CLP exhibits four stages during the time interval from 7.1 to 3.5 Ma, indicating the phased evolution of paleoecology and paleoclimate. From 7.1 to 6.2 Ma, very cold and dry climatic conditions prevailed. From 6.2 to 5.4 Ma, the climate remained cold and dry but was not as dry as the preceding interval, as evidenced by the dominance of CA mollusks and rather more TH species and individuals. From 5.4 to 4.4 Ma, very warm and moist climatic conditions prevailed, evidenced by high values of the TH species and individuals, as well as by the very small numbers of CA species and individuals and their almost complete absence after about

5 Ma. From 4.4 to 3.5 Ma, all of the CA indices increase significantly and remain at a high level, indicating a cooling climate; however, all of the TH indices exhibit relatively high valuesfrom 4.4 to 4.0 Ma, an abrupt decrease from 4.0 Ma and an increase again from 3.7 Ma. The three CA species, *Cathaica pulveraticula*, *Cathaica schensiensis*, *Pupopsis retrodens*, occurred solely during this period and were absent from 7.1 to 5.4 Ma, suggesting that the climate from 4.4 to 3.5 Ma was becoming colder and drier than previous stage.

The very cold and arid climatic conditions, with changes at about 6.2 Ma, are paralleled by a global cooling trend, increased in ice-rafted detrital flux in the Northern Hemisphere [81–85], and the buildup of the Western Antarctica ice sheet [1]. The shift at about 5.3–5.4 Ma roughly corresponds to the onset of the Pliocene when global climate became increasingly warmer than previously. Our results seem to support the suggestion that the warming interval may be related to changes in oceanic circulation and oceanic heat transport [5,94–96]. The 4.4 Ma shift may be related to tectonic events such as the uplift of Tibetan Plateau.

Variations in the diversity of the CA and TH mollusks, *H*(CA) and *H*(TH), are closely related to climatic changes during the Late Miocene to Pliocene. From 7.1 to 5.4 Ma when a cold, dry climate prevailed, the CA group was the more diverse. In contrast, during the relatively warm, wet time interval from 5.4 to 4.4 Ma, the TH terrestrial mollusk populations became more diverse. It should be pointed out that most of the Neogene terrestrial mollusks in the CLP have modern representatives and therefore they have the potential to estimate quantitatively Neogene changes in temperature and precipitation in the CLP. However, such estimates depend upon the development of a training set based on a large number of surface samples. Changes in fossil terrestrial mollusk diversity are significant for the prediction of terrestrial biodiversity changes, and Quaternary loess deposits in Europe and North America have been studied in this context. The 22-Ma loess deposits in China provide an excellent opportunity for understanding long term changes in terrestrial mollusk diversity; and although the present paper focuses on the Late Neogene terrestrial mollusk in the CLP, ongoing studies will focus on other time intervals.

Supporting Information

Table S1 The depth, age, and mollusk data from the Dongwan loess-paleosol sequence. S(CA)–Total number of species of the cold-aridiphilous (CA) mollusk group. *H*(CA)–Diversity of the CA mollusk group. CAMI–Mollusk individuals of the CA mollusk group. S(TH)–Total number of species of the thermo-humidiphilous (TH) mollusk group. *H*(TH)–Diversity of the TH mollusk group. THMI–Mollusk individuals of the TH mollusk group. S(total)–Total number of species of the total mollusk group. *H*(total)–Diversity of the total mollusk group. E(total)–Equitability of the total mollusk group. TMI–Total Mollusk individuals.

Acknowledgments

We thank all of the reviewers for valuable comments, criticisms and suggestions. We are grateful to Dr. Jan Bloemendal for editing the English and for suggestions, Prof. Qingzhen Hao for field assistance and Prof. Shiling Yang for providing us with $\delta^{13}C$ data of soil carbonates from the Lingtai Red Clay sequence.

Author Contributions

Conceived and designed the experiments: FL NW. Performed the experiments: FL NW YP. Analyzed the data: FL NW DDR YD DZ. Contributed reagents/materials/analysis tools: FL NW. Wrote the paper: FL NW DDR.

References

1. Zachos J, Pagani M, Sloan L, Thomas E, Billups K (2001) Trends, rhythms, and aberrations in global climate 65 Ma to present. Science 292: 686–693.
2. Quade J, Cerling TE, Bowaman JR (1989) Development of Asian monsoon revealed by marked ecological shift during the latest Miocene in northern Pakistan. Nature 342: 163–166.
3. Harrison TM, Copeland P, Kidd WSF, Yin A (1992) Raising Tibet. Science 255: 1663–1670.
4. Cerling TE, Harris JM, MacFadden BJ, Leakey MG, Quade J, et al. (1997) Global vegetation change through the Miocene/Pliocene boundary. Nature 389: 153–158.
5. Haug GH, Tiedemann R (1998) Effect of the formation of the Isthmus of Panama on Atlantic Ocean thermohaline circulation. Nature 393: 673–676.
6. Ding ZL, Yang SL (2000) C3/C4 vegetation evolution over the last 7.0 Myr in the Chinese Loess Plateau: evidence from pedogenic carbonate $\delta^{13}C$. Palaeogeogr. Palaeoclimatol. Palaeoecol. 160: 291–299.
7. Zheng HB, Powell CM, An ZS, Zhou J, Dong GR (2000) Pliocene uplift of the northern Tibetan Plateau. Geology 28: 715–718.
8. An ZS, Kutzbach JE, Prell WL, Porter SC (2001) Evolution of Asian monsoons and phased uplift of the Himalayan-Tibetan plateau since Late Miocene times. Nature 411: 62–66.
9. Jiang WY, Peng SZ, Hao QZ, Liu TS (2002) Carbon isotopic records in paleosols over the Pliocene in Northern China: implication on vegetation development and Tibetan uplift. Chin. Sci. Bull. 47: 687–690.
10. Clark MK, House MA, Royden LH, Whipple KX, Burchfiel BC, et al. (2005) Late Cenozoic uplift of southeastern Tibet. Geology 33: 525–528.
11. Fang XM, Yan MD, Voo RV, Rea DK, Song CH, et al. (2005) Late Cenozoic deformation and uplift of the NE Tibetan Plateau: Evidence from high-resolution magnetostratigraphy of the Guide Basin, Qinghai Province, China. Geol. Soc. Am. Bull. 117: 1208–1225.
12. Molnar P (2005) Mio-Pliocene growth of the Tibetan Plateau and evolution of East Asian climate. Palaeontologia Electronica 8: 2A. 1–23.
13. Vandenberghe J, Lu HY, Sun DH, van Huissteden J (2004) The late Miocene and Pliocene climate in East Asia as recorded by grain size and magnetic susceptibility of the Red Clay deposits (Chinese Loess Plateau). Palaeogeogr. Palaeoclimatol. Palaeoecol. 204: 239–255.
14. Wen LJ, Lu HY, Qiang XK (2005) Changes in grain-size and sedimentation rate of the Neogene Red Clay deposits along the Chinese Loess Plateau and implications for the palaeowind system. Sci China Earth Sci 48: 1452–1462.
15. Zhu YM, Zhou LP, Mo DW, Kaakinen A, Zhang ZQ, et al. (2008) A new magnetostratigraphic framework for late Neogene Hipparion Red Clay in the eastern Loess Plateau of China. Palaeogeogr. Palaeoclimatol. Palaeoecol. 268: 47–57.
16. Guo ZT, Peng SZ, Hao QZ, Biscaye PE, An ZS, et al. (2004) Late Miocene-Pliocene development of Asian aridification as recorded in the Red-Earth Formation in the northern China. Glob. Planet. Change 41: 135–145.
17. Wang YX, Yang JD, Chen J, Zhang KJ, Rao WB (2007) The Sr and Nd isotopic variations of the Chinese Loess Plateau during the past 7 Ma: Implications for the East Asian winter monsoon and source areas of loess. Palaeogeogr. Palaeoclimatol. Palaeoecol. 249: 351–361.
18. Chen J, Chen Y, Liu LW, Ji JF, Balsam W, et al. (2006) Zr/Rb ratio in the Chinese loess sequences and its implication for changes in the East Asian winter monsoon strength. Geochim. Cosmochim. Acta 70: 1471–1482.
19. Fan MJ, Song CH, Dettman DL, Fang XM, Xu XH (2006) Intensification of the Asian winter monsoon after 7.4 Ma: Grain-size evidence from the Linxia Basin, northeastern Tibetan Plateau, 13.1 Ma to 4.3 Ma. Earth Planet. Sci. Lett. 248: 186–197.
20. Ding ZL, Xiong SF, Sun JM, Yang SL, Gu ZY, et al. (1999) Pedostratigraphy and paleomagnetism of a 7.0 Ma eolian loess-red clay sequence at Lingtai, Loess Plateau, north-central China and the implications for paleomonsoon evolution. Palaeogeogr. Palaeoclimatol. Palaeoecol. 152: 49–66.
21. Ding ZL, Yang SL, Sun JM, Liu TS (2001) Iron geochemistry of loess and red clay deposits in the Chinese Loess Plateau and implications for long-term Asian monsoon evolution in the last 7.0 Ma. Earth Planet. Sci. Lett. 185: 99–109.
22. Fortelius M, Eronen JT, Jernvall J, Liu L, Pushkina D, et al. (2002) Fossil mammals resolve regional patterns of Eurasian climate change during 20 million years. Evol. Ecol. Res. 4: 1005–1016.
23. Liu LP, Eronen JT, Fortelius M (2009) Significant Mid-Latitude Aridity in the Middle Miocene of East Asia. Palaeogeogr. Palaeoclimatol. Palaeoecol. 279: 201–206.
24. Kaakinen A, Sonninen E, Lunkka JP (2006) Stable isotope record in paleosol carbonates from the Chinese Loess Plateau: implications for late Neogene

paleoclimate and paleovegetation. Palaeogeogr. Palaeoclimatol. Palaeoecol. 237: 359–369.
25. Kaakinen A, Passey BH, Zhang Z, Liu L, Pesonen LJ, et al. (2013) Stratigraphy and Paleoecology of the classical dragon bone localities of Baode County, Shanxi Province. p. 203–217. In Wang X, et al. (eds): Fossil Mammals of Asia: Neogene Biostratigraphy and Chronology. Columbia University Press, New York.
26. Zhang ZQ, Gentry AW, Kaakinen A, Liu LP, Lunkka JP, et al. (2002) Land mammal faunal sequence of the late Miocene of China: new evidence from Lantian, Shaanxi Province. Vert. PalAsiat. 40: 165–176.
27. Passey BH, Ayliffe LK, Kaakinen A, Zhang ZQ, Eronen JT, et al. (2009) Strengthened East Asian summer monsoons during a period of high-latitude warmth? Isotopic evidence from Mio-Pliocene fossil mammals and soil carbonates from northern China. Earth Planet. Sci. Lett. 277: 443–452.
28. Wang Y, Deng T (2005) A 25 m.y. isotopic record of paleodiet and environmental change from fossil mammals and paleosols from the NE margin of the Tibetan Plateau. Earth Planet. Sci. Lett. 236: 322–338.
29. Jia GD, Peng PA, Zhao QH, Jian ZM (2003) Changes in terrestrial ecosystem since 30 Ma in East Asia: Stable isotope evidence from black carbon in the South China Sea. Geology 31: 1093–1096.
30. Steinke S, Groeneveld J, Johnstone H, Rendle-Bühring R (2010) East Asian summer monsoon weakening after 7.5 Ma: Evidence from combined planktonic foraminifera Mg/Ca and $\delta^{18}O$ (ODP Site 1146; northern South China Sea). Palaeogeogr. Palaeoclimatol. Palaeoecol. 289: 33–43.
31. Wei GJ, Li XH, Liu Y, Shao L, Liang X (2006) Geochemical record of chemical weathering and monsoon climate change since the early Miocene in the South China Sea. Paleoceanography 21: PA4214, doi:10.1029/2006PA001300.
32. Chen MH, Wang RJ, Yang LH, Han JX, Lu J (2003) Development of east Asian summer monsoon environments in the late Miocene: radiolarian evidence from Site 1143 of ODP Leg 184. Mar. Geol. 201: 169–177.
33. Wan SM, Li AC, Clift PD, Jiang H (2006) Development of the East Asian summer monsoon: Evidence from the sediment record in the South China Sea since 8.5 Ma. Palaeogeogr. Palaeoclimatol. Palaeoecol. 241: 139–159.
34. Jacques FMB, Shi G, Wang WM (2013) Neogene zonal vegetation of China and the evolution of the winter monsoon. Bull. Geosci. 88: 175–193.
35. Liu YS, Utescher T, Zhou ZK, Sun BN (2011) The evolution of Miocene climates in North China: Preliminary results of quantitative reconstructions from plant fossil records. Palaeogeogr. Palaeoclimatol. Palaeoecol. 304: 308–317.
36. Tang H, Micheels A, Eronen JT, Fortelius M (2011) A regional climate model experiment to investigate the Asian monsoon in the Late Miocene. Clim. Past 7: 847–868.
37. Guo ZT, Ruddiman WF, Hao QZ, Wu HB, Qiao YS, et al. (2002) Onset of Asian desertification by 22 Myr ago inferred from loess deposits in China. Nature 416: 159–163.
38. Sun XJ, Wang PX (2005) How old is the Asian monsoon system?–Palaeobotanical records from China. Palaeogeogr. Palaeoclimatol. Palaeoecol. 222: 181–222.
39. Qiang XK, An ZS, Song YG, Chang H, Sun YB, et al. (2011) New eolian red clay sequence on the western Chinese Loess Plateau linked to onset of Asian desertification about 25 Ma ago. Sci China Earth Sci 54: 136–144.
40. Hao QZ, Guo ZT (2007) Magnetostratigraphy of an early-middle Miocene loess-soil sequence in the western Loess Plateau of China. Geophys. Res. Lett. 34: L18305, doi:10.1029/2003GL031162.
41. Hao QZ, Guo ZT (2004) Magnetostratigraphy of a late Miocene-Pliocene loess-paleosol sequence in the western Loess Plateau in China. Geophys. Res. Lett. 31: L09209, doi:10.1029/2003GL019392.
42. Sun DH, Shaw J, An ZS, Chen MY, Yue LP (1998) Magnetostratigraphy and paleoclimatic interpretation of a continuous 7.2 Ma late Cenozoic eolian sediments from the Chinese Loess Plateau. Geophys. Res. Lett. 25: 85–88.
43. Ding ZL, Sun JM, Yang SL, Liu TS (1998) Preliminary magnetostratigraphy of a thick eolian red clay-Loess sequence at Lingtai, the Chinese Loess Plateau. Geophys. Res. Lett. 25: 1225–1228.
44. Xu Y, Yue LP, Li JX, Sun L, Sun B, et al. (2009) An 11-Ma-old red clay sequence on the Eastern Chinese Loess Plateau. Palaeogeogr. Palaeoclimatol. Palaeoecol. 284: 383–391.
45. An ZS, Huang YS, Liu WG, Guo ZT, Clemens SC, et al. (2005) Multiple expansions of C4 plant biomass in East Asia since 7 Ma coupled with strengthened monsoon circulation. Geology 33: 705–708.
46. Wu NQ, Pei YP, Lu HY, Guo ZT, Li FJ, et al. (2006) Marked ecological shifts during 6.2–2.4 Ma revealed by a terrestrial molluscan record from the Chinese Red Clay Formation and implication for palaeoclimatic evolution. Palaeogeogr. Palaeoclimatol. Palaeoecol. 233: 287–299.
47. Li FJ, Rousseau DD, Wu NQ, Hao QZ, Pei YP (2008) Late Neogene evolution of the East Asian monsoon revealed by terrestrial mollusk record in western Chinese Loess Plateau: from winter to summer dominated sub-regime. Earth Planet. Sci. Lett. 274: 439–447.
48. Zoller L, Semmel A (2001) 175 years of loess research in Germany–long records and "unconformities". Earth-Sci. Rev. 54: 19–28.
49. Liu TS (1985) Loess and the Environment. China Ocean Press, Beijing, 251 p.
50. Rousseau DD, Wu NQ (1997) A new molluscan record of the monsoon variability over the past 130 000 yr in the Luochuan loess sequence, China. Geology 25: 275–278.
51. Rousseau DD, Wu NQ (1999) Mollusk record of monsoon variability during the L_2-S_2 cycle in the Luochuan loess sequence, China. Quat. Res. 52: 286–292.
52. Rousseau DD, Wu NQ, Guo ZT (2000) The terrestrial mollusks as new indices of the Asian paleomonsoons in the Chinese loess plateau. Glob. Planet. Change 26: 199–206.
53. Wu NQ, Rousseau DD, Liu TS (1996) Land mollusk records from the Luochuan loess sequence and their paleoenvironmental significance. Sci. China, Ser. D: Earth Sci. 39: 494–502.
54. Wu NQ, Rousseau DD, Liu XP (2000) Response of mollusk assemblages from the Luochuan loess section to orbital forcing since the last 250 ka. Chin. Sci. Bull. 45: 1617–1622.
55. Wu NQ, Rousseau DD, Liu TS, Lu HY, Gu ZY, et al. (2001) Orbital forcing of terrestrial mollusks and climatic changes from the Loess Plateau of China during the past 350 ka. J. Geophys. Res. 106: 20045–20054.
56. Wu NQ, Liu TS, Liu XP, Gu ZY (2002) Mollusk record of millennial climate variability in the Loess Plateau during the Last Glacial Maximum. Boreas 31: 20–27.
57. Wu NQ, Chen XY, Rousseau DD, Li FJ, Pei YP, et al. (2007) Climatic conditions recorded by terrestrial mollusc assemblages in the Chinese Loess Plateau during marine Oxygen Isotope Stages 12–10. Quat. Sci. Rev. 26: 1884–1896.
58. Li FJ, Wu NQ, Pei YP, Hao QZ, Rousseau DD (2006a) Wind-blown origin of Dongwan late Miocene-Pliocene dust sequence documented by land snail record in western Chinese Loess Plateau. Geology 34: 405–408.
59. Li FJ, Wu NQ, Rousseau DD (2006b) Preliminary study of mollusk fossils in the Qinan Miocene loess-soil sequence in western Chinese Loess Plateau. Sci. China, Ser. D: Earth Sci. 49: 724–730.
60. Hou XY (1983) Vegetation of China with reference to its geographical distribution. Annals of the Missouri Botanical Garden 70: 509–549.
61. Cande SC, Kent DV (1995) Revised calibration of the geomagnetic polarity timescale for the Late Cretaceous and Cenozoic. J. Geophys. Res. 100: 6093–6095.
62. Kukla G, An ZS, Melice JL, Gavin J, Xiao JL (1990) Magnetic susceptibility record of Chinese loess. Trans. R. Soc. Edinb. Earth Sci. 81: 263–288.
63. Puisségur JJ (1976) Mollusques continentaux quaternaires de Bourgogne. Significations stratigraphiques et climatiques. Rapports avec d'autres faunes boréales de France: Université de Dijon Mémoires Géologiques, 3: 241 p.
64. Shannon CE (1948) A mathematical theory of communication. Bell Syst. tech. J. 27: 379–423.
65. Rousseau DD (1992) Terrestrial mollusks as indicators of global aeolian dust fluxes during glacial stages. Boreas 21: 105–109.
66. Rousseau DD, Limondin N, Puissegur JJ (1993) Holocene environmental signals from mollusk assemblages in Burgundy (France). Quat. Res. 40: 237–253.
67. Rousseau DD, Kukla G (1994) Late Pleistocene climate record in the Eustis loess section, Nebraska based on land snail assemblages and magnetic susceptibily. Quat. Res. 42: 176–187.
68. Zhang CF, Wang Y, Deng T, Wang XM, Biasatti D, et al. (2009) C4 expansion in the central Inner Mongolia during the latest Miocene and early Pliocene. Earth Planet. Sci. Lett. 287: 311–319.
69. Rea DK, Snoeckx H, Joseph LH (1998) Late Cenozoic eolian deposition in the North Pacific: Asian drying, Tibetan uplift, and cooling of the northern hemisphere. Paleoceanography 13: 215–224.
70. Eronen JT, Kaakinen A, Liu LP, Passey BH, Tang H, et al. (2014) Here be Dragons: Mesowear and tooth enamel isotopes of the classic Chinese "Hipparion" faunas from Baode, Shanxi Province, China. Ann. Zool. Fennici 51: 227–244.
71. Zheng HB, Powell CM, Rea DK, Wang JL, Wang PX (2004) Late Miocene and mid-Pliocene enhancement of the East Asian monsoon as viewed from the land and sea. Glob. Planet. Change 41: 147–155.
72. Wang PX, Clemens S, Beaufort L, Braconnotd P, Ganssen G, et al. (2005) Evolution and variability of the Asian monsoon system: state of the art and outstanding issues. Quat. Sci. Rev. 24: 595–629.
73. Wu YS (2001) Palynoflora at late Miocene-early Pliocene from Leijiahe of Lingtai, Gansu Province, China. Acta Botanica Sinica 43: 750–756 (in Chinese with English abstract).
74. Wang L, Lu HY, Wu NQ, Li J, Pei YP, et al. (2006) Palynological evidence for Late Miocene-Pliocene vegetation evolution recorded in the red clay sequence of the central Chinese Loess Plateau and implication for palaeoenvironmental change. Palaeogeogr. Palaeoclimatol. Palaeoecol. 241: 118–128.
75. An ZS, Liu TS, Lu YC, Porter SC, Kukla G, et al. (1990) The long-term paleomonsoon variation recorded by the loess-paleosol sequence in Central China. Quat. Int. 7: 91–96.
76. Ding ZL, Liu TS, Rutter NW, Yu ZW, Guo ZT, et al. (1995) Ice-volume forcing of East Asian winter monsoon variations in the past 800,000 years. Quat. Res. 44: 149–159.
77. Hovan SA, Rea DK, Pisias NG, Shackleton NJ (1989) A direct link between the China loess and marine $\delta^{18}O$ records: Aeolian flux to the North Pacific. Nature 340: 296–298.
78. Sun JM, Xu QH, Huang BC (2007) Late Cenozoic magnetochronology and paleoenvironmental changes in the northern foreland basin of the Tian Shan Mountains. J. Geophys. Res. 112: B04107, doi:10.1029/2006JB004653.

79. Sun JM, Zhang ZQ, Zhang LY (2009) New evidence on the age of the Taklimakan Desert. Geology 37: 159–162.
80. Sun DH, Zhang YB, Han F, Zhang Y, Yi ZY, et al. (2011) Magnetostratigraphy and palaeoenvironmental records for a Late Cenozoic sedimentary sequence from Lanzhou, Northeastern margin of the Tibetan Plateau. Glob. Planet. Change 76: 106–116.
81. Jansen E, Sjøholm J (1991) Reconstruction of glaciation over the past 6 Myr from ice-borne deposits in the Norwegian Sea. Nature 349: 600–603.
82. deMenocal PB (1993) Wireline logging on the North Pacific transect. JOIDES J. 19: 29.
83. Wolf-Welling TCW, Cremer M, O'Connell S, Winkler A, Thiede J (1996) Cenozoic Arctic gateway paleoclimate variability: indications from changes in coarse-fraction composition. Proc. Ocean Drill. Program, Sci. Results 151: 515–567.
84. Thiede J, Winkler A, Wolf-Weilling T, Eldholm O, Myhre AM, et al. (1998) Late Cenozoic history of the polar north Atlantic: results from ocean drilling. Quat. Sci. Rev. 17: 185–208.
85. St. John KEK, Krissek LA (2002) The late Miocene to Pleistocene ice-rafting history of southeast Greenland. Boreas 31: 28–35.
86. Hao QZ, Wang L, Oldfield F, Peng SZ, Qin L, et al. (2012) Delayed build-up of Arctic ice sheets during 400000-year minima in insolation variability. Nature 490: 393–396.
87. Kutzbach JE, Guetter PJ, Ruddiman WF, Prell WL (1989) Sensitivity of climate to late cenozoic uplift in southern Asia and American West: Numerical experiments. J. Geophys. Res. 94: 18393–18407.
88. Ruddiman WF, Kutzbach JE (1990) Late Cenozoic plateau uplift and climate change. Trans. R. Soc. Edinb. Earth Sci. 81: 301–314.
89. Liu XD, Yin ZY (2002) Sensitivity of East Asian monsoon climate to the uplift of the Tibetan Plateau. Palaeogeogr. Palaeoclimatol. Palaeoecol. 183: 223–245.
90. Liu XD, Kutzbach JE, Liu ZY, An ZS, Li L (2003) The Tibetan Plateau as amplifier of orbital-scale variability of the East Asian monsoon. Geophys. Res. Lett. 30: 16, 1839, doi:10.1029/2003GL017510.
91. Haywood AM, Dowsett HJ, Valdes PJ, Lunt DJ, Francis JE, et al. (2009) Introduction. Pliocene climate, processes and problems. Phil. Trans. R. Soc. A, 367: 3–17.
92. Rind D, Chandler M (1991) Increased ocean heat transports and warmer climate. J. Geophys. Res. 96: 7437–7461.
93. Raymo ME, Rau G.H (1992) Plio-Pleistocene atmospheric CO_2 levels inferred from POM $\delta^{13}C$ at DSDP Site 607. Eos Trans. AGU 73 (Suppl.), 95.
94. Ravelo AC, Andreasen DH (2000) Enhanced circulation during a warm period. Geophys. Res. Lett. 27: 1001–1004.
95. Cane MA, Molnar P (2001) Closing of the Indonesian seaway as a precursor to east African aridification around 3–4 million years ago. Nature 411: 157–162.
96. Dowsett HJ, Cronin TM, Poore PZ, Thompson RS, Whatley RC, et al. (1992) Micropaleontological evidence for increased meridional heat-transport in the North Atlantic Ocean during the Pliocene. Science 258: 1133–1135.
97. Philander GS, Fedorov AV (2003) Role of tropics in changing the response to Milankovich forcing some three million years ago. Paleoceanography 18: PA1045, doi:10.1029/2002PA000837.
98. Wara MW, Ravelo AC, Delaney ML (2005) Permanent El Nino-like conditions during the Pliocene warm period. Science 309: 758–761.
99. Fedorov AV, Dekens PS, McCarthy M, Ravelo AC, deMenocal PB, et al. (2006) The Pliocene paradox (mechanisms for a permanent El Nino). Science 312: 1485–1489.
100. Haywood AM, Valdes PJ (2004) Modelling Middle Pliocene warmth: contribution of atmosphere, oceans and cryosphere. Earth Planet. Sci. Lett. 218: 363–377.
101. Hall R (2002) Cenozoic geological and plate tectonic evolution of SE Asia and the SW Pacific: Computer-based reconstructions and animations. J. Asian Earth Sci. 20: 353–434.
102. Maier-Reimer E, Mikolajewicz U, Crowley T (1990) Ocean general circulation model sensitivity experiment with an open central American isthmus. Paleoceanography 5: 349–366.
103. Mikolajewicz U, Crowley TJ (1997) Response of a coupled ocean/energy balance model to restricted flow through the central American isthmus. Paleoceanography 12: 429–441.
104. Mikolajewicz U, Maier-Reimer E, Crowley TJ, Kim KY (1993) Effect of Drake and Panamanian gateways on the circulation of an ocean model. Paleoceanography 8: 409–426.
105. Chaisson WP, Ravelo AC (2000) Pliocene development of east-west hydrographic gradient in the equatorial Pacific. Paleoceanography 15: 497–505.
106. Li BH, Wang JL, Huang BQ, Li QY, Jian ZM, et al. (2004) South China Sea surface water evolution over the last 12 Myr: A south-north comparison from Ocean Drilling Program Sites 1143 and 1146. Paleoceanography 19: PA1009, doi:10.1029/2003PA000906.
107. Raymo ME, Ruddiman WF, Froelich PN (1988) Influence of late Cenozoic mountain building on ocean geochemical cycles. Geology 16: 649–653.
108. Raymo ME, Ruddiman WF (1992) Tectonic forcing of late Cenozoic climate. Nature 359: 117–122.

21

Snout Shape in Extant Ruminants

Jonathan P. Tennant[1]*, Norman MacLeod[2]

1 Department of Earth Science and Engineering, Imperial College London, London, United Kingdom, **2** Earth Science Department, The Natural History Museum, London, United Kingdom

Abstract

Snout shape is a prominent aspect of herbivore feeding ecology, interacting with both forage selectivity and intake rate. Previous investigations have suggested ruminant feeding styles can be discriminated via snout shape, with grazing and browsing species characterised by 'blunt' and 'pointed' snouts respectively, often with specification of an 'intermediate' sub-grouping to represent ambiguous feeding styles and/or morphologies. Snout shape morphology is analysed here using a geometric morphometric approach to compare the two-dimensional profiles of the premaxilla in ventral aspect for a large sample of modern ruminant species, for which feeding modes are known from secondary criteria. Results suggest that, when browsing and grazing ruminants are classified ecologically based on a range of feeding style indicators, they cannot be discriminated unambiguously on the basis of snout profile shape alone. Profile shapes in our sample form a continuum with substantial overlap between groupings and a diverse range of morphologies. Nevertheless, we obtained an 83.8 percent ratio of correct post hoc feeding style categorisations based on the proximity of projected profile shapes to group centroids in the discriminant space. Accordingly, this procedure for identifying species whose feeding strategy is 'unknown' can be used with a reasonable degree of confidence, especially if backed-up by additional information. Based on these results we also refine the definitions of snout shape varieties, taking advantage of the descriptive power that geometric morphometrics offers to characterize the morphological disparities observed. The shape variance exhibited by both browsing and grazing ruminants corresponds strongly to body mass, providing further evidence for an interaction between snout shape, feeding style, and body size evolution. Finally, by exploring the role of phylogenetic similarity in snout shape, we find a slight increase in successful categorisation when repeating the analysis with phylogenetic control on the geometric profiles.

Editor: Andrew A. Farke, Raymond M. Alf Museum of Paleontology, United States of America

Funding: These authors have no support or funding to report.

Competing Interests: The authors have declared that no competing interests exist.

* Email: jonathan.tennant10@imperial.ac.uk

Introduction

Members of Ruminantia are even-toed ungulate mammals defined uniquely by possession of a two-step digestion system involving the fermentation chamber in the foregut of the stomach. Some 200 extant species currently recognised [1]. Ruminant feeding styles are reflected in their craniodental and gastrointestinal morphophysiological diversity, and have been conventionally categorised into 'browsers' and 'grazers', with an 'intermediate' sub-category [2–5]. Additionally, browsers are considered obligate non-grazers, but not vice-versa [2]. Some authors further include variants of frugivores, high-level browsers, and fresh grass grazers as independent categories in an attempt to encompass a larger range of feeding styles [3–5]. Variations in feeding style may also occur on different spatial and temporal levels, corresponding to environmental stresses (e.g., drought [6]), and plausibly a hierarchical grazing succession related to species' migration patterns, geomorphology, resource partitioning or forage quality [7–9].

Van Zyl was the first to define an ecological classification scheme for ungulates based explicitly on feeding style [10]. Hofmann extended Van Zyl's definitions to contain a novel qualitative morphological and physiological underpinning [11–16]. Hofmann's modified ungulate feeding classification scheme has been used extensively in vertebrate (palaeo)biology ever since. Recently, availability of software, new data-analysis techniques, and increased computational power have combined to facilitate the use of a wider range of approaches, including geometric morphometrics, that allows us to build on and re-appraise these earlier findings (see Clauss *et al.* [17] and references therein).

Despite a range of morpho-behavioural correlates, the archetypal dichotomy between 'browsers' and 'grazers' is based on a botanical foundation. Browsers typically consume dicotyledonous leaves, stems, twigs and fruits [11,18,19]. Grazers consume monocotyledonous plants, and 'intermediate' feeders vary their consumption preferences depending on season and geography [20,21]. The putative morphological significance of this dietary variation is that the physical, mechanical and biochemical properties of different forage types are adequate to drive and maintain a morpho-functional trichotomy among ruminant species that reflects the physical challenges they face accessing and/or processing different types of forage. It has been argued that the biomechanical properties of different forage types have exerted strong controls on the evolution of the masticatory apparatus and gastrointestinal tract [2], and specifically the reticulorumen physiology [22,23] within ruminants.

The botanical definitions underpinning the classification of browsers and grazers have a complex history, with technical articles containing numerous examples of inconsistent thresholds for distinctions between these classes. Several authors have regarded browsers to be ruminants that consume less than 10 percent grass, and grazers to be those consuming greater than 90 percent grass per annum, with all other species classed as intermediate [24–27]. Conversely, others have selected greater than 75 percent grass consumption per annum as the threshold criterion for their grazer class, and greater than 75 percent browse consumption for the browser class, with little or no empirical evidence or logical rationale provided for these thresholds [4,28–30]. Clauss *et al.* [31] defined grazers as those ungulates consuming greater than 80 percent monocot material, and strict browsers as those with a "very low intake of monocot forage" (p. 399) based on a small sample, similar to a range of comparative studies in which percentage of grass consumed is used as the basis for further investigations into ruminant ecology [2,17,23,32].

In other studies, feeding strategy delimitation has been based purely on qualitative assessments [33], where grazers are classified as those "consuming primarily grasses, sedges and other graminoids" (p. 178). This discordant usage has been summarised partially by Clauss *et al.* [17], with Gordon and Illius [34] finding that different thresholds of classification give different results in ecological analyses. Accordingly, defining these thresholds in congruence with functional or ecological significance remains a problematic issue and one which is only exacerbated when such thresholds are used as a basis for further investigation of ruminant ecology. Additionally, species-level variation in diet and the physiology and morphology of ingestion due to the facultative nature of feeding strategies makes placing discrete parameters around distinct ecological sub-groups challenging [35].

There are numerous morphophysiological parameters that might, in principle, affect digestive rates and masticatory efficiency among ruminant species. Among these, the anterior snout forms the interactive part of the ingestive apparatus that interacts with any and all types of ruminant forage [36]. The anterodorsal section of the snout is formed predominantly by the premaxillae. It has been noted commonly that browsing ruminant species have pointed snouts and grazers a more squared or blunt shape representing a derived cropping condition (e.g., [24,37]). Intermediate feeding styles have been posited to have an intermediate form, considered to conform to a mediolaterally compressed club-like shape [38].

The relationships between the various aspects of herbivore ingestion are complicated, but recent developments have provided considerable insight into the factors that interplay to control efficiency (e.g., Clauss *et al.*, [39]). Snout shape is part of this suite of aspects of herbivore ecology that, combined, determine initial intake rate, chewing efficiency and forage selection ability [20,40,41,42,43]. Traditionally, a more pointed rostrum is associated with increased selection sensitivity, and a blunt rostrum is associated with a less selective cropping process with greater intake [24,39], but these views remain in a state of partial flux [34,44,45]. Codron *et al.* [46] suggested that browsers and grazers vary their diets on a spatiotemporal scale, conforming to earlier studies by Owen-Smith and Du Toit [33 and 47]. Despite this variation, several distinctions have become apparent between browsing and grazing ruminant categories and are supported within a statistical and phylogenetic framework [17].

To date the association between anterodorsal snout morphology and feeding style has not been subjected to any geometric morphometric analysis of pure shape. Fraser and Theodor [43] demonstrated that anterior dentary shape (i.e., the ventral component of the snout) is strongly associated with diet. Furthermore, dentary shape was shown to be a good proxy for premaxilla shape and is strongly deterministic in selectivity during feeding. Such studies highlight the importance of controlling for phylogenetic similarity in tests of functional ecomorphology. We aim to build on such studies by using geometric morphometric techniques focussed exclusively on the dorsal snout, in particular the anterior section of the rostrum formed primarily by the paired premaxillae.

The principle aim of our study is to determine whether empirically assessed patterns of snout shape variation in ruminants support traditional distinctions that have been drawn between 'browser' and 'grazer' categories, and whether a quantitative geometry-based approach allows a more precise morphological definition of these functional categories to be formulated. The secondary aim is an assessment of the extent to which quantitative snout geometry may be used to predict the feeding styles in ruminants of 'unknown' or 'intermediate' feeding style, including fossil specimens for palaeoecology. The statistical null hypothesis under consideration is that snout profile shape exhibits no structured variation such that reliable morpho-functional categorisation is possible.

Furthermore, we investigate the influences of body size on snout morphology. Body size is an important ecological parameter in ruminants, affecting factors such as locomotion, digestive and ingestive efficiency [40,48,49,50], competitive interactions [51], evolutionary and life histories [52–55], biogeography [56,57], and sexual dimorphism [58]. Finally, phylogeny has been repeatedly demonstrated to be an important determinant of phenotypic aspects of ruminant life histories and ecology [59–62]. Accordingly, we explore the role of phylogenetic non-independence on snout morphology similarity using established comparative methods (e.g., [63–66]), and compare the results between a 'raw' geometric morphometric study, and one in which there is a measure of phylogenetic control.

Materials and Methods

Geometric morphometrics involves the multivariate numerical analysis of two- or three-dimensional Cartesian coordinate data, typically defined by discrete, spatially-defined landmarks (i.e., topologically homologous loci on a structure [67]). Zoological studies are increasingly using a wide range of geometric morphometric techniques due to their intrinsic ability to summarize modes of variation in form – and so guide its interpretation – in many different systematic contexts [68], including functional morphology, sexual dimorphism, ontogenic development, and phylogenetic inference. The ruminant specimen-set analysed here consisted of 125 extant species, 119 of which were bovids or cervids as these are the most taxonomically diverse groups. Ecological categorisations were based on a number of sources and independent criteria, provided in Table S1. Categorisation authority was given to more recent studies where possible. Species traditionally classified as 'intermediates' or 'frugivores' were considered to be 'unknown' for the purpose of this analysis. Species with conflicting ecological classifications were additionally classified as unknowns, so that the browser and grazer categories used here were defined by species strictly characterised as browsers and strict grazers in the primary literature. The majority of sampled specimens were housed in the zoology collections at The Natural History Museum (NHM), London, UK, with additional specimens sampled from the Royal Veterinary College (RVC), London, UK. Specimen sex was not taken into account in calculation of the discriminant function, and, where multiple specimens were available, the largest were always selected to maintain consistency in age. The taxonomy follows the NHM labelling system, updated to conform to the species-level taxonomy of Fernández and Vrba [1]. It is assumed here that intraspecific shape differences are of a lower order of magnitude than interspecific shape differences; therefore, only a single specimen per species was used for the investigation.

Snout profile outlines were collected from photographs taken in ventral view with the crosshairs positioned centrally on the sagittal inter-premaxillary suture. The starting point for all the outlines was defined as the point where the suture between the maxilla and premaxilla intersects the left-lateral margin on the ventrolateral surface. This convention ensured that all subsequent semi-landmarks were interpolated to topologically homologous positions with respect to the total set of semi-landmarks used to represent the outline (each semi-landmark has a defined *x–y* position with respect to the co-ordinate system origin). One hundred equally spaced semi-landmarks were collected along each outline, a digitizing resolution sufficient to produce a geometrically faithful representation of the profiles. No semilandmark sliding was allowed for reasons discussed by MacLeod [69]. The raw, untransformed landmark co-ordinate data are provided in data file S1.

These semi-landmark data were subjected to a *Procrustes* (generalised least squares) transformation. *Procrustes* superimposition forms the core for analysis of pure shape by removing the extraneous variation in scale, orientation and position for all specimens' semi-landmark constructions (see [70] and Box 2 of [71]). Optimising the fit of all specimens to each other was achieved by rigid rotation iteration until the distance between successive mean landmark configurations fell below 0.0001. This means that the analysis proceeded in shape space as opposed to form space. The specimens at this stage were sub-divided into their ecological sub-groupings for each subsequent analysis.

Superposed co-ordinate data for defined browsers and grazers were subject to a covariance-based principal components analysis (PCA) [72], which preserves the partial *Procrustes* distances among specimens. Three principal component (PC) axes accounted for 93.4 percent of the total shape variance (Table S1). Accordingly, projected scores on these three PC axes were retained and served as the basis for a secondary discriminant analysis. These principal component scores were then subjected to a canonical variates analysis (CVA) which, unlike PCA, includes the group-level information as an additional variable [73]. This multivariate technique transforms the data to a configuration that achieves the optimal discrimination between group centroids relative to the group dispersion structure [73–75] (S3). A log-likelihood ratio (LLR) test was performed to test group distinctiveness (i.e., the group dispersion structure) of these transformed data, with respect to the sample that defines the discriminant space [76]. The resulting probability estimate represents a validation test of the between-groups covariance structure; i.e., a low probability (<0.05, traditionally) reflects a statistically significant difference in the dispersion structure with respect to the defined groups.

Recently Mitteröcker and Bookstein [77] have questioned the use of CVA in geometric morphometric contexts as it is often the case in such datasets that the sample size is exceeded greatly by the number of variables used to represent form variation among the specimens in the sample (in this case, 200). To address this concern we applied Monte Carlo simulation and bootstrapping variations of CVA, each based on 1,000 pseudo-replicate datasets, to determine whether the between-groups distinctions observed are the product of hyper-dimensionality within the dataset [78]. The Monte Carlo simulation created pseudo-replicate datasets of values drawn randomly from a normal distribution of identical mean and variance to that of the pooled PCA scores. Each pseudo-replicate dataset was then subjected to a CVA and the extent of between groups discrimination summarized via calculation of the LLR index (ϕ) [78]. As all of the values used for the CVA were drawn randomly from the same distribution, this pseudo-replicate ϕ value represent the extent of-groups distinction expected under the null model of no difference between groups other than random sampling error. In addition, the set of 1,000 pseudo-replicate ϕ values was then be tabulated into a frequency distribution and used to assess the statistical significance of observed ϕ value obtained for the investigation dataset. The bootstrapping simulation created pseudo-random datasets drawn randomly from the PCA transformed raw shape data themselves such that the observed group structure was destroyed and, for each pseudo-replicate dataset random agglomeration of species of group sample sizes and variable numbers identical to the observed data were substituted. The distribution of random ϕ values derived from these bootstrapped pseudo-replicate datasets was then compared to the observed ϕ value obtained for the investigation dataset. Passing these sensitivity tests implies that the observed group distributions are the products of some extrinsic group-distinction factor (e.g., biogeography, phylogeny, functional constraints, ecology), as opposed being the result of random sampling or dataset dimensionality issues.

To represent a shape transformation sequence through the data based on hypothetical successive models of the snout profiles in both the principal component space, and a space defined by maximum between-groups shape variation, overlay or 'strobe plot' comparisons of modelled snout shapes were performed [79]. Three principal component axes, with five modelled points per axis, were back-projected into the space defined by the original *Procrustes*-transformed variables. These models represent the two extreme points, the central point, and two medially-interpolated points between these on each PC. The result is a set of non-orthogonal principal component axes oriented with respect to the data within *Procrustes*-scaled landmark data. Each modelled axis was plotted in order to assess, interpret, and illustrate the modes of shape variation represented along each PC axis. We repeated this procedure for the single discriminant axis produced by the CVA, and back-projected this into the PCA space to observe the major mode of between-groups shape variation in the space that defines the sum variation in sample snout shapes.

We additionally performed a test for phylogenetic signal in the principal component scores using Blomberg's k, a commonly used statistic that is independent of sample size and assumes that a trait, in this case snout shape, evolved along a topology under Brownian Motion [80]. We also calculated Pagel's λ which determines whether a structured or non-structured tree topology fits the trait data best [81]. Such practice is becoming increasingly common in ecomorphological analyses, and particularly within ruminants [49,59,60]. While several recent ruminant phylogenies exist for Cervidae [82], Bovidae [83] and all of Ruminantia [1], we opted to use the updated version of the Bininda-Emonds *et al.* mammalian supertree (M. Clauss, 2014, JPT, pers. comm., [84]), pruning both this tree and our data set so that taxonomic lists were congruent ($n = 104$, File S2). Blomberg's k was calculated in the Picante package using the Kcalc() function, and Pagel's λ in the Phytools package using the phylosig() function. We tested for phylogenetic signal using the residuals of a linear regression between body mass and the raw PCA scores. The results of this test prescribe whether or not the analysis needs to be repeated with a measure of phylogenetic control – in this case, using the phyl.pca() function in the Phytools package [85]. Body mass data were extracted from the PanTHERIA database [86] for all sampled species, and log-transformed subsequent to any analysis. All statistical tests were conducted in R v. 3.0.3 ([87]).

The process of dimensionality reduction, discriminant analysis, dispersion structure validation, and model visualisation provides a statistically rigorous protocol for assessing the validity of the ruminant feeding categories, and the morphology of the profiles that define these categories. Additionally, by exploring the effects of phylogeny and body size, we can determine if the pattern of shape variation exhibited by the sampled species is the product of evolutionary similarity, body size, an external ecological factor, such as feeding style, or a combination of all of the above.

Results

Principal Components Analysis

The first two principal component axes explained the overwhelming majority of the sample covariation (89.45 percent; Table S1). These axes can be used to define a low-dimensional shape ordination subspace (Fig. 1A), with an additional 4 percent of the covariance described by PC-3 (Fig. 1B). Grazer-classed species show less variation in this PC-1 and PC-2 subspace compared to browsers. The two groups overlap about the region of the total sample grand mean, but occupy quasi-distinct regions of the PC-1 to 3 subspace. Much of the morphospace occupied by grazing species in PC1–PC2 is defined by several outlier taxa (*Bos taurus*, *Connochaetes gnou*, *C. taurinus* and *Oryx leucoryx*), whereas the majority of this group occupy low negative scores about the grand mean. Grazers occupy more negative regions overall on PC-1 and browsers more positive values on both PC-1 and PC-2. Grazers are strongly constrained along the PC-3 axis, whereas browsers exhibit about twice the range variation in both the positive and negative values.

This dispersion structure implies that the dominant aspects of the shape covariance observed in the sample play a limited role in any between-group separation structure in the sample. Since PC-1 accounts for more than four times the variance as PC-2, this means that much of the shape variation present in the dataset is being determined by something other than distinctions between browsers and grazers, such as phylogeny, body size, or geographical partitioning. Overall the dispersion pattern of browsers is much sparser along PC-1 (defined at the positive-most extremity by *Alces pulmatus* and *Ammodorcas clarkei*, and negatively by *Beatragus hunteri*), whereas the distribution is much more constrained and defined by a higher density of species at extreme ranges on PC-2 (with the extremity defined positively by *Tragulus javanicus* and negatively by and negatively by *Rhynchotragus kirkii*).

The between-groups shape deformation axes were modelled at five coordinate positions along the first three PCA axes (Fig. 2). As can be seen from these models, positive scores on PC-1 are associated with profile shapes exhibiting an elongated rostrum (approximately twice the length compared to the maximal width of the premaxillae), with a convex distal end, and a distinct medial compression at the mid-point of the lateral premaxillae margins. Negative PC-1 scores describe premaxillae that are slightly wider than long, with a distal medial concavity and lateral margins that gradually diverge posteriorly. The anterior end of this model profile is slightly concave. This represents a continuous transformation of lateral broadening and longitudinal contraction from positive to negative score values. This difference in muzzle length between different feeding categories was first noted by Fraser and Theodor [43]. Positive scores on the PC-2 axis describe a sub-triangular geometry, with a convex distal end. The lateral margins of these profile shapes diverge rapidly posteriorly, with a slight lateral contraction that is not distinct as in PC-1. Negative PC-2 scores describe mediolaterally compressed 'club-like' shapes, with a slight anterior concavity similar to PC-1. This axis represents a shape deformation sequence in which the posterior part of the lateral premaxilla narrows in width, and the anterior part expands laterally but compresses longitudinally from positive to negative PC-values. The major mode of change described by PC-3 is from a distally flattened rostrum with strongly laterally convex margins (negative scores) to a distally convex rostrum (positive scores), with a slight component of asymmetrical vergence in the higher values. This pattern of deformation is unusual in that it is occurring in a non-symmetrical mode about the sagittal line of the rostrum. It is likely that this axis is detecting a portion of heterogeneous shape change associated with deformation not removed by the *Procrustes* transformation, and perhaps due to a small degree of warping in the premaxillae from drying involved in the collection and storing process. As this mode of deformation is represented by such a small proportion of the sample variance, we do not consider this to be a major problem with our samples.

Canonical Variates Analysis

The first three PC axes accounted for the majority of the total observed sample shape variance (93.42 percent). Scores of individual semi-landmark shape configurations on each of these axes were parsed into grazers and browsers based on lines of evidence independent of snout morphology and subjected to a CVA. Since only two groups were used for this analysis, a single discriminate axis was defined. A histogram of results for the projection of shape configurations onto this axis is given in Figure 3. Both browsers and grazers occupy relatively broad regions with grazers occupying negative values along CV-1 whereas browsers are distributed more positively (see Table S2 for associated CVA scores).

The overlapping ranges of browsers and grazers implies that the within-groups shape variation is distributed in a manner such that a complete snout profile continuum exists irrespective of whether the shape space is formulated to reflect the major axes of pooled groups variance (Figs. 1A, B) or between-groups distinction (Fig. 3). Nevertheless, quasi-distinct regions within the discriminant space can still be identified. A log-likelihood ratio test confirmed this distribution as not being the product of random variation within the dataset ($p<0.0001$). Monte Carlo simulation and bootstrapping variations of CVA both based on 1,000 pseudoreplicate datasets produced distributions of randomized LLR (ϕ) values that were all well below that of the empirical data ($\phi = 38.13$; $p = 0$ in both circumstances). This indicates that the probability of these groups occupying their positions in the overall CV space as a result of the effect of random sampling of a single, underlying population is well below the traditional 95 percent level of statistical significance. Accordingly, the alternative hypothesis – that the observed magnitude of centroid separation is such that these data were most likely drawn from different shape populations with different characteristics – is supported. Although these snout outline shapes are distributed continuously between ruminant species, there is an underlying trend driven by the different group-based ecological categorisation of these profiles.

Proximity estimates are provided by calculation of a confusion matrix (Table S2), which summarizes the assignment of species with respect to their *a priori*-defined groups based on their distances to the respective group means in the canonical variates space. This result indicates that in over four out of every five cases (83.82% for this dataset), the correct *a posteriori* assignment of each species to its *a priori* designated feeding class, based on

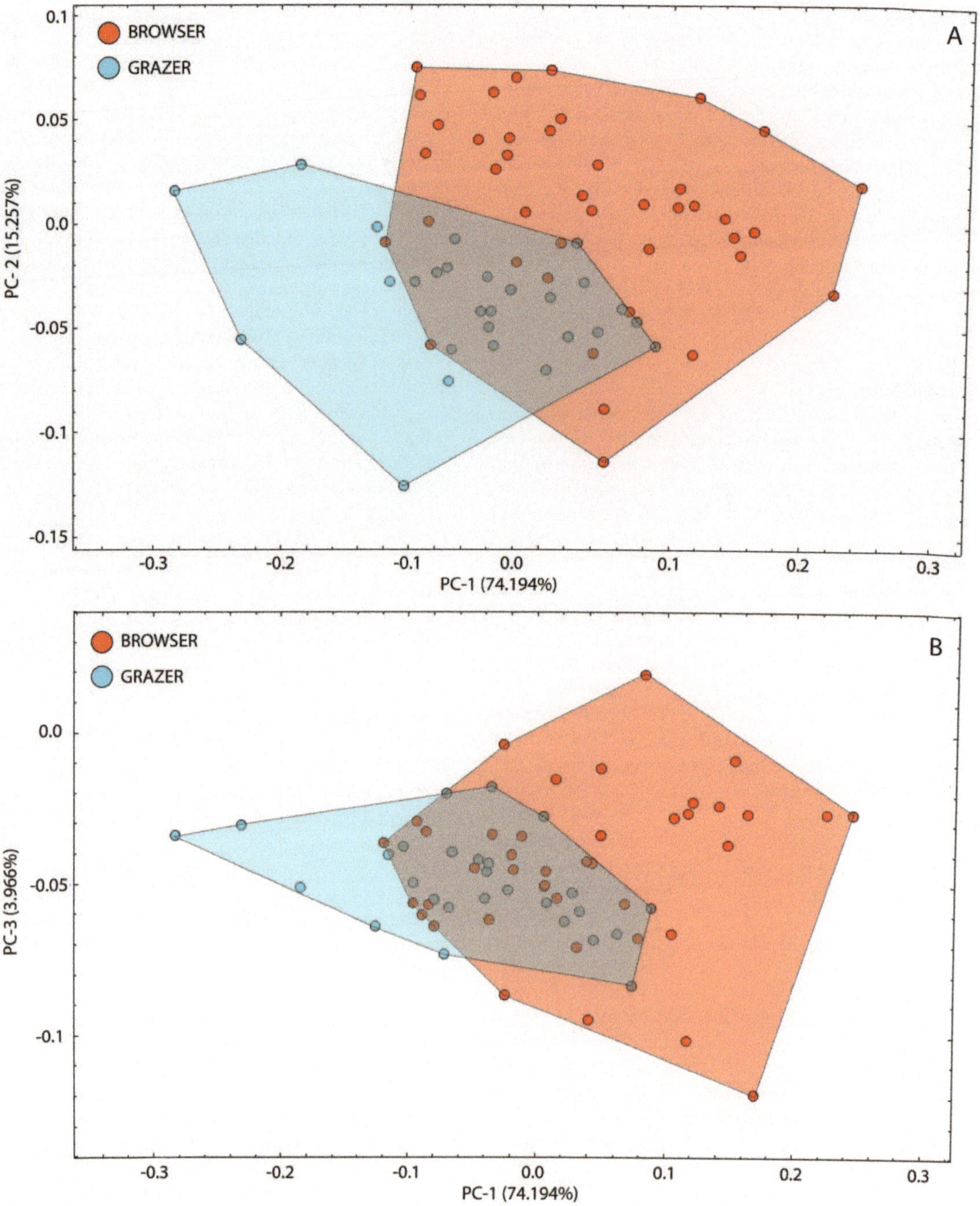

Figure 1. PCA score plots for browsing- and grazing-classed ruminants (A) PC-1 versus PC-2. The convex hulls represent a morphospace constrained by the extreme data points within the range envelope. Ecological classifications and PC scores for the species used to define this space are in Table S1. (B) PC-1 versus PC-3. The browser group occupies a similarly broad range as the PC-1 versus PC-2 plot, with grazers appearing more constrained along the PC-3 axis.

secondary criteria, was possible using premaxilla shape data alone. As such, the division of ruminants in to broad, but non-exclusive, categories of browsers and grazers maintains a high level of statistical support. A jackknifed (leave one out) cross-validation of the performance of this discriminant space produced results similar to the original CVA, with 83.58% of correct *a posteriori* assignment with just a single grazer being identified incorrectly.

The 'unknown' sub-group was projected into this defined space as a way of indicating which of the known groups they belong to, and as such what ecological inferences can be made about them with a confidence of 83.82 percent (Fig. 4). By calculating the distance from each projected point to the known group centroids, we were able estimate the likely candidate group to which these 'unknown' and 'uncertain' species belong. Of the 57 unknown species, 32 are assigned provisionally to the grazer category, and 25 to the browsers. These provisional assignments can be validated by using additional observational data, such as the percentage of grass consumed, or the hypsodonty index, once these are

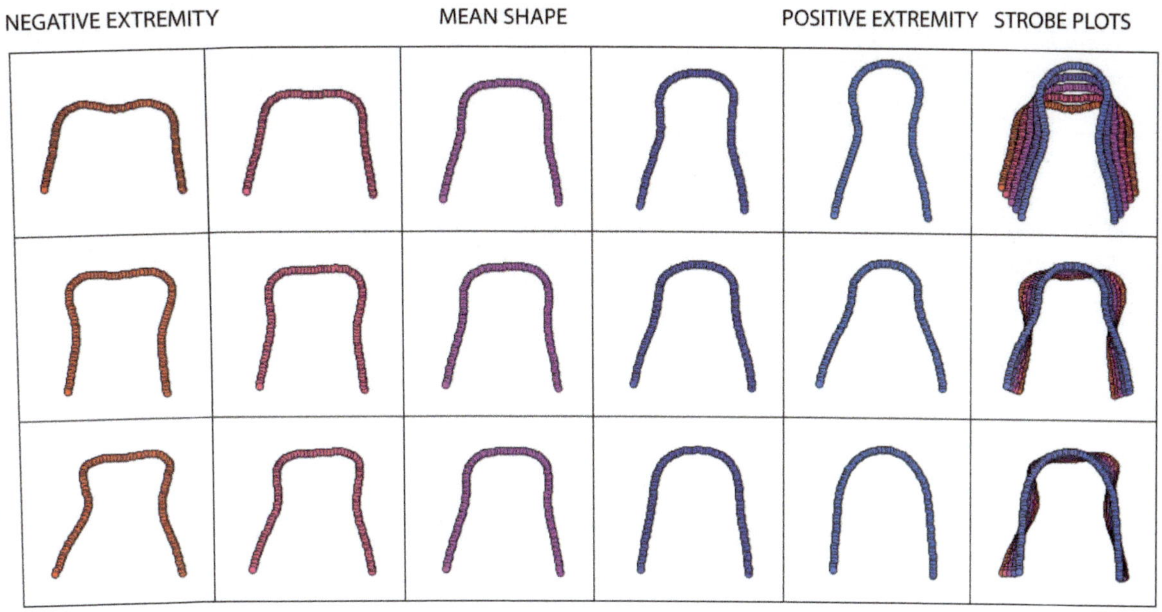

Figure 2. Strobe plots for the axis models associated with PC-1, PC-2 and PC-3. The right hand column is an overlay plot for each model series, showing the progressive deformation along each axis.

rigorously defined as being able to discriminate between the different feeding styles. It is however worth noting that the assignment of several of the taxa that fall outside of any of the known browser- or grazer-defined spaces cannot be justified based on the current analysis.

To interpret the geometric character of optimal between-groups shape discrimination, the single CV axis was modelled at five equally-spaced coordinate positions as with the PC axes. The shape configurations present at these positions along this CV axis were determined in a manner similar to that used to create the PC models (see above), first by back-projecting them into the corresponding PC-space and then reconstructing the semilandmark point configurations at those positions using the method of MacLeod [78,79] (Fig. 5). The pattern of shape variation described by this CV axis incorporates all three of the modes represented by the PC axes described above. It can be regarded as a continuum that shows progressive deformation of the premaxilla from a rostrolaterally broad, moderately laterally convex, and

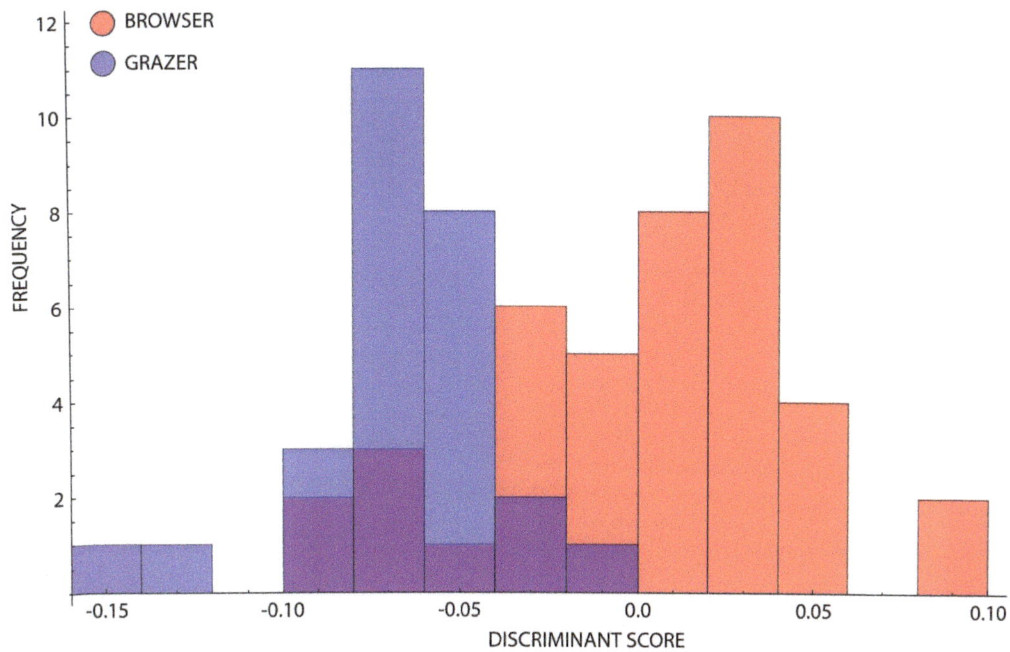

Figure 3. Histogram showing the frequency of occurrences of browser-class and grazer-class species along CV-1. The occupation of quasi-distinct discriminant spaces is clear, with overlap about the grand mean.

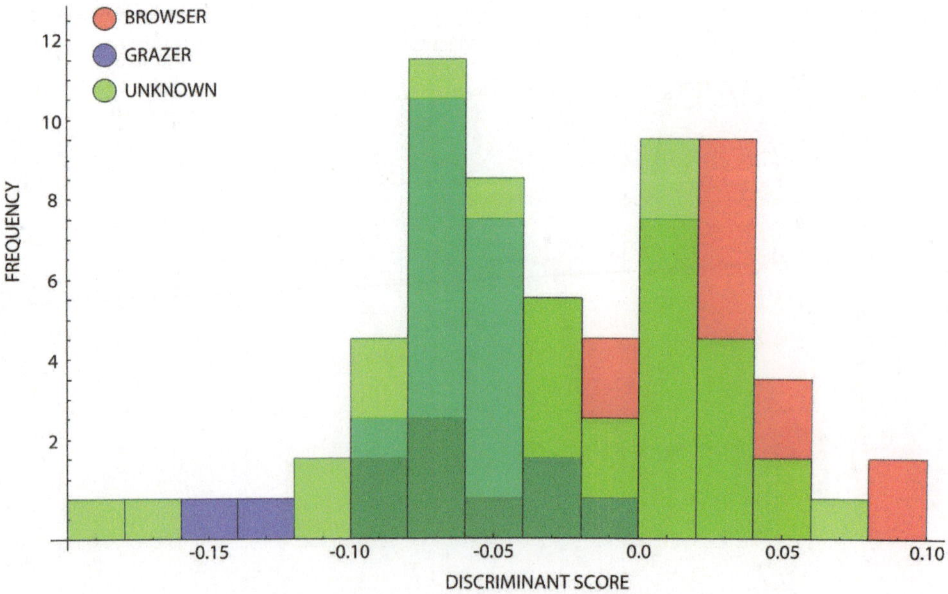

Figure 4. Histogram plot for ruminants classified according to their feeding strategy, with 'unknowns' projected into the space. The unknown-classed species' range plots more negatively on the CV-1 axis, suggesting that there is a 'cryptic' measure of snout shape variance that is not picked up in the traditional browser-grazer dichotomy.

distally depressed geometry (negative scores) into a laterally convex and posteriorly divergent, and distally narrow and pointed shape (positive scores). This model 'strobe plot' provides a more explicit, detailed, and empirically-based visualisation of this broadly pointed-to-blunt shape continuum. The fact that these discriminant axis models are almost identical to those of PC-1, the major axis of shape variation, implies that the shape distinction between browsers and grazers is more significant than any other factor in reflecting the current state of ruminant snout morphological diversity.

Comparison with body mass and phylogeny

We used the Bininda-Emonds *et al.* supertree [82] pruned to include only those taxa that are present in the current sample dataset (n = 104). Using the residuals from a linear regression model of the first two PC axis scores against body mass, we derived Pagel's λ as a measure of phylogenetic signal (PC-1, p = 0.014; PC-2, p = 0.008). We additionally derived the Blomberg's k statistic as a comparison, finding non-significant results for the residuals for both major PC axes (PC-1, p = 0.228; PC-2, p = 0.192). Taken as a whole, however, these results suggest a significant component of phylogenetic signal within these shape data. Accordingly, we calculated independent contrasts for the body mass data and scores for the first two PC axes. A Spearman's rank (p = 0.8613, $\rho = 0.017$) and Kendall's tau test (p = 0.934, $\tau = 0.006$) demonstrates that there is no significant relationship between body mass and PC-1, the primary axis of snout shape variation, within the ruminant data set when phylogenetic independence is controlled for. Similar results were obtained for PC-2 for both Spearman's rank (p = 0.445, $\rho = -0.076$) and Kendall's tau (p = 0.454, $\tau = -0.05$).

Due to the significant phylogenetic signal in the principal component scores, we modified and repeated the previous multivariate analysis procedure using the raw shape variables by performing a phylogenetic PCA on the superimposed *Procrustes* co-ordinates [85]. The results of this extended phylogenetically controlled analysis are given in Table S3. The resulting canonical variates histogram produces similar results to the raw analysis, with overlapping but quasi-distinct discriminant spaces occupied by both browsers and grazers (Fig. 6). It is important to note that the frequency peaks of these distributions are distinct from each other in this space, falling either side of the mean shape. The confusion matrix indicates that the phylogenetically controlled discriminant analysis performs slightly better than the non-controlled analysis in correctly resolving individual species to their category, with 85.45% correct assignments. The jackknifed confusion matrix confirms the stability of this distribution, with a slight reduction to 85.19% correct assignment. The log-likelihood ratio test of this distribution is strongly significant, ($p<0.0001$), and the stability confirmed as before with bootstrapping of the distribution (p = 0.0) and Monte Carlo simulations (p = 0.0).

Projecting the 'unknowns' into this discriminant space as in the raw analysis shows that they overlap both group spaces pervasively, and exceed the browser space on the negative CV-1 axis (Fig. 7). Of the 48 'unknown' species in this slightly reduced dataset, 31 (64.58%) are assigned to the browser group, and 17 to the grazer group (35.42%), which based on the assessment purely with browsers and grazers we can state with approximately an 85% confidence level (Table S3). The overall result is that controlling for phylogenetic similarity in these tests is not sufficient to change the stability of the resulting group dispersion structures, despite their being a high degree of phylogenetic signal in the shape profiles within the sample dataset.

Discussion

Taken as a whole, our results suggest that snout shape is largely sufficient to differentiate between - and so to identify - different feeding styles in ruminants. Initial descriptions of the blunt-pointed dichotomy do indeed represent an aspect of the deformation sequence and describe it in a simple way. While the results of this study largely confirm that of previous research (e.g., [24,36,37,42]), the approach used herein gives analysts access to the total range of shape variation expressed by geometric

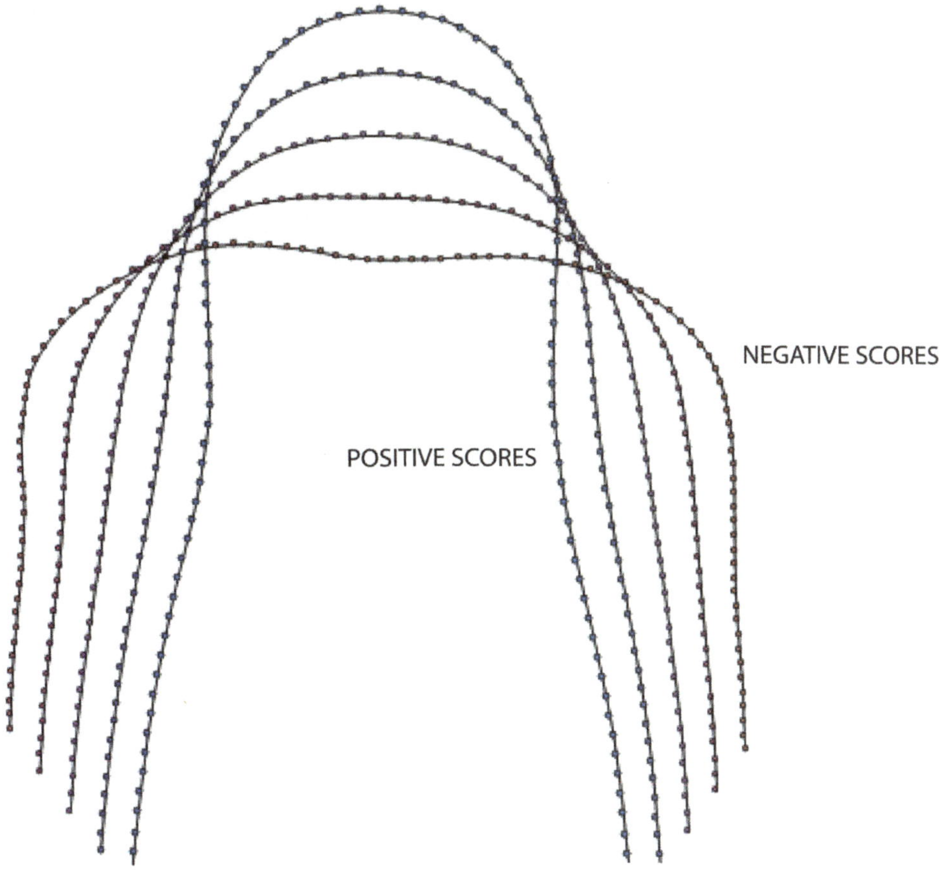

Figure 5. Overlay plot of the CV model axes in PC space for browsers and grazers showing the progressive geometric deformation between modelled axis points. The general profile change is from blunt to pointed, but this excludes some of the subtle profile shape changes.

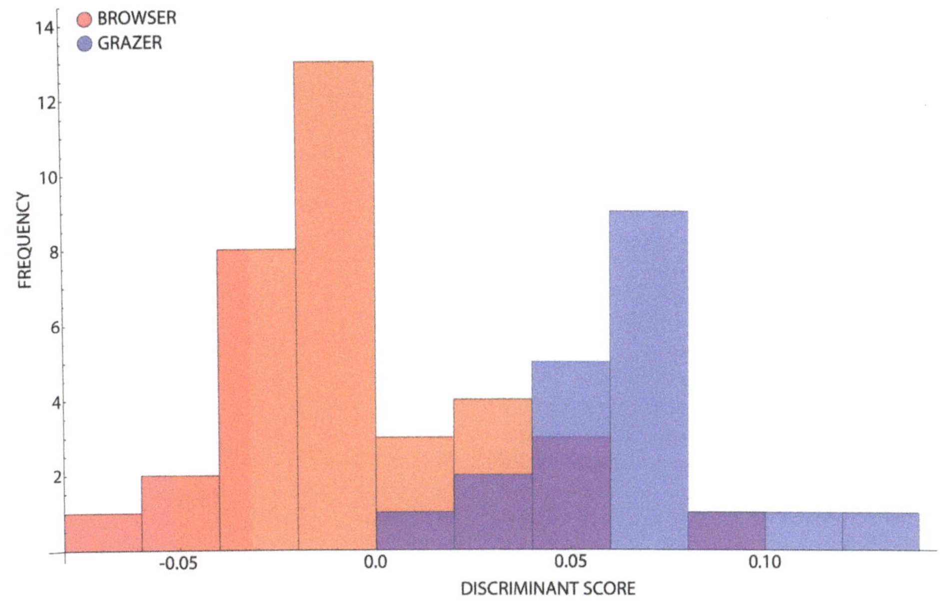

Figure 6. Histogram showing the frequency of occurrences of browser-class and grazer-class species along CV-1, based on a canonical variate decomposition of the phylogenetically controlled principal component scores (PC1, PC2, and PC3). Note that the frequency distributions for each group appear to be superficially taking on that of a normal distribution about different means.

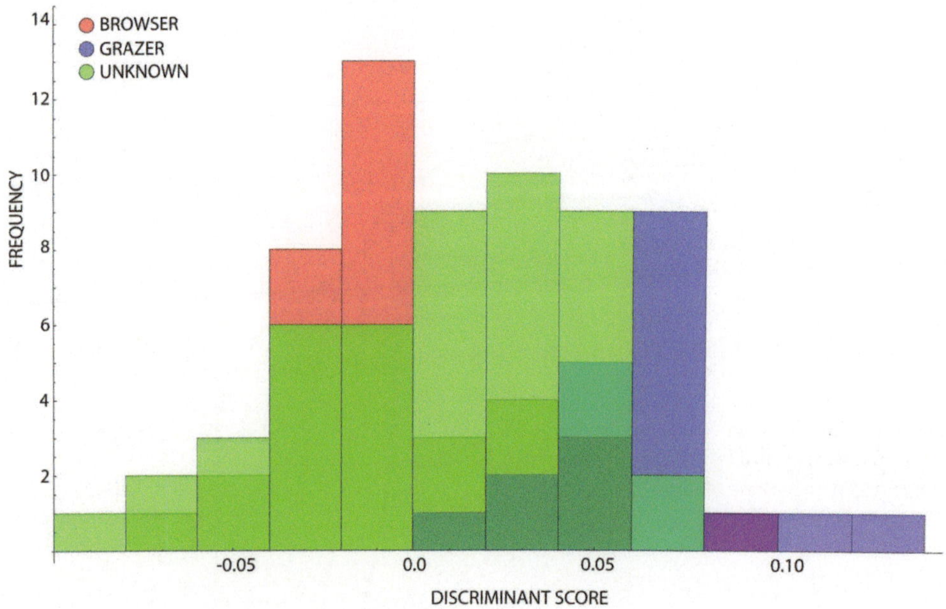

Figure 7. Histogram plot for ruminants classified according to their feeding strategy with 'unknowns' projected into the phylogenetically controlled CV-1 space. Note the high frequency distribution about the grand mean of the browser-grazer defined space.

morphometrics, and so provides a greater quantitative appreciation of the complexity of the shape transformation. It is apparent that ruminants are so morphologically diverse, and have adapted to maximise resource exploitation in their respective ecosystems to such an extent, that they form a continuum of shape variation between 'browser-type' and 'grazer-type' end members. Such a result might be explained by 'constrained divergence', whereby the two groups' rostrum shapes were originally quite similar, but have diverged over evolutionary time.

Previous work assessing the relationship between snout shape and diet (e.g., [36,42]) has largely followed the methodology of Walker [88], using it primarily to aid for inferring the diets for extinct ruminant species. These assessments were based on quantitative interpretation of exemplar taxa, with the method requiring construction of the anterior dorsal snout curve using a cubic spline-fit function to assess intraspecific variation. Typically, authors have selected the 26 degree angle for the whole sample to reconstruct the angle of lateral intersection from the midline, a seemingly arbitrary decision which may not be a consistently strong indicator of snout shape for each specimen. More recent research, such as the comparative study by Fraser and Theodor [43], extended these earlier studies by comparing the utility of numerous snout shape metrics in reconstructing diet, additionally incorporating information from the anterior dentary. We have demonstrated that, using geometric morphometrics, such linear morphometric approaches are largely sufficient in capturing the complexity in shape variation of ruminant snouts. Other authors have identified snout width as a proxy for distal snout shape, with measurements taken at the ventral maxilla-premaxilla intersection on the lateral margin [24,37]. When describing the geometry of complex shapes a single linear metric is usually inadequate as equivalent measurement values can describe completely disparate geometries of varying complexity, and non-comparable function. These authors used this type of measurement, along with the palatal length, to define a 'relative muzzle width ratio', which they used to represent the ratio between body size and the oral aperture, as well as possibly representing oral intake and processing rate. Ratios are poor shape descriptors since all a ratio can represent adequately is an ellipse, if the two measurements represent orthogonal axes, as in the method used by Solounias and Dawson-Saunders [89]. This approach may be sufficient for partially representing extremes of the browser end of the shape spectrum, but can just as easily describe a typically blunt grazing form. The simple fact is that the set of shapes the same ratio can represent is infinite. Hence, ratios can be inappropriate tools for snout shape characterisation (contra [24]). However using linear morphometrics, Fraser and Theodor [43] find a high rate of dietary classification, in particular when the intermediate or mixed feeders are excluded from analyses. The implication of this result is that while outline morphometric methods can describe the complexity of shape variation and confirm the presence (or absence) of shape differences in a more geometric manner, linear distances may be adequate for encapsulating the same amount of ecological information from ruminant snouts, and arguably in a more efficient manner.

The principle focus of this study was to determine whether ruminant snout profile shapes form discrete varieties that covary with feeding strategies (assessed via independent evidence) as had been suggested by numerous previous studies [24,36,37,42]. The corresponding null hypothesis relates to the conclusions of Pérez-Barbéria and Gordon [29], among others, that feeding strategy bears no precise relation to premaxilla morphology. One alternative hypothesis is that the shapes of ruminant premaxilla form a continuum, with characteristic 'browser-type' and 'grazer-type' morphologies comprising end-members, a hypothesis that is increasingly winning support based on a range of detailed investigations. This hypothesis is based on the inference that classifying what are intrinsically morphologically diverse organisms into discrete clusters is problematic and somewhat counter-intuitive, if purely for the purposes of having an antecedent framework onto which new hypotheses of functional morphology can be built. Our results show that, when ruminants are classified ecologically as browsers and grazers based on a range of secondary criteria, they cannot be discriminated completely based on the shape of their premaxillary profile. This result is inconsistent with previous investigations of this issue in which this shape dichotomy was assumed to be absolute based on the sample species used [24,36,37,42]. Premaxilla shape appears to be moderately homoplastic in nature, with a broad range of profile geometries being present in both of the feeding-style sub-groups. Despite exhibiting a degree of shape overlap, these groups retain moderate geometric independence, such that they can be assigned to the correct groups post hoc over 80 percent of the time. While profile-based classification among ruminants is not perfect, it nevertheless has potential to inform studies of fossil ruminants, as it enables quantitative assessment of inferring their ecologies as well as providing a means of quantifying the statistical confidence that can be assigned to these inferences.

The results obtained by our investigations also suggest a possible route of analysis that can be employed in future investigations of functional ecology in ruminants: specifically the use of multivariate ordination analysis combined with tests of statistical confidence to assess the validity of naturally-occurring groups. A similar conclusion was reached by Pérez-Barbería *et al.* [90] that currently accepted boundaries between ruminant feeding strategies remain somewhat arbitrary. One approach to resolving this problem would be to employ a covariate or group of covariates as continuous variables, with thresholds being based on the identification of functionally significant and discrete clusters. However, investigations into this issue so far have found no morphological discrepancies that can explain variation in ruminant digestive efficiency based on digestive, not ingestive, morphology [2,19,32,91]. This perplexing result may, in part, be due to the treatment of species as static entities, when realistically thresholds should be constructed on a sliding scale accounting for population-level ecological, environmental and spatiotemporal variations where appropriate [33,35]. Interpretation of general patterns must also be flexible enough to account for singular exceptions (e.g., frugivores) and are currently insufficient to encapsulate the full diversity of ruminant feeding habits.

Theoretically, a higher food intake rate should covary with the evolution of stronger anatomical structures [92] (e.g., strengthening or fusion of sutures, increased muscle attachment area, decreasing pleurokinesis and increased resistance to strain). This relationship between diet and ecology does not necessarily imply that as snout shape, and hence intake rate, varies, it forces covariation of other morphophysiological parameters. Rather, snout shape constitutes an initial parameter with which other functional domains interact. This morpho-functional relationship was corroborated by Fletcher *et al.* [93], who proposed that the strength of the masticatory apparatus has a functional or adaptational origin, challenging other studies which identified it as being a phylogenetic artefact [29,37,94,95]. This covariation hypothesis requires further investigation, with snout shape being analysed to assess functional significance as a trait affecting both intake rate (volume per unit of time) and selectivity (non-parametric), and plausibly maximum bite size (volume) [96,97]. The results herein imply that snout shape, and thus feeding style, has a strong adaptive component combined with phylogenetic constraint, based on the analysis of a broad range of ruminant species. Finally, our results suggest that major variations in snout shape are related to body size variation, although a directional relationship cannot be established.

Conclusions

Using a two-dimensional representation of the ruminant snout in ventral aspect, we have demonstrated that there is a strong relationship between snout shape and feeding ecology within a highly diverse sample of the major ruminant clades, but only when the data set is restricted to members of the relatively well-defined browser and grazer classes. This between-group discrimination is statistically robust, and supported by recent analyses of the relationship between diet and the shape of the anterior dentary [43]. Snout shape variation is shown to be strongly controlled by phylogenetic similarity, but with this phylogenetic component not affecting the overall dispersal patterns of snout shapes in discriminant space and proportion of successful categorisations. Snout shape variation is also found to be strongly correlated to body size, although this relationship breaks down in a phylogenetically controlled comparison. This corroborates previous hypotheses of relations between feeding style, body size, and ecology, and that while evolutionary similarity is an important component of ecology, snout shape appears to reflect a genuine functional signal.

Based on our results, it is further apparent that previous categorisations, which included putative 'intermediates', or 'unknowns' here, of snout shapes relative to feeding strategy are not fully adequate in their depictions of the full range of exhibited morphological variation (i.e., 'browsers' do not strictly have 'pointed' premaxillae, and 'grazers' do not just have 'blunt' premaxillae as asserted previously by many authors). The geometric complexity of premaxilla morphology is more extensive than this and forms a continuum of shape variation within the modern ruminant fauna. Our results suggest that attempts to place thresholds on other related factors involved in feeding are problematic and quantitative testing is required *a priori* (following the recommendations of Gordon and Illius, [34]).

In light of these results, inferences made by Janis *et al.* [92] - that intake rate forces covariation in the anatomical strength of the mandible - could be explored further to determine the relationship between grazing and browsing ruminants and the relative robustness of the masticatory apparatus. We suggest, in a manner analogous to that of Codron *et al.* [46], that ruminant diets represent a continuum with variation explicitly occurring on a spatiotemporal (geographical and seasonal) scale for all feeding strategies. This requires additional analysis in terms of ruminant phylogenetic affinity, [98,99,100], species' ranges, and functionally significant ecological parameters. Additionally, the role of different ecological categories based on dietary strategies could be explored beyond the traditional browser-intermediate-grazer trichotomy, such as that for African bovids by Gagnon and Chew [101].

The fact that feeding style-based categories were demonstrated to be associated with snout shape in this investigation offers a model for future ecological studies regarding the reconstruction of palaeodiets using a morphometric dataset to delimit and identify extinct browsing and grazing species [36,38]. This aspect of palaeoecology could feasibly be integrated with additional indicators of diet, such as isotopic signatures and microwear in teeth [99,100,102], or the hypsodonty index [103]. Indeed, the incorporation of additional ecological predictors has been demonstrated to increase the accuracy of dietary classification [42].

It is conceivable that our results are the product of a lack of consistency in the ecological definitions of functional feeding groups for ruminants – either in theory or in practice – with respect to other morphophysiological traits. The functional significance of snout shape in relation to bite size, intake rate, body size and selectivity was not addressed explicitly by our investigation. Indeed, our results indicate that closer inspection of these relationships is required. Quantitative metrics describing these ecologically significant parameters should provide a firmer basis for these in (anticipated) future studies [97].

What is undoubtedly necessary in future studies is the dissection of recovered signals to determine what proportion of trait covariation can be explained by phylogenetic relationships [94,95]. Applicable methods include comparative phylogenetic modelling, which has gained increasing interest in the integration of ecology and macroevolution (e.g., [95,104]), and demonstrated in the current study in the context of multivariate statistics. This will facilitate the teasing apart of genuine adaptational signals as opposed to morphological similarity based on common ancestry. Calculation of the rates and direction of snout shape evolution in ruminants, while incorporating fossil data, will be important in elucidating the ecological history of ruminants. Furthermore, if singular or multiple functional traits are found to be phylogenetic artefacts, it may be possible to track the sequence of acquisition, and therefore trace the macroevolutionary and ecological coevolution of ruminants. However, the results obtained here suggest that while phylogeny exhibits a strong control on snout shape in ruminants, it does not affect their ecological classification. Indeed, snout shape and profile-based classification can be explained by a combination of phylogenetic similarity and evolutionary history, body size, and ecology. Finally, in addition to phylogeny, factors such as ontogeny and range size should be scrutinised within a similarly rigorous morphometric-statistical framework to detect potential allometric variation, possible synchronisation of trait acquisition, and evolutionary patterns of character acquisition that might differ between sexes.

Supporting Information

Table S1 Categorical data used for all analyses, PCA eigenvalues, PCA scores, PCA eigenvectors, and body sizes.

Table S2 CVA scores, confusion matrix, jackknifed confusion matrix, browser and grazer distance table, and the unknown projection distance table.

Table S3 Results of the phylogenetically controlled component of this study, including PCA scores, CVA scores, browser and grazer distance table and confusion matrix.

File S1 Raw co-ordinate data for all 125 ruminant species.

File S2 Ruminant phylogeny used for comparative analysis.

Acknowledgments

JT would like to thank Roberto Portela Miguez (Zoology Department, The Natural History Museum, London, UK) and Andrew Crook (Royal Veterinary College, London, UK) for facilitating access to specimens. Marcus Clauss, Danielle Fraser, and one anonymous reviewer provided invaluable assistance on earlier drafts of this manuscript that substantially improved the content.

Author Contributions

Conceived and designed the experiments: JPT NM. Performed the experiments: JPT NM. Analyzed the data: JPT NM. Contributed reagents/materials/analysis tools: JPT NM. Contributed to the writing of the manuscript: JPT NM.

References

1. Fernández MH, Vrba ES (2005) A complete estimate of the phylogenetic relationships in Ruminantia: a dated species-level supertree of the extant ruminants. Biological Reviews 80: 269–302.
2. Clauss M, Lechner-Doll M, Streich WJ (2003) Ruminant diversification as an adaptation to the physiochemical characteristics of forage. A re-evaluation of an old debate and a new hypothesis. Oikos 102: 532–562.
3. Bodmer RE (1990) Ungulate frugivores and the browser-grazer continuum. Oikos 57: 319–325.
4. Mendoza M, Janis CM, Palmqvist P (2002) Characterizing complex craniodental patterns related to feeding behaviour in ungulates: a multivariate approach. Journal of the Zoological Society of London 258: 223–246.
5. Gordon IJ (2003) Browsing and grazing ruminants: are they different beasts? Forest Ecology and Management 181: 13–21.
6. Bourliére F, Hadley M (1970) The ecology of tropical savannahs. Annual Review of Ecology and Systematics 1: 125–152.
7. Gwynne MD, Bell RHV (1968) Selection of vegetation components by grazing ungulates in the Serengeti National Park. Nature 220: 390–393.
8. Fryxell JM (1991) Forage quality and aggregation by large herbivores. The American Naturalist 138(2): 478–498.
9. Voeten MM, Prins HHT (1999) Resource partitioning between sympatric wild and domestic herbivores in the Tarangire region of Tanzania. Oecologia 120: 287–294.
10. Van Zyl JHM (1965) The vegetation of S. A. Lombard Nature Reserve and its utilisation by certain antelopes. Zoologica Africana 1: 55–71.
11. Hofmann RR, Stewart DRM (1972) Grazer or browser: a classification based on the stomach-structure and feeding habits of East African ruminants. Extrait de Mammalia 36(2): 226–240.
12. Hofmann RR (1973) The Ruminant Stomach, Stomach Structure and Feeding Habits of East African Game Ruminants, East African Monographs in Biology, Volume II. East African Literature Bureau, Nairobi, Kenya. 354 p.
13. Hofmann RR (1988) Morphophysiological evolutionary adaptations of the ruminant digestive system. In: Dobson A, Dobson MJ, editors. Aspects of Digestive Physiology in Ruminants, Cornell University Press, Ithaca, New York. pp. 1–19.
14. Hofmann RR (1989) Evolutionary steps of ecophysiological adaptation and diversification of ruminants: a comparative view of their digestive system. Oecologia 78: 443–457.
15. Hofmann RR (1991) Endangered tropical herbivores – their nutritional requirements and habitat demands. In: Ho YW, Wong HK, Abdullah N, Tajuddin ZA, editors. Recent Advances on The Nutrition of Herbivores, Malaysia Society of Animal Production, UPM Serdang. pp. 27–34.
16. Hofmann RR (1999) Functional and comparative digestive system anatomy of Arctic ungulates. Rangifer 20: 71–81.
17. Clauss M, Kaiser T, Hummel J (2008) The morphophysiological adaptations of browsing and grazing mammals, In: Gordon IJ, Prins HHT, editors. The ecology of browsing and grazing, Berlin. pp. 47–88.
18. Jarman PJ (1974) The social organisation of antelope in relation to their ecology. Behaviour 48: 215–267.
19. Janis CM (1995) Correlations between craniodental morphology and feeding behaviour in ungulates: reciprocal illumination between living and fossil taxa. In: Thomason JJ, editor. Functional Morphology in Vertebrate Palaeontology, Cambridge University Press, New York. pp. 76–98.
20. Shipley LA (1999) Grazers and browsers: how digestive morphology affects diet selection. In: Launchbaugh KL, Sanders KD, Mosley JC, editors. Grazing Behaviour of Livestock and Wildlife. Idaho Forest, Wildlife and Range Expeditions Station Bulletin, University of Idaho, Moscow. pp. 20–27.
21. Sanson G (2006) The biomechanics of browsing and grazing. American Journal of Botany 93: 1531–1545.
22. Clauss M, Hume ID, Hummel J (2010) Evolutionary adaptations of ruminants and their potential relevance for modern production systems. Animal 4: 979–992.
23. Clauss M, Hofmann RR, Fickel J, Streich WJ, Hummel J (2009) The intraruminal papillation gradient in wild ruminants of different feeding types: implications for rumen physiology. Journal of Morphology 270: 929–942.
24. Janis CM, Ehrhardt D (1988) Correlation of relative muzzle width and relative incisor width with dietary preference in ungulates. Zoological Journal of the Linnean Society 92: 267–284.
25. Janis CM (1990) Correlation of cranial and dental variables in mammals: a comparison of macropodoids and ungulates. Memoirs of Queensland Museum 281: 349–366.

26. Janis CM, Damuth J, Theodor JM (2000) Miocene ungulates and the terrestrial primary productivity: where have all the browsers gone? Proceedings of the National Academy of Science 97(14): 7899–7904.
27. Pérez-Barbería FJ, Gordon IJ, Nores C (2001) Evolutionary transitions among feeding styles and habitats in ungulates. Evolutionary Biology Research 3: 221–230.
28. Gordon IJ, Illius AW (1996) The nutritional ecology of African ruminants: a reinterpretation. Journal of Animal Ecology 65: 18–28.
29. Pérez-Barbería FJ, Gordon IJ (1999) The functional relationship between feeding type and jaw and cranial morphology in ungulates. Oecologia 118: 157–165.
30. Pérez-Barbería FJ, Gordon IJ, Illius A (2001) Phylogenetic analysis of stomach adaptation in digestive strategies in African ruminants. Oecologia 129: 498–508.
31. Clauss M, Fritz J, Bayer D, Nygren K, Hammer S, et al. (2009) Physical characteristics of rumen contents in four large ruminants of different feeding type, the addax (*Addax nasomaculatus*), bison (*Bison bison*), red deer (*Cervus elaphus*) and moose (*Alces alces*). Comparative Biochemistry and Physiology, Part A 152: 398–406.
32. Sponheimer M, Lee-Thorp JA, DeRuiter D, Smith JM, Van der Merwe NJ, et al. (2003) Diets of Southern African Bovidae: stable isotopic evidence. Journal of Mammalogy 84: 471–479.
33. Owen-Smith N (1997) Distinctive features of the nutritional ecology of browsing versus grazing ruminants. Proceedings of the First International Symposium on Physiology and Ethology of Wild and Zoo Animals 11: 176–191.
34. Gordon IJ, Illius AW (1994) The functional significance of the browser-grazer dichotomy in African ruminants. Oecologia 98: 167–175.
35. Codron D, Clauss M (2010) Rumen physiology constrains diet niche: linking digestive physiology and food selection across wild ruminant species. Canadian Journal of Zoology 88: 1129–1138.
36. Solounias N, Teaford M, Walker A (1988) Interpreting the diet of extinct ruminants: the case of a non-browsing giraffid. Paleobiology 14: 287–300.
37. Gordon IJ, Illius AW (1988) Incisor arcade structure and diet selection in ruminants. Functional Ecology 2: 15–22.
38. Murray MG, Brown D (1993) Niche separation of grazing ungulates in the Serengeti: an experimental test. Journal of Animal Ecology 62: 380–389.
39. Pérez-Barbería FJ, Gordon IJ (1998) The influence of molar occlusal surface area on the voluntary intake, digestion, chewing behaviour and diet selection of red deer (*Cervus elaphus*). Journal of the Zoological Society of London 245: 307–316.
40. Clauss M, Schwarm A, Ortmann S, Streich WJ, Hummel J (2007) A case of non-scaling in mammalian physiology? Body size, digestive capacity, food intake, and ingesta passage in mammalian herbivores. Comparative Biochemistry and Physiology 148: 249–265.
41. Pérez-Barbería F, Gordon IJ (2001) Relationships between oral morphology and feeding style in the Ungulata: a phylogenetically controlled evaluation. Proceedings of the Royal Society of London B 268: 1023–1032.
42. Solounias N, Moelleken SMC (1993) Dietary adaptation of some extinct ruminants determined by premaxillary shape. Journal of Mammalogy 74: 1059–1071.
43. Fraser D, Theodor JM (2011) Anterior dentary shape as an indicator of diet in ruminant artiodactyls. Journal of Vertebrate Paleontology 31: 1366–1375.
44. Robbins TC, Spalinger DE, van Hoven W (1995) Adaptation of ruminants to browse and grass diets: are anatomical-based browser-grazer interpretations valid? Oecologia 103: 208–213.
45. Hofmann RR, Streich WJ, Fickel J, Hummel J, Clauss M (2008) Convergent evolution in feeding types: salivary gland mass differences in wild ruminant species. Journal of Morphology 269: 240–257.
46. Codron D, Lee-Thorp JA, Sponheimer M, Codron J (2007) Nutritional content of savannah plant foods: implications for browser/grazer models of ungulate diversification. European Journal of Wildlife Research 53: 100–111.
47. Du Toit JT (2003) Large herbivores and savannah heterogeneity. In: Du Toit JT, Biggs H, Rogers KH, editors. The Kruger Experience, Island Press. pp. 292–309.
48. Gordon IJ, Illius AW (1996) The nutritional ecology of African ruminants: a reinterpretation. Journal of Animal Ecology 65: 18–28.
49. Clauss M, Nunn C, Fritz J, Hummel J (2009) Evidence for a tradeoff between retention time and chewing efficiency in large mammalian herbivores. Comparative Biochemistry and Physiology, Part A 154: 376–382.
50. Steuer P, Südekum K-H, Müller DWH, Franz R, Kaandorp J, et al. (2011) Is there an influence of body mass on digesta mean retention time in herbivores? Comparative Biochemistry and Physiology, Part A 160: 355–364.
51. Illius AW, Gordon IJ (1992) Modelling the nutritional ecology of ungulate herbivores: evolution of body size and competitive interactions. Oecologia 89: 428–434.
52. Sibly RM, Brown JH (2007) Effects of body size and lifestyle on evolution of mammal life histories. Proceedings of the National Academy of Sciences 104: 17707–17712.
53. Millar JS, Zammuto RM (1983) Life histories of mammals: an analysis of life tables. Ecology 64: 631–635.
54. Isaac NJB, Jones KE, Gittleman JL, Purvis A (2005) Correlates of species richness in mammals: body size, life history, and ecology. The American Naturalist 165: 600–607.
55. Bielby J, Mace GM, Bininda-Emonds ORP, Cardillo M, Gittleman JL, et al. (2007) The fast-slow continuum in mammalian life history: an empirical reevaluation. The American Naturalist 169: 748–757.
56. Du Toit JT, Owen-Smith N (1989) Body size, population metabolism, and habitat specialization among large African herbivores. The American Naturalist 133: 736–740.
57. Sand H, Cederlund G, Danell K (1995) Geographical and latitudinal variation in growth patterns and adult body size of Swedish moose (*Alces alces*). Oecologia 102: 433–442.
58. Pérez-Barbería FJ, Gordon IJ (2000) Differences in body mass and oral morphology between the sexes in the Artiodactyla: evolutionary relationships with sexual segregation. Evolutionary Ecology Research 2: 667–684.
59. Fritz J, Hummel J, Kienzle E, Arnold C, Nunn C, et al. (2009) Comparative chewing efficiency in mammalian herbivores. Oikos 118: 1623–1632.
60. Pérez-Barbería FJ, Elston DA, Gordon IJ, Illius AW (2004) The evolution of phylogenetic differences in the efficiency of digestion in ruminants. Proceedings of the Royal Society of London B 271: 1081–1090.
61. Raia P, Carotenuto F, Eronen JT, Fortelius M (2011) Longer in the tooth, shorter in the record? The evolutionary correlates of hypsodonty in Neogene ruminants. Proceedings of the Royal Society of London B 278: 3474–3481.
62. Sieg AE, O'Connor MP, McNair JN, Grant BW, Agosta SJ, et al. (2009) Mammalian metabolic allometry: do intraspecific variation, phylogeny, and regression models matter? The American Naturalist 174: 720–733.
63. Martins EP, Hansen TF (1997) Phylogenies and the comparative method: a general approach to incorporating phylogenetic information into the analysis of interspecific data. The American Naturalist 149: 646–667.
64. Polly DP, Lawing AW, Fabre A-C, Goswami A (2011) Phylogenetic principal components analysis and geometric morphometrics. Hystrix, the Italian Journal of Mammalogy 24: 33–41.
65. Klingenberg CP, Gidaszewski NA (2010) Testing and quantifying phylogenetic signals and homoplasy in morphometric data. Systematic Biology 59: 245–261.
66. Klingenberg CP, Marugán-Lobón J (2013) Evolutionary covariation in geometric morphometric data: analyzing integration, modularity and allometry in a phylogenetic context. Systematic Biology 62: 591–610.
67. Mitteroecker P, Gunz P (2009) Advances in geometric morphometrics. Evolutionary Biology 36: 235–247.
68. Adams DC, Rohlf FJ, Slice DE (2004) Geometric morphometrics: ten years of progress following the 'revolution'. Italian Journal of Zoology 71: 5–16.
69. MacLeod N (2013) Landmarks and semilandmarks: differences without meaning and meaning without difference. Palaeontological Association Newsletter 82: 32–43.
70. Rohlf FJ, Slice D (1990) Extensions of the *Procrustes* method for the optimal superimposition of landmarks. Systematic Biology 39: 40–59.
71. Klingenberg CP (2010) Evolution and development of shape: integrating quantitative approaches. Nature Reviews: Genetics 11: 623–635.
72. MacLeod N (2005) Principal components analysis (eigenanalysis & regression 5). Palaeontological Association Newsletter 59: 44–57.
73. MacLeod N (2007) Groups II. Palaeontological Association Newsletter 65: 36–49.
74. Campbell NA, Atchley WR (1981) The geometry of canonical variates analysis. Systematic Zoology 30: 268–280.
75. MacLeod N (2007) Groups I. Palaeontological Association Newsletter 64: 35–45.
76. Satorra A, Saris WE (1985) Power of the likelihood ratio test in covariance structure analysis. Psychometrika 50: 83–90.
77. Mitteroecker P, Bookstein F (2011) Linear discrimination, ordination, and the visualization of selection gradients in modern morphometrics, Evolutionary Biology 38: 100–114.
78. MacLeod N (2013) A picture may be worth 1,000 landmarks, Palaeontological Association Newsletter 84: 20–34.
79. MacLeod N (2009) Form and shape models. Palaeontological Association Newsletter 72: 14–27.
80. Zhang W-Q, Zhang M-H (2012) Phylogeny and evolution of Cervidae based on complete mitochondrial genomes. Genetics and Molecular Research 11: 628–635.
81. Blomberg SP, Garland T Jr, Ives AR (2003) Testing for phylogenetic signal in comparative data: behavioral traits are more labile. Evolution 57: 717: 745.
82. Pagel M (1999) Inferring the historical patterns of biological evolution. Nature 401: 877–884.
83. Bibi F (2013) A multi-calibrated mitochondrial phylogeny of extant Bovidae (Artiodactyla, Ruminantia) and the importance of the fossil record to systematics. BMC Evolutionary Biology 13: 166.
84. Bininda-Emonds ORP, Cardillo M, Jones KE, MacPhee RDE, Beck RMD, et al. (2007) The delayed rise of present-day mammals. Nature 446: 507–512.
85. Revell LJ (2009) Size-correction and principal components ofr interspecific comparative studies. Evolution 63: 3258–3268.
86. Jones KE, Bielby J, Cardillo M, Fritza SA, O'Dell J, et al. (2009) PanTHERIA: a species-level database of life history, ecology, and geography of extant and recently extinct mammals. Ecology 90: 2648.
87. R Development Core Team (2008) R: A language and environment for statistical computing. R Foundation for Statistical Computing, Vienna, Austria. ISBN 3-900051-07-0. Available: http://www.R-project.org. Accessed 2014 Oct 15.

88. Walker AC (1984) Extinction in hominid evolution. In: Nitecki MH editor. Extinctions, University of Chicago Press, Chicago. pp. 119–152.
89. Solounias N, Dawson-Saunders B (1988) Dietary adaptations and paleoecology of the Late Miocene ruminants from Pikermi and Samos in Greece. Palaeogeography, Palaeoclimatology, Palaeoecology 65: 149–172.
90. Pérez-Barbería FJ, Elston DA, Gordon IJ, Illius AW (2004) The evolution of phylogenetic differences in the efficiency of digestion in ruminants. Proceedings of the Royal Society of London B 271: 1081–1090.
91. Clauss M, Hofmann RR, Streich WJ, Fickel J, Hummel J (2008) Higher masseter muscle mass in grazing than browsing ruminants. Oecologia 157: 377–385.
92. Janis CM, Constable EC, Houpt KA, Streich WJ, Clauss M (2010) Comparative ingestive mastication in domestic horses and cattle: a pilot investigation. Journal of Animal Physiology and Animal Nutrition 94: 402–409.
93. Fletcher TM, Janis CM, Rayfield EJ (2010) Finite element analysis of ungulate jaws: can mode of digestive physiology be determined? Palaeontologia Electronica 13: 1–15.
94. Figuerido B, Serrano-Alarcón FJ, Slater GJ, Palmqvist P (2010) Shape at the crossroads: homoplasy and history in the evolution of the carnivoran skull towards herbivory. Journal of Evolutionary Biology 23: 2579–2594.
95. Raia P, Carotenuto F, Meloro C, Piras P, Pushkina D (2009) The shape of contention: adaptation, history, and contingency in ungulate mandibles. Evolution 64: 1489–1503.
96. Shipley LA, Gross JE, Spalinger DE, Hobbs NT, Wunder BA (1994) The scaling of intake rate in mammalian herbivores. The American Naturalist 143: 1055–1082.
97. Gordon IJ, Illius AW, Milne JD (1996) Sources of variation in the foraging efficiency of grazing ruminants. Functional Ecology 10: 219–226.
98. Clauss M, Hummel J (2005) The digestive performance of mammalian herbivores: why big may not be that much better. Mammal Review 35: 174–187.
99. Cooper N, Purvis A (2010) Body size evolution in mammals: complexity in tempo and mode. The American Naturalist 175: 727–738.
100. MacFadden J (2000) Mammalian herbivores from the Americas: reconstructing ancient diets and terrestrial communities. Annual Review of Ecology and Systematics 31: 33–59.
101. Gagnon M, Chew AE (2000) Dietary preferences in extant African Bovidae. Journal of Mammalogy 81: 490–511.
102. Codron D, Brink JS, Rossouw L, Clauss M (2008) The evolution of ecological specialization in southern African ungulates: competition- or physical environment turnover? Oikos 117: 344–353.
103. Mendoza M, Palmqvist P (2008) Hypsodonty in ungulates: an adaptation for grass consumption or foraging in open habitats. Journal of Zoology 274: 134–142.
104. Cardillo M, Gittleman JL, Purvis A (2008) Global patterns in the phylogenetic structure of island mammal assemblages. Proceedings of the Royal Society B 275: 1549–1556.

22

Growth-Mortality Relationships in Piñon Pine (*Pinus edulis*) during Severe Droughts of the Past Century: Shifting Processes in Space and Time

Alison K. Macalady[1]*, Harald Bugmann[2]

1 University of Arizona, School of Geography and Development and Laboratory of Tree-Ring Research, Tucson, Arizona, United States of America, **2** Forest Ecology, Department of Environmental Systems Science, ETH Zürich, Zürich, Switzerland

Abstract

The processes leading to drought-associated tree mortality are poorly understood, particularly long-term predisposing factors, memory effects, and variability in mortality processes and thresholds in space and time. We use tree rings from four sites to investigate *Pinus edulis* mortality during two drought periods in the southwestern USA. We draw on recent sampling and archived collections to (1) analyze *P. edulis* growth patterns and mortality during the 1950s and 2000s droughts; (2) determine the influence of climate and competition on growth in trees that died and survived; and (3) derive regression models of growth-mortality risk and evaluate their performance across space and time. Recent growth was 53% higher in surviving vs. dying trees, with some sites exhibiting decades-long growth divergences associated with previous drought. Differential growth response to climate partly explained growth differences between live and dead trees, with responses wet/cool conditions most influencing eventual tree status. Competition constrained tree growth, and reduced trees' ability to respond to favorable climate. The best predictors in growth-mortality models included long-term (15–30 year) average growth rate combined with a metric of growth variability and the number of abrupt growth increases over 15 and 10 years, respectively. The most parsimonious models had high discriminatory power (ROC>0.84) and correctly classified ~70% of trees, suggesting that aspects of tree growth, especially over decades, can be powerful predictors of widespread drought-associated die-off. However, model discrimination varied across sites and drought events. Weaker growth-mortality relationships and higher growth at lower survival probabilities for some sites during the 2000s event suggest a shift in mortality processes from longer-term growth-related constraints to shorter-term processes, such as rapid metabolic decline even in vigorous trees due to acute drought stress, and/or increases in the attack rate of both chronically stressed and more vigorous trees by bark beetles.

Editor: Hormoz BassiriRad, University of Illinois at Chicago, United States of America

Funding: AKM was supported by a US Department of Energy Graduate Research Environmental Fellowship (GREF) and the GK-12 Graduate Fellows Program of the National Science Foundation. HB was supported by a Haury Fellowship from the University of Arizona, Laboratory of Tree-Ring Research. This work is also a contribution to the Swiss National Center of Competence in Research (NCCR) 'Climate', funded by the Swiss National Science Foundation. The funders had no role in study design, data collection and analysis, decision to publish, or preparation of the manuscript.

Competing Interests: The authors have declared that no competing interests exist.

* E-mail: amacalad@email.arizona.edu

Introduction

Climate-related tree mortality has been documented in forests around the world and may be intensifying in some regions due to rising temperature and enhanced drought [1–6]. Forests play an important role in regulating the earth's energy, carbon and water cycles [7], and increases in tree mortality rates or rapid collapses in forest cover could have major implications for ecosystems and ecosystem services [8–10]. Yet, our ability to predict future forest dynamics is limited by gaps in our understanding of tree death, and associated uncertainty about how represent tree mortality in vegetation models [6,11–15].

The processes underlying drought-associated tree mortality are particularly unclear [16–20]. Leading hypotheses suggest that tree mortality may arise from two interrelated mechanisms: (1) carbon starvation, i.e., water stress causes trees to close their stomata, thus reducing photosynthesis and constraining the availability of carbon necessary for maintaining metabolic functions or fending off insects and pathogens; and (2) hydraulic failure, i.e., the collapse of the water-conducting system in the xylem [16,21]. Much recent drought-mortality research has built on this framework in an experimental context to detect failure points within the linked hydraulic, photosynthetic and carbon transport systems of trees subject to drought, with the aim of clarifying mechanisms and identifying physiological and climatological thresholds beyond which death occurs (cf. [20,22,23]). However, our understanding of tree mortality remains inadequate for projecting the impacts of climate change on forests [6]. An important set of knowledge gaps relates to understanding *variability* in interacting drought-mortality processes and thresholds across landscapes and through time, and determining the time scales that are most important for understanding mortality risk, including the influence of previous events on the status and resistance of trees [22,24,25].

Tree ring studies offer an excellent way to complement more physiologically detailed but temporally and spatially limited experimental studies of drought-associated tree mortality. Radial stem growth is, at least in the short term, a low priority for a tree's allocation of available carbon [26,27]. Thus tree growth is often a sensitive indicator of changes in a tree's carbon balance due to environmental or tree intrinsic factors. In the context of drought-associated mortality, for example, if trees close their stomata to reduce the risk of desiccation and hydraulic failure, carbon uptake and presumably radial growth are reduced. Hypothesized relationships between growth, a tree's carbon budget, and mortality are reflected by mortality algorithms in many forest models, where recent (1–3 year) growth is a basis for determining the risk of death in a given time-step [11,15]. Tree rings have been used to test and improve the empirical basis for such algorithms (cf. [28–32]). However, these studies have focused mostly on sporadic, individual-tree mortality, rather than on tree die-off associated with severe droughts. Furthermore, a paucity of long-term datasets has hindered the evaluation of the temporal and spatial generality of relationships between growth and the likelihood of death, although stable relationships are critical to the incorporation of growth-mortality algorithms in dynamic vegetation models [24,33]. A number of studies have explored growth rates in trees that eventually died during drought to test hypotheses about the physiological mechanisms of mortality [34–43], but to our knowledge no studies have quantitatively assessed the importance of multiple growth variables and time scales for shaping mortality risk during drought-related die-off.

In the semi-arid conifer forests of western North America, prolonged drought and heat have interacted with bark beetles to produce mortality across millions of hectares during the last 20 years [5,44–46]. In the piñon-juniper woodlands of the southwestern USA (SW), drought and the activity of bark beetles led to the widespread death of piñon pines (*Pinus edulis* Engelm. and *P. monophylla* Torr. & Frem) [5,44,47]. It has been suggested that several aspects of the 2000s drought and associated die-off were novel, and that unusually warm conditions caused elevated mortality rates that anticipate 'global change' conditions [5,44,48,49]. However, tree mortality was also widespread in the SW during the 1950s [50,51]. Although cooler than the 2000s drought across much of New Mexico [49], instrumental and tree-ring based records indicate that the 1950s drought was one of the most severe and protracted of the past 500 years [5]. Exceptional preservation of long-dead trees in these landscapes provides a unique opportunity to compare tree-ring growth and mortality patterns during the two mortality episodes, allowing not only for an evaluation of long- and short-term factors operating during widespread drought-associated die-off, but also an assessment of the generality of mortality processes and growth-mortality relationships during droughts with distinct climatic patterns.

In this study we draw on recent and archived tree-ring collections and stand structural measurements from sites spanning a latitudinal gradient in New Mexico, USA, to investigate drought-related mortality of piñon pine (*P. edulis* Engelm.). Specifically, we (1) analyze *P. edulis* growth patterns and mortality during the 1950s and 2000s droughts; (2) determine the influence of climate and competition on growth in trees that died vs. those that survived; and (3) derive regression models of growth-mortality risk and evaluate their performance across space and time.

Materials and Methods

Study Areas

We assessed *P. edulis* mortality close to lower treeline at four study areas that span gradients in climate and stand composition in New Mexico (Tables 1, 2; Fig. S1). Each area (TRP, BNM, WRK, SEV) is located on relatively gentle terrain at the middle-to-high end of the local elevation range of piñon, where the species co-occurs with one or more *Juniperus* species (*J. monosperma* and *J. scopulorum*). We refer to the study areas by acronym, and add 2000 or 1950 after each to distinguish between the sampling representing the two drought-mortality events.

Ethics Statement and Data Availability

This research was performed at the Sevilleta National Wildlife Refuge, Bandelier National Monument and the Carson National Forest. Necessary permissions were obtained from the National Park Service, National Fish and Wildlife Service and National Forest Service. All data are available upon request.

Field Sampling and Laboratory Methods

2000s mortality. In 2008–2010, we selected 10–30 recently dead (bark and fine branches remaining) "target" trees at each of the four study areas (Tables 1, 2). Living trees were selected as "control cases" to compare with each dead tree based on proximity, similarity in micro-topography, tree diameter and overall stature [29,31]. Mortality processes can differ between adult and juvenile trees [52]; we selected trees greater than 9cm diameter at root collar (DRC) in order to focus our study on the mortality of mature trees [53]. Very few mature piñon trees were alive at BNM in 2010, thus 30 dead target trees were selected at this site for comparison to live and dead trees at the other sites.

Two increment cores were extracted from each living target tree at breast height (135 cm), and a cross-section was taken from each paired dead tree. In the laboratory, cores and cross-sections were prepared, crossdated and ring widths were measured using standard dendrochronological techniques [54]. We confirmed visual crossdating statistically using the computer program COFECHA [55]. In total, 167 ($n = 98$ dead and $n = 69$ live) target trees were sampled and successfully crossdated. One live target tree at TRP and two dead trees at BNM could not be crossdated, and were dropped from the study. For the characterization of stand structure and spatial patterns of mortality, we measured the diameter and noted the status (live or dead) and species of each tree (>1 cm DBH) within a 7.5 m radius plot centered on each live and dead target tree.

1950s mortality. SEV and BNM are associated with a previous study for which long-dead trees were measured and sampled along with living neighbors (Allen, Betancourt and Swetnam, *unpublished data*). All living trees, snags and downed remnants within two 0.5 ha plots at BNM and SEV were measured, and each piñon tree was sampled for dendroecological purposes. These 0.5 ha plots fall within the larger sampling areas for each site described above. Each core and cross section was prepared and visually crossdated as above in order to determine inside and outside ring dates for each tree. To investigate growth-mortality relationships associated with the 1950s drought, we selected target trees from the archived specimens of this sampling campaign (all housed at the Laboratory of Tree-Ring Research, University of Arizona). We identified dead specimens with good preservation (e.g. where the last year of radial growth could reliably be determined) that were ≥9 cm DRC and had outer ring dates between 1940 and 1960. By many definitions, the 1950s drought actually stretched from the mid-1940s through the beginning of the 1960s [5], and so we allowed selection of trees with outer ring dates slightly preceding the first significant drought year in the 1940s (1946). Dead tree samples were measured and checked for dating errors as described above. We then identified and measured cores or cross-sections from trees ≥9 cm DRC that survived through the early 1960s. Growth increments from these

Table 1. Location, characteristics and sample sizes for each study site.

Study Area	Location	Elevation (m)	Annual Precip. (mm)	Annual Temp. (C°)	Mortality event	Site Acronym	Live Target Trees	Dead Target Trees
Tres Piedras, NM	36.34°N 105.93°W	2100	312	8.7	2000s	TRP2000	29	30
White Rock, NM	35.81°N 106.24°W	1990	340	10.1	2000s	WRK2000	10	10
Bandelier Natl. Monument, NM	35.76°N 106.27°W	1940	340	10.1	2000s	BNM2000	0	28
					1950s	BNM1950	22	23
Sevilleta Natl. Wildlife Refuge, NM	34.34°N 106.55°W	2050	387	10.6	2000s	SEV2000	30	30
					1950s	SEV1950	27	26

samples were used to estimate tree diameter in the year 1960. Trees from this pool were matched as much as was possible to dead "case" trees based on the 1960-diameter estimate and 0.5 ha study plot. The pool of suitable survivor trees was limited by the fact that larger trees appear to have been preferentially killed during the 1950s drought at these sites. Ultimately we selected 25 dead and 26 surviving target trees for SEV1950, and 23 dead and 22 surviving trees for BNM1950, for a total of 96 trees, with the 1950s dataset containing slightly larger dead than surviving trees (Table 2). However, all trees met our size criteria (DRC ≥9 cm) and were estimated to be at least 65 years old by the 1950s drought.

Target Tree Characteristics

Outside ring dates were recorded as the best-available approximation of the death date for each dead tree. Direct observations of dying trees at BNM2000 suggest that tree-ring estimates and actual death dates agree within a year or two (C. Allen, *personal communication*). Many insects and diseases are known to affect piñon pine, but *Ips confusus* LeConte – the piñon *Ips* bark beetle – has been associated with the most severe damage, and is known to attack both living and recently dead trees [56,57]. We thus documented evidence of *Ips* attack for each dead target tree by noting whether an individual contained *Ips* beetle galleries on the sampled portion of the tree bole and/or whether the sample contained blue stain – a fungal pathogen introduced by bark beetles – in the sapwood [58].

Woodland Structure and Spatial Patterns of Mortality

We calculated density and basal area of live and dead trees within neighborhood plots and used quasi-binomial regression to assess relationships between woodland structure and mortality severity (the percentage of recently dead trees) during the 2000s mortality event. We assessed fine-scale spatial patterning of recent mortality by testing for differences in stand composition and mortality-severity around live versus dead target trees.

It was not possible to make the same assessments of woodland structure prior to the 1950s mortality episode because of unknown tree locations within the 0.5 ha plots, but we made conservative estimates of piñon mortality severity using the dendrochronologically determined birth and death dates of the trees sampled on the 0.5 ha plots (cf. above).

The rings of juniper trees at our sites cannot be reliably crossdated due to many false and missing rings and lack of circuit uniformity, and therefore no assessments of juniper size, structure and mortality were made.

Growth and Growth Indices

We calculated basal area increments (BAI), relative basal area increments (RelBAI) and ring width indices (RWI) from raw ring-widths (RW) for use in subsequent analyses. Basal area increments ($cm^2\ yr^{-1}$) were calculated from ring widths ($mm\ yr^{-1}$) for each tree radius using the inside-out method:

$$BAI_t = \frac{\pi * \left(\sum_{t}^{i} RW + d\right)^2 - \pi * \left(\sum_{t-1}^{i} RW + d\right)^2}{100} \quad (1)$$

where d is an estimate of the distance from the first measured ring to the pith [59], i is the first year of growth in the time series, and t is the current year of growth.

RelBAI ($cm^2\ yr^{-1}\ yr^{-1}$) was calculated by dividing basal area increments for each year by the previous year's total basal area

Table 2. Mortality severity and stand characteristics for each study site.

Site	Piñon mortality (%)	Juniper Mortality (%)	Tree Density (stems/ha)	Piñon Density (stems/ha)	Total Basal Area (m^2/ha)	Piñon Basal Area (m^2/ha)
TRP2000	64.0 (4.1)	2.9 (1.7)	426.8 (36.8)	333.8 (36.9)	8.8 (0.9)	6.1 (0.7)
WRK2000	82.1 (7.5)	0.2 (0.2)	605.5 (78.1)	206.6 (34.4)	7.0 (1.1)	3.0 (0.6)
BNM2000	99.6 (0.4)	2.6 (1.0)	808.4 (71.5)	282.9 (41.0)	9.7 (1.1)	3.5 (0.7)
SEV2000	19.9 (3.2)	3.1 (1.0)	726.2 (49.0)	331.1 (31.3)	11.2 (0.8)	3.1 (0.3)
BNM1950	64.5 (7.5)	-	-	262.1 (26.5)	-	-
SEV1950	46.5 (8.5)	-	-	432.0 (61.0)	-	-

Standard errors for each measurement are indicated in parentheses. For 2000s sites, the mean and standard error were calculated based on 7.5m neighborhood plot data. For 1950s sites, the mean and standard error are from measurements at two separate 0.5-hectare (ha) plots. Tree density and basal area reflect pre-mortality conditions. No estimates of juniper mortality, total tree density or basal area were made for the 1950s sites due to lack of dendroecological data for juniper and lack of tree size reconstructions for all trees within each plot.

[60]. We utilized the C-method to generate ring-width indices [61]. This method transforms ring widths by dividing individual series by a curve that reflects the biological expectation of constant annual basal area increment for each tree. The C-method thus standardizes individual ring width series to a common mean and variance, but unlike other standardization methods, it allows individual index series to retain low-frequency variability and trends due to, for example, injury, senescence, competition, and climatic influences. Measurements from multiple radii were averaged to generate single records of RW, BAI, RelBAI and RWI for each tree.

Although BAI is often considered to be a more biologically meaningful growth metric than raw ring-widths or ring-width indices [61,29,62,39], we utilized RW for building models of growth-based mortality risk and assessing growth relationships to climate and competition, and RWI for the calculation of average growth chronologies. RWI was used in chronologies in order to minimize the influence of particular trees with high mean growth and variance, and to highlight changes in the *trajectory* versus the average growth rate in dead versus surviving trees. We utilized RW in quantifying growth-mortality relationships because, unlike RWI, it retains gross differences in growth rates between live and dead trees, and contrary to expectation, we found less pronounced size-related trends in piñon RW compared to BAI and RelBAI (Fig. S2). Although data from our neighborhood plots indicate that tree size was likely a predisposing factor for mortality at our 2000s sites (Fig. S3), our study design is not well-suited to quantify the combined influence of tree size and growth on mortality risk, because (1) the 1950s dataset is slightly biased towards larger dead trees due to preservation issues; and (2) average sampled tree size (and age) was not stratified between study sites (Table 2). Thus even though tree size and age matched relatively well between living and dead trees and target trees shared similar dominant or co-dominant status, we sought to choose a growth metric that is least sensitive to tree size in order to make a conservative estimation of the growth-mortality relationships. For the sake of comparison, we also analyzed basic growth differences between live and dead trees using all metrics, and generated growth-mortality models using BAI, RelBAI and RWI as outlined below. These models had slightly different predictor variables and performance, but did not lead to different conclusions and thus are not shown.

We generated four types of indices from growth time series to develop a pool of predictor variables of mortality risk: average growth, growth variability, growth trend and the frequency of abrupt growth changes. Average growth [16,26] was calculated as the mean of annual growth measurements over $k=3$, 5, 7, 10, 15…50 years. Growth variance has been documented as a factor that influences predisposition to mortality in semi-arid woodlands [35,36]. We chose mean sensitivity, a statistic of year-to-year growth variability that reflects both the variance and the first-order autocorrelation of the time series [63], because compared to standard deviation (Table 2) or first-order autocorrelation (not shown) it differed more strongly between living and dead trees. Mean sensitivity was calculated from growth time series according to Eq. 2 in Biondi and Quedan [64], where k = the length of the tree-ring series $t=1,2,\ldots,k$ = year in the tree-ring series:

$$MS=\frac{k}{k-1}\frac{\sum_{t=2}^{k}|RW_t-RW_{t-1}|}{\sum_{t=1}^{k}RW_t} \quad (2)$$

Growth trend was calculated as the slope of the linear regression of growth fitted across all years within different time intervals [29]. Both mean sensitivity and growth trend were calculated over $k=5$, 10, 15…50 years.

Several studies have noted a preponderance of abrupt growth declines in trees predisposed to die [31,34], and we also noted differences in the frequency of growth releases in our dataset. We defined abrupt growth changes as 50% reductions or increases in growth averaged over a 10 year period as compared to average growth over the previous 10 years. We counted the number of such changes over $k=5$, 10, 15…50 years.

For the calculation of all predictor variables, live tree growth was truncated at the last year of the corresponding dead tree pair.

Mortality Modeling

We used linear mixed effects logistic regression to relate growth indices to tree status (live or dead) [65,66]. We followed the general procedure of Das et al. [31], modified for mixed effects modeling, to identify the most parsimonious model structure. We designated study site/period as a random effect. For the fixed effects, we: (1) generated models with only one covariate using a temporal range of growth indices from each growth category (average growth, growth trend, mean sensitivity, and abrupt growth changes). We did not include indices after 35 years because multiple trees with shorter crossdated growth records dropped out of the predictor pool after this point, making it difficult to compare models; (2) used Akaike's Information Criterion (AIC) to compare the support for each model [67]; and (3) created a suite of models with multiple fixed effect predictors using the variables from the three single-variable models with the lowest AIC score in each growth category and/or with differences in AIC scores below 2 [67]. Independent variables were transformed if Wald tests indicated non-linear relationships [68]. Random effects were dropped from each model if AIC scores and likelihood ratio tests on nested models indicated that a simpler model structure was more parsimonious [69]. The three models with the lowest AIC scores overall are presented along with the best-ranked single variable models for comparison.

Model Diagnostics, Validation and Interpretation

We computed a variety of diagnostic and validation statistics to aid in the interpretation of best-ranked logistic models. Correct classification rates were calculated from confusion matrices generated by a bootstrapped internal validation routine (1000 iterations) in which models were fit repeatedly with a random sub-sample containing 60% of the data and validated on the remaining 40% [29]. Trees were classified as living if their survival probability was above an empirically determined threshold [70]. We also externally validated best-ranked models on the dataset from BNM2000, which contained only dead trees, did not match the case-control study design, and thus could not be used in model building. However we expected that the best models would correctly classify the majority of dead trees at BNM2000 if mortality processes and thresholds were similar here compared with other sites. The Area Under the Receiver Operating Characteristic curve (ROC) is a threshold-independent measure of model discrimination, where a value of 0.5 suggests no discrimination and values above 0.8 suggest excellent discrimination between live and dead trees [71]. Odds ratios were calculated from regression coefficients to assess changes in relative mortality risk associated with changes in growth. Odds ratios indicate a *change* in the likelihood of mortality given a meaningful change in the predictor variables. For example, an odds ratio of 2.0 associated with a 0.1 mm increase in average growth can be interpreted as a doubling of the likelihood of survival with each 0.1 mm growth increase, all else being equal.

Effects of Climate and Competition on Tree Growth and Mortality

We fit a separate set of linear mixed-effect models to make *post-hoc* assessments of how two factors - climate and competition - influenced growth in trees destined to die and survive drought-mortality events [41]. For target trees at TRP, WRK and SEV, we modeled ring width as a function of cool season precipitation (previous September through May) (PPT_{cool}), early summer (May–July) average vapor pressure deficit (VPD_{MJJ}), a continuous index of competitive pressure (CI), and a categorical variable representing tree status (Live or Dead). The seasonal climate variables were chosen based on initial comparisons between mean-value chronologies and PRISM climate model output (4 km resolution) [72] (Text S1). PRISM data were chosen instead of weather station data because PRISM data better explained the variability in tree growth chronologies (not shown).

Indices of competitive pressure (CI) were derived from the neighborhood plot data (cf. above). A distance- and size-weighted index, calculated using only conspecific (e.g. piñon) neighbors was used for modeling growth, as it yielded the strongest and most consistent correlations with recent growth:

$$CI = \sum_{j=1}^{n} \frac{DBH_j^2 / DBH_i^2}{DIST_{ij}} \tag{3}$$

where $j = 1,\ldots,n$ are competitor trees, i is the target tree, DBH is tree diameter and DIST is the distance between target and competitor trees [29]. For analyses of the 1950s sites, competition was not included as a covariate because neighborhood data were not available. Growth, climate and competition variables were converted to z-scores specific to site and the modeled period.

Model fitting reflected our goals of assessing the effects of climate and competition on tree growth, and testing whether trees destined to die and survive responded differently to climate and competition in the years leading up to mortality. First, through exploratory analyses, we noted that the relationship between ring width and climate is sometimes slightly curvilinear, and models without higher-order climate terms contained highly skewed residuals. Thus, following [69] we started with a 'beyond-optimal' model that included as fixed effects both linear and quadratic climate variables along with tree status, competition, and site. Because of the potential complexity of interactions in a model with a high number of predictors, we restricted the 'beyond-optimal' model to 2 and 3-way interactions that represented interpretable biological processes [69]. We did not include interactions between climate variables, as our initial analyses indicated that they were small. Interactions between quadratic climate terms and other predictors were not included, either. Tree ID was added as a random effect to account for non-independence of growth within individual trees. More complicated random effects structures (for example, allowing different growth trends over years for different trees, or nesting Tree ID within site) were rejected, as they did not improve the full model. Residual autocorrelation of growth between years and heterogeneity of variance in residuals were accounted for by adding model correlation structures and variance weights [69,65,41].

We used an iterative, AIC-based backwards selection to sequentially drop terms from the 'beyond-optimal' model [69]. We developed a final model structure using growth data that started in 1960 and 1910 for the 2000s and 1950s datasets, respectively. We retained this structure for models of growth over different time periods to allow for the straightforward comparison of coefficients. Residuals were inspected to ensure that assumptions about residual independence, heterogeneity and normality were adequately met. Final models were fit using the restricted maximum likelihood criteria. Growth data from BNM2000 were not included in model fitting, as there were no living trees.

Unless otherwise noted, all statistical tests were performed in R v3.0.1 [73]. We used the *dplR* library v1.5.6 for BAI and RWI calculations and dendrochronology statistics [74]. Generalized linear mixed modeling was performed using functions from the *lme4* library v0.99999911-6 [75]. Linear mixed models of tree growth were fit using the library *nlme* v3.1-110 [65].

Results

Spatial and Temporal Mortality Patterns

The severity of piñon mortality during the recent (2000s) drought ranged from 20% to 99% across our study sites, and was least severe at SEV2000 (Table 2). Piñon mortality associated with drought in the 1940s and 1950s was also severe at both SEV1950 and BNM1950 (45%–65%), with slightly higher mortality at BNM1950 (Table 2). These measurements are based on 7.5 m neighborhood plot data around each target tree (2000s drought) or dead and live trees in 0.5 ha plots (1950s drought), and thus reflect mortality severity at our study sites only. However, for the 2000s drought the patterns in our data conform to other published studies with more extensive sampling, in which the 2000s mortality was found to be greater in northern versus south-central New Mexico [76].

Outside ring dates correspond well with periods of decadal drought, but mortality was more or less synchronous depending on site and mortality episode (Fig. 1). Fine-scale spatial patterning of mortality at the 2000s study sites varied along the latitudinal gradient, with a non-significant positive relationship between tree density and mortality severity at TRP2000 grading into a weakly significant negative relationship between density and mortality severity at SEV2000 (Fig. S4). Finer-scale clumping of dead piñon was also characteristic of mortality at the northern sites in the 2000s (TRP2000 and WRK2000), with more piñon trees and more dead piñon trees around dead versus living target trees (not significant at the 0.05 level at WRK2000). No such clustering existed at SEV2000 (Table S1). Drought-associated mortality was generally concentrated in medium-sized to larger trees at the 2000s sites, based on size-class distributions of living and dead piñon trees measured in neighborhood plots (Fig. S3).

Evidence of Insect Attack

Evidence on sampled cross-sections points to the almost ubiquitous presence of *Ips* beetles during both mortality events. Seventy-nine percent of dead trees in the 2000s dataset contained *Ips* galleries on sampled bole cross-sections, and 94% had sapwood colored by blue stain fungus. At least one type of evidence was present on 97% of dead samples. All samples with no evidence of successful bark beetle attack were from the SEV2000 site, where 10% of dead trees had no indication of activity. In our 1950s dataset, 67% of samples exhibited *Ips* galleries on sampled cross-sections, and 100% contained blue stain fungus.

Differences in Growth between Dead and Surviving Trees

Averaged across all sites, trees that survived drought had higher recent growth rates (53% higher 3-yr average RW, $p<0.001$) and lower average mean sensitivity (18% lower 20-yr mean sensitivity, $p<0.001$) than dead trees. Average recent growth and mean sensitivity exhibited the same direction of difference between live and dead trees at all sites where both live and dead trees were sampled, although differences were only significant at SEV2000 and the 1950s sites (Table 3; Figs. S5–S6). Significant differences in growth trends and abrupt growth changes were present between live and dead trees at some sites, but they were less consistent in magnitude and direction (Table 3; Figs. S7–S8). Differences in average growth and mean sensitivity extended back many decades at some sites, yet they were generally minor and insignificant earlier in the life of trees (Figs. S5–S6). At some sites with stronger differences between live and dead trees, a divergence in growth occurred after previous severe and/or protracted drought intervals (e.g., after the 1950s for 2000s sites, and after a drought at the turn of the century for the 1950s sites; cf. Fig. 1).

Growth-based Models of Mortality Risk

The best logistic growth-mortality models were highly significant and resulted in good discrimination between live and dead trees, with ROC scores above 0.80 and correct classification rates slightly above 70% (Table 4). Models containing multiple and longer-term growth variables featured substantially lower AIC scores and higher discrimination statistics than models that included only recent growth as a predictor. The best-ranked model included an average growth variable, a measure of year-to-year growth variability, and a term characterizing abrupt growth increases (Table 4). The relative survival probability associated with each predictor varied depending on site (Fig. 2). The direction of growth-mortality relationships was consistent across sites included in the model-building dataset. However, the strength

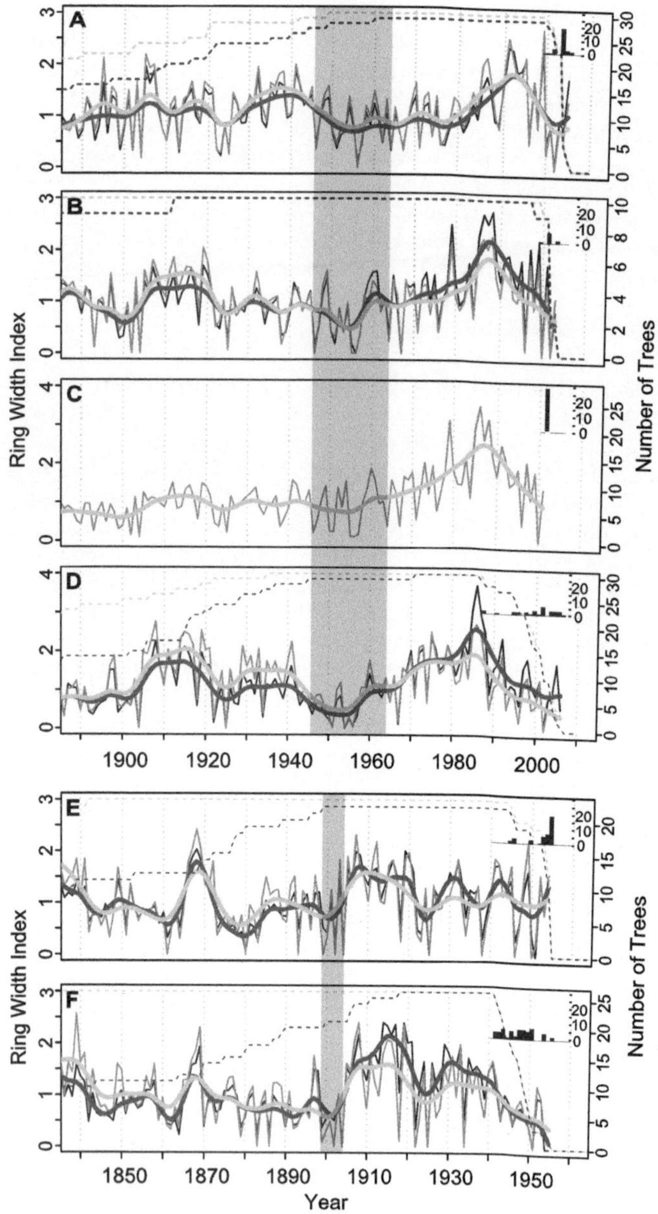

Figure 1. Growth chronologies and death dates from piñon target trees. Live (black) and dead (grey) tree ring-width index chronologies for TRP2000 (A), WRK2000 (B), BNM2000 (C), SEV2000 (D), BNM50 (E), and SEV50 (F). Tukey's biweight robust mean was used to calculate chronology values from individual index series. A smoothing spline (df = 40) (thicker lines) is overlain on the annual mean value chronologies (thinner lines). A horizontal dashed line indicates the number of trees contributing to chronologies in each year. Bar plots of outside ring dates for dead trees at each site are shown in the small panels within each larger time series panel. The transparent grey boxes show SW drought events (as defined in [5]) preceding the 2000s (A–D) and 1950s (E–F) mortality events. The period 1945–1964 was the sixth strongest drought event since 1000 A.D., and the period 1899–1904 was the seventeenth strongest.

Table 3. Average size and growth characteristics of live and dead target trees.

	TRP2000		BNM2000		WRK2000		SEV2000		BNM1950		SEV1950	
	Live	Dead	Live	Dead	Live	Dead	Live	Dead	Live	Dead	Live	Dead
Diameter (cm dbh)	15.9	16.1	-	13.6	17.4	17.2	11	11.9	**8.8**	**12.1**	**7.7**	**10.7**
Recent average growth												
3yr RW (mm/yr)	0.328	0.234	-	0.476	0.357	0.248	**0.412**	**0.157**	**0.439**	**0.239**	**0.358**	**0.131**
3yr RWI	0.824	0.765	-	1.186	1.337	1.036	**1.329**	**0.686**	1.146	1.102	**1.013**	**0.662**
3yr BAI (cm^2/yr)	1.771	1.284	-	1.948	2.005	1.431	**1.389**	**0.602**	1.162	0.927	**0.830**	**0.466**
3yr RelBAI (cm^2/yr/yr)	0.009	0.006	-	0.016	0.010	0.006	**0.017**	**0.006**	**0.017**	**0.007**	**0.020**	**0.005**
Long-term average growth												
20yr RW (mm/yr)	0.609	0.514	-	0.859	0.460	0.348	**0.578**	**0.319**	**0.442**	**0.236**	**0.521**	**0.220**
20yr RWI	1.488	1.614	-	1.954	1.668	1.406	**1.805**	**1.341**	1.200	1.087	**1.366**	**1.077**
20yr BAI (cm^2/yr)	2.840	2.290	-	2.958	2.286	1.915	**1.738**	**1.184**	1.015	0.886	1.103	0.784
20yr RelBAI (cm^2/yr/yr)	0.020	0.017	-	0.037	0.013	0.009	**0.027**	**0.012**	**0.020**	**0.008**	**0.031**	**0.008**
Growth variability												
20yr Mean Sensitivity (RW)	0.509	0.563	-	0.501	0.578	0.591	**0.419**	**0.536**	**0.521**	**0.729**	**0.478**	**0.694**
20yr Mean Sensitivity (RWI)	0.515	0.569	-	0.508	0.581	0.596	**0.423**	**0.544**	**0.528**	**0.732**	**0.477**	**0.693**
20yr Mean Sensitivity (BAI)	0.523	0.577	-	0.515	0.588	0.598	**0.426**	**0.540**	**0.538**	**0.733**	**0.475**	**0.692**
20yr Standard Deviation (RW)	0.310	0.282	-	0.432	0.229	0.178	**0.295**	**0.205**	**0.232**	**0.158**	**0.261**	**0.139**
20yr Standard Deviation (RWI)	0.783	0.898	-	1.014	0.828	0.724	0.915	0.742	0.651	0.739	0.667	0.669
20yr Standard Deviation (BAI)	1.401	1.252	-	1.489	1.132	0.947	0.944	0.879	0.589	0.596	0.568	0.482
Growth trend												
20yr Trend (RW)	−0.02	−0.02	-	−0.04	−0.01	−0.01	−0.02	−0.02	**−0.01**	**0.00**	−0.02	−0.01
20yr Trend (RWI)	−0.03	−0.06	-	−0.07	−0.04	−0.04	−0.06	−0.08	−0.01	0.00	−0.04	−0.03
20yr Trend (BAI)	−0.04	−0.06	-	−0.07	−0.04	−0.05	−0.05	−0.07	0.00	0.00	−0.03	−0.03
Abrupt growth changes												
20yr Increases (RW)	6.41	5.43	-	3.68	4.70	3.90	3.70	2.27	1.00	0.91	0.70	1.35
20yr Increases (RWI)	7.41	6.10	-	5.18	5.30	5.00	4.60	2.80	1.41	1.65	1.52	1.54
20yr Increases (BAI)	8.34	6.70	-	8.18	5.70	5.20	**7.43**	**3.43**	3.50	1.87	**4.15**	**2.12**

Significant differences between live and dead trees within sites are indicated by boldface type ($p<0.05$, Student's t-test). No live trees were sampled for this study at BNM2000.

Table 4. Best-ranked growth-mortality models.

Model Form		*Model Diagnostics and Validation*						
Fixed Effects	**Random Effects**	**Δ AIC**	**ROC**	**ROC$_{boot}$**	**Dead Trees CCR**	**Live Trees CCR**	**All Trees CCR**	**kappa**
log(RW30) + Sens15+ AbruptIncrease10	1+ log(RW30) | site	0.00	0.847	0.844	72.6%	72.7%	72.6%	0.453
log(RW30) + Sens20+ AbruptIncrease10	1+ log(RW30) | site	2.20	0.846	0.848	71.5%	71.4%	71.5%	0.429
log(RW25) + Sens15+ AbruptIncrease10	1+ log(RW25) | site	2.80	0.842	0.838	73.5%	71.5%	72.5%	0.450
Best 2-variable								
log(RW30) + Sens15	1+ log(RW30) | site	5.90	0.834	0.826	71.6%	73.4%	72.5%	0.450
Best 1-variable								
Average growth								
log(RW20)	1+ log(RW20) | site	12.90	0.816	0.812	71.4%	69.4%	70.4%	0.408
Growth sensitivity								
Sens15	-	31.20	0.746	0.736	67.3%	67.5%	67.4%	0.348
Growth trend								
Trend15	-	71.90	0.584	0.580	48.9%	57.5%	53.2%	0.064
Abrupt growth changes								
AbruptIncrease10	-	73.00	0.512	0.517	53.4%	43.3%	48.4%	−0.032
Recent Average Growth								
log(RW3)	-	20.90	0.773	0.782	72.4%	72.9%	72.6%	0.453

Variables include average growth (RW), mean sensitivity (Sens), growth trend (Trend), and abrupt growth changes (AbruptIncreases), with the number of years over which variables were averaged indicated after variable type. The best single-variable models in different growth categories, along with a model containing recent average growth as the only predictor variable (log(RW3)), are shown for comparison. ΔAIC is the difference in AIC between the best-ranked model and the model shown in each table row, with smaller values indicating more parsimonious model fit. ROC is a threshold independent measure of model discrimination, where 0.5 suggests no discrimination and values above 0.8 suggest excellent discrimination. Correct classification rates (CCR) are based on a bootstrapped internal validation with 1000 iterations in which 60% of the data was used for model fitting and 40% was used for model validation. Trees were classified as living if model output was greater than the empirically defined threshold [70]. ROC$_{boot}$ is an average of the ROC statistics generated in the model-fitting portion of the bootstrapping routine. The kappa statistic measures the proportional improvement of the model classification over a random assignment of tree status [99], and was also estimated by taking an average of kappa statistics generated in the bootstrapping routine.

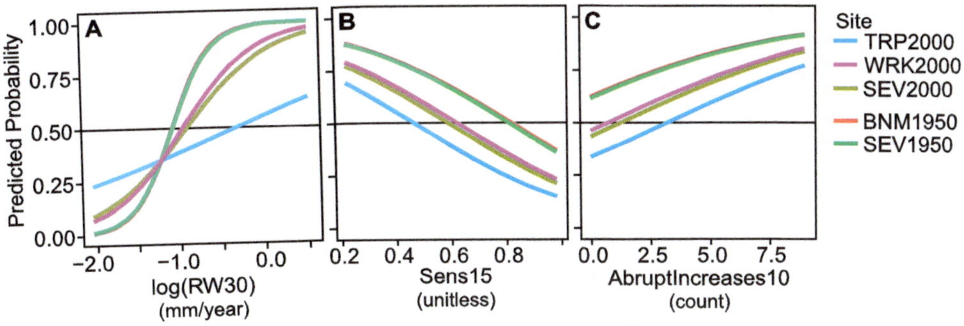

Figure 2. Predicted probabilities of mortality associated with the best-ranked logistic regression model. The figure shows predicted survival probabilities associated with the best-ranked growth-mortality model in Table 4. Values for mean sensitivity over 15 years (Sens15) and the count of abrupt growth increases over 10 years (AbruptIncreases10) are held at their mean for the dataset in (A); AbruptIncreases10 and log(RW30) (average 30-year growth rate) are held at their mean in (B); log(RW30) and Sens15 are held at their mean in (C).

of growth-mortality relationships was weaker overall for the 2000s sites (and at TRP2000 in particular), and the threshold of mortality varied. Specifically, trees at the 1950s sites exhibited lower growth and higher growth variability than trees at the 2000s sites while still featuring survival probabilities above the empirically determined mortality threshold in our model (0.497) (Fig. 2).

Odds ratios for the best-ranked model indicate that, averaged across sites, a 0.1 mm increase in RW averaged over 30 years increases the relative odds of survival by 1.32, all else being equal. Likewise, a 0.1 increase in mean sensitivity over a 15-year period decreases survival odds by 0.72, and one additional growth increase leads to an increase in survival probability by a factor of 1.2. Model coefficients and 95% confidence intervals generated in the bootstrapping routine indicates that all terms with the exception of the random slope term for average growth are significant at the 95% level (Table S2). We nonetheless retained the random slope term, given consistent improvements in AIC and significant increases in the log-likelihood compared to models without this term [69].

Although the best-ranked models produced satisfactory correct classification rates when evaluated internally, they performed poorly when externally validated on the BNM2000 dataset, significantly under-predicting mortality (only 12% of dead trees were classified correctly) (Table 5). Best-ranked models also performed unevenly when internally validated at the site level (Table 5). Site-specific correct classification rates indicate good model performance at the 1950s sites and SEV2000, but correct classification rates were not much better at TRP2000 and WRK2000 than if trees were classified by chance. Fitting growth-mortality models only on data from those individual sites did not dramatically improve this outcome (not shown).

Effects of Climate and Competition on Growth and Mortality

We used linear mixed-effects models to make *post-hoc* assessments of the influence of climatic and competitive factors on radial growth in trees that survived and eventually died. These models confirm that growth over the decades prior to drought-mortality events was different depending on eventual tree status (live/dead), significantly and positively related to precipitation (PPT_{cool}), and negatively related to growing-season VPD (VPD_{MJJ}), though the effect of tree status and the slope of the growth response to climate varied across sites (Tables 6–7; Figs. 3–4, S9–S10). A tree's growth was related to eventual tree status more strongly and consistently in 1950s and SEV2000 trees, confirming a generally stronger growth-mortality signal when compared to WRK2000 and TRP2000.

Interaction terms between tree status, PPT_{cool} and VPD_{MJJ} provide evidence of a differential response to climate amongst surviving and dying trees (Tables 6–7). Survivors from both the 2000s and 1950s exhibited a generally greater response (steeper slope) to precipitation than dying trees, driven by an enhanced growth response during wet years (Tables 6–7; Figs. 3A,C,E, 4A,C). Interactions between tree status and VPD_{MJJ} indicate that surviving trees also usually had a greater growth response to VPD_{MJJ}, driven by enhanced growth during years when

Table 5. Site-specific correct classification rates for best-ranked growth-mortality models.

	Correct Classification Rates		
Site	**Dead Trees**	**Live Trees**	**All Trees**
TRP2000	60.0–63.3%	62.1%	61.0–62.7%
WRK2000	60.0%	60.0–70.0%	60.0–65.5%
BNM2000	12.0%	-	12.0%
SEV2000	**73.3%**	**80.0–83.3%**	**76.7–78.3%**
BNM1950	**82.6%**	**95.5%**	**88.8%**
SEV1950	**90.1%**	**77.3%**	**84.1%**

Each column shows the range of correct classification rates for the three best-ranked general mortality models. Bold typeface highlights correct classification rates that are consistently above 70%.

Table 6. The modeled relationship between tree growth, status, competition and climate in trees from 2000s sites.

	TRP2000				*WRK2000*				*SEV2000*			
Model Terms	**β**	**SE**	**t-value**	**P-value**	**β**	**SE**	**t-value**	**P-value**	**β**	**SE**	**t-value**	**P-value**
Intercept	**−0.51**	**0.06**	**−7.84**	**<0.001**	**−0.47**	**0.15**	**−3.19**	**0.002**	**−0.49**	**0.08**	**−5.95**	**<0.001**
PPT_{Cool}	**0.28**	**0.01**	**19.09**	**<0.001**	**0.26**	**0.03**	**9.20**	**<0.001**	**0.24**	**0.02**	**14.86**	**<0.001**
$(PPT_{Cool})^2$	**0.03**	**0.01**	**4.05**	**<0.001**	**−0.07**	**0.01**	**−6.95**	**<0.001**	**−0.04**	**0.01**	**−5.01**	**<0.001**
VPD_{MJJ}	**−0.13**	**0.01**	**−8.89**	**<0.001**	**−0.17**	**0.03**	**−5.49**	**<0.001**	**−0.08**	**0.01**	**−5.51**	**<0.001**
$(VPD_{MJJ})^2$	**0.02**	**0.01**	**4.29**	**<0.001**	**0.06**	**0.01**	**4.58**	**<0.001**	**−0.03**	**0.01**	**−3.34**	**0.001**
CI	**−0.11**	**0.04**	**−2.50**	**0.014**	−0.10	0.11	−0.91	0.362	**−0.15**	**0.06**	**−2.47**	**0.015**
Live Status	0.14	0.09	1.55	0.124	0.30	0.24	1.24	0.217	**0.63**	**0.14**	**4.59**	**<0.001**
CI:Live Status	−0.02	0.06	−0.38	0.707	−0.02	0.06	−0.38	0.706	−0.02	0.06	−0.38	0.707
PPT_{Cool}:CI	**−0.02**	**0.01**	**−3.01**	**0.003**	−0.02	0.03	−0.73	0.466	−0.02	0.01	−1.73	0.084
PPT_{Cool}:Live Status	0.03	0.02	1.38	0.169	**0.10**	**0.05**	**2.00**	**0.045**	**0.08**	**0.03**	**2.65**	**0.008**
VPD_{MJJ}:CI	0.01	0.01	1.26	0.207	0.01	0.01	1.26	0.207	0.01	0.01	1.26	0.207
VPD_{MJJ}:Live Status	0.01	0.02	0.27	0.790	−0.03	0.05	−0.60	0.549	**−0.10**	**0.03**	**−3.73**	**<0.001**
VPD_{MJJ}:CI:Live Status	**0.03**	**0.01**	**2.40**	**0.017**	**0.03**	**0.01**	**2.40**	**0.017**	**0.03**	**0.01**	**2.40**	**0.017**

Summary of the linear mixed-effects model with the formula RW~PPT_{Cool} + $(PPT_{Cool})^2$+ VPD_{MJJ} + $(VPD_{MJJ})^2$+ CI + Tree Status + Site + CI:Tree Status + PPT_{Cool}:CI + PPT_{Cool}:Tree Status + VPD_{MJJ}:CI + VPD_{MJJ}:Tree Status + PPT_{Cool}:Site + $(PPT_{Cool})^2$:Site + VPD_{MJJ}:Site + $(VPD_{MJJ})^2$:Site + CI:Site + Tree Status:Site + PPT_{Cool}:CI:Site + PPT_{Cool}:Tree Status:Site + VPD_{MJJ}:Tree Status:Site + VPD_{MJJ}:CI:Tree Status, random = (~1 | TreeID). A correlation term and variance weights were also included in the model in order to account for residual autocorrelation of growth between years and variance heterogeneity of residuals by TreeID and across fitted values [65]. Growth was modeled from 1980, with the end of the modeled period varying depending on the outer growth year of the dead tree in each pair. Model parameters with estimates of ($p<0.05$) are in boldface type. The reference level for Tree Status is Dead. Contrasts were applied to calculate coefficients and significances associated with each site.

Table 7. The modeled relationship between tree growth, status, and climate in trees from 1950s sites.

Model Terms	*BNM50*				*SEV50*			
	β	SE	t-value	P-value	β	SE	t-value	P-value
(Intercept)	**−0.34**	**0.06**	**−5.47**	**<0.001**	**−0.43**	**0.05**	**−9.10**	**<0.001**
PPT_{Cool}	**0.40**	**0.02**	**20.31**	**<0.001**	**0.11**	**0.01**	**8.33**	**<0.001**
$(PPT_{Cool})^2$	**−0.10**	**0.01**	**−13.48**	**<0.001**	**−0.10**	**0.01**	**−13.48**	**<0.001**
VPD_{MJJ}	**−0.13**	**0.01**	**−8.79**	**<0.001**	**−0.09**	**0.01**	**−6.10**	**<0.001**
$(VPD_{MJJ})^2$	−0.02	0.01	−1.24	0.2149	−0.01	0.01	−0.99	0.320
Live Status	**0.53**	**0.07**	**7.41**	**<0.001**	**0.53**	**0.07**	**7.41**	**<0.001**
PPT_{Cool}:Live Status	**0.08**	**0.03**	**2.53**	**0.0116**	0.01	0.02	0.42	0.6774
VPD_{MJJ}:Live Status	**−0.10**	**0.03**	**−3.65**	**<0.001**	**−0.10**	**0.02**	**−4.45**	**<0.001**

Summary of the linear mixed-effects model with the formula RW~PPT_{Cool} + $(PPT_{Cool})^2$+ VPD_{MJJ} + $(VPD_{MJJ})^2$+ Tree Status + Site + PPT_{Cool}:Tree Status + VPD_{MJJ}:Tree Status + PPT_{Cool}:Site + $(PPT_{Cool})^2$:Site + $(VPD_{MJJ})^2$:Site + Tree Status:Site + PPT_{Cool}:Tree Status:Site + VPD_{MJJ}:Tree Status:Site, random = (~1 | TreeID). A correlation term and variance weights were also included in the model in order to account for residual autocorrelation of growth between years and variance heterogeneity of residuals by TreeID and across fitted values [65]. Growth was modeled from 1930, with the end of the modeled period varying depending on the outer growth year of the dead tree in each pair. Model parameters with estimates of (p<0.05) are in boldface type. The reference level for Tree Status is Dead. Contrasts were applied to calculate coefficients and significances associated with each site.

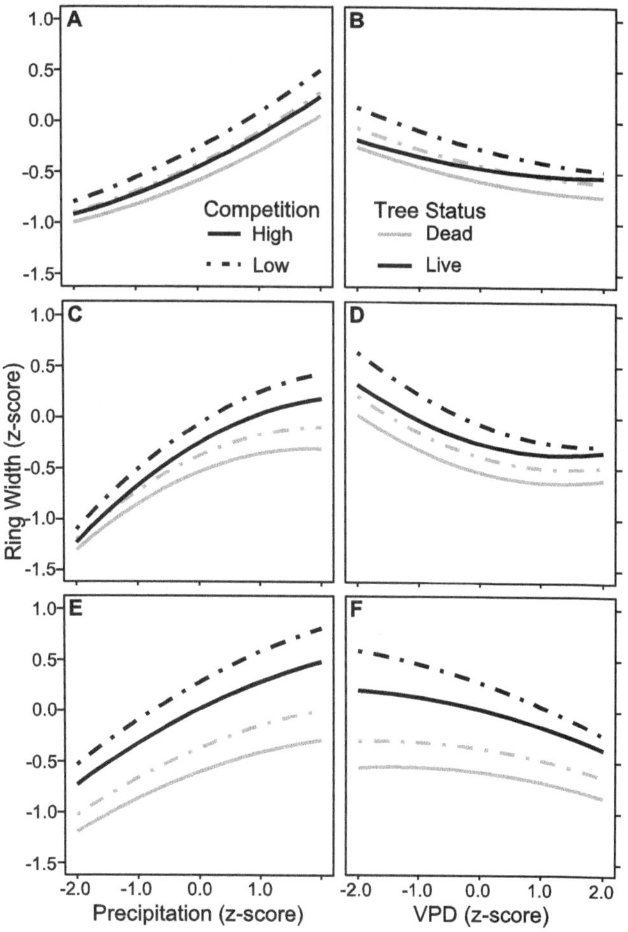

Figure 3. The predicted effects of precipitation (PPT_{cool}), vapor pressure deficit (VPD_{MJJ}) and competition (CI) on growth in surviving and dying trees from 2000s sites. The relationships reflect the model shown in Table 6. High and low competition levels are set to 75th and 25th percentiles of CI, respectively, with the predicted effects shown separately for TRP2000 (A, B), WRK2000 (C, D) and SEV2000 (E, F).

atmospheric vapor demand was low (Tables 6–7; Figs. 3B,D,F, 4B,D). However, the interactions between tree status and climate were weaker and less consistent at TRP2000 (Table 6; Figs. 3B, S9).

Competition played a complex yet significant role in modulating tree growth. Growth was negatively affected by the presence of conspecific neighbors, and competition also reduced the ability of trees to grow well during years with abundant PPT_{cool}, although this effect was weaker at WRK2000 and SEV2000 than at TRP2000 (Table 6; Figs. 3A,C,E, S9). Competition also modulated the response of trees to VPD_{MJJ}, but this effect was contingent on tree status, with live trees with low CI best able to exploit years with low atmospheric demand, and the response of dying trees to VPD_{MJJ} not significantly affected by competition (Table 6; Figs. 3B,D,F, S9).

The magnitude and significance of the growth predictors shifted slightly through time, but the direction of effects remained generally consistent (Figs. S9, S10).

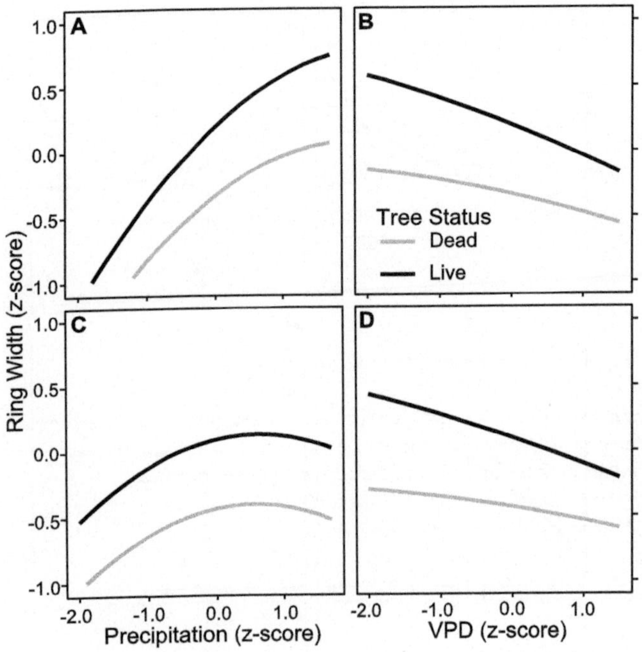

Figure 4. The predicted effects of precipitation (PPT_{cool}) and vapor pressure deficit (VPD_{MJJ}) on growth in surviving and dying trees from 1950s sites. The relationships reflect the model shown in Table 7. Predicted effects are shown separately for BNM50 (A, B) and SEV1950 (C, D).

Discussion

Long-term Factors Predispose Trees to Die during Drought Events

Radial tree growth serves as a proxy for tree status in the years preceding death, integrating the effects of drought, injuries, and tree structural characteristics on the carbon dynamics that can lead to mortality during prolonged and severe droughts. We observed consistently lower average growth in trees that died versus trees that survived, and significantly greater year-to-year growth variability. The latter may be related to carbon reserves and thus the relative capacity of trees to buffer themselves against inter-annual swings in resource availability [39,77], although we cannot prove a physiological link via our data. Lower average growth and higher mean sensitivity among dying piñon are consistent with previous studies in other semi-arid forests [35,36,42], and with the hypothesis that piñon death during the droughts of the 1950s and 2000s was related at least in part to constraints on carbon uptake and/or storage, leading to lower growth and, ultimately, the inability to meet metabolic requirements or repel attacking insects [15,16,39].

More surprising than the observed growth differences *per se* is the fact that at some sites these differences extend over multiple decades, and yet were not present early in the growth records of trees (Figs. 1, S5–S8). This is consistent with the decline-disease theory of tree death [52,78], and suggests that long-term processes or the contingent effects individuals' response to previous events are underlying at least a significant portion of the mortality during two large, seemingly sudden die-off events. We propose two non-exclusive processes that are consistent with our observations.

First, growth in surviving and dying trees appears to diverge most strongly after previous record-setting decadal droughts, at least at some sites (Fig. 1). Some trees may have reacted to these droughts by shedding leaf or root mass [21,79,80], other physiological adjustments [21], and/or they may have sustained injuries such as loss of xylem conductance [25,81,82]. As a result of such responses, some trees became more vulnerable to mortality during subsequent drought [78]. This is consistent with findings on much shorter time scales in *P. sylvestris*, where drought and herbivory-associated reductions in leaf area reduced carbon uptake and reserves and influenced mortality risk during a subsequent drought year [21,83]. It is also consistent with the inclusion of the abrupt growth increase term in our best growth-mortality model (Table 4), which reflects how *recovery* from periods of lower growth (and presumably higher stress) is important to mortality risk beyond the influence of average growth rate alone.

Second, the mortality of neighbors during previous decadal drought may have freed survivors from competition, allowing some to recover faster and/or boost their productivity. The lack of significant differences between spatial neighborhoods around dead and surviving trees at SEV2000 and WRK2000 (Table S1) – sites that exhibited the strongest growth divergences – along with weak relationships between mortality severity and tree density or basal area in neighborhood plots (Fig. S4), suggests that overall tree density was not the most important factor driving mortality risk among the trees in our study. However, the negative influence of conspecific neighbors on piñon growth, and the negative interaction between tree climate response and competition suggests that, although the effect may be complex, competition contributes to growth trajectories, and by extension, likely influences drought-mortality risk (Fig. 3).

The decades-long divergences in growth between dying and surviving trees that we observed are in agreement with those documented in a few other long-term studies [34,36–38], and suggest that understanding tree recovery after drought may be critical to understanding the full impacts of drought-mortality events and anticipating future tree mortality. For example, would the 2000s die-off have been worse if the drought had occurred a decade sooner, when fewer trees had sufficiently recovered from the 1950s drought? Was the severity of mortality during the 2000s drought contingent upon the character and timing of the 1950s drought and the climate in the following years? Past divergences in growth also suggest that there is now a new pool of vulnerable trees that were injured but not killed during the early 2000s drought [25], further highlighting that potential changes in the *frequency* of drought may dictate the *severity* of future die-off events, along with changes in drought intensity and duration.

These results and interpretations are consistent with the view of extreme ecological events put forth by Gutchick and BassiriRad [84], in which the consequences of such events are hypothesized to become most evident during a long recovery period. Among our trees, structural adjustments or injuries caused by previous severe drought (as hypothesized above) might have had a genetic component [84], and the associated fitness costs in terms of lost growth potential proved fatal, even if decades later. Thus, drought-mortality processes may be more fully understood if, in addition to quantifying the instantaneous effects of climate on tree carbon and hydraulic dynamics, we expand the consideration of the controls on prolonged recovery from severe drought events, which may determine how individuals respond to environmental variability years to decades later [21,84].

Dying Trees Exhibit a Differential Response to Climate

The significantly different growth rates of dying vs. surviving trees leading up to the 2000s and 1950s droughts are partly due to differential responses to precipitation and VPD (Tables 6–7; Figs. 3–4, Figs. S9–S10). McDowell et al. [39] found that growth of *P. ponderosa* that died during the 2000s drought was more responsive to a drought index than growth of surviving trees, with dying trees growing less during the drier years leading up to mortality. They suggest that this is consistent with trees predisposed to die by low leaf-level gas exchange and carbon uptake driven by chronic water stress. However, Ogle et al. [35] found that growth in mature drought-killed piñon in Arizona was less responsive to drought variability than in survivors, and Millar et al. found that limber pine (*P. flexilus*) [42] and whitebark pine (*P. albicaulis*) [43] that died were less responsive to decreasing water deficit than survivors, at least at high temperatures.

We found that trees predisposed to die exhibited higher mean sensitivity (e.g. higher year-to-year growth variability, Table 2). Previous researchers have suggested that high mean sensitivity reflects greater limitation by inter-annual swings in climate or other environmental variables (e.g. [77] and see above). Although dying trees grew less than survivors during hotter, drier years (Figs. 3–4), our models suggest that growth in dying piñon was generally *less* responsive to the overall range of PPT_{cool} than in survivors (Tables 6–7; Figs. 3–4, S9–S10). The response of dying trees to VPD_{MJJ} was also generally less pronounced than among survivors, though this effect was reduced or negligible at TRP2000. Importantly, these differences were driven by enhanced growth of surviving trees during wet or cool, rather than dry or hot years (Figs. 3–4). This suggests that, in addition tree response to drought stress *per se*, the ability to maximize photosynthesis and growth during years with abundant water supply and low VPD may be an important aspect of tree survival during subsequent severe or prolonged drought. Recent evidence generated by precision dating of ^{14}C in carbon within and respired by trees points to the utilization of years-to-decades-old stored carbohydrates for functions such as dormant-season metabolism, defense and repair [85–88]. Trees that survived may have been able to store excess carbon from enhanced photosynthesis during wet periods, and to use this carbon for defense and metabolism during subsequent drought periods when growth and photosynthesis were severely constrained [89].

Competition Reduces Tree Growth and Modulates Trees' Ability to Respond to Favorable Climate

High stand density has been found to increase the likelihood of mortality via competition in many forests [29,52,78]. However, recent studies of drought-associated mortality patterns in southwestern woodlands have documented variable relationships between tree density (or basal area) and the severity of mortality within a stand or site [53,76,90–92]. The proportion of dead trees in the neighborhood plots at our study sites was also inconsistently related to tree density or basal area (Fig. S3). However, growth of piñon trees across sites was negatively influenced by the presence of conspecific neighbors, with CI also reducing the ability of trees to take advantage of wet and/or cool conditions over the decades before drought (Table 6; Figs. 3, S9). Similar models fit with a CI that was calculated using neighbors of all species produced less consistent responses across sites (not shown), suggesting differences in the competitive effects of conspecific vs. heterospecific neighbors. Thus our results suggest that managing competition in forests is likely to be important to promote resistance to mortality during drought (cf. [93,94]). At the same time, it is important to note the apparent complexity involved, with species mix in addition to overall tree density needing consideration, as well as the potential role of stand structural characteristics beyond their influence on tree vigor (cf. [95]). Ultimately, further study is required to resolve the role of competition and tree density on mortality.

Modeling Drought-associated Tree Mortality Using Growth-based Predictor Variables

Do simple metrics of tree productivity and carbon balance reflect the complex physiological, structural and life history aspects of individual trees that lead to their mortality during severe drought events? To our knowledge, our study represents the first to comprehensively assess and quantify growth-mortality relationships in the context of widespread drought-associated die-off, and it is one of only a few studies to look at the stability of growth-mortality relationships across space and time (but see [24]). Our models correctly classified ca. 70% of the trees across sites and drought events, which is comparable to or slightly below correct classification rates in other growth-mortality studies [29,31]. Thus, there is promise in using relatively simple, growth-based empirical approaches for assessing drought-mortality risk, at least for certain species or functional groups, even when drought and associated insect activity causes rapid and widespread mortality.

However, we found variable growth-mortality relationships and thresholds between sites and drought events, and uneven model performance at the site level. Although individual sites did not exhibit opposite relationships between growth and mortality, as with background mortality in some European forests [96], the northern sites (TRP, WRK, BNM) were characterized by weaker associations between growth and mortality (Table 5; Fig. 2). This may represent a shift from mortality factors associated with chronic constraints on overall tree growth, to factors associated with shorter-term and/or exogenous factors. These factors include rapid hydraulic failure or carbon starvation in some more vigorous trees due to acute drought stress [40,97], and/or the build-up of bark beetle populations that were able to overcome the resistance of trees with relatively higher growth rates [98]. Many bark beetles favor more vigorous trees with larger food stores, but they are only able to overcome the defenses of such trees at higher population densities [98]. The build-up of relatively large *Ips* populations may have been favored at some sites during the comparatively hot 2000s drought [49], as beetle development is accelerated by warm conditions [45].

Regardless of the underlying physiological causes, growth-mortality models calibrated on empirical data from the 1950s drought alone would have under-predicted mortality during the 2000s drought, thus underscoring the problems with projecting future mortality rates using empirical relationships established based on one drought event or experiment (see also [23]). If bark beetle attack is an underlying driver, integrating information on bark beetle dynamics with empirical indicators of tree physiological stress will be important for improving the predictive capacity of mortality models. Furthermore, our best growth-mortality models included longer-term, less simplistic growth metrics, suggesting that the time scales and/or cumulative processes considered by many current models should be extended (cf. [31]).

Conclusions

Understanding the processes that underlie drought-related tree mortality is critical for anticipating future forest dynamics and associated feedbacks to the earth system, and for developing management plans that enhance the robustness and resilience of forests to climate change. Our study documented high levels of piñon mortality during both the 2000s and 1950s droughts, with almost ubiquitous evidence of bark beetle activity on dead trees. More synchronous mortality was observed at northern sites in New Mexico during the 2000s event. Dying trees generally had lower average growth rates and greater year-to-year growth variability than trees that survived, but early in their life, these differences were not evident. Instead, decades-long growth divergences between surviving and dying trees suggest that recent growth differences are related to the response and recovery of trees to previous severe droughts, at least at some sites. This pattern further suggests that a pool of trees that survived the early 2000s drought may now be particularly vulnerable during future droughts. These trees should be investigated in more detail to reveal the processes that influence their recovery [84].

We show that tree growth response to climate is an important predisposing factor underlying mortality during widespread, drought- and insect-related mortality events. In particular, our results suggest that tree response to wet/cool years, in addition to the response to drought years, may be an important aspect of vulnerability. The growth response of surviving trees during very wet/cool years in the decades preceding mortality events likely enhanced their carbon reserves, which was important for withstanding subsequent drought and insect attack. The competitive environment also influenced tree growth and the ability of trees to respond favorably to wet conditions, suggesting that controlling tree density is likely to enhance tree resistance to mortality during drought. However, conspecific vs. heterospecific competitive effects appear to be different, and should therefore be considered in detail.

The discriminatory ability of logistic growth-mortality functions underscores the potential of simple empirical approaches to represent mortality risk in models of vegetation dynamics, even in the context of widespread mortality events associated with drought and bark beetles. However, incorporating multiple and longer-term aspects of tree growth and life history is important for fully capturing mortality risk. Furthermore, shifting growth-mortality relationships across space and time point to the challenges associated with calibrating mortality algorithms. Although we cannot fully explain the weakening of growth-mortality relationships at the northern study sites during the 2000s drought, one consistent hypothesis is that bark beetle dynamics played a more important role in the recent die-off event, shifting the physiological basis for mortality and highlighting the need for further study of tree-insect dynamics to improve the prediction of tree mortality during drought.

Supporting Information

Figure S1 Study area locations within New Mexico, USA. The distribution of *P. edulis* is shown in light gray.

Figure S2 Relationships between tree diameter and radial growth as represented by different growth metrics. Only live tree data is shown. A linear regression line with 95% confidence intervals is plotted over the raw data for relative basal area increments (RelBAI) (A), raw ring widths (RW) (B), basal area increments (BAI) (C), and ring width indices (RWI) (D). Trends are significant at the 95% level for (A) and (C), but not significant for (B) and (D). Growth is represented by an average of the most recent 3-years in each growth record.

Figure S3 Size-class distributions of *Pinus edulis* that died and survived the 2000s drought. Distributions are derived from measurements made in 7.5m neighborhood plots around each target tree. Bars in each histogram represent 2.5cm size classes, based on tree diameters at breast height (DBH). Data are pooled across all 2000s study sites in (A), and shown separately for TRP2000 (B), BNM2000 (C), WRK2000 (D) and SEV2000 (E). Size-class distributions for live and dead trees are significantly different overall ($p<0.0001$, Kolmogorov-Smirnov test), for TRP2000 ($p=0.0313$), WRK2000 ($p=0.0006$), and SEV2000 (0.0014). There were not enough living trees to test for differences at BNM2000.

Figure S4 The relationship between tree density and mortality severity. Each point represents the ratio of dead *Pinus edulis* (PIED) versus total tree density in 7.5m neighborhood-plots surrounding target trees at 2000s study sites ((A) TRP2000 (n = 60), (B) WRK2000 (n = 20) and (C) SEV2000 (n = 60)). A linear regression line is shown to provide a visual estimate of the relationship. Quasi-binomial regression was used to statistically assess the direction, magnitude and significance of depicted relationships. For TRP2000 (A), $0.00091x + 0.18240$, $p=0.19$. For WRK2000 (B), $y=0.00246x+0.06733$, $p=0.17$. For SEV2000 (C), $y=-.00107x-0.22008$, $p=0.021$.

Figure S5 Box and whisker plots of ring widths averaged over different time intervals. Live tree growth is truncated at the outside ring date of the dead tree in the tree pair. Boxes drawn around time intervals on the x-axis denote significant differences between live and dead trees ($p<0.05$, Student's t-test).

Figure S6 Box and whisker plots of growth mean sensitivity averaged over different time intervals. Live tree growth is truncated at the outside ring date of the dead tree in the tree pair. Boxes drawn around time intervals on the x-axis denote significant differences between live and dead trees ($p<0.05$, Student's t-test).

Figure S7 Box and whisker plots of tree growth trends averaged over different time intervals. Live tree growth is truncated at the outside ring date of the dead tree in the tree pair. Boxes drawn around time intervals on the x-axis denote significant differences between live and dead trees ($p<0.05$, Student's t-test).

Figure S8 Box and whisker plots of the number of abrupt growth increases averaged over different time intervals. Live tree growth is truncated at the outside ring date of the dead tree in the tree pair. Boxes drawn around time intervals on the x-axis denote significant differences between live and dead trees ($p<0.05$, Student's t-test).

Figure S9 Coefficients from models relating growth to climate, tree status and competition for trees that died and survived the 2000s drought. Coefficients are from linear mixed-effects models relating growth (RW) to precipitation (PPT_{cool}), vapor pressure deficit (VPD_{MJJ}), competition (CI), and tree status (L/D). Coefficients were calculated separately for TRP2000 (A), WRK2000 (B), and SEV2000 (B). All predictor variables were converted to z-scores prior to modeling, allowing for a direct comparison of coefficients between models. Growth was modeled starting from five different dates leading up to the 2000s drought-mortality event (bar colors). The end of the modeled period varied depending on tree pair, with growth in surviving trees truncated at the outer year of growth in the corresponding dead tree.

Figure S10 Coefficients from models relating growth to climate and tree status for trees that died and survived the 1950s drought. Coefficients are from linear mixed-effects models relating growth (RW) to precipitation (PPT_{cool}), vapor pressure deficit (VPD_{MJJ}), and tree status (L/D). Coefficients were calculated separately for BNM1950 (A) and SEV50 (B). All predictor variables were converted to z-scores prior to modeling, allowing for a direct comparison of coefficients between models. Growth was modeled starting from five different dates leading up to the 1950s drought-mortality event (bar colors). The end of the modeled period varied depending on tree pair, with growth in surviving trees truncated at the outer year of growth in the corresponding dead tree.

Table S1 Fine scale spatial patterning of mortality at 2000s sites. Significant differences in tree density and basal area in neighborhood plots surrounding dead versus living target trees are in boldface type ($p<0.05$, Student's t-test). PIED is *Pinus edulis*. JUMO is *Juniperus monosperma*.

Table S2 Bootstrapped estimates and confidence intervals for model terms in the best-ranked growth-mortality model. The model formula is Tree Status ~log(RW30) + Sens15 + AbruptIncreases10 + (1 + log(RW30) | Site), with validation statistics shown in Table 4. Variables include average growth (RW), mean sensitivity (Sens), and the number of abrupt growth increases (AbruptIncreases), with the number of years over which variables were averaged indicated after variable type. Bootstrapped estimates were generated by fitting models to 1000 samples drawn from the calibration data. The *Estimates* columns represent model coefficients for fixed effects and standard deviations for random effects.

Text S1 Justification for the choice of climate predictors in linear mixed-effects models.

Acknowledgments

We thank Craig D. Allen, Julio L. Betancourt and Thomas J. Swetnam for access to archived samples, constructive discussions and review of an earlier draft of the manuscript. The comments of Vincent Gutschick and an anonymous reviewer greatly improved the manuscript. We appreciate project support, feedback and review of an earlier draft of the manuscript from Nate G. McDowell and Connie Woodhouse. Christof Bigler and Dean Billiheimer helped with statistical methods. We are grateful for field assistance from Kay Beeley, Collin Haffey, Matthias Kläy, Derek Murrow and Greg Pederson.

Author Contributions

Conceived and designed the experiments: AKM HB. Performed the experiments: AKM HB. Analyzed the data: AKM HB. Contributed reagents/materials/analysis tools: AKM HB. Wrote the paper: AKM HB.

References

1. Van Mantgem PJ, Stephenson NL, Byrne JC, Daniels LD, Franklin JF, et al. (2009) Widespread Increase of Tree Mortality Rates in the Western United States. Science 323: 521–524. doi:10.1126/science.1165000.
2. Phillips OL, Aragão LEOC, Lewis SL, Fisher JB, Lloyd J, et al. (2009) Drought Sensitivity of the Amazon Rainforest. Science 323: 1344–1347. doi:10.1126/science.1164033.
3. Peng C, Ma Z, Lei X, Zhu Q, Chen H, et al. (2011) A drought-induced pervasive increase in tree mortality across Canada's boreal forests. Nature Climate Change 1: 467–471. doi:10.1038/nclimate1293.
4. Carnicer J, Coll M, Ninyerola M, Pons X, Sánchez G, et al. (2011) Widespread crown condition decline, food web disruption, and amplified tree mortality with increased climate change-type drought. Proceedings of the National Academy of Sciences: 201010070. doi:10.1073/pnas.1010070108.
5. Williams AP, Allen CD, Macalady AK, Griffin D, Woodhouse CA, et al. (2013) Temperature as a potent driver of regional forest drought stress and tree mortality. Nature Climate Change 3: 292–297. doi:10.1038/nclimate1693.
6. Allen CD, Macalady AK, Chenchouni H, Bachelet D, McDowell N, et al. (2010) A global overview of drought and heat-induced tree mortality reveals emerging climate change risks for forests. Forest Ecology and Management 259: 660–684. doi:10.1016/j.foreco.2009.09.001.
7. Bonan G (2008) Carbon cycle: Fertilizing change. Nature Geoscience 1: 645–646. doi:10.1038/ngeo328.
8. Kurz WA, Dymond CC, Stinson G, Rampley GJ, Neilson ET, et al. (2008) Mountain pine beetle and forest carbon feedback to climate change. Nature 452: 987–990.
9. Breshears DD, López-Hoffman L, Graumlich LJ (2011) When ecosystem services crash: preparing for big, fast, patchy climate change. Ambio 40: 256–263.
10. Reichstein M, Bahn M, Ciais P, Frank D, Mahecha MD, et al. (2013) Climate extremes and the carbon cycle. Nature 500: 287–295. doi:10.1038/nature12350.
11. Bugmann H (2001) A review of forest gap models. Climatic Change 51: 259–305. doi:10.1023/A:1012525626267.
12. Keane RE, Austin M, Field C, Huth A, Lexer MJ, et al. (2001) Tree mortality in gap models: application to climate change. Climatic Change 51: 509–540.
13. Fisher R, McDowell N, Purves D, Moorcroft P, Sitch S, et al. (2010) Assessing uncertainties in a second-generation dynamic vegetation model caused by ecological scale limitations. New Phytologist 187: 666–681.
14. Seidl R, Fernandes PM, Fonseca TF, Gillet F, Jönsson AM, et al. (2011) Modelling natural disturbances in forest ecosystems: a review. Ecological Modelling 222: 903–924. doi:10.1016/j.ecolmodel.2010.09.040.
15. McDowell NG, Beerling DJ, Breshears DD, Fisher RA, Raffa KF, et al. (2011) The interdependence of mechanisms underlying climate-driven vegetation mortality. Trends in Ecology & Evolution 26: 523–532. doi:10.1016/j.tree.2011.06.003.
16. McDowell N, Pockman WT, Allen CD, Breshears DD, Cobb N, et al. (2008) Mechanisms of plant survival and mortality during drought: why do some plants survive while others succumb to drought? New Phytologist 178: 719–739.
17. Sala A, Piper F, Hoch G (2010) Physiological mechanisms of drought-induced tree mortality are far from being resolved. New Phytologist 186: 274–281. doi:10.1111/j.1469-8137.2009.03167.x.
18. McDowell NG (2011) Mechanisms linking drought, hydraulics, carbon metabolism, and vegetation mortality. Plant Physiology 155: 1051.
19. Choat B, Jansen S, Brodribb TJ, Cochard H, Delzon S, et al. (2012) Global convergence in the vulnerability of forests to drought. Nature 491: 752–755. doi:10.1038/nature11688.
20. Zeppel MJ, Adams HD, Anderegg WR (2011) Mechanistic causes of tree drought mortality: recent results, unresolved questions and future research needs. New Phytologist 192: 800–803.
21. Bréda N, Huc R, Granier A, Dreyer E (2006) Temperate forest trees and stands under severe drought: a review of ecophysiological responses, adaptation processes and long-term consequences. Annals of Forest Science 63: 625–644.
22. Anderegg WR, Kane JM, Anderegg LD (2013) Consequences of widespread tree mortality triggered by drought and temperature stress. Nature Climate Change 3: 30–36.
23. McDowell NG, Fisher RA, Xu C, Domec JC, Hölttä T, et al. (2013) Evaluating theories of drought-induced vegetation mortality using a multimodel–experiment framework. New Phytologist 200: 304–321.
24. Wunder J (2007) Conceptual advancement and ecological applications of tree mortality models based on tree-ring and forest inventory data [Ph.D. dissertation]. Zurich, Switzerland: ETH Zurich.
25. Anderegg WRL, Plavcová L, Anderegg LDL, Hacke UG, Berry JA, et al. (2013) Drought's legacy: multiyear hydraulic deterioration underlies widespread aspen forest die-off and portends increased future risk. Global Change Biology 19: 1188–1196. doi:10.1111/gcb.12100.
26. Waring R (1987) Characteristics of trees predisposed to die. BioScience 37: 569–574.
27. Dobbertin M (2005) Tree growth as indicator of tree vitality and of tree reaction to environmental stress: a review. European Journal of Forest Research 124: 319–333. doi:10.1007/s10342-005-0085-3.
28. Wyckoff PH, Clark JS (2002) The relationship between growth and mortality for seven co-occurring tree species in the southern Appalachian Mountains. Journal of Ecology 90: 604–615.
29. Bigler C, Bugmann H (2003) Growth-dependent tree mortality models based on tree rings. Canadian Journal of Forest Research 33: 210–221.
30. Bigler C, Bugmann H (2004) Assessing the performance of theoretical and empirical tree mortality models using tree-ring series of Norway spruce. Ecological Modelling 174: 225–239.
31. Das AJ, Battles JJ, Stephenson NL, Van Mantgem PJ (2007) The relationship between tree growth patterns and likelihood of mortality: a study of two tree species in the Sierra Nevada. Canadian Journal of Forest Research 37: 580–597.
32. Wunder J, Reineking B, Bigler C, Bugmann H (2008) Predicting tree mortality from growth data: how virtual ecologists can help real ecologists. Journal of Ecology 96: 174–187.

33. Hawkes C (2000) Woody plant mortality algorithms: description, problems and progress. Ecological Modelling 126: 225–248.
34. Pedersen BS (1998) The role of stress in the mortality of midwestern oaks as indicated by growth prior to death. Ecology 79: 79–93.
35. Ogle K, Whitham TG, Cobb NS (2000) Tree-ring variation in pinyon predicts likelihood of death following severe drought. Ecology 81: 3237–3243.
36. Suarez ML, Ghermandi L, Kitzberger T (2004) Factors predisposing episodic drought-induced tree mortality in Nothofagus– site, climatic sensitivity and growth trends. Journal of Ecology 92: 954–966. doi:10.1111/j.1365-2745.2004.00941.x.
37. Bigler C, Bräker OU, Bugmann H, Dobbertin M, Rigling A (2006) Drought as an inciting mortality factor in Scots pine stands of the Valais, Switzerland. Ecosystems 9: 330–343.
38. Bigler C, Gavin DG, Gunning C, Veblen TT (2007) Drought induces lagged tree mortality in a subalpine forest in the Rocky Mountains. Oikos 116: 1983–1994.
39. McDowell N, Allen CD, Marshall L (2010) Growth, carbon-isotope discrimination, and drought-associated mortality across a Pinus ponderosa elevational transect. Global Change Biology 16: 399–415.
40. Levanic T, Cater M, McDowell NG (2011) Associations between growth, wood anatomy, carbon isotope discrimination and mortality in a Quercus robur forest. Tree physiology 31: 298–308. doi:10.1093/treephys/tpq111.
41. Hereş A-M, Martínez-Vilalta J, López BC (2012) Growth patterns in relation to drought-induced mortality at two Scots pine (Pinus sylvestris L.) sites in NE Iberian Peninsula. Trees 26: 621–630. doi:10.1007/s00468-011-0628-9.
42. Millar CI, Westfall RD, Delany DL (2007) Response of high-elevation limber pine (Pinus flexilis) to multiyear droughts and 20th-century warming, Sierra Nevada, California, USA. Canadian Journal of Forest Research 37: 2508–2520. doi:10.1139/X07-097.
43. Millar CI, Westfall RD, Delany DL, Bokach MJ, Flint AL, et al. (2012) Forest mortality in high-elevation whitebark pine (Pinus albicaulis) forests of eastern California, USA; influence of environmental context, bark beetles, climatic water deficit, and warming. Canadian Journal of Forest Research 42: 749–765. doi:10.1139/x2012-031.
44. Breshears DD, Cobb NS, Rich PM, Price KP, Allen CD, et al. (2005) Regional vegetation die-off in response to global-change-type drought. Proceedings of the National Academy of Sciences of the United States of America 102: 15144.
45. Bentz BJ, Régnière J, Fettig CJ, Hansen EM, Hayes JL, et al. (2010) Climate change and bark beetles of the western United States and Canada: direct and indirect effects. BioScience 60: 602–613.
46. Anderegg WR, Berry JA, Smith DD, Sperry JS, Anderegg LD, et al. (2012) The roles of hydraulic and carbon stress in a widespread climate-induced forest die-off. Proceedings of the National Academy of Sciences 109: 233–237.
47. Raffa KF, Aukema BH, Bentz BJ, Carroll AL, Hicke JA, et al. (2008) Cross-scale Drivers of Natural Disturbances Prone to Anthropogenic Amplification: The Dynamics of Bark Beetle Eruptions. BioScience 58: 501. doi:10.1641/B580607.
48. Adams HD, Guardiola-Claramonte M, Barron-Gafford GA, Villegas JC, Breshears DD, et al. (2009) Temperature sensitivity of drought-induced tree mortality portends increased regional die-off under global-change-type drought. Proceedings of the National Academy of Sciences 106: 7063.
49. Weiss JL, Castro CL, Overpeck JT (2009) Distinguishing Pronounced Droughts in the Southwestern United States: Seasonality and Effects of Warmer Temperatures. Journal of Climate 22: 5918–5932. doi:10.1175/2009JCLI2905.1.
50. Allen CD, Breshears DD (1998) Drought-induced shift of a forest–woodland ecotone: rapid landscape response to climate variation. Proceedings of the National Academy of Sciences of the United States of America 95: 14839.
51. Swetnam TW, Betancourt JL (1998) Mesoscale disturbance and ecological response to decadal climatic variability in the American Southwest. Journal of Climate 11: 3128–3147.
52. Franklin JF, Shugart HH, Harmon ME (1987) Tree Death as an Ecological Process. BioScience 37: 550–556. doi:10.2307/1310665.
53. Floyd ML, Clifford M, Cobb NS, Hanna D, Delph R, et al. (2009) Relationship of stand characteristics to drought-induced mortality in three Southwestern piñon-juniper woodlands. Ecological Applications 19: 1223–1230.
54. Stokes MA, Smiley TL (1968) An introduction to tree-ring dating. Tucson, AZ: University of Arizona Press. 73 p.
55. Holmes RL (1983) Computer-assisted quality control in tree-ring dating and measurement. Tree-Ring bulletin 43: 69–78.
56. Wood S (1982) The Bark and Ambrosia Beetles of North and Central America (Coleoptera: Scolytidae), a Taxonomic Monograph. Great Basin Naturalist Memoirs, Brigham Young University Number 6.
57. Rogers TJ (1993) Insect and disease associates of the piñon-juniper woodlands. Gen. Tech. Rep. RM-236. Santa Fe, NM: US Department of Agriculture, Forest Service, Rocky Mountain Forest and Range Experiment Station. 42–62.
58. Perkins DL, Swetnam TW (1996) A dendroecological assessment of whitebark pine in the Sawtooth-Salmon River region, Idaho. Canadian Journal of Forest Research 26: 2123–2133.
59. Applequist M (1958) A simple pith locator for use with off-center increment cores. Journal of Forestry 56: 141.
60. Wunder J, Reineking B, Matter JF, Bigler C, Bugmann H (2007) Predicting tree death for Fagus sylvatica and Abies alba using permanent plot data. Journal of Vegetation Science 18: 525–534.
61. Biondi F, Qeadan F (2008) A theory-driven approach to tree-ring standardization: defining the biological trend from expected basal area increment. Tree-Ring Research 64: 81–96.
62. Bigler C, Bugmann H (2004) Predicting the time of tree death using dendrochronological data. Ecological Applications 14: 902–914.
63. Strackee J, Jansma E (1992) The statistical properties of mean sensitivity–a reappraisal. Dendrochronologia 10: 121–135.
64. Biondi F, Qeadan F (2008) Inequality in paleorecords. Ecology 89: 1056–1067. doi:10.1890/07-0783.1.
65. Pinheiro JC, Bates DM (2000) Mixed-Effects Models in S and S-Plus. Springer. 560 p.
66. Agresti A (2013) Categorical data analysis. 3rd ed. Hoboken, NJ: John Wiley & Sons. 744 p.
67. Burnham KP, Anderson D (2002) Model Selection and Multi-Model Inference: A Practical Information-Theoretic Approach. 2nd ed. New York, NY: Springer-Verlag New York. 496 p.
68. Harrell FE (2001) Regression modeling strategies: with applications to linear models, logistic regression, and survival analysis. New York, NY: Springer-Verlag New York. 568 p.
69. Zuur AF, Ieno EN, Walker NJ, Saveliev AA, Smith GM (2009) Mixed effects models and extensions in ecology with R. Ney York, NY: Springer-Verlag New York. 580 p.
70. Fawcett T (2003) ROC graphs: Notes and practical considerations for data mining researchers. Technical Report HPL-2003-4. Palo Alto, CA, USA: HP Laboratories. Available: http://www.hpl.hp.com/techreports/2003/HPL-2003-4.pdf.
71. Hosmer DW, Lemeshow S, Sturdivant RX (2013) Applied logistic regression. 3rd ed. Hoboken, NJ: John Wiley & Sons. 528 p.
72. Daly C, Neilson RP, Phillips DL (1994) A Statistical-Topographic Model for Mapping Climatological Precipitation over Mountainous Terrain. Journal of Applied Meteorology 33: 140–158. doi:10.1175/1520-0450(1994)033<0140:ASTMFM>2.0.CO;2.
73. R Core Team (2013) R: A language and environment for statistical computing. Vienna, Austria: R Foundation for Statistical Computing. Available: http://www.R-project.org/.
74. Bunn A (2008) A dendrochronology program library in R (dplR). Dendrochronologia 26: 115–124. doi:10.1016/j.dendro.2008.01.002.
75. Bates DM, Maechler M, Bolker B (2013) lme4: Linear-mixed Effects Models Using S4 Classes.
76. Clifford MJ, Royer PD, Cobb NS, Breshears DD, Ford PL (2013) Precipitation thresholds and drought-induced tree die-off: insights from patterns of Pinus edulis mortality along an environmental stress gradient. New Phytologist 200: 413–421. doi:10.1111/nph.12362.
77. Fritts HC, Smith DG, Cardis JW, Budelsky CA (1965) Tree-ring characteristics along a vegetation gradient in northern Arizona. Ecology: 394–401.
78. Manion PD (1991) Tree disease concepts. 2nd ed. Prentice Hall. 422 p.
79. Poyatos R, Aguadé D, Galiano L, Mencuccini M, Martínez-Vilalta J (2013) Drought-induced defoliation and long periods of near-zero gas exchange play a key role in accentuating metabolic decline of Scots pine. New Phytologist 200: 388–401. doi:10.1111/nph.12278.
80. Landhäusser S, Lieffers V (2012) Defoliation increases risk of carbon starvation in root systems of mature aspen. Trees 26: 653–661. doi:10.1007/s00468-011-0633-z.
81. Plaut JA, Wadsworth WD, Pangle R, Yepez EA, McDowell NG, et al. (2013) Reduced transpiration response to precipitation pulses precedes mortality in a pinon–juniper woodland subject to prolonged drought. New Phytologist 200: 375–387.
82. Hacke UG, Stiller V, Sperry JS, Pittermann J, McCulloh KA (2001) Cavitation Fatigue. Embolism and Refilling Cycles Can Weaken the Cavitation Resistance of Xylem. Plant Physiology 125: 779–786.
83. Galiano L, Martínez-Vilalta J, Lloret F (2011) Carbon reserves and canopy defoliation determine the recovery of Scots pine 4 yr after a drought episode. New Phytologist 190: 750–759. doi:10.1111/j.1469-8137.2010.03628.x.
84. Gutschick VP, BassiriRad H (2003) Extreme events as shaping physiology, ecology, and evolution of plants: toward a unified definition and evaluation of their consequences. New Phytologist 160: 21–42.
85. Guérard N, Maillard P, Bréchet C, Lieutier F, Dreyer E (2007) Do trees use reserve or newly assimilated carbon for their defense reactions? A 13C labeling approach with young Scots pines inoculated with a bark-beetle-associated fungus (Ophiostoma brunneo ciliatum). Annals of Forest Science 64: 601–608. doi:10.1051/forest:2007038.
86. Carbone MS, Czimczik CI, Keenan TF, Murakami PF, Pederson N, et al. (2013) Age, allocation and availability of nonstructural carbon in mature red maple trees. New Phytologist: n/a–n/a. doi:10.1111/nph.12448.
87. Richardson AD, Carbone MS, Keenan TF, Czimczik CI, Hollinger DY, et al. (2013) Seasonal dynamics and age of stemwood nonstructural carbohydrates in temperate forest trees. New Phytologist 197: 850–861.
88. Gaylord ML, Kolb TE, Pockman WT, Plaut JA, Yepez EA, et al. (2013) Drought predisposes piñon-juniper woodlands to insect attacks and mortality. New Phytologist 198: 567–578. doi:10.1111/nph.12174.
89. Breshears DD, Myers OB, Meyer CW, Barnes FJ, Zou CB, et al. (2008) Tree die-off in response to global change-type drought: mortality insights from a decade of plant water potential measurements. Frontiers in Ecology and the Environment 7: 185–189.

90. Negron JF, Wilson JL (2003) Attributes associated with probability of infestation by the piñon ips, Ips confusus (Coleoptera: Scolytidae), in piñon pine, Pinus edulis. Western North American Naturalist 63: 440–451.
91. Greenwood DL, Weisberg PJ (2008) Density-dependent tree mortality in pinyon-juniper woodlands. Forest Ecology and Management 255: 2129–2137.
92. Santos MJ, Whitham TG (2010) Predictors of Ips confusus Outbreaks During a Record Drought in Southwestern USA: Implications for Monitoring and Management. Environmental Management 45: 239–249. doi:10.1007/s00267-009-9413-6.
93. Linares JC, Camarero JJ, Carreira JA (2010) Competition modulates the adaptation capacity of forests to climatic stress: insights from recent growth decline and death in relict stands of the Mediterranean fir *Abies pinsapo*. Journal of Ecology 98: 592–603. doi:10.1111/j.1365-2745.2010.01645.x.
94. Ruiz-Benito P, Lines ER, Gómez-Aparicio L, Zavala MA, Coomes DA (2013) Patterns and Drivers of Tree Mortality in Iberian Forests: Climatic Effects Are Modified by Competition. PLoS ONE 8: e56843. doi:10.1371/journal.pone.0056843.
95. Fettig CJ, Klepzig KD, Billings RF, Munson AS, Nebeker TE, et al. (2007) The effectiveness of vegetation management practices for prevention and control of bark beetle infestations in coniferous forests of the western and southern United States. Forest Ecology and Management 238: 24–53.
96. Wunder J, Brzeziecki B, Żybura H, Reineking B, Bigler C, et al. (2008) Growth–mortality relationships as indicators of life-history strategies: a comparison of nine tree species in unmanaged European forests. Oikos 117: 815–828.
97. Sthultz CM, Gehring CA, Whitham TG (2009) Deadly combination of genes and drought: increased mortality of herbivore-resistant trees in a foundation species. Global Change Biology 15: 1949–1961. doi:10.1111/j.1365-2486.2009.01901.x.
98. Boone CK, Aukema BH, Bohlmann J, Carroll AL, Raffa KF (2011) Efficacy of tree defense physiology varies with bark beetle population density: a basis for positive feedback in eruptive species. Canadian Journal of Forest Research 41: 1174–1188.
99. Fleiss JL, Levin B, Paik MC (2013) Statistical methods for rates and proportions. 3rd ed. Hoboken, NJ: John Wiley & Sons. 800 p.

PERMISSIONS

All chapters in this book were first published in PLOS ONE, by The Public Library of Science; hereby published with permission under the Creative Commons Attribution License or equivalent. Every chapter published in this book has been scrutinized by our experts. Their significance has been extensively debated. The topics covered herein carry significant findings which will fuel the growth of the discipline. They may even be implemented as practical applications or may be referred to as a beginning point for another development.

The contributors of this book come from diverse backgrounds, making this book a truly international effort. This book will bring forth new frontiers with its revolutionizing research information and detailed analysis of the nascent developments around the world.

We would like to thank all the contributing authors for lending their expertise to make the book truly unique. They have played a crucial role in the development of this book. Without their invaluable contributions this book wouldn't have been possible. They have made vital efforts to compile up to date information on the varied aspects of this subject to make this book a valuable addition to the collection of many professionals and students.

This book was conceptualized with the vision of imparting up-to-date information and advanced data in this field. To ensure the same, a matchless editorial board was set up. Every individual on the board went through rigorous rounds of assessment to prove their worth. After which they invested a large part of their time researching and compiling the most relevant data for our readers.

The editorial board has been involved in producing this book since its inception. They have spent rigorous hours researching and exploring the diverse topics which have resulted in the successful publishing of this book. They have passed on their knowledge of decades through this book. To expedite this challenging task, the publisher supported the team at every step. A small team of assistant editors was also appointed to further simplify the editing procedure and attain best results for the readers.

Apart from the editorial board, the designing team has also invested a significant amount of their time in understanding the subject and creating the most relevant covers. They scrutinized every image to scout for the most suitable representation of the subject and create an appropriate cover for the book.

The publishing team has been an ardent support to the editorial, designing and production team. Their endless efforts to recruit the best for this project, has resulted in the accomplishment of this book. They are a veteran in the field of academics and their pool of knowledge is as vast as their experience in printing. Their expertise and guidance has proved useful at every step. Their uncompromising quality standards have made this book an exceptional effort. Their encouragement from time to time has been an inspiration for everyone.

The publisher and the editorial board hope that this book will prove to be a valuable piece of knowledge for researchers, students, practitioners and scholars across the globe.

LIST OF CONTRIBUTORS

Martin Aberhan
Museum für Naturkunde, Leibniz Institute for Evolution and Biodiversity Science, Berlin, Germany

Wolfgang Kiessling
Museum für Naturkunde, Leibniz Institute for Evolution and Biodiversity Science, Berlin, Germany
GeoZentrum Nordbayern, Paläoumwelt, Universität Erlangen2 Nürnberg, Erlangen, Germany

James L. Rainford and Peter J. Mayhew
Department of Biology, University of York, York, United Kingdom

Michael Hofreiter
Department of Biology, University of York, York, United Kingdom
Faculty of Mathematics and Natural Sciences, Institute for Biochemistry and Biology, University of Potsdam, Potsdam, Germany

David B. Nicholson
Department of Biology, University of York, York, United Kingdom
Department of Natural Sciences, National Museums Scotland, Edinburgh, United Kingdom
Department of Earth Sciences, The Natural History Museum, London, United Kingdom

Euan D. Reavie
Center for Water and the Environment, Natural Resources Research Institute, University of Minnesota Duluth, Duluth, Minnesota, United States of America

Adam J. Heathcote
Groupe de Recherche Interuniversitaire en Limnologie, Département des Sciences Biologiques, Université du Québecà Montréal, Montréal, Québec, Canada

Victoria L. Shaw Chraïbi
Department of Earth and Atmospheric Sciences, University of Nebraska-Lincoln, Lincoln, Nebraska, United States of America

Joshua R. Thienpont, Elisa S. Cheng and John P. Smol
Paleoecological Environmental Assessment and Research Lab (PEARL), Department of Biology, Queen's University, Kingston, Ontario, Canada

Steven V. Kokelj
Northwest Territories Geoscience Office, Government of the Northwest Territories, Yellowknife, Northwest Territories, Canada

Cyndy Desjardins, Linda E. Kimpe and Jules M. Blais
Department of Biology, University of Ottawa, Ottawa, Ontario, Canada

Jennifer B. Korosi
Paleoecological Environmental Assessment and Research Lab (PEARL), Department of Biology, Queen's University, Kingston, Ontario, Canada
Department of Biology, University of Ottawa, Ottawa, Ontario, Canada

Michael F.J. Pisaric
Department of Geography and Environmental Studies, Carleton University, Ottawa, Ontario, Canada
Department of Geography, Brock University, St. Catharines, Ontario, Canada

Mariana C. Teixeira, Sidinei M. Thomaz, Thaisa S. Michelan, Roger P. Mormul and Thamis Meurer
José Vitor B. Fasolli, Márcio J. Silveira Universidade Estadual de Maringá, Nupélia, Maringá-PR, Brazil

Juan C. Larrasoaña
Instituto Geológico y Minero de España, Unidad de Zaragoza, Zaragoza, Spain
Research School of Earth Sciences, The Australian National University, Canberra, Australia

Andrew P. Roberts
Research School of Earth Sciences, The Australian National University, Canberra, Australia

Eelco J. Rohling
Research School of Earth Sciences, The Australian National University, Canberra, Australia
School of Ocean and Earth Science, University of Southampton, National Oceanography Centre, Southampton, United Kingdom

Mariana G. Bender and Sergio R. Floeter
Departamento de Ecologia e Zoologia, Universidade Federal de Santa Catarina, Florianó polis, SC, Brazil

Cassiano Monteiro-Netto, Carlos E. L. Ferreira and Gustavo R. Machado
Departamento de Biologia Marinha, Universidade Federal Fluminense, Niteroi, RJ, Brazil

Paulo José de Azevedo Silva
Fundação Instituto de Pesca, Arraial do Cabo, RJ, Brazil

Osmar J. Luiz
Department of Biological Sciences, Macquarie University, Sydney, NSW, Australia

Antony G. Brown
Palaeoenvironmental Laboratories (PLUS), University of Southampton, Southampton, United Kingdom

Laura S. Basell
School of Geography, Archaeology and Palaeoecology, Queen's University Belfast, Belfast, Northern Ireland

Sian Robinson
MRC Lifecourse Epidemiology Unit, Southampton General Hospital, University of Southampton, Southampton, United Kingdom

Graham C. Burdge
Institute of Developmental Sciences Building, Faculty of Medicine, University of Southampton, Southampton, United Kingdom

Patricia Sánchez-Baracaldo
School of Geographical Sciences, University of Bristol, Bristol, United Kingdom

Gavin H. Thomas
Department of Animal and Plant Sciences, University of Sheffield, Sheffield, South Yorkshire, United Kingdom

Timme H. Donders and Friederike Wagner-Cremer
Palaeoecology, Department of Physical Geography, Faculty of Geosciences, Utrecht University, Laboratory of Palaeobotany and Palynology, Utrecht, The Netherlands

Kimberley Hagemans
Palaeoecology, Department of Physical Geography, Faculty of Geosciences, Utrecht University, Laboratory of Palaeobotany and Palynology, Utrecht, The Netherlands
Department of Environmental Sciences, Copernicus Institute, Faculty of Geosciences, Utrecht University, Utrecht, The Netherlands

Stefan C. Dekker
Department of Environmental Sciences, Copernicus Institute, Faculty of Geosciences, Utrecht University, Utrecht, The Netherlands

Letty A. de Weger
Department of Pulmonology, Leiden University Medical Centre, Leiden, The Netherlands

Pim de Klerk
Botany section, Staatliches Museum für Naturkunde Karlsruhe, Karlsruhe, Germany

Lee-Ann C. Hayek
Smithsonian Institution Mathematics and Statistics NMNH MRC-121, Washington D.C., United States of America

Brent Wilson
Petroleum Geoscience Programme, Department of Chemical Engineering, The University of the West Indies, Saint Augustine, Trinidad, Trinidad and Tobago

Catalina Pimiento
Florida Museum of Natural History, Gainesville, Florida, United States of America; Department of Biology, University of Florida, Gainesville, Florida, United States of America; Smithsonian Tropical Research Institute, Balboa, Panama

Christopher F. Clements
Institute of Evolutionary Biology and Environmental Studies, The University of Zurich, Zurich, Switzerland

Washington Luiz Silva Vieira, Kleber Silva Vieira, Rômulo Pantoja Nóbrega and Paulo Fernandes Guedes Pereira Montenegro
Laboratório de Ecofisiologia Animal, Departamento de Sistemática e Ecologia, Universidade Federal da Paraíba, João Pessoa, PB, Brazil

Gentil Alves Pereira Filho
Museu de Zoologia da Universidade de São Paulo, Ipiranga, São Paulo, SP, Brazil

Gindomar Gomes Santana
Programa de Pós-Graduaçãoem Ecologia e Conservação (PPGEC)/Departamento de Biologia, Universidade Estadual da Paraíba, Campina Grande, Paraíba, Brazil

Rômulo Romeu Nóbrega Alves
Programa de Pós-Graduação em Ecologia e Conservação (PPGEC)/Departamento de Biologia, Universidade Estadual da Paraíba, Campina Grande, Paraíba, Brazil
Departamento de Biologia, Universidade Estadual da Paraíba, Campina Grande, Paraíba, Brazil

Waltécio Oliveira Almeida
Departamento de Química Biológica, Centro de Ciências Biológicas e da Saúde, Universidade Regional do Cariri - URCA, Campus do Pimenta, Crato, CE, Brazil

Alexandre Vasconcellos
Departamento de Sistemática e Ecologia, Universidade Federal da Paraíba, João Pessoa, PB, Brazil

Jian-Guo Huang, Andria Dawson and Philip G. Comeau
Department of Renewable Resources, University of Alberta, Edmonton, Alberta, Canada

Kenneth J. Stadt
Forest Management Branch, Sustainable Resource Development, Edmonton, Alberta, Canada

Věra Pavelková Řičánková, Jan Robovský and Jan Riegert
Department of Zoology, Faculty of Science, University of South Bohemia, České Budějovice, Czech Republic

Guillermo Rodríguez-Gómez, Ana Mateos and Jesús Rodríguez
Paleofisiología y Ecología Social de homínidos, Centro Nacional de Investigación sobre la Evolución Humana (CENIEH), Burgos, Spain

Jesús Angel Martín-González
Departamento de Matemáticas y Computación, Universidad de Burgos, Burgos, Spain, and temporarily assigned to CENIEH, Burgos, Spain

Ruth Blasco
The Gibraltar Museum, Gibraltar, Gibraltar

Jordi Rosell
Àrea de Prehistória, Universitat Rovira i Virgili (URV), Tarragona, Spain
Institut Catalàde Paleoecologia Humana i Evolució Social (IPHES), Tarragona, Spain

Xingzhao Liu, Kun Li, Weiqiang Xu and Junhua Yan
South China Botanical Garden, Chinese Academy of Sciences, Guangzhou, China

Wei Meng
Paulownia Research and Development Centre, China Forestry Administration, Zhengzhou, China

Guohua Liang
State Key Laboratory of Conservation and Utilization of Subtropical Agro-Bioresources, South China Agricultural University, Guangzhou, China

Liujing Huang
Guangdong Provincial Academy of Environmental Science, Guangzhou, China

Courtney E. Gorman and Jennifer A. Schweitzer
Dept. of Ecology and Evolutionary Biology, University of Tennessee, Knoxville, Tennessee, United States of America

Joseph K. Bailey
Dept. of Ecology and Evolutionary Biology, University of Tennessee, Knoxville, Tennessee, United States of America
School of Plant Science, University of Tasmania, Hobart, Tasmania, Australia

Brad M. Potts
School of Plant Science, University of Tasmania, Hobart, Tasmania, Australia

Carrie L. Tyler and Michał Kowalewski
Florida Museum of Natural History, University of Florida, Gainesville, Florida, United States of America

Fengjiang Li, Naiqin Wu, Yajie Dong and Dan Zhang
Key Laboratory of Cenozoic Geology and Environment, Institute of Geology and Geophysics, Chinese Academy of Sciences, Beijing, China

Denis-Didier Rousseau
Laboratoire de Meteorologie Dynamique, UMR INSU-CNRS 8539 & CERES-ERTI, Ecole Normale Superieure, Paris, France
Lamont-Doherty Earth Observatory of Columbia University, Palisades, New York, United States of America

Yunpeng Pei
School of the Earth Sciences and Resources, China University of Geosciences, Beijing, China

Jonathan P. Tennant
Department of Earth Science and Engineering, Imperial College London, London, United Kingdom

Norman MacLeod
Earth Science Department, The Natural History Museum, London, United Kingdom

Alison K. Macalady
University of Arizona, School of Geography and Development and Laboratory of Tree-Ring Research, Tucson, Arizona, United States of America

Harald Bugmann
Forest Ecology, Department of Environmental Systems Science, ETH Zürich, Zürich, Switzerland

Index